VOLUME ONE HUNDRED AND SEVENTY

PROGRESS IN MOLECULAR BIOLOGY AND TRANSLATIONAL SCIENCE

Computational Approaches for Understanding Dynamical Systems: Protein Folding and Assembly

VOLUME ONE HUNDRED AND SEVENTY

PROGRESS IN MOLECULAR BIOLOGY AND TRANSLATIONAL SCIENCE

Computational Approaches for Understanding Dynamical Systems: Protein Folding and Assembly

Edited by

BIRGIT STRODEL

Institute of Biological Information Processing (IBI-7: Structural Biochemistry), Forschungszentrum Jülich, Jülich, Germany; Institute of Theoretical and Computational Chemistry, Heinrich Heine University Düsseldorf, Düsseldorf, Germany

BOGDAN BARZ

Institut für Physikalische Biologie, Heinrich Heine University Düsseldorf, Düsseldorf, Germany; Institute of Biological Information Processing (IBI-7: Structural Biochemistry), Forschungszentrum Jülich, Jülich, Germany

ELSEVIER

ACADEMIC PRESS

An imprint of Elsevier

Academic Press is an imprint of Elsevier
50 Hampshire Street, 5th Floor, Cambridge, MA 02139, United States
525 B Street, Suite 1650, San Diego, CA 92101, United States
The Boulevard, Langford Lane, Kidlington, Oxford OX5 1GB, United Kingdom
125 London Wall, London EC2Y 5AS, United Kingdom

First edition 2020

Notices
Knowledge and best practice in this field are constantly changing. As new research and experience broaden our understanding, changes in research methods, professional practices, or medical treatment may become necessary.

Practitioners and researchers must always rely on their own experience and knowledge in evaluating and using any information, methods, compounds, or experiments described herein. In using such information or methods they should be mindful of their own safety and the safety of others, including parties for whom they have a professional responsibility.

To the fullest extent of the law, neither the Publisher nor the authors, contributors, or editors, assume any liability for any injury and/or damage to persons or property as a matter of products liability, negligence or otherwise, or from any use or operation of any methods, products, instructions, or ideas contained in the material herein.

ISBN: 978-0-12-821135-9
ISSN: 1877-1173

For information on all Academic Press publications
visit our website at https://www.elsevier.com/books-and-journals

Publisher: Zoe Kruze
Acquisitions Editor: Ashlie M. Jackman
Editorial Project Manager: Joanna Collett
Production Project Manager: James Selvam
Cover Designer: Greg Harris

Typeset by SPi Global, India

Contents

Contributors

Sandro Bottaro
Structural Biology and NMR Laboratory & Linderstrøm-Lang Centre for Protein Science, Department of Biology, University of Copenhagen, Copenhagen, Denmark; Atomistic Simulations Laboratory, Istituto Italiano di Tecnologia, Genova, Italy

Nicolae-Viorel Buchete
School of Physics; Institute for Discovery, University College Dublin, Dublin, Ireland

Yang Cao
State Key Laboratory of Chemical Oncogenomics, School of Chemical Biology and Biotechnology, Peking University Shenzhen Graduate School, Shenzhen, China

Cezary Czaplewski
Faculty of Chemistry, University of Gdańsk, Gdańsk, Poland

Philippe Derreumaux
Laboratory of Theoretical Chemistry; Faculty of Pharmacy, Ton Duc Thang University, Ho Chi Minh City, Vietnam

Ron Elber
Oden Institute for Computational Engineering and Sciences; Department of Chemistry, University of Texas at Austin, Austin, TX, United States

Arman Fathizadeh
Oden Institute for Computational Engineering and Sciences, University of Texas at Austin, Austin, TX, United States

Szilard N. Fejer
Provitam Foundation, Department of Research and Development, Sfântu Gheorghe, Romania; University of Pécs, Department of Chemistry, Faculty of Sciences, Pécs, Hungary

Robert Ganzynkowicz
Faculty of Chemistry, University of Gdańsk, Gdańsk, Poland

Artur Giełdoń
Faculty of Chemistry, University of Gdańsk, Gdańsk, Poland

Łukasz Golon
Faculty of Chemistry, University of Gdańsk, Gdańsk, Poland

Wei Han
State Key Laboratory of Chemical Oncogenomics, School of Chemical Biology and Biotechnology, Peking University Shenzhen Graduate School, Shenzhen, China

Jochen S. Hub
Theoretical Physics and Center for Biophysics, Saarland University, Saarbrücken, Germany

Agnieszka Karczyńska
Faculty of Chemistry, University of Gdańsk, Gdańsk, Poland

Paweł Krupa
Institute of Physics, Polish Academy of Sciences, Warsaw, Poland

Andreas Haahr Larsen
Structural Biology and NMR Laboratory & Linderstrøm-Lang Centre for Protein Science, Department of Biology; Structural Biophysics, Niels Bohr Institute, Faculty of Science, University of Copenhagen, Copenhagen, Denmark; Structural Bioinformatics and Computational Biochemistry Unit, Department of Biochemistry, University of Oxford, Oxford, United Kingdom

Justin A. Lemkul
Department of Biochemistry, Virginia Tech, Blacksburg, VA, United States

Qinghua Liao
Science for Life Laboratory, Department of Chemistry-BMC, Uppsala University, Uppsala, Sweden

Kresten Lindorff-Larsen
Structural Biology and NMR Laboratory & Linderstrøm-Lang Centre for Protein Science, Department of Biology, University of Copenhagen, Copenhagen, Denmark

Agnieszka G. Lipska
Faculty of Chemistry, University of Gdańsk, Gdańsk, Poland

Adam Liwo
Faculty of Chemistry, University of Gdańsk, Gdańsk, Poland; School of Computational Sciences, Korea Institute for Advanced Study, Seoul, Republic of Korea

Jennifer Loschwitz
Institute of Theoretical and Computational Chemistry, Heinrich Heine University Düsseldorf, Düsseldorf; Institute of Biological Information Processing (IBI-7: Structural Biochemistry), Forschungszentrum Jülich, Jülich, Germany

Emilia A. Lubecka
Institute of Informatics, Faculty of Mathematics, Physics, and Informatics, University of Gdańsk, Gdańsk, Poland

Maciej Maciejczyk
Department of Physics and Biophysics, Faculty of Food Science, University of Warmia and Mazury in Olsztyn, Olsztyn, Poland

Dominique Madern
Univ. Grenoble Alpes, CEA, CNRS, IBS, Grenoble, France

Mariusz Makowski
Faculty of Chemistry, University of Gdańsk, Gdańsk, Poland

Sandipan Mohanty
Institute for Advanced Simulation, Jülich Supercomputing Centre, Forschungszentrum Jülich, Jülich, Germany

Magdalena A. Mozolewska
Institute of Computer Science, Polish Academy of Sciences, Warsaw, Poland

Brajesh Narayan
School of Physics; Institute for Discovery, University College Dublin, Dublin, Ireland

Phuong H. Nguyen
CNRS, Université de Paris, UPR 9080, Laboratoire de Biochimie Théorique; Institut de Biologie Physico-Chimique-Fondation Edmond de Rothschild, PSL Research University, Paris, France

Olujide O. Olubiyi
Institute of Biological Information Processing (IBI-7: Structural Biochemistry), Forschungszentrum Jülich, Jülich, Germany; Department of Pharmaceutical Chemistry, Faculty of Pharmacy, Obafemi Awolowo University, Ile-Ife, Nigeria

Simone Orioli
Structural Biology and NMR Laboratory & Linderstrøm-Lang Centre for Protein Science, Department of Biology; Structural Biophysics, Niels Bohr Institute, Faculty of Science, University of Copenhagen, Copenhagen, Denmark

Chetan S. Poojari
Theoretical Physics and Center for Biophysics, Saarland University, Saarbrücken, Germany

Adam K. Sieradzan
Faculty of Chemistry, University of Gdańsk, Gdańsk, Poland; School of Computational Sciences, Korea Institute for Advanced Study, Seoul, Republic of Korea

Fabio Sterpone
CNRS, Université de Paris, UPR 9080, Laboratoire de Biochimie Théorique; Institut de Biologie Physico-Chimique-Fondation Edmond de Rothschild, PSL Research University, Paris, France

Birgit Strodel
Institute of Theoretical and Computational Chemistry, Heinrich Heine University Düsseldorf, Düsseldorf; Institute of Biological Information Processing (IBI-7: Structural Biochemistry), Forschungszentrum Jülich, Jülich, Germany

Xuan Tang
State Key Laboratory of Chemical Oncogenomics, School of Chemical Biology and Biotechnology, Peking University Shenzhen Graduate School, Shenzhen, China

Stepan Timr
CNRS, Université de Paris, UPR 9080, Laboratoire de Biochimie Théorique; Institut de Biologie Physico-Chimique-Fondation Edmond de Rothschild, PSL Research University, Paris, France

Miao Yuan
State Key Laboratory of Chemical Oncogenomics, School of Chemical Biology and Biotechnology, Peking University Shenzhen Graduate School, Shenzhen, China

Ye Yuan
School of Physics; Institute for Discovery, University College Dublin, Dublin, Ireland

Preface

Biomolecular simulations have the unique ability to explore the dynamics of biological systems at atomistic resolution and can therefore reveal biomolecular mechanisms difficult to observe with experimental techniques. The advantages of molecular simulations, in combination with clever biophysical and biochemical experiments, have enabled and will continue to enable great progress in providing a molecular understanding of functional and disease-related biological mechanisms. As proteins are central to the majority of biomolecular mechanisms, gaining a detailed understanding of their structure, dynamics, and function is key for apprehending life. The book "Computational Approaches for Understanding Dynamical Systems: Protein Folding and Assembly" from the series *Progress in Molecular Biology and Translational Science* brings together some of the most important advances in the field of computational modeling and simulations of proteins, with emphasis on two main mechanisms that are essential to life: protein folding and assembly.

A fundamental ingredient for classical molecular simulations of proteins is the so-called force field which defines basic interactions between any atoms present in the modeled biological system. The more than four decades of developments that led to the current state of atomistic protein force fields are described in the first chapter of the book. Of equal complexity has been the development of coarse-grained force fields that enable simulations of biological systems of larger sizes and longer times as the number of degrees of freedom in the model is reduced, yet at the cost of reduced resolution. The second chapter provides theoretical and analytical steps for deriving coarse-grained models able to predict protein structures and to simulate protein folding. An important application of molecular dynamics (MD) simulations is the interpretation of experimental measurements and integration of experimental data into simulations in order to improve the biophysical models. This is the closest partnership between computational and experimental techniques studying protein dynamics and has a very promising future, as described in Chapter 3. Transitions between protein conformations usually require overcoming free energy barriers, which is not a trivial task computationally if the free energy barriers become large. This leads to considerable gaps between the timescales reached by classical MD simulations and those governing biological processes. Chapter 4 provides an overview of the various enhanced sampling methodologies that have been developed to fill in these gaps and also gives an introduction to free energy calculations. Chapter 5 complements this topic by reviewing two increasingly popular methods that allow the exploration of longer timescales by MD simulations, Markov models and the Milestoning method. Both methods make use of ensembles of short trajectories to compute long-time kinetics and thermodynamics, opening new avenues for the investigation of biomolecular mechanisms.

Without protein folding it would be difficult to imagine the biological life as it is. This mechanism that manages to create a unique protein structure or fold based purely on the amino acid sequence has been intensively studied for decades. Once folded, proteins have generally very stable conformations, unless they become denatured due to chemical agents, radiation, or high temperature. However, some proteins preserve their fold and function even at high temperatures and are highly adapted to these extreme conditions. Recent computational studies contributed greatly to understanding the rational behind the stability of thermophilic proteins at high temperatures and are described in Chapter 6. A special class of proteins are membrane proteins which are vital to the functioning of cells as they are involved in protein-lipid interactions governing processes such as cell-cell recognition, molecular transport across membranes, and signal transduction, to name a few. Chapter 7 demonstrates that MD simulations have greatly contributed to our understanding of membrane proteins, while also providing a comprehensive overview of methods and software that are available for studying protein-membrane systems.

Next to protein folding, protein assembly is equally important for keeping a cell functioning by building short- and long-range structures and mediating biomolecular interactions. Protein assembly, however, can also be related to diseases, as it is the case in viral capsid formation or the aberrant aggregation of misfolded proteins. The assembly of viral capsids is a complex process and involves the association of proteins with different geometric shapes into a large and stable structure able to carry the genetic material of the virus. This process has been studied computationally and is described in Chapter 8. A particular case of disease-related assembly involves so-called amyloid proteins which initially form small oligomers, thought to be responsible for the death of cells, and later aggregate into highly ordered amyloid fibrils that can extend for micrometers. The assembly of amyloid peptides has been successfully investigated computationally at both the atomistic and coarse-grained levels. Such simulations are presented in Chapter 9, and are related to the amyloid-β peptide involved in the development of Alzheimer's disease which has been the topic of intense study by both experiments and simulations.

Recently, it has been discovered that the surface of amyloid fibrils can act as a catalyst for oligomer and fibril formation via a so-called secondary nucleation mechanism. This mechanism is interconnected with fibril elongation, and together, they lead to a considerable acceleration of the aggregation process. Chapter 10 presents simulation approaches to model these molecular events. The final chapter of this book also addresses the problem of protein aggregation and liquid-liquid phase separation, or coacervation, yet using a simulation approach alternative to MD simulations, namely Monte Carlo simulations. A short introduction to the theory of Monte Carlo simulations is provided, followed by reviewing several studies falling into the topic of the chapter.

We hope that this book will be a great guide and source of information regarding molecular simulations in life science, not only for computational researchers but also for curious experimentalists. We would like to thank all the authors for their contributions to this book and to the series editor Prof. David Teplow for making this book possible.

Birgit Strodel
Bogdan Barz

CHAPTER ONE

Pairwise-additive and polarizable atomistic force fields for molecular dynamics simulations of proteins

Justin A. Lemkul*
Department of Biochemistry, Virginia Tech, Blacksburg, VA, United States
*Corresponding author: e-mail address: jalemkul@vt.edu

Contents

Abstract

Protein force fields have been undergoing continual development since the first complete parameter sets were introduced nearly four decades ago. The functional forms that underlie these models have many common elements for the treatment of bonded and nonbonded forces, which are reviewed here. The most widely used force fields to date use a fixed-charge convention in which electronic polarization effects are treated via a mean-field approximation during partial charge assignment. Despite success in modeling folded proteins over many years, the fixed-charge assumption has limitations that cannot necessarily be overcome within their potential energy equations. To overcome these limitations, several force fields have recently been derived that explicitly treat electronic polarization effects with straightforward extensions of the potential

Progress in Molecular Biology and Translational Science, Volume 170
ISSN 1877-1173
https://doi.org/10.1016/bs.pmbts.2019.12.009

energy functions used by nonpolarizable force fields. Here, we review the history of the most popular nonpolarizable force fields (AMBER, CHARMM, OPLS, and GROMOS) as well as studies that have validated them and applied them to studies of protein folding and misfolding. Building upon these force fields are more recent polarizable interaction potentials, including fluctuating charge models, POSSIM, AMOEBA, and the classical Drude oscillator. These force fields differ in their implementations but all attempt to model electronic polarization in a computationally tractable manner. Despite their recent emergence in the field of protein folding, several studies have already applied these polarizable models to challenging problems in this domain, including the role of polarization in folding free energies and sequence-specific effects on the stability of α-helical structures.

1. Introduction

The field of molecular modeling and simulation has become integral to the investigation of biomolecules—proteins, nucleic acids, carbohydrates, and lipids, among others. Simulations offer the ability to interrogate molecular systems at atomistic resolution in ways that most experimental techniques cannot. As such, theoretical investigations stand on their own in generating hypotheses and proposing mechanistic details of biological phenomena but are also intimately related to widely used experimental techniques. The quality of any biomolecular simulation derives primarily from two attributes—the exhaustiveness of sampling and the quality of the underlying physical model, called a force field, which governs how all atoms interact. As such, choosing a suitable force field is perhaps the most important element of designing a simulation. It must be applicable to all elements of the system and must be of sufficiently high quality to address the scientific question(s) at hand.

To date, most biomolecular force fields have relied upon an approximation to assign partial charges to all of the atoms in the system, that the medium in which the molecules will be solvated is water. While this approximation generally works well for most globular proteins, there are instances in which considering changes in polarity or electric field are important, and as such, the distribution of charges must change to address the physical phenomena being modeled. For instance, it has long been recognized that electronic polarization is critical to describing enzyme catalysis,[1–3] thus requiring quantum mechanical (QM) or hybrid quantum mechanical/molecular mechanical (QM/MM) approaches. Thus, the development of explicitly polarizable, purely classical force fields is desirable and has become a focus of the molecular modeling community in recent years.

Here, we review the concepts underlying atomistic force fields for simulations of proteins, and discuss successes and challenges in modeling protein dynamics, with a specific emphasis on protein folding and misfolding. Section 2 will introduce the concept of force fields, with an overview of the mathematical basis underlying the most commonly used functional forms. Section 3 will discuss nonpolarizable force fields, which have been routinely employed in protein simulations for the better part of four decades. Section 4 will discuss more recent developments in force field technology, with the advent of force fields that explicitly model electronic polarization. Section 5 concludes with an outlook on the next challenges that the field will attempt to resolve.

2. Force field overview

The foundational concept underlying all atomistic molecular dynamics (MD) simulations is that, given a set of coordinates, $\vec{R}$, one can compute the potential energy, $U\left(\vec{R}\right)$. One possibility is to use quantum mechanical (QM) calculations but doing so is prohibitively expensive for systems as large as condensed-phase biomolecules like proteins. Instead, the typical approach is to use a parametric equation based on classical mechanics. The forces on each atom can subsequently be computed by taking the derivative of this function with respect to the coordinates. Thus, a force field is composed of two elements: (1) the functional form used to compute $U\left(\vec{R}\right)$ and (2) the parameters (numerical constants) used in this functional form.

A force field functional form is a combination of simple mathematical functions, each describing a different type of interaction between two or more atoms. A typical functional form is shown in Eq. (1):

$$U\left(\vec{R}\right) = \frac{1}{2}\sum_{Bonds} k_b(b-b_0)^2 + \frac{1}{2}\sum_{Angles} k_\theta(\theta-\theta_0)^2 + \sum_{Dihedrals} k_\phi[1+\cos(n\phi-\delta)] + \sum_{LJ}\varepsilon_{ij}^{min}\left[\left(\frac{R_{ij}^{min}}{r_{ij}}\right)^{12} - 2\left(\frac{R_{ij}^{min}}{r_{ij}}\right)^{6}\right] + \sum_{Coulomb}\frac{q_iq_j}{4\varepsilon_0\varepsilon r_{ij}} \tag{1}$$

2.1 Bonded interactions

Bonded energy terms include those interactions that exist between atoms that are covalently linked, including bond stretching, valence angle bending, and dihedral rotation. Bond stretching and angle bending (the first two terms on the right-hand side of Eq. 1) are typically represented by harmonic functions. The required parameters for both of these energy terms are a force constant and an equilibrium bond or angle value. The force constants (k_b and k_θ for bonds and angles, respectively) define the stiffness of the harmonic oscillations. The equilibrium values (b_0 and θ_0 for bonds and angles, respectively) define the geometry at which the energy associated with the bond or angle is zero. These values are generally taken from optimized QM geometries or crystallographic data and reflect the only energy minimum in the potential energy equation for these interactions. Force constants can be derived from QM vibrational analysis, spectroscopic data, and/or by fitting to the potential energy profile as a function of perturbing the target bond or angle away from its equilibrium value. The equilibrium bond length or valence angle can similarly be taken from an optimized QM geometry or from crystal structures.

Dihedral rotation (the third term on the right-hand side of Eq. 1) is often expressed as a cosine series, reflecting the periodic nature of rotations about covalent bonds, such that one or more energy minimum will exist over a period of 360-degree rotation (Fig. 1). The rotation around a covalent bond is a function of noncovalent interactions between atoms separated by three bonds, so-called "1–4 pairs." Steric effects give rise to local energy minima and maxima (Fig. 1). Dihedral terms are typically employed in classical force fields to account for inaccuracies in the treatment of nonbonded terms (see Section 2.2). The parameters required for dihedral energy terms are the force constant (k_ϕ), multiplicity (n), and the phase angle (δ). Force constants dictate the barrier heights between energy minima. The multiplicity refers to the number of energy minima over 360° and the phase angle is the offset that indicates where the energy minima are. A phase angle of 180° means there will be an energy minimum at 0°, and a phase angle of 0° corresponds to an energy minimum at 180°. There may be multiple dihedral terms associated with each rotatable bond, depending on the complexity of the potential energy surface. Each of these terms can have their own set of parameters, and those with larger force constants have a greater impact on the energy surface.

Another energy term is often used, which is called an "improper" dihedral. This nomenclature comes from the fact that the interaction is among four atoms and therefore the angle between two planes, but in this case, the energy term is associated with out-of-plane deformation around a planar center, such as a carbon atom in an aromatic ring or a peptide bond.

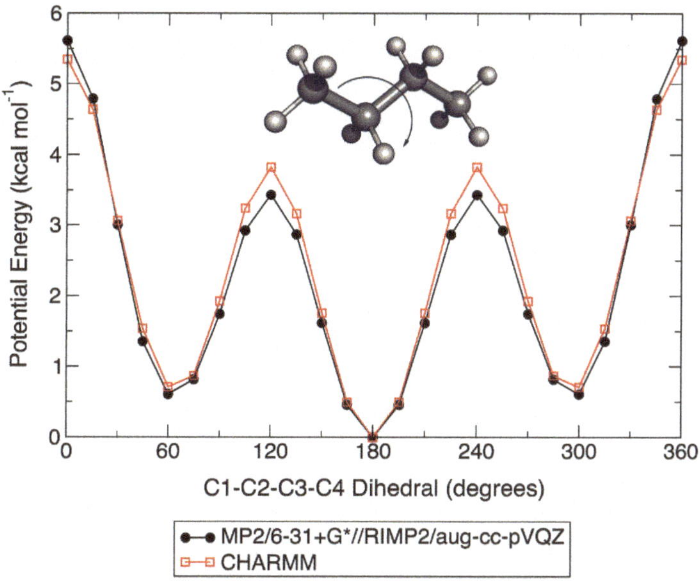

Fig. 1 Potential energy scan of the dihedral rotation about the central C—C bond in butane. The dihedral was fixed at values in increments of 15° and other degrees of freedom were allowed to relax. QM optimization was performed using a MP2/6-31+G* model chemistry, with subsequent single-point energy evaluation with a RIMP2/aug-cc-pVQZ model chemistry. Corresponding molecular mechanical energies were obtained with the CHARMM force field.[4]

Unlike "proper" dihedrals, improper dihedrals are typically treated as harmonic interactions, with the functional form given in Eq. (2):

$$U\left(\vec{R}\right) = \sum_{Impropers} k_{\varphi}(\varphi - \varphi_0)^2 \tag{2}$$

where k_{φ} is the force constant and φ_0 is the equilibrium improper dihedral, typically 0°, corresponding to planarity. Improper dihedral terms are not assigned to all planar groups; in some cases, the "proper" dihedral angles are sufficiently balanced to yield planarity at equilibrium and model the out-of-plane fluctuations accurately.

Together, the bonded energy terms describe the internal energy among covalently bonded atoms, as a function of the bond lengths (b), valence angles (θ), dihedral angles (ϕ), and often improper dihedral angles (φ). More complex potentials exist for describing bonded interactions, including a Morse potential that can model bond breaking, but this energy function is not commonly employed in simulations of proteins. Additionally, there are other force fields that include cross-terms that describe the interplay of bonds, angles, and dihedrals using higher-order mathematical functions, but these force fields are not commonly applied to proteins and therefore will only be described in the relevant case of the AMOEBA polarizable force field (see Section 4.2).

2.2 Nonbonded interactions

The second type of interactions among atoms in an MD simulation is among atoms that are not covalently bonded; as such, these interactions are called "nonbonded interactions," and are typically described by two terms in the energy function, van der Waals interactions and electrostatic interactions. Computing forces arising from nonbonded interactions is typically the most expensive part of an MD simulation, therefore computationally efficient functions are required.

The most common expression for calculating van der Waals interactions is via the Lennard-Jones (LJ) 12-6 potential (Eq. 1). In this construction, repulsion between atoms in close proximity is described by an r^{-12} term and the attractive portion of the interaction has an r^{-6} dependence. As such, at very close distance, the r^{-12} term dominates and the atoms are repelled from one another, emulating the interaction of electron clouds in real atoms, which cannot overlap due to the Pauli exclusion principle. The decay of the LJ potential with r^{-6} dependence allows for truncation of van der Waals interactions at relatively short distances to decrease computational burden since interactions at longer distances are negligible on a per-atom pair basis and methods exist to correct for missing dispersion interactions.[5,6]

The LJ potential requires the assignment of two parameters for each atom type in a force field, which describe how favorable the interatomic interaction is and at what distance the energy minimum occurs. The distance of minimum energy is denoted R_{min}, and the magnitude of the well depth at this distance is denoted ε (Fig. 2). Some force fields may express R_{min} differently, in terms of σ, which is defined as the distance at which the LJ potential between two atoms is zero. R_{min} and σ can easily be interconverted via the relationship

$$\sigma = R_{min} \times 2^{-1/6} \tag{3}$$

LJ parameters are determined for each atom type, but how are heterogeneous interactions among atoms described? Each force field has a "combination rule" (alternatively "mixing rule") for how the individual

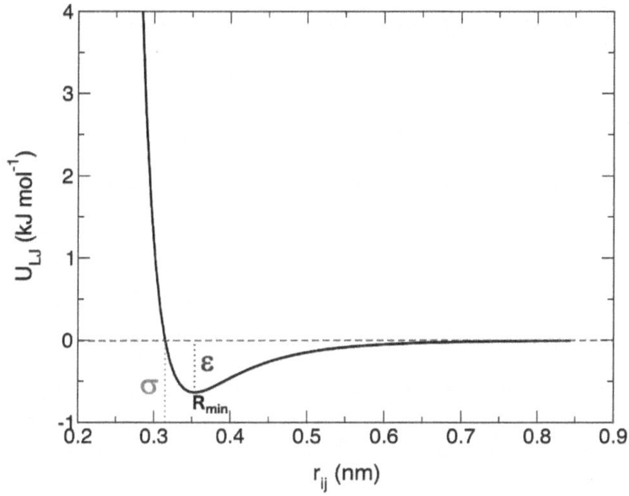

Fig. 2 Potential energy as a function of interatomic distance, r_{ij}, for the Lennard-Jones 12–6 potential. The curve shown was generated using the values of R_{min} and ε from the oxygen atom in the TIP3P water model.[7] The value of σ corresponds to the interatomic distance at which U_{LJ} is zero and ε is the well depth at the distance of minimum energy, R_{min}.

ε and R_{min} of each atom type are combined to describe the interaction between different types. For two given atom types i and j with parameters (ε_{ii}, R^{ii}_{min}) and (ε_{jj}, R^{jj}_{min}), their combined interaction can be determined using arithmetic or geometric means of the parameters. A force field that uses strictly arithmetic means would determine these values as:

$$\varepsilon_{ij} = \frac{1}{2}\left(\varepsilon_{ii} + \varepsilon_{jj}\right) \tag{4}$$

$$R^{ij}_{min} = \frac{1}{2}\left(R^{ii}_{min} + R^{jj}_{min}\right) \tag{5}$$

A force field that relies on purely geometric means (such as OPLS, which typically reports σ instead of R_{min}, see Section 3.4) would calculate these terms as:

$$\varepsilon_{ij} = \sqrt{\varepsilon_{ii}\varepsilon_{jj}} \tag{6}$$

$$\sigma_{ij} = \sqrt{\sigma_{ii}\sigma_{jj}} \tag{7}$$

The final type of combination rule is a mixture of geometric and arithmetic means, where the well depth (ε_{ij}) is determined via geometric mean

(Eq. 6) and the value of R_{min}^{ij} is determined via arithmetic mean (Eq. 4). This combination rule is called "Lorentz-Berthelot" and is used by the AMBER and CHARMM force fields (see Sections 3.1 and 3.3).

Occasionally, combination rules fail to accurately describe the strength of LJ interactions between atom types due to their simple nature and the underlying assumptions of the suitability of the LJ equation. In this case, most force fields allow for pair-specific values of ε_{ij} and R_{min}^{ij} that override the combination rule values. Such corrections may be referred to as "off-diagonal" LJ terms (since they arise from overriding the values that would be precomputed in the matrix of atom types) or "NBFIX" for "nonbonded fix," nomenclature that is specific to the CHARMM force field (see Section 3.3).

Electrostatic interactions comprise the remainder of the nonbonded interactions in a simulation system and are typically described by Coulomb's Law (rightmost term in Eq. 1). In pairwise-additive force fields, each atom is assigned a partial charge based on the electron distribution in a molecule; typically charge assignment is carried out according to QM target data such as molecular dipole moments and electrostatic surface potentials (discussed in Section 3 in the context of force field parametrization). Charges are then assigned assuming some level of overestimation of the gas-phase dipole moment to account for polarization effects in the aqueous phase, for which most biomolecular force fields are parametrized. Thus, these force fields do not consider electronic polarization explicitly, rather they estimate it according to a mean-field approximation assuming a constant dielectric medium surrounding the molecule(s) of interest.

Force fields that explicitly model electronic polarization include additional energy terms or particles that represent electronic degrees of freedom, allowing the electronic structure of a molecule to change in response to alterations of the local electric field. Such changes may arise from partitioning into a different medium or interactions with other molecules of differing polarity. Thus, these force fields include multi-body contributions to electrostatic interactions; that is, the relaxation of induced dipoles is not explicitly modeled in pairwise-additive force fields. Polarizable force fields are therefore non-additive; if a molecule is removed from the system, the dipoles of the other species will be aligned differently and will have different magnitudes, leading to different interaction energies among the remaining molecules. In contrast, with a nonpolarizable (additive) force field, removal of any one species will have no effect on the interaction energies between other species. An example of this phenomenon will be given in Section 4.

2.3 Types of atomistic force fields

Two subsets of atomistic force fields are routinely used in protein simulations. The first is called an "all-atom" force field, which is the most common. As the name implies, these force fields explicitly represent all atoms in the system, including all hydrogen atoms. The other subset of atomistic force fields are "united-atom" parameter sets, and in this convention, aliphatic hydrogen atoms are encompassed by the carbon atom to which they are connected, with the LJ parameters and mass of the carbon atom adjusted to implicitly account for the presence of the hydrogen atoms. Polar hydrogen atoms are included to accurately represent hydrogen bonding interactions that are critical to biomolecular simulations. United-atom force fields are advantageous because they reduce the total number of particles in the system, therefore speeding up the simulation. All-atom force fields may be preferable for accurately modeling aromatic interactions, which have specific orientation effects to their interactions that are not accounted for by a simple spherical representation of an extended carbon atom.[8–10]

3. Pairwise-additive force fields

Many of the most popular nonpolarizable force fields used today have their origins in foundational work conducted nearly 50 years ago. Some of the earliest examples include the use of crystallographic and spectroscopic data to assign bonded parameters (equilibrium geometries and force constants) and use of simple QM calculations for charge assignment. Work by the Lifson group on amides[11] and the Scheraga group on polypeptides[12] stand as some of the earliest examples of successful calculations of small-molecule and polypeptide properties using classical force fields. In 1977, the first simulation of a protein (bovine pancreatic trypsin inhibitor, BPTI) was carried out in Martin Karplus' group using a simplified model of the protein backbone and disulfide linkages.[13] This model was subsequently expanded to include parameters for amino-acid side chains in 1979.[14] Other groups subsequently expanded and adapted these early parameter sets for their own investigations, yielding new families of molecular mechanics force fields that are still in use today. This section will describe the most commonly applied nonpolarizable (pairwise-additive electrostatic) force fields, including their general parametrization strategies and applications to the study of dynamics and folding of proteins. The development history of the nonpolarizable, atomistic force fields described in this chapter is shown in Fig. 3.

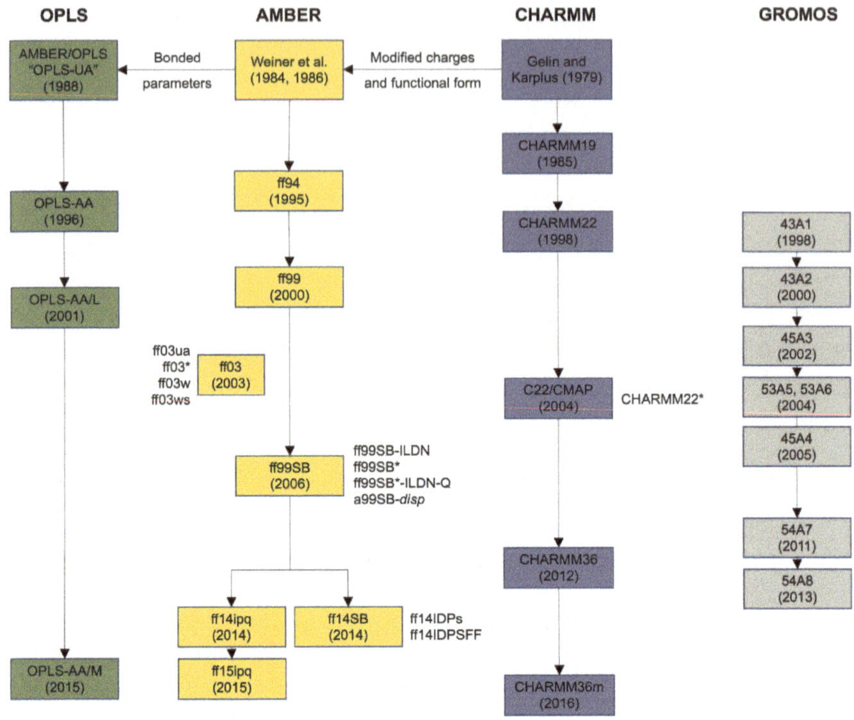

Fig. 3 Force field development history for the AMBER, CHARMM, OPLS, and GROMOS atomistic force fields. Vertical progression reflects major reparametrization efforts and core parameter sets are boxed. Additional force fields appear alongside the parameter set from which they were derived. The common origins of the CHARMM, AMBER, and OPLS force fields is indicated by the parameters transferred or differences between them.

3.1 AMBER

The AMBER (Assisted Model Building with Energy Refinement) force field has its origins in the early 1980s,[15,16] using a modified parameter set derived by Gelin and Karplus.[14] Changes to the original parameter set included an explicit energy term associated with hydrogen bonding and charge derivation using Mulliken population analysis. Additional refinement of the force field, which included data from several spectroscopic methods, prior molecular mechanics calculations, and refinement of charges from more rigorous electrostatic surface potential calculations yielded a united-atom parameter set for proteins and nucleic acids reported by Weiner et al.[17] and the subsequent all-atom version in 1986.[18] Since this time, AMBER force fields have generally been all-atom rather than united-atom, with nomenclature that indicates the year the force field was produced. The construction of the name for an AMBER force field parameter set generally follows the pattern of "ff" (force field) + XX (two-digit year) + some descriptor of modifications, if applicable.

AMBER parametrization methodology, as with most force fields, relies on a combination of empirical and QM target data. Charge assignment in the AMBER family of force fields relies on the use of the electrostatic surface potential (ESP) using a method that enforces symmetry in equivalent atoms/groups via a restraint in the fitting algorithm, called RESP. The model chemistry underlying most AMBER ESP and geometry optimizations is HF/6-31G* as this method overestimates molecular dipole moments by 10–20% due to the absence of electron correlation and dispersion effects, therefore implicitly accounting for polarization effects in the aqueous phase, which is desirable when deriving a biomolecular force field. Torsional scans and relative energies of model compounds are carried out at higher levels of theory, at least MP2/6-31G* but often using larger basis sets or more complex electronic structure methods. Validation of AMBER parameter sets involves comparison of MD outcomes to crystallographic data, free energies of solvation (originally assuming the use of the TIP3P water model, but as will be discussed below, improved water models may also be combined with the AMBER force field), and conformational energies from QM calculations.

Building off of the all-atom force field by Weiner et al., Cornell et al.[19] proposed a revised force field (called "ff94") that involved a reparametrization of charges, removal of the explicit hydrogen-bonding energy term, and inclusion of torsional terms for ϕ and ψ in the peptide backbone due to the availability of QM energy data for dipeptides. Partial atomic charges on small peptides were fit using a RESP method to avoid conformation-specific charge assignment that was encountered when simply fitting to the QM ESP; that is, when considering only one conformation of a molecule, asymmetries often arise in the charges assigned to each atom. The RESP method enforces symmetry to chemically equivalent atoms and groups that overcomes this problem, with the goal of more physically realistic conformational sampling. The suitability of the RESP charge assignment was subsequently evaluated by Wang et al.,[20] leading to a reparametrization of dihedral parameters to improve conformational energies of model compounds that give rise to proteins and nucleic acids. This parameter set, ff99, has become the basis for all subsequent versions of the AMBER force field.

A significant departure from the previous AMBER parametrization methodology was made with the release of ff03,[21] in which conformational energy profiles and ESP were generated in the presence of a continuum solvent ($\varepsilon = 4.0$) rather than in vacuo, as previous force fields had done. The authors chose a low dielectric constant for their continuum model, rather than the dielectric constant of water ($\varepsilon = 78.4$) to mimic the protein interior and to avoid potentially over-polarizing the force field. The combination of a low dielectric and the Hartree-Fock electronic structure method was determined to be an adequate model for charge assignment for solution-phase dynamics. Geometry optimizations were performed with the HF/6-31G** model chemistry, with subsequent single-point energy calculations (performed on multiple conformations of each dipeptide) employing the B3LYP/cc-pVTZ model chemistry. Dihedral parameters for backbone torsions were derived using a 2D grid of ϕ and ψ in the full range of 360°, in increments of 30°. A united-atom version of ff03, ff03ua, was later produced by Yang et al.[22]

Given the dramatic differences in the methods applied by Duan et al. from ff94 and ff99, ff03 is often considered a distinct force field rather than an extension or revision of ff99, and development of the core AMBER force field has proceeded from ff99 rather than ff03, though some modifications to ff03 have been proposed. Best and Hummer determined that ff03 over-stabilizes α-helices and subsequently proposed a revision to the backbone torsions, calling the resulting force field ff03*.[23] The ff03* force field produced more accurate helical content in the model Ac-(AAQAA)$_3$-NH_2 peptide at ambient temperature relative to ff03, which was almost entirely helical. Moreover, the effect of the parameter modification on globular proteins was shown to be minimal as the backbone order parameters and residual dipolar couplings (RDC) remained in good agreement with both experimental data and the original ff03 parameter set. An examination of helix formation employed the Lifson-Roig model, which describes helix-coil equilibrium in terms of a nucleation parameter, v, that is the equilibrium constant between helix and coil for a given residue, and an extension parameter, w, that describes the equilibrium constant for converting a disordered residue at the end of a helix into a helical one. The simulations performed by Best and Hummer revealed that the ff03* parameters produced a w value that is too small and is insufficiently sensitive to temperature relative to experimental data; the v value is too large and therefore indicates inadequate cooperativity in helix formation. These deficiencies may reflect an intrinsic limitation in nonpolarizable force fields, in that they lack hydrogen bonding directionality and a polarization enhancement as more hydrogen bonds form, though the effect could potentially be captured by augmenting the force field with off-center "lone pair" virtual sites to account for directionality and possibly by adding an energy term to account for hydrogen bonding cooperativity.

Since ff03* was unable to reproduce temperature dependence of α-helical content in the Ac-$(AAQAA)_3$-NH_2 peptide, Best and Mittal proposed the use of the TIP4P/2005 water model[24] instead of TIP3P and a modification of the force constant on the ψ backbone dihedral from 1.5 kJ mol^{-1} in ff03* to 0.75 kJ mol^{-1} in the new force field, which they called ff03w.[25] The phase angle, δ, for ψ in both ff03* and ff03w was set to 285.5°, whereas most force fields use values of 0° or 180°. A description of such cases and their implications will be provided below in discussion on the CHARMM force field in Section 3.3. By evaluating similar properties as in the case of ff03*, Best and Mittal obtained better agreement in Lifson-Roig parameters for Ac-$(AAQAA)_3$-NH_2 and similar agreement with chemical shifts and RDC data in folded proteins. Thus, it was concluded that minor refinement of backbone torsions in concert with a more accurate water model could be well suited to examining helix-coil equilibrium. Expanding upon this concept, Best et al. proposed in 2014 that a scaling factor could be applied to protein-water interactions to improve the balance of protein-protein and protein-water interactions, thereby increasing the accuracy of the resulting conformational ensemble.[26] They combined their previous ff03w parameter set with a scaling factor, γ, that is applied to the well depth of the LJ interaction between all protein atoms (ε_i) and the oxygen atom of water (ε_O):

$$\varepsilon' = \gamma\sqrt{\varepsilon_O \varepsilon_i} \tag{8}$$

where γ was set to 1.1 by Best et al., thereby increasing the attraction between water and protein atoms by 10% without changing intra-protein interactions, yielding a parameter set they called ff03ws. This simple adjustment resulted in more expanded structures of intrinsically disordered proteins that were in agreement with small-angle X-ray scattering (SAXS) profiles and Förster resonance energy transfer (FRET) data. Additionally, small-molecule hydration free energy values were improved, and there were no adverse effects on folded protein ensembles. These outcomes emphasize the importance of balance in force fields and inclusion of hydration in parametrization, as overly strong intramolecular interactions will adversely affect conformational sampling and properties of peptides in solution. The ff03 series of parameter sets was recently evaluated in a systematic fashion by Zerze et al. in an examination of unfolded and disordered proteins.[27] By analyzing disordered proteins of different size and composition, they were able to determine the tendencies for different force fields to favor compact versus elongated conformations as well as underlying secondary structure preferences. The authors used a well-known relationship to describe the collapse of a polymer chain in terms of its radius of gyration (R_g) as a function of the number of constituent monomers, N:

$$R_g = kN^{\nu} \tag{9}$$

where k is a prefactor and ν is the scaling exponent. The larger the value of ν, the more extended the chain is. Collapsed chains have values of ν on the order of 0.5 in water, whereas denatured structures are characterized by ν of 0.6 and higher. Zerze et al. note that most atomistic force fields yield values of ν approximately 0.3–0.4, indicating that the structural ensembles of disordered peptides are too compact relative to experimental evidence. The ff03* and ff03w force fields produced ν values (0.38 and 0.39, respectively) that indicate their ensembles are too compact, while ff03ws produced $\nu=0.53$, in good agreement with structural data on disordered states. Interestingly, an examination of islet amyloid polypeptide (IAPP), *Thermatoga maratima* cold-shock protein (CSP), and TAR-DNA binding protein (TDP) with each of these three force fields revealed that despite very different levels of compaction produced by each force field, secondary structure content was comparable. This outcome is sensible in light of the fact that the ψ dihedral parameters are shared by all three force fields in an effort to obtain better balance between helical and extended states. Further validation against chemical shifts and scalar ^{3}J-couplings (both of which also report on local structure) showed similar results among the parameter sets. Thus, the conclusion of the work is that it is important to consider multiple types of experimental data in parametrizing and refining atomistic force fields. While improved agreement with quantities related to local secondary structure may be encouraging, it is still possible that the model does not capture global aspects of protein structure.

Given that ff03 was a substantial theoretical departure from previous AMBER force fields, most revisions to the AMBER force field have used ff99 as the basis for further development. The first major revision to ff99 occurred in 2006, when Hornak et al. produced the ff99SB ("SB" for

Stony Brook University, where the Simmerling group is based) parameter set.[28] Having observed incorrect balance in the conformational sampling of previous ff94 and ff99 variants, Hornak et al. reparametrized the backbone ϕ and ψ torsions using high-level QM energies of glycine and alanine tetrapeptides, a more rigorous assessment than previous efforts using dipeptides and limited conformations. Hornak et al. determined that previous efforts to correct the imbalance in secondary structure propensities suffered from limitations in the manner in which the backbone torsion parameters were fit. For all non-glycine amino acids, a second set of dihedral parameters affects ϕ and ψ rotation, ϕ' (C-N-Cα-Cβ) and ψ' (Cβ-Cα-C-N). Thus, parameter fitting to ϕ and ψ alone for alanine-based peptides that were carried out by many groups will result in incorrect sampling behavior when applied to glycine, which lacks ϕ' and ψ'. After parameter fitting, the new backbone dihedral parameters were employed in simulations of glycine and alanine tetrapeptides (Gly_3 and Ala_3) using ff99, ff99SB, and several other ff99 variants that preceded the study by Hornak et al. The sampling in (ϕ,ψ) space was compared against a survey of the Protein Data Bank (PDB); ff99SB significantly improved agreement for Gly_3 and Ala_3 over previous force fields, with results comparable to ff03. For Ala_3, ff94 and ff99 showed a pronounced preference for sampling α_R conformations, consistent with observations that these force fields over-stabilize helices. The authors subsequently performed a rigorous test of the force field by evaluating the energy of longer peptides with known structures. Lower deviations from the experimental structure should correspond to lower energy and indeed ff99SB performed well in this regard, suggesting that the parameter fitting to glycine and alanine tetrapeptides was transferable to other polypeptides and proteins. Finally, as in other evaluations of force field revisions, Hornak et al. analyzed order parameters in lysozyme and ubiquitin, finding that ff99SB produced the best correlation between experimental and computed order parameters between the two proteins. Together, the data demonstrated that careful reparametrization of the backbone dihedral terms in the ff99 force field could be undertaken in a consistent manner that does not bias the force field toward particular conformations as other previous attempts had, and the ff99SB force field became the recommended AMBER parameter set at that time.

Several revisions to ff99SB were subsequently proposed to further improve this force field. The first of these occurred in 2006, when Lindorff-Larsen et al. proposed several revisions to amino-acid side-chain torsions.[29] These revisions specifically targeted the χ_1 dihedral in isoleucine,

χ_2 in leucine, and both χ_1 and χ_2 in aspartate and asparagine, hence the force field was named ff99SB-ILDN, according to the one-letter codes for these amino acids. After parameter refitting, MD simulations of helical Ac-A_4XA_4-amide peptides (X = one of the 20 canonical amino acids) were performed and the side-chain rotamer distributions were extracted and compared against distributions obtained from the PDB. Application of ff99SB-ILDN led to dramatic improvement in χ_1 of isoleucine and leucine, and modest improvement for aspartate and asparagine. Following this evaluation, a more stringent test against solution NMR data (^{3}J scalar couplings and RDC) was performed using four proteins as test cases (bovine pancreatic trypsin inhibitor, GB3, lysozyme, and ubiquitin). Though both ff99SB and ff99SB-ILDN produced stable structures with low backbone root-mean-square deviation (RMSD), ff99SB-ILDN reduced deviations in scalar ^{3}J-couplings associated with χ_1 sampling in the targeted amino acids. RDC data were available for GB3 and lysozyme, and again the RMSD between computed and experimental couplings was dramatically improved using the modified parameter set, reducing the RDC RMSD in lysozyme by half relative to ff99SB. The results of this refinement and validation clearly demonstrate the importance of using experimental solution-phase data in addition to conformational populations in validating protein force fields. Rauscher et al. recently evaluated the performance of ff99SB-ILDN in simulations of intrinsically disordered proteins (IDPs), finding that this parameter set produced peptide conformations that were too compact and that the hydrophobic effect is too strong with this force field.[30] These results indicate that despite the notable improvements in ff99SB-ILDN relative to ff99SB, additional refinement would be necessary to make this parameter set suitable for studies on protein folding.

In addition to revising ff03 to produce ff03*, Best and Hummer also revised ff99SB by adjusting the ψ dihedral parameters ($k_\psi = 0.1788$ kJ mol^{-1}, $\delta = 105.4°$), assessing this revised parameter set in the same manner as described above for ff03*.[23] The revised ff99SB* performed comparably to ff03*, demonstrating that even simple revisions to single torsions can have pronounced impact on disparate parameter sets that were derived in different ways. Not long after, Best et al. also proposed a revision to ff99SB-ILDN, which they called ff99SB*-ILDN-Q,[31] which included the ψ parameter revisions of Best and Hummer[23] and the χ_1 and χ_2 revisions of Lindorff-Larsen et al. with an additional update to partial charges (hence "Q" at the end of the force field name to reflect the notation of charges in the Coulomb equation) in lysine, arginine, protonated histidine, aspartate, and glutamate. In ff99, the partial charges of backbone atoms were constrained to be uniform across amino acids to simplify the fitting process, however, in the approach taken by Best et al., charged residues could adopt different charges in backbone atoms, distinguishing them from neutral amino acids. The changes to the aspartate charges necessitated a reparametrization of the ff99SB-ILDN χ_1 for this residue. The ff99SB*-ILDN-Q produced helical propensities that were in better agreement with experimental data, while preserving the structural integrity of folded proteins (ubiquitin and lysozyme in this case).

In a very recent development, the ff99SB*-ILDN-Q force field served as the starting point for a refinement by Robustelli et al., who modified this parameter set and the TIP4P-D water model[32] to yield a well-balanced force field to describe both folded and disordered proteins.[33] Their strategy was to minimally perturb the existing parameters and better capture the conformational sampling of a wide variety of proteins, in terms of NMR observables and global properties from SAXS data. The TIP4P-D LJ parameters were adjusted to better balance intra- and inter-molecular interactions in the proteins, the attraction between amide carbonyl oxygen and nitrogen atoms was refined to better model hydrogen bonding strength within the polypeptide backbone, several backbone and side-chain torsions were adjusted, and charges were refined in aspartate, glutamate, and arginine. Together, these refinements led to a force field the authors called a99SB-*disp*, so named for the adjustments to the dispersion strength of the TIP4P-D water model. The a99SB-*disp* force field was subsequently evaluated against a benchmark set of 21 different proteins, ranging from short peptides to globular structures. Notably, a99SB-*disp* retained the ability to model the folded proteins while simultaneously improving on agreement with disordered proteins, though some difficulties were still noted in capturing the folding behavior of $(AAQAA)_3$, CLN025, and Trp-cage,[33] suggesting that additional refinement will still be necessary to accurately simulate this wide range of proteins.

The next substantial revision in the ff99 series was a modification of backbone and side-chain dihedral parameters in the ff99SB parameter set, called ff14SB, which sought to address deficiencies in side-chain sampling and under-stabilization of α-helical structures.[34] A key aspect to the approach taken by Maier et al. was the use of multiple backbone conformations in model peptides, rather than relying only on gas-phase minima. In the fitting protocol for ff14SB, both α ($\phi = -65°$, $\psi = -45°$) and β ($\phi = -135°$, $\psi = 135°$) geometries were used in scanning side-chain rotational energy profiles. This consideration is significant as there is known coupling between torsions in peptides and proteins, whereas previous versions of the force field (ff94 and ff99) had assumed all dihedrals could be parametrized independently. As such, the development of ff14SB was important in reducing or removing the dependence of side-chain sampling on backbone geometry. Another important advancement was the use of full amino acids in QM calculations, rather than small-molecule analogs as had been used in ff99 and ff99SB; the coupling of dihedrals may not be adequately treated if all relevant degrees of freedom are not considered in tuning dihedral parameters. Lastly, new atom types were introduced into the AMBER force field to allow independent tuning of backbone and side chain terms, as well as decoupling from the nucleic acid force field, which shared some of the same standard parameters. While the goal of any force field is transferability between species, compromises are often made in the name of accuracy and the introduction of new atom types resulted in a larger number of parameters but greater overall accuracy in the protein force field.

The ff14SB force field introduced a single modification of the backbone ϕ dihedral parameters as well as several side-chain dihedrals; no adjustment was made to ψ as in other ff99 revisions discussed above. Side-chain dihedral parameters were adjusted based on over 15,000 dipeptide conformations with energies evaluated with the MP2/6-31+G** model chemistry. The combination of the revised side chain dihedral parameters and the ϕ dihedral adjustment led to the final force field. Importantly, Maier et al. also tested the use of only the side-chain parameter modifications in a force field they denoted ff14SBonlysc. Doing so allowed them to demonstrate specific enhancements that arose from more accurate side-chain parameters. The final parameter set was chosen by demonstrating improvement in the agreement of Ala_5 scalar couplings and helical content in two model peptide systems for which helical content in solution is known. The ff99SB parameter set uniformly yielded helical content that was about 25% of what is expected from experiments; use of ff14SB produced near-quantitative agreement with experimental data. Crucially, it was shown that ff14SBonlysc nearly doubled the helical content of ff99SB, demonstrating that improved treatment of amino-acid side chains influences backbone conformational sampling. It is therefore important to carefully consider the unification of various modifications; for example, Maier et al. note that the correction of Best and Hummer that produced ff99SB*[23] may have led to improvement by cancellation of errors in backbone and side-chain terms. Applying an additional correction to side-chain dihedrals in this case may lead to worse agreement. Therefore, considering the coupling of dihedrals in protein dynamics is crucial to producing an accurate force field model.

The ff14SB force field was further validated against a β-hairpin peptide (CLN025, an engineered version of the fast-folding chignolin) and four folded proteins (GB3, ubiquitin, lysozyme, and bovine pancreatic trypsin inhibitor). In the case of CLN025, ff14SB favored the native-like fold by 2.9 kcal mol^{-1} relative to ff99SB, which sampled a larger number of conformations in which the N- and C-terminal β-strands were shifted out of register by one amino acid. ff14SB also produced a greater variety of unfolded structures throughout 4-μs simulations and produced lower nuclear Overhauser effect (NOE) deviations than ff99SB. In the case of the four folded proteins, ff14SB, ff14SBonlysc, and ff99SB-ILDN performed comparably in regard to side-chain scalar couplings, and all were substantially better than ff99SB. In case of solvent-exposed α-helical residues with only χ_1 dihedrals, ff14SB was the best in terms of agreement with experimental scalar couplings. This outcome further emphasizes the importance of including multiple backbone conformations during fitting, as ff99SB-ILDN used only β conformations and ff14SB used both α and β. Since its publication, ff14SB has become the recommended AMBER force field in the ff99 lineage for simulations of peptides and proteins. A recent comparison of a dozen widely used protein force fields revealed that ff14SB, ff99SB-ILDN, and ff99SB*-ILDN were among the best at reproducing side-chain rotamer populations in ubiquitin and GB3 when compared to NMR data.[35] Thus, the recent strategies in refining these force fields have proven robust for sampling subtle details of protein dynamics.

It should be noted here that the ff14SB parameter set has been further modified specifically for use in simulations of IDPs. Song et al. introduced the ff14IDPs force field,[36,37] and subsequently ff14IDPSFF,[38] specifically for use in simulations of IDPs. Song et al. added energy correction terms in the style of CHARMM CMAP corrections (see Section 3.3) to change the backbone sampling specifically for residues identified to be enriched in disordered protein sequences. This specific parametrization target differs from other force fields that seek to recover the probabilities of (ϕ,ψ) pairs of a full Ramachandran plot that includes folded proteins. IDPs simulated with ff14IDPs had more extended structures with less α-helicity, and overall produced more diverse conformational ensembles. Simulations of ubiquitin and lysozyme (100 ns in length) suggested folded proteins were still stable when using ff14IDPs, though longer tests may be needed to conclusively determine the suitability of an IDP force field for folded proteins. In the next step of this IDP-specific force field development, Song et al. proposed backbone correction terms for all 20 canonical amino acids, not just those believed to induce disorder, and the resulting parameter set for all amino acids was subsequently named ff14IDPSFF.[38] The updated ff14IDPSFF parameter set produced more diverse conformational ensembles than ff14SB and ff14IDPs and allowed peptides to sample even more extended states, as quantified by radius of gyration.

The ff14SB and ff14IDPs force fields were recently evaluated by Krupa et al. in a systematic comparison of the amyloid β-peptide (Aβ) via continuous, unbiased MD and temperature replica-exchange MD (REMD).[39] The ff14IDPs force field performed the best in terms of agreement with NMR chemical shifts, but produced similar distributions of radius of gyration compared to its parent force field, ff14SB. Simulations with ff14IDPs also produced similar secondary structure content as the old ff99SB force field, which overestimated β-strand content relative to experimental measurements.[40,41] The authors noted that whereas the other force fields examined (which included several CHARMM parameter sets, see Section 3.3), use of ff14IDPs in REMD simulations led to no enhancement of conformational sampling, suggesting that this force field does not respond quickly enough to changes in temperature that induce local structural transitions. However, ff14IDPs performed very well in continuous MD simulations, indicating that investigators may not need to run expensive REMD simulations to obtain good sampling, reflecting a benefit to the use of ff14IDPs. It is also important to note that it is somewhat difficult to assess the quality of ff14IDPs against other parameter sets in the context of IDPs, because other force fields are designed to be broadly applicable to many different classes of folded and unfolded proteins, whereas ff14IDPs was specifically designed for IDPs.

3.2 AMBER-derived implicitly polarized force fields

Recently, the Case group has led the development of a modification to the AMBER force field, called the IPolQ model,[42] that applies implicitly polarized charges with the goal of improving the representation of electrostatics in the fixed-charge convention by specifically targeting the polarization response of protein functional groups in a reaction field in the partial charge assignment. Rather than modifying dihedral terms as many previous force field refinements have employed, the IPolQ model seeks to better address the effects of electronic polarization without incurring the computational expense of modeling this property explicitly. The IPolQ model was constructed by importing LJ parameters from the ff99SB parameter set, while recomputing charges in vacuum and in the presence of a reaction field that models water. The final charge set is the average of these two charge sets, reflecting an approximate treatment of electronic polarization. The initial IPolQ model was iteratively refined by computing charges in conjunction with ff99SB LJ parameters to reproduce hydration free energies of amino-acid analogs; this fitting protocol resulted in some modifications to the LJ parameters of several atom types as directly importing them from ff99SB, which uses a different set of charges, led to good agreement that was substantially improved upon refinement of LJ radii.

A crucial component of the IPolQ parametrization methodology is the calculation of ESP around model compounds, in keeping with the AMBER force field convention. Multiple ESP were computed for different conformations of dipeptides in vacuum, and one conformation of a dipeptide in the presence of a reaction field generated by placement of perturbing charges. The generation of this reaction field was novel in the context of force-field parametrization. Simulations of fixed dipeptides were performed in a bath of TIP4P-Ew water molecules[43] to obtain the solvent distributions around the dipeptides. Perturbing charges were then placed to model water distribution during the QM calculations to generate the target ESP. The final parameter set developed by Cerutti et al. showed important differences with respect to ff94 and ff03, namely in that polar protein functional groups were predicted to be more polar in IPolQ as a consequence of the implicit polarization underlying the charge model. Some deviations in IPolQ ESP were noted at hydroxyl and carbonyl oxygen moieties, specifically in the locations where lone pairs would be located. This observation suggested that this initial IPolQ model could be improved with the inclusion of virtual sites off of the atomic sites.

From the initial IPolQ model, Cerutti et al. developed the ff14ipq force field for proteins in 2014, as the previous model did not refine dihedral parameters to capture proper sampling of the polypeptide backbone and amino-acid side chains. In producing the ff14ipq parameter set, Cerutti et al. sought to solve two issues, one related to the compatibility of the assigned charges with the dihedral parameters, and the second regarding problems with the force field model in reproducing high-energy features of QM potential energy surfaces. Targeting QM potential energy surfaces for dihedral rotation in the gas phase in the presence of condensed-phase charges can lead to problems in transferability, as the balance of forces in the gas and aqueous phases will be different. Therefore, in refining the IPolQ method towards building the ff14ipq force field, a new method for deriving dihedral parameters using gas-phase target data was needed.

To do so, the final polarized charges, Q^{IPol}, were expressed as a perturbation of vacuum charges, Q^{vac}, via:

$$Q^{IPol} = Q^{vac} + \Delta Q \tag{10}$$

where ΔQ is a perturbation to the charges that reflects a change in dipole moment upon transfer to aqueous solvent. Thus, the dihedral parameters could be fit to the QM potential energy surface using Q^{vac} and transferred for use with Q^{IPol} by constraining ΔQ to be small during parameter fitting. To overcome the issue of high-frequency motions impacting the potential energy surfaces and thus fitting to lower-frequency motions in model compounds, Cerutti et al. developed a method by which dipeptide and tetrapeptide simulations were performed and an ensemble of conformations was extracted. These configurations were then subjected to single-point QM energy calculations. By doing so, configurations that are most relevant in MD simulations are prioritized and serve as new fitting data. The final modification made in ff14ipq parameter refinement was the use of off-diagonal LJ terms to describe interactions with water. In the original IPolQ parametrization, the radii of several atoms had to be increased to improve hydration free energies, but these larger radii led to problems in dihedral parameter fitting. Thus, the increased radii were employed only in computing interactions with the TIP4P-Ew oxygen atom, thus allowing the intramolecular LJ interactions (primarily 1–4 interactions that affect dihedral rotation most strongly) were computed with shorter radii. The ff14ipq was shown to model ϕ/ψ distributions well in model peptides and produce good agreement with NMR J-couplings in Ala_5, and while most globular proteins remained stable, several issues emerged. In general, polar interactions were somewhat too strong, leading to very tight salt bridges, over-stabilization of the GB1 hairpin, and potentially backbone interactions that were too strong. These deficiencies motivated the refinement of ff14ipq to yield ff15ipq in 2016.[44]

The parametrization of ff15ipq involved new angle and dihedral terms, a reparametrization of charges, and new atom types in different classes of amino acids (N, Cα, or Cβ) to allow for more specific dihedral refinement. A major change in the force field was the use of the SPC/E_b water model[45] in place of TIP4P-Ew; this change was motivated by (1) its lower computational cost due to having only three sites instead of four and (2) it was parametrized to correctly model rotational diffusion of proteins in aqueous solution. A particular advantage of the second point is that the use of SPC/E_b

allows for accurate comparisons of fast motions in proteins, particularly backbone dynamics, that are not possible when using water models with incorrect diffusion behavior, like TIP3P that is used for AMBER force field development. The change in water model required a complete reparametrization of charges on the protein moieties as the charge distribution from TIP4P-Ew that was used to generate the reaction field in QM calculations could no longer be used.

Debiec et al. demonstrated that cation-anion pairing in solution of amino-acid side chain analogs were too strong with ff14ipq (as well as in other state-of-the-art nonpolarizable force fields); the source of this problem was the use of off-diagonal LJ terms to describe ff14ipq protein-water interactions. The increased radii on some protein atoms led to weakened affinity with water, causing desolvation and contact ion pairing among charged residues. To overcome this issue, the authors discarded the pair-specific LJ parameters and instead increased the radii selectively on N and H atoms, leading to improved agreement with experiment. The ff15ipq parameter set was overall the best among nonpolarizable force fields in describing the probability of ion pairing in solution, however, none of these force fields was as accurate as two polarizable models, AMOEBA and Drude, which will be discussed in Sections 4.2 and 4.3.

Validation simulations with ff15ipq included small β-hairpins and short α-helices, globular proteins including the Trp-cage miniprotein, and intrinsically disordered peptides from p53 and the S-peptide (simulated alone in solution and in complex with their respective binding partners to test the ability to adopt α-helical conformations). The inclusion of IDPs in the validation set but not during force field parametrization and refinement represents a stringent test of the force field that differs from the approaches described above for IDP-specific force fields, which directly targeted IDP properties. Debiec et al. demonstrated that ff15ipq produced good balance between sampling α-helical and extended β-strand conformations. The K19 and $(AAQAA)_3$ model peptides showed temperature-dependent unfolding of their α-helical structures, with K19 maintaining slightly more α-helical character at ambient temperature than $(AAQAA)_3$, as expected. The authors noted, however, that this observation may be due to intrinsic differences in lysine and glutamine residues that arose from parametrization; lysine has a more prominent energy minimum in the α region of the Ramachandran surface, and its backbone carbonyl oxygen charge is ~0.05 *e* more negative than in glutamine, thus altering the propensity to form hydrogen bonds. Nevertheless, given the multiple folding and unfolding events in each trajectory, it was determined that ff15ipq does not show a bias toward α-helical structures. Simulations on β-hairpins GB1, chignolin, and hyper-stable chignolin derivative CLN025 again revealed qualitative agreement in terms of the relative stabilities of chignolin and CLN025, but CLN025 was too stable beyond its melting temperature, remaining folded for approximately 80% of simulation time at 370 K, when experimental evidence suggests it should be roughly 25% folded at this temperature. Together, the simulations of small α-helices and β-hairpins do not indicate any systematic bias toward a given secondary structure, a major improvement over earlier parameter sets described above.

Globular proteins were stable with ff15ipq over very long (10-μs) MD simulations, some of the most rigorous validation simulations performed on any proteins to date. Debiec et al. noted the importance of long simulations in assessing force field quality, as some conformational changes took several microseconds to observe. The Trp-cage miniprotein remained stable as in previous AMBER parameter sets and backbone sampling of proline-containing dipeptides (a specific target of dihedral refinement) was noted to be in good agreement with experimental evidence. In other proteins, some deviations in side-chain sampling were noted in branched-chain amino acids, indicating that further refinement would be appropriate.

In the final set of simulations, the ff15ipq force field was applied to an intrinsically disordered region of p53 and the S-peptide. When p53 binds MDM2, it converts from a random coil to an α-helix. S-peptide binding to S-protein induces the same type of conformational change. Over the course of 10-μs simulations of each peptide alone and in complex with its respective binding partner, ff15ipq generally produced the expected behavior. Simulations of p53 alone transiently sampled α-helical states but S-peptide sampled none. Both peptides retained their α-helical structure when simulated in complex with their binding partners, though some inter-protein contacts were lost in each case and S-peptide transiently sampled partially disordered states. Overall, these results are encouraging and emphasize the importance of proper charge assignment, consistent development of internal (bonded) force field terms, and consideration for balance of solute-solute and solute-solvent interactions. The implicitly polarized charge model represents an important step forward in fixed-charge force field parametrization.

3.3 CHARMM

CHARMM (Chemistry at HARvard Macromolecular Mechanics) is another all-atom force field that is commonly employed in simulating proteins.

Its parametrization methodology relies primarily on first matching quantities computed using QM calculations (optimized geometries, molecular dipole moments, rotational energy profiles, and interactions with water) and comparing the physical properties of the resulting molecules produced by the force field with available experimental data (crystal geometries, heats of vaporization and sublimation for solids, enthalpy of vaporization and density for neat liquids, and conformational sampling of biomolecules). QM optimizations are typically carried out at the MP2/6-31G* level of theory (6-31+G* to include diffuse functions for molecular ions), and explicit water interaction energies are computed using the HF/6-31G* (or 6-31+G*) model chemistry; the use of the Hartree-Fock method is for the same reason as described above in the discussion on AMBER force fields, to approximate polarization response in solution. In assigning partial charges, polarization effects are also implicitly accounted for by targeting the MP2 molecular dipole moment, overestimated by approximately 20%. The target QM-optimized distances between water and the relevant interacting group are shortened by 0.2 Å and the interaction energy scaled by a factor of 1.16 to mimic the same effect and to achieve balance among solute-solute and solute-solvent interactions, which is critical for the accuracy of any force field.[4,46] Such scaling is only applied to neutral compounds; in the case of molecular ions, no scaling is performed on the target data. Scaling of the distance is necessary to account for missing dispersion interactions using the Hartree-Fock model and to yield an appropriate density in the TIP3P water model,[7,47] which is the model for which the CHARMM force field is parametrized for use.

Each version of the CHARMM force field is indicated by a number that corresponds to the version of the CHARMM software with which the parameter set was released. The first parameter set, CHARMM19,[48,49] was a united-atom force field, thus including only polar hydrogen atoms with "extended" atom types for aliphatic carbon atoms. CHARMM19 included topologies and parameters for the 20 canonical amino acids, water, heme, and several heme ligands. The next version was called CHARMM22,[4] which was an all-atom parameter set that was subsequently revised in 2004 to include polypeptide backbone energy corrections called "CMAP,"[50,51] and the force field name became CHARMM22/CMAP. Due to the coupled nature of the ϕ and ψ torsions, it is difficult to capture the relative energies of each conformation with the simple cosine energy terms described above. Therefore, a CMAP term is applied as an additional energy term to the CHARMM functional form, which is similar to the one shown in Eq. (1). The correction that is applied at each (ϕ,ψ) point is the difference in QM and MM energies of model dipeptides. The fineness of the grid is 15°, and corrections applied to values that do not fall exactly on the specified points are interpolated in a grid-based manner. The application of CMAP to the CHARMM energy function was subsequently shown to improve the relative energetics of different backbone conformations in model dipeptides and also to reduce deviations in backbone geometries of a range of globular proteins. However, MacKerell et al. noted that while it is possible to reproduce the QM energy surface with the MM force field using the CMAP corrections, doing so led to deviations in α-helical and β-sheet structure, requiring empirical modifications to recover better condensed-phase behavior. The conclusion from this observation is that while corrections can be made to additive force fields, the polarization response in the gas phase and in solution will be different, therefore a purely QM-based parametrization scheme for a nonpolarizable force field will not yield perfect agreement with conformational sampling in solution.

Subsequent MD studies using the CHARMM22/CMAP force field revealed that it over-stabilized α-helical structures,[52,53] prompting Piana et al. to propose a revision to several force field terms as part of a systematic evaluation of AMBER and CHARMM parameter sets their ability to produce correctly folded structures of the villin headpiece.[54] The authors proposed the elimination of the CMAP terms from CHARMM22/CMAP for all residues except glycine and proline, and replacing the CMAP with new parameters for the ϕ and ψ dihedrals. Additionally, to correct over-stabilization of salt bridges (which ultimately impacts conformational sampling and would restrict accessible conformational space in folding proteins), Piana et al. reparametrized the charges on the side-chain atoms of aspartate, glutamate, and arginine, as well as changing the side-chain dihedral parameters for aspartate. The combination of these revisions yielded a force field that the authors called CHARMM22*. The authors subsequently evaluated AMBER force fields ff03[21] and ff99SB-ILDN[29] and CHARMM force fields CHARMM22/CMAP[4,50,51] and CHARMM22*. It should be noted that Piana et al. refer to CHARMM22/CMAP as "CHARMM27," as is common in the literature. The CMAP revisions to CHARMM22 appeared in version 27 of CHARMM and were incorporated into force field files with the same name, hence the colloquial adoption of the force field name. Piana et al. noted that all the tested force fields could reproduce the correct folded state of the villin headpiece, and over the course of simulations in excess of 100 μs, each force field produced trajectories during which the protein underwent reversible folding events, therefore allowing calculation of the folding rate. Both AMBER ff03 and CHARMM22/CMAP exhibited higher thermal stability due to their known problem of over-stabilizing α-helices. Therefore, both of these force fields yielded structures that had comparably large values of α-helical content in unfolded states and therefore predict different folding pathways via the nucleation of helical elements in different orders. Although CHARMM22* improved upon helical content and yielded good results in terms of folding rate and thermodynamics, it was also shown that it could not reproduce the dependence of helical content in the $(AAQAA)_3$ peptide as a function of temperature, remaining largely flat across a wide temperature range.[54]

Lindorff-Larsen et al. subsequently used CHARMM22* in a landmark study on simulating the folding of small proteins that are structurally diverse,[55] expanding upon the work of Piana et al. and serving as a more rigorous assessment of the CHARMM22* force field. Lindorff-Larsen et al. also produced trajectories up to 1 ms in length, at least 1000 times longer than what is normally considered state of the art. By doing so, they were able to interrogate protein folding in an exhaustive manner and again compute folding trajectories as well as folding rates. They found that they could generally obtain folded structures within 2-Å Cα RMSD to experimentally determined structures, though exceptions were noted for larger proteins and those with greater loop content, as these regions are difficult to resolve experimentally and also to simulate. For each of the 12 proteins examined, a dominant folding pathway emerged as a function of the nucleation of secondary structure elements, which formed before longer-range (tertiary) contacts were established. As part of a comparison of several force fields and combinations with different water models, Rauscher et al. investigated the CHARMM22* force field to determine the agreement of simulation ensembles of disordered peptides with experimental data such as radius of gyration, hydrodynamic radius, and SAXS profiles.[30] CHARMM22* (when used with the CHARMM-modified TIP3P water model) performed the best in reproducing these quantities, as well as chemical shifts but not RDC data.

The parametrization of the CHARMM22* force field made use of some unconventional force field terms, including force constants with negative values and phase angles that deviated from 0° or 180°. Mathematically, nothing prevents either of these conventions from being used, however, it is generally preferred within the CHARMM force field to use positive force constants (therefore simply inverting the phase angle) and to use 0° and 180° as the only phase angles to preserve symmetry. Using other values may create problems when attempting to apply the parameters to D-amino acids, for instance, because the potential energy surface will no longer simply be the inverse of that of the L-amino acid.

The CHARMM22/CMAP force field was subsequently revised in 2012 to yield the CHARMM36 force field,[56] which addressed several of the problems of the previous version. The CMAP parameters were refined for glycine and proline against more accurate QM potential energy surfaces, and the CMAP terms for other amino acids were adjusted based on solution NMR data including peptides with little secondary structure. This approach corrected the tendency of CHARMM22/CMAP to adopt too much α-helical structure. Side-chain χ_1 and χ_2 dihedrals were also reparametrized using 2D QM potential energy surfaces of dipeptides in the α_R, β, and α_L backbone geometries. The resulting force field was compared against solution NMR data, including RDC, scalar ^{3}J-couplings, and chemical shifts, as well as survey data from the PDB to compare populations of side-chain conformations. After the refinement, backbone sampling (as defined by a root-mean-square Q-factor), was improved in CHARMM36 over CHARMM22/CMAP for α_R, PPII, and α_L regions, and performed similarly in β and coil regions. Notably, CHARMM36 was better than AMBER ff99SB for all conformations except β and coil, though the difference in the latter was negligible. CHARMM36 was also among the best-performing force fields in the systematic analysis of side-chain sampling by Petrović et al.[35]

Despite this improvement, in simulations of intrinsically disordered peptides, it was observed that CHARMM36 over-stabilized arginine-serine (RS) polypeptides in α_L conformations compared to available experimental data,[30] motivating a minor revision to the CHARMM36 parameter set. The origin of the α_L over-sampling problem was actually not related to backbone terms directly; instead, a reduction in the LJ radius of β carbons included in the alkane parameter set of CHARMM36 reduced steric clashes with the backbone and allowed for more favorable sampling of the α_L geometry. The smaller radius is desirable for improved agreement of alkane properties and transferability between molecules, therefore recalibration of the CMAP terms to account for this change was needed. Huang et al. used a reweighting scheme to adjust the CMAP energy surface in the α_L region to correct this deficiency. Along with an NBFIX for guanidinium-carboxylate interactions to improve salt bridge populations, the force field revisions yielded what is called the "CHARMM36m" parameter set.[57] A particularly notable outcome of the CHARMM36m simulations was dramatically improved agreement between experimental and computed SAXS profiles. The CHARMM36m parameter set reduced χ^2 from 0.63 with CHARMM36 to 0.12.[57] Given that the adjustments to the CMAP energy surface were confined to the αL region to disfavor the adoption of these conformations, the structural properties of globular, folded proteins are unaffected and the CHARMM36m force field is suitable for simulations of either folded or unfolded proteins.

The ability of CHARMM36 and CHARMM36m to simulate an IDP was assessed by Krupa et al. in their systematic comparison of recent AMBER and CHARMM parameter sets. Both of these parameter sets produced the most expanded Aβ structures with α-helical and β-sheet content that was in good agreement with experimental data in both continuous MD and REMD simulations. Notably, CHARMM36m produced the expected antiparallel β-sheet that involved residues 16–21 and 29–36. The refined CHARMM36m force field produced NMR chemical shifts that were in slightly better agreement with experimental data than the previous CHARMM36 force field, and both of these parameter sets produced comparable agreement with AMBER ff99SB and ff14SB parameter sets. As noted above, ff14IDPs produced the best agreement in terms of this property, suggesting that though CHARMM36m is capable of reproducing many key properties of Aβ, additional refinement of the CHARMM36m force field may be necessary to more accurately model difficult IDP systems.

3.4 OPLS

The OPLS (Optimized Potential for Liquid Simulations) force field, like AMBER and CHARMM, began as a united-atom force field (OPLS-UA)[58] and later converted to an all-atom form (OPLS-AA).[59] The theory underlying the OPLS force field is that since proteins share many functional groups that are similar to (or the same as) organic liquids, optimizing the behavior of pure liquids in simulations should be a robust starting point for the construction of a protein force field. Agreement between computed and experimental enthalpy of vaporization and densities of pure organic liquids would give confidence that the nonbonded parametrization is reliable, and further augmenting this approach with hydration free energies (when available) could serve as a test of the ability of these parameters to interact adequately with water for the purposes of simulating proteins in aqueous solution. The OPLS parametrization methodology relies on QM calculations of relative conformational energies and interactions between organic molecules and water, typically making use of the 6-31G* basis set to include polarization effects. The original OPLS-UA parameter set introduced by Jorgensen and Tirado-Rives[58] demonstrated remarkable agreement of enthalpy of vaporization and molecular volume (thus density) for a class of neutral organic compounds that underlie proteins. Since molecular ions cannot be simulated as pure liquids, these compounds were parametrized using QM interaction energies with water. While these interaction energies were somewhat underestimated relative to the QM calculations, they were generally in good agreement with the limited experimental data at the time and hydration structure around solvated species confirmed the utility of the OPLS-UA parameters. The OPLS parameters were originally developed for use with the TIP4P water model,[7,60] but were also shown to produce good behavior when combined with TIP3P[7] and SPC.[61]

The earliest OPLS parameter set, OPLS-UA, has a connection with the AMBER united-atom parameter set of Weiner et al. discussed above.[17] Bonded parameters were taken from the AMBER force field and coupled with the newly defined OPLS nonbonded parameters (including 65 atom types to describe proteins), to create what Jorgensen and Tirado-Rives called "AMBER/OPLS" at the time (now referred to as OPLS-UA). Unlike the united-atom AMBER force field, the OPLS potential function included no explicit hydrogen bonding term and did not make use of lone pairs, thus the description of hydrogen bonding was quite different from the AMBER interaction potential of the time. Even with a simpler functional form, it was shown that OPLS-UA produced comparable results to the AMBER force field, viewed at the time as the best parameter set for protein simulations. The OPLS-UA force field was shown to reproduce relative conformational energies of butane, methylethylether, and glycine and alanine dipeptides with similar quality to the AMBER united-atom force field, suggesting that the combination of these parameter sets was suitable for use in proteins. Of note was the modification of the scaling factor for 1–4 interactions, which was set to 0.5 in AMBER for both electrostatic and LJ interactions. However, when increasing radii for extended carbon atoms, the LJ scaling factor was reduced to 0.125; it is this combination of scaling factors (0.5 for electrostatics and 0.125 for LJ) that was applied in OPLS-UA. The OPLS-UA force field was subsequently applied to energy minimizations of crystals of cyclic peptides and the protein crambin. The RMSD of each of these systems was generally lower with OPLS-UA than AMBER, and other aspects of geometries such as deviations in backbone torsional geometry and side chain torsions were comparable between both force fields. Thus, Jorgensen and Tirado-Rives concluded that the OPLS-UA force field, even with its more simplified potential energy function, was a suitable model of proteins that was competitive with AMBER, the best united-atom force field of the time.

The first all-atom form of the OPLS force field (OPLS-AA) was introduced by Jorgensen et al.[59] The change from a united-atom to all-atom representation in the OPLS potential was motivated by early studies that showed inclusion of explicit aliphatic hydrogen atoms allowed for more flexibility in charge assignment and dihedral refinement, leading to a substantial improvement in hydration free energies of alkanes.[62] Thus, the OPLS-AA force field was reparametrized by deriving new nonbonded parameters (charges and LJ) and re-deriving dihedral parameters by targeting conformational energy profiles of 50 molecules at the RHF/6-31G* level of theory. Bond stretching and angle bending parameters were generally retained from their values in AMBER. Of note is the dihedral functional form used in OPLS-AA, which is a Fourier series, given by:

$$U_{dihedral} = \sum_i \frac{V_1^i}{2}[1 + \cos(\phi_i + f_i1)] + \frac{V_2^i}{2}[1 + \cos(2\phi_i + f_i2)] + \frac{V_3^i}{2}[1 + \cos(3\phi_i + f_i3)] \quad (11)$$

where V_n are the coefficients of the Fourier series, $f1$, $f2$, and $f3$ are the phase angles (functionally equivalent to δ in Eq. 1) and ϕ_i is the instantaneous value of dihedral angle *i*. Dihedral parameter fitting was undertaken for a series of organic molecules, with parameters shared between molecules fit at the same time to avoid biasing the force field terms. Relative energies of dipeptides were compared to RHF/6-31G* QM energies, which were also found to be comparable to higher-level calculations, suggesting that the less expensive QM calculations were sufficiently robust to define the energy differences among peptide backbone geometries. Parameter fitting for amino-acid χ_1 torsions was performed using dipeptides with planar backbones, and all other side-chain dihedral parameters were taken from the analogous small-molecule parameters.

The OPLS-AA force field was evaluated by performing Monte Carlo simulations of 30 different organic liquids and comparing their structural and thermodynamic properties, including enthalpy of vaporization, density, heat capacity at constant pressure, and compressibility, to available experimental data. In the OPLS-AA simulations, all bonds were allowed to be flexible, which represented a departure from the simpler approach with OPLS-UA in which all bonds were held rigid. The differences in such treatments were determined to be negligible by Jorgensen et al. Both the OPLS-UA and OPLS-AA force fields produced excellent agreement with experimental properties of pure liquids, with errors generally on the order of 2% or less, though Jorgensen et al. noted some areas of future improvement, including an incorrect ordering of density and enthalpy of vaporization for propanol and 2-propanol that may be alleviated with more specific charge assignment rather than the transferable approach taken in constructing OPLS-AA. Specific improvements were noted for sulfur-containing compounds (which are important for proteins containing cysteine and methionine), as the more rigorous parametrization of OPLS-AA led to better agreement in thermodynamic properties.

Building upon the 1996 OPLS-AA parameters from Jorgensen et al., a more complete and accurate OPLS-AA protein force field was introduced in 2001 by Kaminski et al.[63] This new force field version aimed to refine protein-specific dihedral parameters by fitting to higher quality QM data. This OPLS version was named "OPLS-AA/L," with the L denoting the use of localized MP2 (LMP2) as the electronic structure method in the underlying QM calculations. Specifically, conformational energies were computed using the LMP2/cc-pVTZ(-f) level of theory on model compounds and dipeptides optimized using the HF/6-31G** model chemistry. For charged side chains, a special technique using a continuum solvent model was employed to avoid overly strong interactions between the side chains and backbone, an approach reminiscent of the continuum model employed in producing the AMBER ff03 force field.[21] Eliminating conformers that would not occur in the aqueous phase was an important consideration in deriving usable parameters for these species.

The new target data and fitting protocol led to dihedral refinements of backbone ϕ and ψ torsions as well as χ_1 in many amino acids. The alanine dipeptide was used as the model for refining the backbone dihedrals and these parameters were applied to all amino acids. To demonstrate transferability, Kaminski et al. compared QM conformational energies of the alanine tetrapeptide with those obtained with OPLS-AA/L and found substantial improvement in the RMS error of the energies and comparable deviations in the backbone dihedrals when compared to the previous OPLS-AA parameter set. Adjustments were made to the cysteine side chain charges and sulfur LJ parameters to improve QM dimerization energies, which were too low using the OPLS-AA force field, and similar modifications were made to methionine sulfur and carbon atoms. The new force field parameters retained or improved upon the quality of liquid structural and thermodynamic properties in methanethiol and ethanethiol. The relative conformational energies of all amino acids were generally improved, though the increase in accuracy of OPLS-AA/L over OPLS-AA was less pronounced for charged amino acids, reflecting the difficult nature of parametrizing these species, some of which had few target conformers to use during fitting. Nevertheless, the dihedral reoptimization that was undertaken was a major improvement in the OPLS force field series.

Beauchamp et al. recently assessed the performance of OPLS-AA/L in a systematic evaluation of recent protein force fields to determine how well they can reproduce NMR chemical shifts and scalar J-couplings.[64] By performing simulations of dipeptides, tripeptides, alanine tetrapeptide, and ubiquitin, the authors demonstrated that while some recent AMBER variants (specifically tuned to reproduce NMR data) performed well, the OPLS-AA/L force field performed poorly in these tests, with large errors compared to all but the earliest AMBER parameter sets, ff94 and ff96. Simulations of OPLS-AA/L in explicit solvent using the SPC water model fared the best in this examination. Around the same time, Lindorff-Larsen et al. performed a similar assessment to determine how well protein force fields reproduce NMR J-couplings, order parameters, and RDC in folded proteins, the fraction of folded structures in the model $(AAQAA)_3$ helix and CLN025 β-hairpin, and in the folding of villin and the FiP35 WW domain.[65] Again, deficiencies were found in the agreement between OPLS-AA/L and NMR observables, with errors comparable to those of the CHARMM22 force field for both ubiquitin and GB3. OPLS-AA/L also underestimated the α-helical content of $(AAQAA)_3$ at low temperature and showed no temperature dependence. Similarly, OPLS-AA/L predicted CLN025 to be less folded below its melting temperature and too ordered above the melting temperature, reflecting a similar lack of temperature dependence. Together, the studies of Beauchamp et al. and Lindorff-Larsen et al. point to inadequacies in the conformational sampling produced by OPLS-AA/L, suggesting that additional refinement, particularly in dihedral terms to improve agreement with the NMR observables, was needed with this parameter set.

OPLS-AA/L has also been evaluated in the context of other challenging systems including IDPs and Aβ. Gerben et al. included OPLS-AA/L in a systematic evaluation of recent force fields including AMBER, CHARMM, and GROMOS (see Section 3.5) variants on the conformational ensembles of Aβ.[66] OPLS-AA/L produced short α-helices in over 66% of all its structures, with an overall average 12% helical content, in reasonable agreement with experimental data, though the population of β-strand structures was somewhat low. OPLS-AA/L also produced the most compact structures as determined by radius of gyration, but the average value was in line with experimental measurements. In comparing OPLS-AA/L to NMR data, the force field was among the best in reproducing Cα and Hα chemical shifts, but tended to have relatively large error (as determined by RMSD) in backbone scalar J-coupling data. In a more recent study, Smith et al. examined a fragment of Aβ,[67] encompassing residues 21–30, and found that OPLS-AA/L produced the highest β-strand structure when used with TIP4P, but this secondary structure preference was diminished in TIP3P and SPC, suggesting a distinct dependence on the water model. OPLS-AA/L also yielded the highest number of intrapeptide hydrogen bonds, suggesting a preference for compaction in aqueous solution rather than elongation. Such compaction was also noted by Carballo-Pacheco and Strodel in a 2017 REMD study on $A\beta_{42}$;[68] OPLS-AA/L again produced the lowest radius of gyration, and also yielded modest agreement with most NMR observables, with CHARMM22* producing the best agreement.

Raucher et al. used OPLS-AA/L in conjunction with the ABSINTH continuum solvent model[69] in an evaluation of IDPs with available SAXS and NMR data. Under these conditions, simulations with OPLS-AA/L yielded polypeptide radius of gyration values that were similar to those of CHARMM22* but exhibited more prevalent intra-peptide hydrogen bonding, similar to the observation of Smith et al. in their simulations of the Aβ fragment. Ultimately, the conformations produced by OPLS-AA/L in this implicit solvent model were too expanded compared to experimental data, though it appears the force field may manifest different behavior when combined with an explicit solvent representation, as noted above.

To address the problems noted above related to torsions, the OPLS-AA/L force field was refined by Robertson et al., yielding OPLS-AA/M.[70] The main goal of this version of the OPLS force field was to use even more rigorous QM calculations to refine backbone and side-chain dihedral parameters, thereby improving conformational sampling in proteins. The authors employed density functional theory (DFT) in their geometry optimizations and single-point energy evaluations using functionals that include dispersion correction;[71–73] optimizations were performed with the ωB97X-D/6-311+G(d,p) model chemistry and subsequent single-point energy calculations were performed at the B2PLYP-D3BJ/aug-cc-pVTZ level of theory. Scans of the side-chain χ_1 and χ_2 torsions were performed with the backbone fixed in both α_R and β-strand geometries. A key improvement in the force field parametrization of OPLS-AA/M was the use of Boltzmann weighting in the error function between QM and MM energies for the peptide conformations subjected to dihedral parameter fitting, which the authors found to substantially improve the quality of their parameter fitting. Model alanine and proline tetrapeptides produced more reliable target QM conformational energy data, as previous efforts using Hartree-Fock methods may have yielded flawed geometries due to a lack of electron correlation. Notably, the RMS error in energy for the alanine tetrapeptide with the OPLS-AA/M force field was better than all the tested AMBER ff99 variants and was comparable to the polarizable AMOEBA force field (see Section 4.2), suggesting that the new OPLS force field was competitive with even more rigorous force field models that include electronic polarization. The OPLS-AA/M force field was validated by simulating an alanine pentapeptide (Ala_5), glycine tripeptide (Gly_3), and all other non-proline amino acids as dipeptides for a total of 600 ns. The Ala_5 simulations produced increased PPII conformations compared to the previous OPLS-AA and OPLS-AA/L parameter sets, in agreement with experimental data, and both Ala_5 and Gly_3 exhibited better agreement with experimental scalar J-couplings. Dipeptide simulations of the other amino acids revealed dramatically better agreement with side-chain rotamer populations taken from denatured ubiquitin and protein G as well as a protein coil library. Robertson et al. specifically noted that the new parameter set eliminated extreme values that were observed with OPLS-AA and OPLS-AA/L that indicated 100% or 0% sampling of different side-chain conformations. Agreement between scalar J-couplings for backbone and side-chain torsions was substantially improved for the dipeptides, as well as ubiquitin and GB3. Together, these data indicated a major improvement in the OPLS force field, and that Boltzmann weighting can greatly increase the quality of dihedral parameters, yielding improved agreement with high-level QM conformational energy data and solution NMR data that report on backbone and side-chain sampling.

The most recent revision to the OPLS protein force field was undertaken by Harder et al. in 2016, with their introduction of the OPLS3 force field that addressed both small molecules and proteins.[74] This force field is not a direct modification of OPLS-AA/M described above, but is related to the previous developments in OPLS force field parameters for small molecules, denoted OPLS2.0[75] and OPLS2.1,[76] which have used semiempirical charge assignment in their parametrization of small molecules. OPLS3 was derived in a compatible manner, starting with the nonbonded parameters of OPLS-AA/L and re-deriving backbone and side-chain dihedral parameters in model dipeptides and tetrapeptides. Similar to the approach taken in the parametrization of OPLS-AA/M, DFT was used for geometry optimization, using the B3LYP/6-31G* level of theory, followed by single-point energy calculations using a LMP2/cc-pVTZ(-f) model chemistry. Refitting dihedral parameters to these energy surfaces for alanine, glycine, and proline dipeptides reduced the RMS error relative to OPLS2.1. Side-chain torsions were refit similarly by fixing the dipeptide backbone in either α or β geometries, and the resulting dihedral angle distributions were compared against a survey of the PDB. Here again, OPLS3 outperformed OPLS2.1, though a comparison with OPLS-AA/M would be interesting in the future to determine which fitting method produced the best results in this regard. OPLS-AA/M was validated against solution data and OPLS3 against the PDB, using high-resolution crystal structures, which are generally dehydrated to some extent. As a final test, OPLS3 was employed in simulations of proteins in solution, and the α-helical content of model peptides was shown to be improved for the K19, $(AAQAA)_3$, and CLN025 peptides, though the α-helical content was overestimated in K19. A series of globular proteins was also evaluated, with OPLS3 generally having the lowest RMSD when compared to OPLS2.1, but CHARMM22* and AMBER ff99SB-ILDN also produced comparable results.

3.5 GROMOS

The GROMOS (Groningen Molecular Simulation) force field is among the most widely used atomistic force fields that has utilized a united-atom representation throughout its entire development history. Unlike the other force fields discussed above, which originated as united-atom parameter sets and moved to full, all-atom representation, GROMOS has retained the simpler representation of aliphatic groups and therefore requires fewer particles in simulations that make use of it. Assignment of partial charges in the GROMOS parametrization process originated from molecular orbital methods employed by Poland and Scheraga;[77] more recently, DFT calculations have been used to initially assign partial charges,[78] which can be adjusted empirically (D. Geerke, personal communication) to target condensed-phase properties such as hydration free energies, enthalpies of vaporization, and densities of pure liquids.

The functional form of the GROMOS force field differs from those of the other atomistic force field described in its treatment of bond stretching.

Rather than using a harmonic function, the GROMOS force field makes use of a fourth-order (quartic) potential, which can be expressed as:

$$U_{bond} = \frac{1}{4} k_b \left(b^2 - b_0^2\right)^2 \tag{12}$$

where b, b_0, and k_b are the bond length, equilibrium bond length, and force constant, respectively.

The GROMOS force field lineage is tied to the releases of the GROMOS software, which underwent significant changes between GROMOS87[79] and GROMOS96 versions. Here, we focus on the GROMOS96 force field parameter sets as these are still in frequent use today. GROMOS96 parameter sets are named by first denoting the number of available atom types, followed by a letter indicating whether the parameter set is for use in the condensed (A) or gas (B) phase, and then a version number. The B versions of the force field arise from adjusting charges and protonation states to yield amino acids in their uncharged forms, and are intended for vacuum simulations. As most protein folding simulations are performed in the condensed phase (typically aqueous solution), we will focus here on the development of the A versions of each GROMOS parameter set.

One of the first widely used GROMOS96 parameter set for protein simulations was known as 43A1,[80] indicating that there were 43 atom types, the parameter set was for the condensed phase, and it was version 1 of the force field. Version 43A1 represents the first version that made use of the functional form still in use today by the GROMOS force field. Daura et al. noted that physical properties such as liquid density and enthalpy of vaporization were strongly dependent upon the cutoff value used for van der Waals interactions. To be computationally more efficient, cutoffs as short as 0.8 nm were in use at the time, but to achieve better agreement with experimental data, longer cutoffs on the order of 1.4–1.6 nm were shown to be necessary. As such, the aliphatic groups in the GROMOS force field were reparametrized under this new convention, leading to the 43A1 parameter set. Dihedral parameters and 1–4 interactions in aliphatic chains were subsequently refined to yield better torsion angle distributions in short alkanes, resulting in the 43A2 parameter set.[81] Longer alkanes (n-heptane and longer) were shown to have densities that were too high compared to experiments, leading to a further revision of alkane parameters in the 45A3 parameter set,[82] which was subsequently used to parametrize lipids,[83] carbohydrates,[84] and nucleotides[85] to yield the 45A4 parameter set. While these improvements and additions do not directly pertain to protein force fields, mention of them here is important in this review for completeness of the GROMOS96 version history.

The 53A5 and 53A6 versions of the GROMOS force field[86] represented a major change in the force field, including a complete reparametrization of nonbonded terms (LJ and charges) related to protein functional groups. The goal of these parameter sets was to reproduce experimental solvation free energies in cyclohexane (53A5) and water (53A6). LJ parameters for atom types in amino-acid side-chain analogs were adjusted from their 45A4 values to more accurately reproduce solvation free energies in cyclohexane. This parametrization could be performed independently of charge fitting because in the united-atom GROMOS convention, the CH_2 groups in cyclohexane are uncharged, therefore there is no electrostatic contribution to the solvation free energy. After refitting LJ parameters, charges were reparametrized by targeting hydration free energies to yield the 53A6 parameter set. The approach taken by Oostenbrink et al. was the first instance of a biomolecular force field explicitly seeking to reproduce partitioning behavior between polar and nonpolar media. Modeling such behavior accurately is important in simulations of proteins at membrane-water interfaces or within membranes, and in simulations of protein folding with the goal of balancing protein-protein and protein-water interactions, particularly with respect to interactions among the nonpolar amino acids that form the core of most globular proteins.

The 53A6 parameter set was validated in a subsequent study by Oostenbrink et al. using short (5-ns) simulations of lysozyme and longer (25- or 100-ns) simulations of a β-dodecamer peptide in methanol and water.[87] The latter case represents a strict test of this force field, which was explicitly parametrized to perform well in solvents of differing polarity, as noted above. These simulations were compared to analogous ones performed using the 45A3 parameter set that preceded 53A6 in the development history. Over the course of 5 ns, the lysozyme structure was generally more stable than in the 3.5-ns simulation using 45A3, as quantified by backbone RMSD. NOE violations were improved using 53A6, particularly when excluding the C-terminal Leu129, which was very flexible in the simulation. Excluding Leu129 from the analysis of the simulation using the 45A3 parameter set did not improve the results, indicating that the bulk of the improvement in NOE agreement came from the remainder of the protein. The β-dodecamer peptide is expected to be helical in methanol and unfolded in water. With the 45A3 parameter set, the opposite held true;

the peptide was generally unfolded in methanol and primarily helical in water, though the RMSD time series indicated unfolding events towards the end of the 100-ns trajectory. With the 53A6 parameter set, the β-dodecamer was more stable in methanol, though again a spike in RMSD was noted at ~75 ns that did not relax back to the native structure. In water, only 25 ns of MD simulation were performed with the 53A6 force field, though the β-dodecamer rapidly unfolded. Interestingly, the authors report NOE violations for the β-dodecamer that are improved using 53A6 for the peptide in methanol but were of worse quality in water compared to 45A3. They concluded that agreement was reasonable and that the 53A6 parameter set represented an improvement over the previous protein parameter set, 45A3, however, acknowledging that short α-helices may be imbalanced. Several years later, in comparing free energy surfaces of peptide backbone sampling, Best et al. noted a very low free energy barrier between α-helical and PPII states in Ala_5 when using the GROMOS 53A6 parameter set,[56] further suggesting that α-helical states are under-stabilized with this force field.

It was confirmed by Schmid et al. that the 53A6 parameter set did not model the balance of secondary structures well, particularly in that α-helices were under-stabilized due to excess repulsion between backbone carbonyl and amide groups. To address this problem, backbone ϕ and ψ dihedral terms were changed by targeting the sampling of backbone conformations observed in crystal structures. Further, the LJ interaction between the backbone carbonyl oxygen and amide nitrogen was made less repulsive, yielding the 54A7 parameter set.[88] The new parameter set was validated using 20-ns MD simulations of four proteins (lysozyme, Fox1 RNA binding domain, chorismite mutase, and the GCN4-p1 peptide), unlike the validation of 53A6 that included only lysozyme and GCN4-p1. The application of the GROMOS 54A7 parameter set to these proteins generally resulted in lower deviations from the experimental structures relative to 53A6, though the difference was minimal in the case of lysozyme and a kink emerged in the second α-helix of chorismite mutase towards the end of the 20-ns MD trajectory. In lysozyme, some features of the secondary structure were lost, including the loss of short 3_{10} helices in favor of longer α-helices, which Schmid et al. noted could be an indication of slight over-stabilization of α-helices. The NOE violations in the three globular proteins (lysozyme, Fox1, and chorismite mutase) were generally lower than in the simulations performed with 53A6. The GCN4-p1 peptide unfolded and refolded during a 50-ns MD simulation with the 54A7 parameter set, primarily sampling α- and π-helical structures. The resulting conformational ensemble was in worse agreement with experimental NOE data than the 53A6 parameter set, however, the authors noted that NOE data are not sensitive enough to distinguish between these two helix types. In general, though α-helical structures were better preserved in globular proteins using the updated 54A7 parameter set, additional refinement was likely needed to more accurately describe protein folding and better balance the populations of α-, π-, and 3_{10}-helical structures.

Reif et al. undertook a refinement of the charged amino acid analogs in the 54A7 parameter set, recomputing partial charges and adjusting the LJ parameters of methyl groups in charged amines and leading to the 54A8 parameter set.[89] The refinement that led to the 53A6 parameter set addressed only neutral side-chain analogs, leaving partial charges for the charged amino acids essentially unchanged since 1984. Reif et al. used hydration free energies of these molecular ions as a target for their refinement, finding that small changes in the partial charges led to dramatically improved hydration free energies and better agreement in the peak locations in ion-water radial distribution functions (RDFs) in 54A8 over 54A7. The 54A8 parameter set was subsequently applied to protein simulations the following year in another study by Reif et al.[90] The test set of proteins in these simulations was the same as in 54A7 discussed above, but also included the B1 domain of protein G, *Escherichia coli* cold-shock protein CspA, and hyperthermophilic protein Sac7a from *Sulfolobus acidocaldarius*. Each simulation was performed for 20 ns as in previous GROMOS validation studies, except for the GCN4-p1 peptide, which was simulated for 100 ns. The stability of each protein was comparable to the previous results with the 54A7 force field, which was to be expected given that the only changes made in parametrizing the 54A8 force field were adjustments to charged side chains. Agreement with NMR data on these proteins was also similar between the 54A7 and 54A8 parameter sets. In all, the adjustments to the charged side chains introduced in the 54A8 parameter set appear to be compatible with the previous refinements that gave rise to the 54A7 parameter set and produce reasonable structural ensembles for proteins in water on the time scale of 10 of nanoseconds.

GROMOS force fields have been used widely in application studies on protein folding and misfolding. We will focus the present discussion on several studies that have applied the GROMOS force field to the challenging case of the Aβ peptide. It is important to note that other studies beyond those described here have been performed on Aβ using the GROMOS force field, which often cite the GROMOS software manual instead of the specific reference for the force field, leaving it ambiguous as to which parameter set was applied during the simulation. As a result, it is difficult to contextualize these studies in this comparative discussion. We have chosen to focus on a few studies that use recent GROMOS96 parameter sets in simulating challenging systems like Aβ to comment on GROMOS96 force field quality and the conformational preferences of different parameter sets in the GROMOS96 lineage. Early simulation studies made use of the 43A1 parameter set, with encouraging results. Tomaselli et al. demonstrated that $A\beta_{42}$ was helical at 300 K but became disordered with partial β-strand content at 340 K.[91] Kittner and Knecht used the same force field to study a fragment of Aβ encompassing residues 25–35 to examine the conformational sampling of dimers using replica-exchange MD simulations of 100 ns in length.[92] They found that β-hairpin and U-shaped, disordered structures dominated the conformational ensemble. Together, these outcomes suggested that the 43A1 parameter set was capable of modeling helix-coil transitions and that perhaps the α-to-β transition of Aβ could be modeled accurately.

More recent studies have focused on comparing the GROMOS parameter sets against each other to determine if the sequence of revisions described above has led to accurate modeling of Aβ. Olubiyi and Strodel compared the 43A2 and 53A6 parameter sets in 1.5-μs MD simulations of $A\beta_{40}$ and $A\beta_{42}$, finding that whereas 43A2 stabilized helices to a greater extent, disorder and β-strand content dominated in simulations with 53A6.[93] Moreover, simulations performed with the 53A6 parameter set led to better reproduction of NMR chemical shifts, particularly in the case of $A\beta_{42}$. Gerben et al. came to a similar conclusion, comparing the 53A6 and 54A7 parameter sets.[66] The total helical content in $A\beta_{40}$ increased from 4% to 20% in 53A6 and 54A7, respectively, and β-strand content concomitantly decreased from 32% to 14%. Both force fields produced outcomes that are in reasonable agreement with experimental data, so based on secondary structure content alone, it could not be determined which force field was better in this regard. Comparison of Cα and Hα chemical shifts revealed that 53A6 produced the best agreement with experimental data, and scalar J-coupling constants showed similar results. Thus, even though the 53A6 parameter set underestimates α-helical stability, in simulations of disordered proteins like Aβ, it may still serve as a useful model for the disordered conformational ensemble. Smith et al. subsequently came to the same conclusion regarding a fragment of Aβ encompassing residues 21–30, that force fields like GROMOS 53A6 and OPLS-AA/L that suppress helical structures in favor of turns and β-strands are superior to AMBER and CHARMM parameter sets.[67] It should be noted that neither Gerben et al. nor Smith et al. analyzed the most current AMBER and CHARMM force fields available at the time in their studies, which may have been a limitation of the force fields supported by the GROMACS software.[94] As a final example, Carballo-Pacheco et al. studied the aggregation of the $A\beta_{16-22}$ fragment with different force fields. The GROMOS96 54A7 parameter set produced the fastest and most favorable formation of hexameric aggregates.[95] Hexamers produced using 54A7 had the highest β-strand content and manifested irreversible aggregation; that is, once formed, the aggregates never broke apart. This outcome is indicative of binding that is too strong relative to experiments, in which Aβ aggregates over a matter of hours. Thus, the 54A7 force field may over-stabilize the aggregation of $A\beta_{16-22}$, suggesting that additional refinement of protein-protein and protein-water interactions may be appropriate. In general, for a difficult peptide like Aβ, the GROMOS force fields may not quantitatively reproduce equilibrium behavior or unfolding kinetics, but the ensemble properties are in good agreement with available NMR data, suggesting that the latest GROMOS parameter sets may be appropriate for use in simulations of disordered peptides.

4. Polarizable force fields

Electronic polarization refers to the reorientation of an electron cloud in response to a change in the surrounding electric field. Electronic degrees of freedom were, for many years, not commonly represented in an explicit fashion in MD simulations; these phenomena were left to QM calculations. However, given the size of most proteins, applying QM calculations is infeasible. Thus, the potential energy functions in molecular mechanics calculations were typically expressed in terms of nuclear coordinates and electronic polarization effects were only approximated by a mean-field assumption of the solvent.[96] As discussed above, partial charges were assigned to each atom in a molecule assuming a uniform surrounding medium, typically water.

Recent developments in force field technology and simultaneous enhancements in software algorithms and hardware capabilities have made more computationally intense simulations tractable. In this section, we address common methods of modeling electronic polarization in classical MD simulations. The most popular methods to model electronic polarization are fluctuating charge models, induced dipole and multipole models, and the classical Drude oscillator, also called "charge on a spring." Each of these methods will be discussed in turn in this section, including an overview of their development, successes and limitations, and applications to protein dynamics.

4.1 Fluctuating charge models

Electronic polarization corresponds to the change in electron distribution around a molecule in response to an electric field. As such, a simple method for modeling this phenomenon is to allow the partial charges assigned to each atom in the system to change over time in response to changes in geometry and interactions with different species in the system. Thus, the point charges become dynamical variables such that the topology assigned to each molecule is not constant. Such force fields are called "fluctuating charge" or "charge equilibration" models. These models derive from the principle that charge will redistribute among atoms to equalize electronegativity at each point, as this property reflects the equilibration of chemical potential. In fluctuating charge simulations, the calculation of electrostatic energy differs from Coulomb's Law introduced in Eq. (1). For an N-atom system, the electrostatic energy is expressed as:

$$U_{elec} = \sum_{i=1}^{N} \chi_i^0 q_i + \frac{1}{2}\sum_{i=1}^{N}\sum_{j=1}^{N} \eta_{ij} q_i q_j + \sum_{i=1}^{N} \Phi_i q_i \tag{13}$$

where q_i and q_j represent partial charges on atoms i and j, respectively, as in Eq. (1). χ_i^0 values are the electronegativities of each atom, η_{ij} is a hardness factor describing the resistance to flow of charge between atoms i and j, and Φ_i is the external electrostatic potential at the location of atom i. In parametrizing a fluctuating charge model, the values of χ_i^0 and η_{ij} are tunable parameters. The use of the hardness term describes a screened Coulomb potential for nearby atoms and is applied to 1–2, 1–3, and 1–4 atomic sites; interactions beyond three bonds in the topology are computed by a normal Coulomb potential. In this way, charge can flow between covalently bonded atoms but also reflect a linear $1/r_{ij}$ dependence beyond ~2.5 Å.[97] The fluctuating charge model has the advantage that its formalism accounts for higher-order multipoles that dipole polarizable models neglect,[98] but suffers from the limitation that it cannot model out-of-plane polarization response (e.g., in aromatic rings) without the use of auxiliary charged sites (virtual sites or "lone pairs").

Fluctuating charge models exist for water and other small molecules,[98–100] and the most fully developed fluctuating charge model for protein simulations, called CHARMM-FQ, was introduced by Patel et al.[97,101] as an extension of the CHARMM22 nonpolarizable force field. The authors implemented the model in the CHARMM program via extended Lagrangian dynamics, whereby the partial atomic charges are propagated in time and are constrained to being transferred only within a given molecule. Although charge transfer between molecules is not impossible, it becomes computationally more demanding and could lead to unphysical, net partial charges on molecules.

Patel et al. parametrized CHARMM-FQ by establishing a linear response model for the change in charge distribution as a function of the introduction of an external electric potential, allowing them to fit the atomic hardness parameters:

$$\Delta\overline{Q} = -\overline{\overline{\eta}}^{-1}\overline{\phi} \tag{14}$$

The hardness parameters of the CHARMM-FQ model were then fit using a series of molecules resembling protein functional groups by computing the change in CHelpG charges via DFT calculations as a function of the presence or absence of 30 different perturbing fields generated by a dipolar probe that simulates water. Therefore, the CHARMM-FQ model sought to reproduce the polarization response to solvation in an aqueous medium, with the ultimate goal of applying the parameters to a full protein. After the hardness parameters were fit, atomic electronegativities were determined by reproducing dipole moments of gas-phase optimized geometries. Fitting was performed on a set of small molecules and alanine dipeptide and then the parameters were validated by applying them to other molecules that mimic protein moieties (acetic acid, benzene, dimethylsulfide, etc.) that were not part of the training set and computing dipole moments and molecular polarizabilities. Good agreement was obtained for most molecules except for planar species, for which the fluctuating charge model cannot reproduce out-of-plane polarization response, as noted above. Molecular polarizabilities with the CHARMM-FQ model were slightly overestimated, but strict agreement with this property may not be necessary to achieve good condensed-phase behavior. For example, the TIP4P-FQ water model[99] underestimates the gas-phase molecular polarizability but reproduces many condensed-phase properties of water. The final parametrization step was to balance solute-solute and solute-water interactions. The energies and geometries (hydrogen-bonding distances and angles) of small molecule-water dimers were computed and compared to gas-phase QM dimers and to assess solute-solute interactions, simulations of bulk liquids or crystals (for indole and imidazole) were performed and compared against experimental densities, heats of vaporization, crystal volumes and lattice parameters, and heats of sublimation. The errors associated with these quantities using the CHARMM-FQ model were comparable to those produced by the nonpolarizable CHARMM22 force field.

CHARMM-FQ simulations of pure liquids produced sharper RDF peaks than CHARMM22, suggesting a stronger electrostatic response, especially among hydrogen-bonding molecules. This phenomenon was validated by computing the pair interaction energy in both force fields; CHARMM-FQ was more favorable by 2.0 kcal mol^{-1}, owing to the polarization response, which was considerable. Aliphatic species had increased dipole moments (on the order of ~0.2 D) in the condensed phase compared to in vacuum, and polar species like methanol exhibited a larger dipole moment change, approximately 1 D. These observations emphasize the importance of electronic polarization response as a function of environment. The final validation simulation of small molecules using CHARMM-FQ was *N*-methylacetamide (NMA) in water. The CHARMM-FQ model produced a larger dipole moment than previous calculations, by nearly 2 D. The outcome suggested that explicit inclusion of solvation energies may be an important component of parameter fitting, to account for the cooperative polarization effects in the aqueous phase.

To overcome this over-polarization problem in the construction of the CHARMM-FQ force field for proteins,[101] Patel et al. scaled atomic hardness factors by 1.15. Previous studies had shown that the polarization response in the condensed phase is somewhat lower than in the gas phase,[102,103] therefore targeting vacuum polarizability data may lead to an overly large polarization response in solution. The scaled hardness values led to better agreement with experimental molecular polarizabilities for a set of small molecules (amides, acetic acid, and imidazole). The scaled hardness factors were used for protein moieties and the bonded parameters were taken from the CHARMM22 force field, with a reparametrization of the CMAP energy surface based on the alanine dipeptide. Short (1-ps), gas-phase MD simulations of 22 different proteins yielded stable structures, with most backbone RMSD values falling in the range of 1.5–2.0 Å, suggesting no major deficit in the CHARMM-FQ protein force field. Some RMSD values were larger than 2.5 Å, which reflects a comparatively large structural change in a very short amount of time.

Condensed-phase simulations of proteins with CHARMM-FQ also enabled Patel et al. to compute the change in atomic charge as a function of moving from vacuum to an aqueous medium. Whereas hydrogen atoms remained near their gas-phase values, carbon, nitrogen, and oxygen exhibited large deviations from gas-phase charges, reflecting the inherent change in electronic structure of any molecule as it is exposed to a polarizing medium like water. Fluctuations in atomic charges over time were on the order of 0.01–0.05 e, which Patel et al. noted were consistent with QM/MM simulations. Water molecules around the polypeptide backbone of six different proteins exhibited increases in their molecular dipole moments on the order of 0.2 D relative to bulk water, with the sharpest enhancements coming among water molecules directly hydrogen-bonded to the peptide bond carbonyl oxygen atom. In summary, CHARMM-FQ was the first polarizable CHARMM force field model to be applied to proteins, yielding expected differences in charges as a function of solvation, dipole moments of water at increasing distance from the polypeptide backbone, and differences in charge distribution along the sequence of each of the different proteins that were assessed. The computational efficiency of the model was somewhat lower than a nonpolarizable force field (approximately 2–4 times slower) but overall, the simulations were computationally tractable.

4.2 Induced dipole and multipole models

Induced dipole models represent electronic polarization via the use of fixed atomic charges and inducible dipoles on polarizable atoms, such that the resulting dipole moment of atom i, $\boldsymbol{\mu}_i$, is linearly proportional to the electric field at that location, $\boldsymbol{E}_i$, according to:

$$\boldsymbol{\mu}_i = \alpha_i \boldsymbol{E}_i^0 \tag{15}$$

where α_i is the polarizability of atom i and $\boldsymbol{E}_i^0$ is the electric field arising from the permanent charges in the system. Computing the electrostatic potential in such a polarizable system requires the self-consistent relaxation of the point dipoles in the electric field, comprised of the permanent charges and the other induced dipoles in the system:

$$\boldsymbol{\mu}_i = \alpha_i \boldsymbol{E}_i^0 + \alpha_i \sum_{j \neq i} \boldsymbol{T}_{ij} \alpha_j \boldsymbol{E}_j^0 \tag{16}$$

where T_{ij} is a tensor of dipole-dipole interactions between point inducible dipoles i and j. These calculations are carried out iteratively to achieve self-consistency at each simulation step, requiring considerable computational expense.

The POSSIM (POlarizable Simulations with Second order Interaction Model) is an induced dipole force field for proteins developed by Kaminski and co-workers. POSSIM originated with a simple polarizable force field (PFF) model in 2002[104] and was subsequently refined. Having developed backbone parameters using alanine dipeptide[105] and small molecules mimicking amino-acid side chains,[106,107] the current version of the POSSIM force field was produced in 2014.[108] POSSIM models electrostatics via screened pairwise-additive charge-charge interactions (with 1–2 and 1–3 pairs excluded and 1–4 pairs shielded by a factor of 0.5). POSSIM uses the second-order expression for modeling dipole-dipole interactions given in Eq. (16). Li et al. parametrized dihedral terms for side-chain torsions (χ_1, χ_2, etc.) against high-level QM energies and applied them to the parameters taken from alanine dipeptide and model compounds in previous work to give the full POSSIM protein force field. The POSSIM force field was evaluated against energies and geometries of four conformations of dipeptides for each of the canonical amino acids, and it was demonstrated that the agreement was as good as, if not better, than OPLS-AA and PFF. A Monte Carlo simulation of a collagen-like triplex demonstrated that POSSIM yielded reasonable structures with an average RMSD <1 Å, comparable to OPLS-AA, suggesting that the force field adequately models proteins in solution. POSSIM was recently shown to very accurately compute pKa shifts in amino acids,[109] which may be significant in simulating protein microenvironments that induce such shifts.

The AMOEBA (Atomic Multipole Optimized Energetics for Biomolecular Applications) force field[110–113] combines the induced dipole model with higher-order multipole terms for an even more accurate representation of electronic polarization. That is, partial charges are replaced by permanent atomic multipoles, including monopole (partial charge), dipole vector, and quadrupole tensor. Dipole and quadrupole moments are defined in the local molecular reference frame rather than absolute Cartesian coordinates; therefore, the description of these permanent multipoles is constructed with respect to the participating atoms. Electronic polarization is represented by inducible dipoles, which interact with the permanent multipoles and the other inducible dipoles in the system. Such a convention represents a critical improvement in the treatment of electronic polarization, allowing for better reproduction of molecular polarizability, directionality of hydrogen bonding, and response to environments of differing polarity. In the AMOEBA convention, molecules are divided into polarization groups, chemical moieties without rotatable bonds. The permanent atomic multipoles only polarize the inducible dipoles outside each polarization group, while inducible dipoles interact with all atoms in the system, irrespective of group assignment.

There are several elements of the AMOEBA functional form that are important to note as they differ from the nonpolarizable force fields discussed above. The AMOEBA force field includes explicit energy terms associated with coupling between bond stretching and angle bending, accounting for anharmonicity as well as explicit treatment of out-of-plane deformation of valence angles. That is, for two bonds (b and b') that share an atom and thus comprise a valence angle, the cross term describing the coupling of their stretching and bending is given as:

$$U_{b\theta} = k_{b\theta}[(b - b_0) + (b' - b_0')](\theta - \theta_0) \tag{17}$$

Van der Waals interactions in the AMOEBA force field do not make use of the 12-6 LJ potential shown in Eq. (1); instead, a buffered 14-7 form[114] is used:

$$U_{vdw} = \varepsilon_{ij}\left(\frac{1.07}{\rho_{ij} + 0.07}\right)^7\left(\frac{1.12}{\rho_{ij}^7 + 0.12} - 2\right) \tag{18}$$

where ε_{ij} is the well depth as in the LJ equation and ρ_{ij} is the ratio between the distance between atoms i and j and the minimum energy distance for the interaction. The buffered 14-7 form has the benefit of being "softer" than the LJ repulsive region but also introduces a steep repulsion at very short distance. Additionally, while the van der Waals parameters of non-hydrogen atoms are assigned to be coincident with the nuclear coordinates, for hydrogen atoms, the center of the van der Waals interaction is located along the X—H bond, which models van der Waals anisotropy along the bond vector for these atoms.

To avoid over-polarization in the AMOEBA force field, a damping function proposed by Thole[115] is used. This damping function effectively "smears" the atomic multipole moments, via the following expression:

$$\rho = \frac{3a}{4\pi}\exp\left[-a\left(\frac{r_{ij}}{\sqrt[6]{\alpha_i\alpha_j}}\right)^3\right] \tag{19}$$

The value of *a* in Eq. (19) determines the strength of the damping; Thole originally proposed a value of 0.572, which has been modified in the AMOEBA force field to be 0.39 for all dipole-dipole interactions.[110]

In 2010, small molecules in the AMOEBA force field were compared against experimental hydration free energies and relative binding free energies from the 2009 Statistical Assessment of the Modeling of Proteins and Ligands (SAMPL) competition. AMOEBA performed well in the assessment of hydration free energies, with an RMS error of only 0.68 kcal mol^{-1} across a diverse set of 30 compounds, many of which mimic protein side chains. Such performance is generally better than with additive force field models, which tend to have errors on the order of 1 kcal mol^{-1} or greater. Similarly, AMOEBA performed favorably in relative binding free energy calculations, particularly for very polar compounds like sugars. AMOEBA did not fare as well for halogenated species, which have highly polarizable halogen atoms that must be parametrized carefully. The final assessment of AMOEBA in 2010 was of a rudimentary protein force field that was applied to several small proteins (20–72 amino acids); over short simulations (3–19.6 ns), the structures were generally stable, with Cα RMSD values <2.0 Å. The only deviation of note was in GB3, which was not converged during the simulation but showed instability due to hydration of the hydrophobic core. However, the simulation was only 3 ns in length, requiring further investigation into the effects of this phenomenon.

Shi et al. introduced a refined AMOEBA protein force field, called AMOEBA-2013,[111] named for the year it was published. Backbone parameters were parametrized against high-level QM ESP and potential energy surfaces for model peptides containing alanine, glycine, or proline. Trial simulations of alanine tripeptide (Ala_3) in solution were included during parameter fitting and compared against a potential of mean force (PMF) generated from PDB statistics, according to $-\ln(P)$, where P is the probability of a backbone (ϕ,ψ) combination. By doing so, the authors were able to incorporate empirical structural data into the parametrization of the protein backbone in the AMOEBA model. The combination of a polarizable electrostatic model, gas-phase QM conformational energies and ESP, and empirical PDB data proved to be a robust approach in producing transferable parameters, as was shown by performing simulations of Ala_5. These simulations also produced (ϕ,ψ) combinations that were in good agreement with the most favorable regions of the PDB PMF. The properties of alanine tetrapeptides and dipeptides containing the other canonical amino acids were validated by comparing the components of the molecular dipole moment and ESP for three conformations of each dipeptide, none of which were part of the parametrization set of conformations. Remarkably, the AMOEBA-2013 force field produced near-perfect agreement with QM values in these properties, the best of any protein force field to date. The quality of the AMOEBA-2013 force field for the protein backbone was assessed against eight different scalar J-coupling values for Ala_5, with a χ^2 of 0.994 and RMS difference of only 0.33 Hz. When using parameters derived from only gas-phase QM data (thus excluding the PDB PMF), the agreement was substantially worse, emphasizing the importance of including empirical conformational data in the parametrization process.

Having demonstrated good agreement with simple model peptide data, Shi et al. assessed the AMOEBA-2013 force field in simulations of the $Ac\text{-}(AAQAA)_3\text{-}NH_2$ model peptide and 10 full proteins. AMOEBA-2013 produced reasonable α-helical content of $Ac\text{-}(AAQAA)_3\text{-}NH_2$ at 303 K, though a few residues had overestimated propensity for forming α-helices. The authors did not address the temperature-sensitivity of α-helix formation. Simulations of 10 full-length proteins (crambin, Trp-cage, villin headpiece, ubiquitin, GB3 domain, RD1 antifreeze protein, SUMO-2 domain, BPTI, FK binding protein, and lysozyme) were then performed to assess their stability using AMOEBA-2013. Over the course of 30-ns MD simulations, all the proteins remained stable, with an average backbone RMSD of just over 1 Å. Further, the flexibility of ubiquitin and lysozyme were compared against order parameters from NMR. While agreement was generally good, several loop regions in the proteins were more flexible in the AMOEBA-2013 simulations than the NMR data suggest. Nevertheless, the absence of any major deviations suggests that the computed ensembles of the proteins are in excellent agreement with available experimental data. As a final test, Shi et al. compared side-chain J-coupling NMR data against their simulation outcomes for BPTI, GB3 domain, ubiquitin, and lysozyme. The results were comparable in quality to the AMBER ff99SB and ff99SB-ILDN parameter sets. Importantly, the latter two force fields were explicitly parametrized against these NMR data, but AMOEBA-2013 was not. This outcome demonstrates that a carefully parametrized polarizable model that includes both QM and empirical data can produce side-chain dihedral sampling that is comparable to nonpolarizable models that were specifically parametrized to reproduce these quantities. Overall, the AMOEBA-2013 represents a robust polarizable force field model for protein simulations, though longer simulations should be performed in future studies to confirm the findings presented by Shi et al.

Recently, Célerse et al. examined the importance of induced polarization in the unfolding of peptides and proteins as well as the dissociation of a protein-protein complex.[116] In their simulations, they used AMOEBA-2013 (which they denoted as AMOEBAPRO13) as well as a recent revision to the AMOEBA force field that includes nucleic acids, called AMOEBABIO18.[117] These results were compared against current nonpolarizable force fields. The free energy barrier to unfolding a model Ala_{10} peptide in vacuum was much lower using both of the AMOEBA variants than any of the nonpolarizable force fields. This outcome was not surprising, as the partial charges assigned in the nonpolarizable models are designed to be used in the condensed phase and therefore likely model hydrogen bonding as being too strong in vacuum. The polarizable models, however, are designed to respond to different environments and are parametrized against gas-phase dipole moments, polarizabilities, and ESP and should therefore serve as better models of peptides and proteins in vacuum. The differences between polarizable and nonpolarizable force fields were less pronounced in the condensed phase in examining the unfolding of ubiquitin. Both AMOEBAPRO13 and AMOEBABIO18 produced results in line with the nonpolarizable force fields, though there was a small but non-negligible discrepancy between the two AMOEBA variants. Given that AMOEBABIO18 produced a small standard deviation of work across multiple steered MD trajectories, the authors suggested its results are likely to be more reliable. In the last test case, Célerse et al. induced the dissociation of the CD2-CD58 complex, which is held together by salt bridges. Ahmed et al. previously found that most nonpolarizable force fields overestimate salt-bridge affinity (ff15ipq being the exception),[118] so a test of polarizable force fields in this context is very important. Given the large size of the fully solvated CD2-CD58 complex, the comparison was done between only CHARMM22 and AMOEBABIO18. The two force fields produced similar free energy profiles but different mechanisms. The CHARMM22 simulations involved first breaking of several salt bridges at the CD2-CD58 interface, followed by reorientation of other residues to engage in salt bridges in the partially dissociated state, then dissociation of CD2 without any substantial unfolding. The AMOEBABIO18 simulations showed similar salt bridge dynamics but CD58 partially disordered during the dissociation process, though a minor population of trajectories followed the same path as was observed with CHARMM22. The AMOEBABIO18 results, while more variable, ultimately led to a free energy barrier that is in better agreement with the experimentally determined dissociation enthalpy, thus the authors concluded that despite variations, the AMOEBABIO18 results were a better model of reality. These simulations also exposed an important methodological consideration. Pulling simulations with polarizable force fields require a slower application of the biasing potential. Whereas water molecules reorient quickly in nonpolarizable simulations to solvate exposed ionic groups, water reorientation is slower in polarizable simulations. The TIP3P water model is known to diffuse too fast, whereas polarizable water models reproduce the experimental diffusion coefficient more accurately. Thus, the pulling velocity must be chosen to be suitable with these slower, but ultimately more accurate, dynamics.

4.3 The classical Drude oscillator

Electronic polarization can also be represented via explicit particles in the simulation system that represent electronic degrees of freedom. This approach is known by several names, such as the charge-on-a-spring model, shell model, or the classical Drude oscillator, so named for Drude et al., who first proposed the convention.[119] Hereafter, this approach is referred to as the classical Drude oscillator model. The auxiliary particles are negatively charged and attached to their parent atoms via harmonic springs using the same bond energy function shown in Eq. (1). In principle, the polarization response of a given atom can then be parametrized as a function of either the charge ascribed to the Drude oscillator (auxiliary particle), the bond force constant, or both.

The most complete Drude force field for proteins and other biomolecules has been developed by MacKerell, Roux, and co-workers, and has been reviewed previously.[120,121] A brief overview of this force field will be provided here. At the time of the writing of this chapter, the current version of the protein force field is Drude-2013, so named for the year it was published.[122] Other elements of the force field, such as nucleic acids, were published later and therefore are named for the year they were publicly released (e.g., Drude-2017 for nucleic acids[123–125]). Given this convention, in discussing the principles of this force field here, it will simply be referred to as "the Drude force field," recognizing, however, that other Drude polarizable models exist for different types of molecules.

In the Drude force field, only non-hydrogen atoms are treated as polarizable (Fig. 4). Any polarization response attributed to hydrogen atoms is incorporated into the connected heavy atom as part of the parametrization

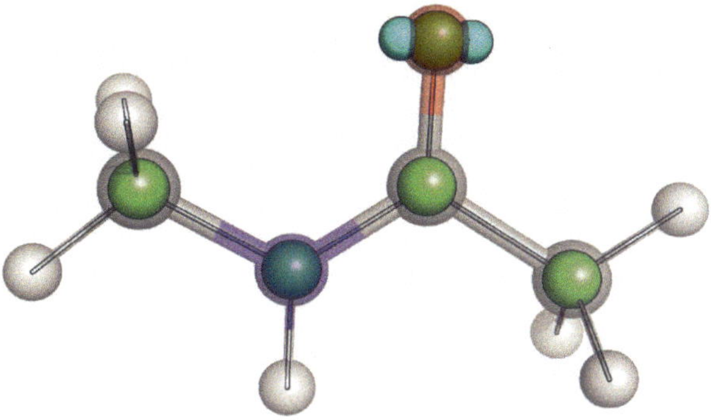

Fig. 4 The structure of *N*-methylacetamide, a model compound commonly used for parametrizing the protein backbone. Real atoms are shown as translucent spheres and colored by element (H = white, C = gray, N = blue, O = red). Drude oscillators and lone pairs (virtual sites) are shown as small green and cyan spheres, respectively.

protocol. The charge on each Drude oscillator is assigned according to the atomic polarizability of the atom, according to Eq. (20):

$$\alpha = \frac{q_D^2}{k_D} \tag{20}$$

where α is the atomic polarizability, q_D is the charge on the Drude oscillator, and k_D is the bond force constant between the Drude oscillator and its parent atom. As noted above, both q_D and k_D could be adjusted to properly model polarization response. In the Drude force field, k_D is fixed for all Drude-atom bonds and α is subject to empirical refinement. Starting values for α are taken from the atomic hybrid polarizabilities proposed by Miller,[126] and adjusted to match target data (discussed below).

In the Drude force field, neighboring dipoles explicitly contribute to the energy function. In additive force fields, interactions within three bonds are normally excluded, and first- or second-neighbor effects are implicitly represented in the parametrization of bond and angle force field terms. In contrast, the Drude force field includes 1–2 and 1–3 dipole-dipole interactions, which are damped via a screening function proposed by Thole,[115] shown in Eq. (21):

$$S_{ij}(r_{ij}) = 1 - \left[1 + \frac{ar_{ij}}{2(\alpha_i\alpha_j)^{1/6}}\right] exp\left[\frac{-ar_{ij}}{(\alpha_i\alpha_j)^{1/6}}\right] \tag{21}$$

where α_i and α_j are the atomic polarizabilities of atoms i and j, respectively, and the constant a determines the strength of the screening. The value of a is also a freely tunable parameter, with a default value of 2.6, similar to the value proposed by Thole (2.089). In practice, the value of a is the sum of individual Thole screening factors for each atom, e.g., $a = a_i + a_j$ for atoms i and j, each of which can be adjusted during parametrization.[127] This function is conceptually identical to the smearing function used by the AMOEBA force field described above; it differs only in its mathematical form. The screening function is incorporated into the energy function as a prefactor for the electrostatic interactions among atoms and Drude oscillators involved in 1–2 and 1–3 interactions. The benefit of including screened 1–2 and 1–3 interactions is an improved representation of molecular polarizability.[121]

In addition to Drude oscillators, atoms in the Drude force field may also have virtual sites linked to them.[128] These particles also carry negative charge and represent lone pairs on hydrogen bond acceptors (carbonyl oxygen atoms, unprotonated nitrogen atoms in heteroaromatic rings, etc.). As such, they are positioned away from the nucleus to which they are attached and have their coordinates updated relative to the constructing atoms during MD simulations. The inclusion of these virtual sites improves the ability of the force field to produce quadrupole moments and electrostatic surface potentials in agreement with QM target data. In conjunction with these virtual sites, the polarizability of a given atom may be treated by expanding the isotropic polarizability into a tensor to allow for anisotropic treatment of the atomic polarizability. That is, the self-polarization energy attributed to the displacement of the Drude oscillator from its parent atom, **d**, when treated isotropically:

$$U_{self}(\mathbf{d}) = \frac{1}{2} k_D \mathbf{d}^2 \tag{22}$$

becomes

$$U_{self} = \frac{1}{2}\left(\left[K_{11}^D\right] d_1^2 + \left[K_{22}^D\right] d_2^2 + \left[K_{33}^D\right] d_3^2\right) \tag{23}$$

for polarizable atoms that are anisotropically polarizable. The Drude bond force constant, k_D, is expanded to a tensor, K^D, with off-diagonal elements assigned a zero value. The Drude-atom displacement, **d**, is expressed in its vector components (d_1, d_2, d_3) that correspond to the local molecular frame. That is, the axes used in the computation of the anisotropic self-polarization energy are not the Cartesian axes but are instead defined by the local molecular geometry.

Parametrization of the Drude force field requires the determination of all bonded parameters as in additive force fields, LJ parameters, assignment of partial charges on each atom (e.g., the total charge on a Drude-atom pair), atomic polarizabilities, per-atom Thole screening factors, and the diagonal elements of the force constant tensor in the case of anisotropically polarizable atoms. Like many other force fields, target data for parametrization of the Drude force field include both QM and empirical, condensed-phase data. A model compound or set of model compounds is constructed to assemble the biomolecule of interest. A QM geometry optimization is performed, and the geometry is used to optimize the internal parameters to recover the same geometry with the Drude force field. The dipole moment, molecular polarizability, and ESP are subsequently targeted for parametrization of all electrostatic terms (charges, α, Thole screening factors, K^D, and virtual site charges and positions, if applicable). The ESP is generated using a Connelly surface of the optimized geometry at increasing distance from the model compound and is computed in the presence and absence of a field of perturbing charges (+0.5 *e*). In this way, the molecular dipole moment and orientational dependence of the molecular polarizability can be captured in detail.

After achieving satisfactory agreement with the QM target data of the isolated model compound(s), the parameter set is validated against available condensed-phase data. These data may include crystal parameters (molecular volumes and lattice parameters), heat of sublimation (for solids), and enthalpy of vaporization (for neat liquids), among other properties for which experimental data exist. This assessment determines if the polarization response is adequately modeled in solids or liquids and is refined if satisfactory agreement is not obtained. At this stage, interactions with individual water molecules are also assessed with QM calculations and the Drude parameter set. The optimized geometries and interaction energies between the model compound and one water molecule at a time (in different poses around all relevant functional groups) are used to validate the existing parameters or serve as additional target data in the event that deficiencies are identified. By doing so, the Drude force field seeks to accurately capture fine details of hydration, which are critical for simulations of biomolecules in aqueous solution. Agreement of properties in a variety of environments (neat liquids, crystals, and in water) is expected to be strong given that the force field is polarizable; that is, the electronic structure of the molecule(s) being assessed should change in response to differences in the dielectric properties of each medium.

To illustrate the non-additive effects in polarizable MD simulations, we present here an example of interactions among the octahedrally coordinated water molecules in the first shell around a Mg^{2+} ion, which is strongly polarizing due to its considerable charge density. In a nonpolarizable force field, the interaction energy between any two molecules will be the same regardless of the presence of Mg^{2+} or any other water molecule. Using the CHARMM force field and TIP3P water model, we demonstrate this property by computing the interaction energy (IE) between water molecules 1 and 2 before and after deletion of the Mg^{2+} ion from the $[Mg(H_2O)_6]^{2+}$ complex. Results are shown in Table 1, and it is clear that the IE is identical at 2.02 kcal mol^{-1}; the net repulsion between these water molecules is offset by the strong attraction between the ion and first-shell waters when the Mg^{2+} is present. In contrast, with the Drude force field, after Mg^{2+} is deleted and the Drude oscillators allowed to relax to reflect the dipole reorientation upon removal of the ion, the net repulsion between SWM4-NDP water molecules is greatly reduced, from 3.45 to 0.67 kcal mol^{-1}.

The change in IE is also coupled to a dramatic change in molecular dipole moment of these water molecules. In the presence of Mg^{2+}, the polarizable water molecules have a dipole moment of 2.88 D, an increase from the bulk value of 2.46 D.[129] The SWM4-NDP water molecules have their dipoles aligned toward the Mg^{2+} ion in the center of the complex (Fig. 5A and C). After removal of the Mg^{2+} ion and relaxation of the dipoles, the Drude oscillator on each water molecule moves toward the oxygen atom and inverts its orientation (Fig. 5B and D), yielding dipole moments of 1.45 D for each water molecule, lower than the gas-phase value of 1.85 D, reflecting the net repulsion that exists between neighboring water molecules in this configuration. The TIP3P water molecules in the additive calculation have fixed dipole moments of 2.35 D and are, by definition, incapable of responding to such changes in the molecular system.

Table 1 Interaction energies (IE, kcal mol^{-1}) between two diagonal equatorial water molecules in an $[Mg(H_2O)_6]^{2+}$ octahedral complex before and after deletion of the Mg^{2+} ion.

	TIP3P(1)–TIP3P (2)	SWM4(1)–SWM4 (2)
Before Mg^{2+} deletion	2.02	3.45
After Mg^{2+} deletion	2.02	0.67
ΔIE	0	−2.78

In the polarizable system, Drude oscillators were allowed to relax following the deletion of the ion but nuclei were kept fixed.

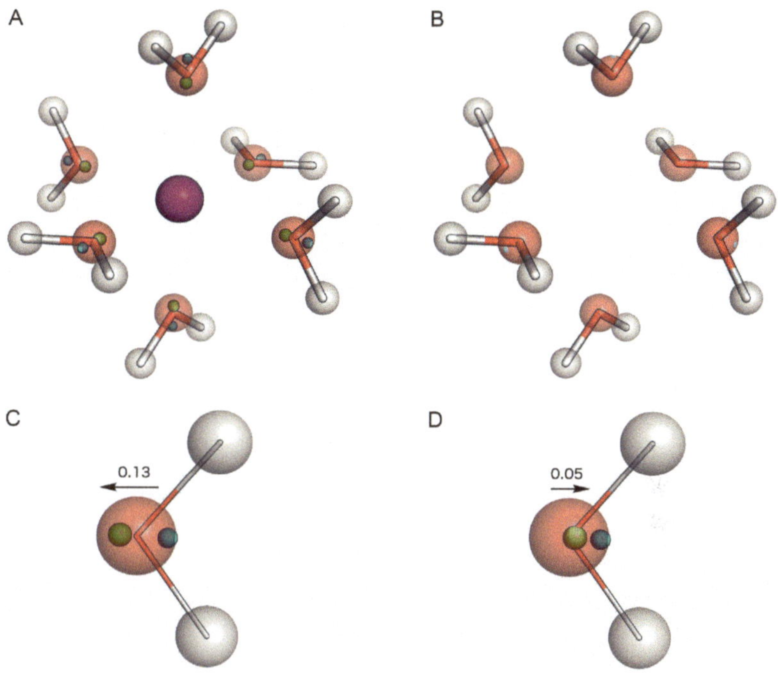

Fig. 5 Multi-body effects in water interactions in the first solvation shell around a Mg^{2+} ion. (A) Structure of the $[Mg(H_2O)_6]^{2+}$ complex and (B) structure of water molecules after removal of the Mg^{2+} ion and relaxation of the water dipoles. Detailed views of the length (in Å) and directionality of the Drude-oxygen bonds in the full complex (C) and after removal of Mg^{2+} (D). Atoms are colored by element (O=red, H=white), Drude oscillators and lone pairs are shown as small green and cyan spheres, respectively. Calculations were performed with the Drude force field.[121,130]

Having briefly described the fundamental concepts of the Drude oscillator model and its parametrization protocol, we now discuss applications of this force field to the study of protein biophysics, dynamics, and folding. The validation of the Drude-2013 protein force field included simulations of a variety of polypeptides and proteins.[122] Analysis of these trajectories revealed interesting dipole moment properties of amino-acid side chains and first-shell water molecules. Specifically, Lopes et al. noted that tryptophan residues in lysozyme had dipole moments that were intrinsically larger than those of additive simulations using CHARMM36, and that the side chains had greater variation in their dipole moments. This observation is important as it reflects the sensitivity of the Drude-2013 force field; amino acids are capable of responding to subtle changes in their environment (neighboring amino acids, water, etc.) and the simulations can produce entirely different properties than observed in additive simulations. In a related observation, Lopes et al. reported that water molecules in the first solvation shell around charged residues such as glutamate and lysine had larger dipole moments than bulk water. Thus, simulations that rely on rigid water models with fixed charges may be missing subtle details of biomolecular hydration that are important for defining the balance among interactions (intramolecular and intermolecular) in the simulation system.

Ngo et al. recently examined the properties of peptide solvation using the Drude force field.[131] They found that the Drude force field reproduced hydration free energies of amino acid side-chain analogs better than the nonpolarizable CHARMM36 force field, with the exception of acidic residues, which may be over-polarized and thus their interactions with water are somewhat too strong. As a result, the hydration free energy of acetate (the side chain analog of glutamate and aspartate) was too favorable by approximately 7 kcal mol^{-1}. This deficiency was addressed by a refinement of the acetate parameters by Lin et al., who reproduced its hydration free energy with an error of only -0.3 kcal mol^{-1}.[132] The improved accuracy of the Drude force field over CHARMM36 in modeling hydration free energy is likely a result of the ability of the dipole moment of the small molecules to change between the gas and condensed phases, a property that pairwise-additive force fields cannot manifest.

In addition, Ngo et al. assessed the conformational sampling of peptides, interactions with ions, the strength of peptide-water, and associated hydrogen-bonding dynamics. Interestingly, they found that the Drude force field accessed different conformations than the CHARMM36 force field, some of which may be induced by cations, though a more thorough examination of this phenomenon is required. They found that the polarizable SWM4-NDP water model[129] used with the Drude force field generally depolarizes around protein surfaces but interacts more favorably with the nearby amino acids than TIP3P does with the CHARMM36 force field. Moreover, the water-water interactions at the protein surface in TIP3P are consistently favorable, whereas they are repulsive in the Drude force field. As a result, simulations with CHARMM36 have shorter hydrogen-bonding lifetimes than with the Drude force field since waters are more quickly exchanged. In simulations with the Drude force field, the favorably associated water molecules at the protein surface are retained for longer periods of time and are not replaced by bulk water. Kinetics of water exchange will have important consequences for protein dynamics and conformational change, and the results of Ngo et al. suggest that polarizable simulations hold great promise for more accurately modeling these subtle effects.

Building upon the observations made in the development of the Drude-2013 force field, Huang et al. reported on microsecond simulations of ubiquitin.[133] This study represents an important development in simulation technology, demonstrating that microsecond simulations are practical when using a polarizable force field, which are inherently more computationally demanding than additive simulations. Moreover, the outcomes of the simulation showed that slow transitions in amino-acid side chain rotations may have functional significance in protein dynamics. The authors highlighted the case of Gln41, which samples both solvent-facing and solvent-occluded orientations. When located in a hydrophobic pocket in the ubiquitin structure, the side chain of Gln41 depolarizes such that its side chain dipole moment is ~4 D, and when exposed to water, the dipole moment increases to ~5.6 D. These conformational transitions occur over several hundred nanoseconds, emphasizing the importance of long simulations to adequately sample these relatively slow dynamics. Additionally, the dramatic change in dipole moment as a function of microenvironment demonstrates that polarizable force fields are essential for capturing changes in electronic structure that correspond to changes in environment. Such behaviors will impact protein folding, partitioning into membranes, and binding to small molecules. Additive simulations will fail to capture such behavior, as the charges assigned to each molecule are optimized for interaction with water and may therefore disfavor partitioning into more hydrophobic environments.

Another interesting observation made by Huang et al. regards the nature of the protein interior. Early in the development of the Drude force field, it was determined that the inclusion of electronic polarization is necessary to capture subtle differences in dielectric constants of alkanes.[134] Whereas the additive CHARMM force field produced nearly uniform, underestimated dielectric constants for all alkanes, irrespective of length (from propane to decane), the Drude force field was able to better reproduce experimental dielectric constants. Though these values only range from 1.75 to 2.0, the ability to capture subtle differences reflects a strength of polarizable simulations. Building upon these early findings, Huang et al. analyzed the dielectric constants of six different proteins, decomposing these quantities into the interior and only the hydrophobic core. Again, the Drude and CHARMM36 force fields produced dramatically different results, with the Drude force field yielding an average optical dielectric constant of 2.0 (in agreement with the commonly used value).[133] Decomposition of the protein structures into interior (SASA $<50\%$) and hydrophobic core (SASA $<20\%$) further revealed that the CHARMM force field again produced very low dielectric constant values, whereas the Drude force field produced a wider range of values, all of which were systematically larger than those of CHARMM. Thus, from small molecules (alkanes) to full proteins, it appears that the inclusion of electronic polarization is required for accurately modeling hydrophobic microenvironments.

Building on simulations of folded, globular proteins, the Drude force field has been employed in additional studies that examine helix-coil transitions in polypeptides, properties of β-hairpins, and dynamics of amyloid peptides, which are known to unfold and aggregate. Simulating such proteins accurately requires that the underlying force field be carefully balanced to give accurate representations of the strength of hydrophobic interactions, intramolecular hydrogen bonding, and hydration. In 2014, Huang and MacKerell applied Hamiltonian replica exchange simulations to investigate the conformational ensemble of the $(AAQAA)_3$ peptide,[135] which is often used as a test case in force field validation due to the availability of helical content data as a function of temperature from NMR experiments.[136] This peptide is only ~20% α-helical at room temperature. Huang and MacKerell used Hamilton replica exchange MD simulations to simulate $(AAQAA)_3$ over a range of temperatures, finding that the Drude force field, not CHARMM36, could reproduce the helical content as a function of temperature. Central to these findings was the cooperativity of backbone peptide-bond dipole moments, quantified by the dipole enhancement. As the amide groups that comprise peptide bonds engage in hydrogen bonds, they polarize along the direction of the hydrogen bond. It is this phenomenon, the alignment of individual dipoles, that gives rise to helical macrodipoles. The enhancement in the dipole moment of each peptide bond can be computed by calculating the corresponding intrinsic dipole moment (from an alanine dipeptide with the same ϕ and ψ backbone dihedrals) and intramolecular dipole moments (by reoptimizing the Drude oscillators in the absence of solvent). By doing so, Huang and MacKerell demonstrated that residues engaging in α-helices, defined as 3 or more residues with (ϕ,ψ) pairs in the α_R region, had peptide bond dipole moments that were enhanced by approximately 1 D. Such enhancement leads to cooperative folding of α-helical turns and therefore a more accurate representation of helix-coil equilibrium that non-polarizable force fields may not achieve in their current functional forms.

Another force field benchmark system is the GB1 hairpin, taken from the C-terminal residues 41–56 in the B1 domain of streptococcal protein G. Hazel et al. used the CHARMM36, CHARMM22*, and

Drude-2013 force fields in replica exchange-umbrella sampling (REUS) simulations to compute folding free energies of the GB1 hairpin.[137] They found that CHARMM36 dramatically over-stabilized the hairpin structure, and while both CHARMM22* and Drude-2013 were within experimental error, the Drude force field favored unfolded states with few hydrogen bonds, properties that are inconsistent with experimental evidence. The authors determined that the backbone parameters in Drude-2013 favored solvation and peptide-peptide interactions were too weak. An increase in the polarizability of the N atom in the amide backbone improved the folding free energy landscape. Additional revisions to the polarizability of hydroxyl groups in serine and threonine led to better hydration properties. Notably, both the CHARMM22* and Drude force fields accurately modeled the weakly formed Asp47-Lys50 salt bridge in the turn of the GB1 hairpin; CHARMM36 over-stabilized this interaction, contributing to the overly favorable folding free energy. Thus, Hazel et al. concluded that the Drude force field accurately models α-helices but may be somewhat less accurate for β-sheets, but a small reparametrization of the backbone is sufficient to bring the force field into agreement with experimental data.

Two recent studies have used the Drude force field to examine the dynamics of amyloidogenic proteins.[138,139] In the first,[138] we examined a fragment of Aβ that is central to Alzheimer's disease; the fragment encompassed residues 15–27, including the central hydrophobic cluster ($L_{17}VFFA_{21}$) and residues 22 and 23, the sites of familial mutations that enhance aggregation and lead to a more severe phenotype. Initially built as an ideal α-helix, the wild-type Aβ and four mutants (E22G, E22K, E22Q, and D23N) all unfolded, though the pathway for each was different. After resolving the structure of $A\beta_{40}$ in 1998, Coles et al. proposed that the acidic dyad of Glu22 and Asp23 caused unfolding of the peptide; that is, the proximity of two negative charges in an α-helix would lead to unwinding due to charge repulsion.[140] Our study confirmed that this is the case of wild-type Aβ, but even more interestingly, the proximity of Glu22 and Asp23 caused a dramatic increase in the side-chain dipole moment of Asp23, perturbing the local electric field and inverting the direction of peptide-bond dipole moments. In a concerted fashion, the α-helical structure was lost as a result of weakened hydrogen bonding that propagated through the entire helical structure. Mutant variants of Aβ showed similar behaviors, such that perturbations to hydrogen bonding networks and salt bridge formations led to unfolding in a mutation-specific manner. Therefore, polarizable MD simulations revealed that it is not simply charge repulsion that leads to unfolding of Aβ; instead, the structure is sensitive to perturbations in hydrogen bonding networks that lead even mutant forms (in which charges are neutralized or inverted) to unfold for different reasons.

Building upon this work, we recently investigated the properties of five different amyloid fibrils—two variants of the microtubule-associated protein tau, the two-layer S-shaped $A\beta_{42}$ fibril, and parallel and antiparallel forms of $IAPP_{21-27}$.[139] The goal of this study was to determine properties of different fibril microenvironments and the impact that induced polarization has on them, ultimately revealing insights into how the fibrils form. Our simulations revealed several interesting features about these fibrils. Both the tau paired helical filament (PHF) and in the $A\beta_{42}$ fibril bound water molecules in internal cavities. The water molecules in tau were concentrated in a narrow, hydrophobic pocket, causing the water molecules to depolarize and align themselves with the fibril axis to maximize hydrogen bonding. At the same time, a buried serine residue depolarized and formed hydrogen bonds with these water molecules. In the $A\beta_{42}$ fibril, which has larger internal cavities near the filament interface, water molecules also depolarized but remained fluid; they did not preferentially align along any axis. As such, internal water molecules may contribute to stabilizing hydrophobic-hydrophilic interfaces by depolarizing, but their fluidity depends on the nature of the given fibril. We also examined salt bridge properties, noting that the tau PHF has a "polar zipper" motif with several charged and polar residues that are buried, which would likely incur a substantial desolvation penalty. While solvent-exposed lysine residues showed dramatic shifts in their side-chain dipole moments as a function of engagement in salt bridges, the buried Lys370 was intrinsically more polarized and the difference in its side-chain dipole moment was smaller. This result suggests that buried charged residues in such an environment may polarize to a greater extent to form strong interactions that help to drive their burial from solvent. Lastly, we noted that glycine has a flexible electronic structure and often occurs at sequence and structural positions between hydrophobic and hydrophilic motifs. We noted glycine dipole moments that ranged from 2.2 to 4.7 D as a function of the surrounding environment. These properties had an important effect in the $IAPP_{21-27}$ fibril, which naturally adopts an antiparallel architecture. When converted to a parallel form by inverting the orientation of every-other β-strand, the peptide-bond dipole moment between Gly23 and Ala24 dropped sharply, weakening inter-strand hydrogen bonding and the parallel fibril was less stable. The perturbation was a function of perturbed packing of bulky hydrophobic residues and the plasticity of the electronic structure of glycine.

Together, these recent studies using the Drude polarizable force field on proteins suggest that there are important properties of protein folding, dynamics, and assembly that are impacted by electronic polarization. While some improvements and refinements should be made to the force field (noted above), the Drude model has potential to reveal many important properties about a variety of protein systems.

5. Conclusions and future directions

Here, we have reviewed the development history of the most popular nonpolarizable force fields for protein simulations (AMBER, CHARMM, OPLS, and GROMOS) as well as several important applications in the context of protein folding and dynamics. Over time, greater computational power, advancements in theoretical method development, and increased experimental data with greater precision have led to more robust target data and sophisticated force field parameter-fitting algorithms. These nonpolarizable force fields have each undergone substantial changes over their respective histories, yet all may be reaching a limit in accuracy that is intrinsic to their pairwise-additive functional forms.

Classical force fields that explicitly model electronic polarization are newer developments in simulation technology but have benefitted from the "lessons learned" over the course of the development of their nonpolarizable predecessors. These force fields have rapidly matured and represent the next generation of biomolecular simulation technology, though even these more elegant models require additional refinement and more thorough examination through application studies of protein dynamics. The availability of highly optimized software and GPU hardware has made polarizable MD simulations computationally tractable, despite the overhead associated with polarizable functional forms and/or additional particles. AMOEBA simulations are approximately 1–2 orders of magnitude slower than nonpolarizable simulations,[141] whereas simulations with the Drude force field are about 3–4 times slower than nonpolarizable simulations.[142,143] The AMOEBA force field is supported by the AMBER,[144,145] Tinker,[146] Tinker-HP,[147] and OpenMM[141,148] simulation programs, and the Drude model is available in CHARMM,[149,150] NAMD,[151,152] OpenMM,[142,148] and GROMACS.[94,143]

In summary, the field of protein simulations benefits from a number of robust models from which investigators can choose when conducting their studies. While no single force field universally captures all aspects of protein conformational sampling in folded and unfolded states, the latest version of each protein force field described here can be well suited to many different applications. The broad availability of polarizable force fields in many common simulation software packages will enable investigators to tackle emerging challenges in protein folding, ligand binding, partitioning, and QM/MM modeling that are likely to benefit from these more elegant electrostatic models.

References

1. Warshel A, Levitt M. Theoretical studies of enzymic reactions: dielectric, electrostatic and steric stabilization of the carbonium ion in the reaction of lysozyme. *J Mol Biol.* 1976;103(2):227–249.
2. Warshel A, Sharma PK, Kato M, Parson WW. Modeling electrostatic effects in proteins. *Biochim Biophys Acta.* 2006;1764(11):1647–1676.
3. Warshel A, Sharma PK, Kato M, Xiang Y, Liu H, Olsson MHM. Electrostatic basis for enzyme catalysis. *Chem Rev.* 2006;106(8):3210–3235.
4. MacKerell Jr AD, Bashford D, Bellott M, et al. All-atom empirical potential for molecular modeling and dynamics studies of proteins. *J Phys Chem B.* 1998;102(18): 3586–3616.
5. Wu X, Brooks BR. Isotropic periodic sum: a method for the calculation of long-range interactions. *J Chem Phys.* 2005;122(4):044107.
6. Shirts MR, Mobley DL, Chodera JD, Pande VS. Accurate and efficient corrections for missing dispersion interactions in molecular simulations. *J Phys Chem B.* 2007;111(45): 13052–13063.
7. Jorgensen WL, Chandrasekhar J, Madura JD, Impey RW, Klein ML. Comparison of simple potential functions for simulating liquid water. *J Chem Phys.* 1983;79(2): 926–935.
8. Burley SK, Petsko GA. Aromatic-aromatic interaction: a mechanism of protein structure stabilization. *Science.* 1985;229(4708):23–28.
9. Burley SK, Petsko GA. Weakly polar interactions in proteins. In: *Advances in Protein Chemistry.* Academic Press; 1988:125–189. vol. 39.
10. Lybrand T, Kollman P. Molecular mechanical calculations on the interaction of ethidium cation with double-helical DNA. *Biopolymers.* 1985;24(10):1863–1879.
11. Warshel A, Levitt M, Lifson S. Consistent force field for calculation of vibrational spectra and conformations of some amides and lactam rings. *J Mol Spectrosc.* 1970; 33(1):84–99.
12. Momany FA, McGuire RF, Burgess AW, Scheraga HA. Energy parameters in polypeptides. VII. Geometric parameters, partial atomic charges, nonbonded interactions, hydrogen bond interactions, and intrinsic torsional potentials for the naturally occurring amino acids. *J Phys Chem.* 1975;79(22):2361–2381.
13. McCammon JA, Gelin BR, Karplus M. Dynamics of folded proteins. *Nature.* 1977;267:585–590.
14. Gelin BR, Karplus M. Side-chain torsional potentials: effect of dipeptide, protein, and solvent environment. *Biochemistry.* 1979;18(7):1256–1268.
15. Blaney JM, Weiner PK, Dearing A, et al. Molecular mechanics simulation of protein-ligand interactions: binding of thyroid hormone analogues to Prealbumin. *J Am Chem Soc.* 1982;104(23):6424–6434.
16. Wipff G, Dearing A, Weiner PK, Blaney JM, Kollman PA. Molecular mechanics studies of enzyme-substrate interactions: the interaction of L- and D-*N*-Acetyltryptophanamide with α-chymotrypsin. *J Am Chem Soc.* 1983;105(4):997–1005.

17. Weiner SJ, Kollman PA, Case DA, et al. A new force field for molecular mechanical simulation of nucleic acids and proteins. *J Am Chem Soc*. 1984;106(3):765–784.
18. Weiner SJ, Kollman PA, Nguyen DT, Case DA. An all atom force field for simulations of proteins and nucleic acids. *J Comput Chem*. 1986;7(2):230–252.
19. Cornell WD, Cieplak P, Bayly CI, et al. A second generation force field for the simulation of proteins, nucleic acids, and organic molecules. *J Am Chem Soc*. 1995;117(19):5179–5197.
20. Wang J, Cieplak P, Kollman PA. How well does a restrained electrostatic potential (RESP) model perform in calculating conformational energies of organic and biological molecules? *J Comput Chem*. 2000;21(12):1049–1074.
21. Duan Y, Wu C, Chowdhury S, et al. A point-charge force field for molecular mechanics simulations of proteins based on condensed-phase quantum mechanical calculations. *J Comput Chem*. 2003;24(16):1999–2012.
22. Yang L, Tan C-h, Hsieh M-J, et al. New-generation Amber united-atom force field. *J Phys Chem B*. 2006;110(26):13166–13176.
23. Best RB, Hummer G. Optimized molecular dynamics force fields applied to the helix-coil transition of polypeptides. *J Phys Chem B*. 2009;113(26):9004–9015.
24. Abascal JLF, Vega C. A general purpose model for the condensed phases of water: TIP4P/2005. *J Chem Phys*. 2005;123:234505.
25. Best RB, Mittal J. Protein simulations with an optimized water model: cooperative helix formation and temperature-induced unfolded state collapse. *J Phys Chem B*. 2010;114(46):14916–14923.
26. Best RB, Zheng W, Mittal J. Balanced protein-water interactions improve properties of disordered proteins and non-specific protein association. *J Chem Theory Comput*. 2014;10(11):5113–5124.
27. Zerze GH, Zhang W, Best RB, Mittal J. Evolution of all-atom protein force fields to improve local and global properties. *J Phys Chem Lett*. 2019;10:2227–2234.
28. Hornak V, Abel R, Okur A, Strockbine B, Roitberg A, Simmerling C. Comparison of multiple Amber force fields and development of improved protein backbone parameters. *Proteins: Struct Funct Bioinform*. 2006;65(3):712–725.
29. Lindorff-Larsen K, Piana S, Palmo K, et al. Improved side-chain torsion potentials for the Amber ff99SB protein force field. *Proteins: Struct Funct Bioinform*. 2006, 78(8):1950–1958.
30. Rauscher S, Gapsys V, Gajda MJ, Zweckstetter M, de Groot BL, Grubmüller H. Structural ensembles of intrinsically disordered proteins depend strongly on force field: a comparison to experiment. *J Chem Theory Comput*. 2011;11(11):5513–5524.
31. Best RB, de Sancho D, Mittal J. Residue-specific α-helix propensities from molecular simulation. *Biophys J*. 2012;102(6):1462–1467.
32. Piana S, Donchev AG, Robustelli P, Shaw DE. Water dispersion interactions strongly influence simulated structural properties of disordered protein states. *J Phys Chem B*. 2015;119(16):5113–5123.
33. Robustelli P, Piana S, Shaw DE. Developing a molecular dynamics force field for both folded and disordered protein states. *Proc Natl Acad Sci USA*. 2018;115(21): E4578–E4766.
34. Maier JA, Martinez C, Kasavajhala K, Wickstrom L, Hauser KE, Simmerling C. ff14SB: improving the accuracy of protein side chain and backbone parameters from ff99SB. *J Chem Theory Comput*. 2015;11(8):3696–3713.
35. Petrović D, Wang X, Strodel B. How accurately do force fields represent protein side chain ensembles? *Proteins: Struct Funct Bioinform*. 2018;86(9):935–944.
36. Song D, Wang W, Ye W, Ji D, Luo R, Chen H-F. *ff14IDPs* force field improving the conformational sampling of intrinsically disordered proteins. *Chem Biol Drug Des*. 2017;89(1):5–15.

37. Wang W, Ye W, Jiang C, Luo R, Chen H-F. New force field on modeling intrinsically disordered proteins. *Chem Biol Drug Des*. 2014;84(3):253–269.
38. Song D, Luo R, Chen H-F. The IDP-specific force field *ff14IDPSFF* improves the conformer sampling of intrinsically disordered proteins. *J Chem Inf Model*. 2017;57(5):1166–1178.
39. Krupa P, Huy PDQ, Li MS. Properties of monomeric Aβ42 probed by different sampling methods and force fields: role of energy components. *J Chem Phys*. 2019; 151(5):055101.
40. Kirkitadze MD, Condron MM, Teplow DB. Identification and characterization of key kinetic intermediates in amyloid β-protein fibrillogenesis. *J Mol Biol*. 2001; 312(5):1103–1119.
41. Ono K, Condron MM, Teplow DB. Structure-neurotoxicity relationships of amyloid β-protein oligomers. *Proc Natl Acad Sci U S A*. 2009;106(21):14745–14750.
42. Cerutti DS, Rice JE, Swope WC, Case DA. Derivation of fixed partial charges for amino acids accommodating a specific water model and implicit polarization. *J Phys Chem B*. 2013;117(8):2328–2338.
43. Horn HW, Swope WC, Pitera JW, et al. Development of an improved four-site water model for biomolecular simulations: TIP4P-Ew. *J Chem Phys*. 2004;120(20): 9665–9678.
44. Debiec KT, Cerutti DS, Baker LR, Gronenborn AM, Case DA, Chong LT. Further along the road less traveled: AMBER ff15ipq, an original protein force field built on a self-consistent physical model. *J Chem Theory Comput*. 2016;12(8):3926–3947.
45. Takemura K, Kitao A. Water model tuning for improved reproduction of rotational diffusion and NMR spectral density. *J Phys Chem B*. 2012;116(22):6279–6287.
46. MacKerell Jr AD, Karplus M. Importance of attractive van der Waals contribution in empirical energy function models for the heat of vaporization of polar liquids. *J Phys Chem*. 1991;95(26):10559–10560.
47. Jorgensen WL. Optimized intermolecular potential functions for liquid alcohols. *J Phys Chem*. 1986;90(7):1276–1284.
48. Reiher III WE. *Theoretical Studies of Hydrogen Bonding*. Cambridge, MA: Chemistry, Harvard University; 1985.
49. Neria E, Fischer S, Karplus M. Simulation of activation free energies in molecular systems. *J Chem Phys*. 1996;105:1902–1921.
50. MacKerell Jr AD, Feig M, Brooks III CL. Extending the treatment of backbone energetics in protein force fields: limitations of gas-phase quantum mechanics in reproducing protein conformational distributions in molecular dynamics simulations. *J Comput Chem*. 2004;25:1400–1415.
51. MacKerell Jr AD, Feig M, Brooks III CL. Improved treatment of the protein backbone in empirical force fields. *J Am Chem Soc*. 2004;126(3):698–699.
52. Best RB, Buchete N-V, Hummer G. Are current molecular dynamics force fields too helical? *Biophys J*. 2008;95(1):L07–L09.
53. Freddolino PL, Park S, Roux B, Schulten K. Force field bias in protein folding simulations. *Biophys J*. 2009;96(9):3772–3780.
54. Piana S, Lindorff-Larsen K, Shaw DE. How robust are protein folding simulations with respect to force field parameterization? *Biophys J*. 2011;100(9):L47–L49.
55. Lindorff-Larsen K, Piana S, Dror RO, Shaw DE. How fast-folding proteins fold. *Science*. 2011;334(6055):517–520.
56. Best RB, Zhu X, Shim J, et al. Optimization of the additive CHARMM all-atom protein force field targeting improved sampling of the backbone φ, ψ and side-chain χ_1 and χ_2 dihedral angles. *J Chem Theory Comput*. 2012;8(9):3257–3273.
57. Huang J, Rauscher S, Nawrocki G, et al. CHARMM36m: an improved force field for folded and intrinsically disordered proteins. *Nat Methods*. 2016;14(1):71–73.

58. Jorgensen WL, Tirado-Rives J. The OPLS potential function for proteins. Energy minimization for crystals of cyclic peptides and crambin. *J Am Chem Soc.* 1988;110(6): 1657–1666.
59. Jorgensen WL, Maxwell DS, Tirado-Rives J. Development and testing of the OPLS all-atom force field on conformational energetics and properties of organic liquids. *J Am Chem Soc.* 1996;118(45):11225–11236.
60. Jorgensen WL, Madura JD. Temperature and size dependence for Monte Carlo simulations of TIP4P water. *Mol Phys.* 1985;56(6):1381–1392.
61. Berendsen HJC, Postma JPM, van Gunsteren WF, Hermans J. Interaction models for water in relation to protein hydration. In: Pullman B, ed. *Intermolecular Forces*. Dordrecht: Holland; 1981:331.
62. Kaminski GA, Duffy EM, Matsui T, Jorgensen WL. Free energies of hydration and pure liquid properties of hydrocarbons from the OPLS all-atom model. *J Phys Chem.* 1994;98(49):13077–13082.
63. Kaminski GA, Friesner RA, Tirado-Rives J, Jorgensen WL. Evaluation and reparametrization of the OPLS-AA force field for proteins via comparison with accurate quantum chemical calculations on peptides. *J Phys Chem B.* 2001; 105(28):6474–6487.
64. Beauchamp KA, Lin Y-S, Das R, Pande VS. Are protein force fields getting better? A systematic benchmark on 524 diverse NMR measurements. *J Chem Theory Comput.* 2012;8(4):1409–1414.
65. Lindorff-Larsen K, Maragakis P, Piana S, Eastwood MP, Dror RO, Shaw DE. Systematic validation of protein force fields against experimental data. *PLoS One.* 2012;7(2):e32131.
66. Gerben SR, Lemkul JA, Brown AM, Bevan DR. Comparing atomistic molecular mechanics force fields for a difficult target: a case study on the Alzheimer's amyloid β-peptide. *J Biomol Struct Dyn.* 2014;32(11):1817–1832.
67. Smith MD, Rao JS, Segelken E, Cruz L. Force-field induced Bias in the structure of $A\beta_{21-30}$: a comparison of OPLS, AMBER, CHARMM, and GROMOS force fields. *J Chem Inf Model.* 2015;55(12):2587–2595.
68. Carballo-Pacheco M, Strodel B. Comparison of force fields for Alzheimer's $A\beta_{42}$: a case study for intrinsically disordered proteins. *Protein Sci.* 2017,26(2).174–185.
69. Vitalis A, Pappu RV. ABSINTH: a new continuum solvation model for simulations of polypeptides in aqueous solution. *J Comput Chem.* 2009;30(5):673–699.
70. Robertson MJ, Tirado-Rives J, Jorgensen WL. Improved peptide and protein torsional energetics with the OPLS-AA force field. *J Chem Theory Comput.* 2015;11(7):3499–3509.
71. Chai J-D, Head-Gordon M. Long-range corrected hybrid density functional with damped atom-atom dispersion corrections. *Phys Chem Chem Phys.* 2008; 10(44):6615–6620.
72. Grimme S. Semiempirical hybrid density functional with perturbative second-order correlation. *J Chem Phys.* 2006;124(3):034108.
73. Grimme S, Ehrlich S, Goerigk L. Effect of the damping function in dispersion corrected density functional theory. *J Comput Chem.* 2011;32(7):1456–1465.
74. Harder E, Damm W, Maple J, et al. OPLS3: a force field providing broad coverage of drug-like small molecules and proteins. *J Chem Theory Comput.* 2016;12(1):281–296.
75. Shivakumar D, Harder E, Damm W, Friesner RA, Sherman W. Improving the prediction of absolute solvation free energies using the next generation OPLS force field. *J Chem Theory Comput.* 2012;8(8):2553–2558.
76. Wang L, Wu Y, Deng Y, et al. Accurate and reliable prediction of relative ligand binding potency in prospective drug discovery by way of a modern free-energy calculation protocol and force field. *J Am Chem Soc.* 2015;137(7):2695–2703.

77. Poland D, Scheraga HA. Energy parameters in polypeptides. I. Charge distributions and the hydrogen bond. *Biochemistry*. 1967;6(12):3791–3800.
78. Malde AK, Zuo L, Breeze M, et al. An automated force field topology builder (ATB) and repository: version 1.0. *J Chem Theory Comput*. 2011;7(12):4026–4037.
79. Hermans J, Berendsen HJC, van Gunsteren WF, Postma JPM. A consistent empirical potential for water-protein interactions. *Biopolymers*. 1984;23(8):1513–1518.
80. Daura X, Mark AE, van Gunsteren WF. Parametrization of aliphatic CH_n united atoms of GROMOS96 force field. *J Comput Chem*. 1998;19(5):535–547.
81. Schuler LD, van Gunsteren WF. On the choice of dihedral angle potential energy functions for *n*-alkanes. *Mol Sim*. 2000;25(5):301–319.
82. Schuler LD, Daura X, van Gunsteren WF. An improved GROMOS96 force field for aliphatic hydrocarbons in the condensed phase. *J Comput Chem*. 2001;22(11): 1205–1218.
83. Chandrasekhar I, Kastenholz M, Lins RD, et al. A consistent potential energy parameter set for lipids: dipalmitoylphosphatidylcholine as a benchmark of the GROMOS96 45A3 force field. *Eur Biophys J*. 2003;32(1):67–77.
84. Lins RD, Hünenberger PH. A new GROMOS force field for hexopyranose-based carbohydrates. *J Comput Chem*. 2005;26(13):1400–1412.
85. Soares TA, Hünenberger PH, Kastenholz MA, et al. An improved nucleic acid parameter set for the GROMOS force field. *J Comput Chem*. 2005;26(7):725–737.
86. Oostenbrink C, Villa A, Mark AE, van Gunsteren WF. A biomolecular force field based on the free enthalpy of hydration and solvation: the GROMOS force-field parameter sets 53A5 and 53A6. *J Comput Chem*. 2004;25(13):1656–1676.
87. Oostenbrink C, Soares TA, van der Vegt NFA, van Gunsteren WF. Validation of the 53A6 GROMOS force field. *Eur Biophys J*. 2005;34(4):273–284.
88. Schmid N, Eichenberger AP, Choutko A, et al. Definition and testing of the GROMOS force-field versions 54A7 and 54B7. *Eur Biophys J*. 2011;40:843–856.
89. Reif MM, Hünenberger PH, Oostenbrink C. New interaction parameters for charged amino acid side chains in the GROMOS force field. *J Chem Theory Comput*. 2012;8(10):3705–3723.
90. Reif MM, Winger M, Oostenbrink C. Testing of the GROMOS force-field parameter set 54A8: structural properties of electrolyte solutions, lipid bilayers, and proteins. *J Chem Theory Comput*. 2013;9(2):1247–1264.
91. Tomaselli S, Esposito V, Vangone P, et al. The α-to-β conformational transition of Alzheimer's Aβ-(1-42) peptide in aqueous media is reversible: a step by step conformational analysis suggests the location of β conformation seeding. *Chembiochem*. 2006;7(2):257–267.
92. Kittner M, Knecht V. Disordered versus fibril-like amyloid β (25-35) dimers in water: structure and thermodynamics. *J Phys Chem B*. 2010;114(46):15288–15295.
93. Olubiyi OO, Strodel B. Structures of the amyloid β-peptides $A\beta_{1-40}$ and $A\beta_{1-42}$ as influenced by pH and a D-peptide. *J Phys Chem B*. 2012;115(10):3280–3291.
94. Abraham MJ, Murtola T, Schulz R, et al. GROMACS: high performance molecular simulations through multi-level parallelism from laptops to supercomputers. *SoftwareX*. 2015;1–2:19–25.
95. Carballo-Pacheco M, Ismail AE, Strodel B. On the applicability of force fields to study the aggregation of amyloidogenic peptides using molecular dynamics simulations. *J Chem Theory Comput*. 2018;14(11):6063–6075.
96. MacKerell Jr AD. Empirical force fields for biological macromolecules: overview and issues. *J Comput Chem*. 2004;25(13):1584–1604.
97. Patel S, Brooks III CL. CHARMM fluctuating charge force field for proteins: I parameterization and application to bulk organic liquid simulations. *J Comput Chem*. 2004;25:1–16.

98. Rick SW, Berne BJ. Dynamical fluctuating charge force fields: the aqueous solvation of amides. *J Am Chem Soc.* 1996;118:672–679.
99. Rick SW, Stuart SJ, Berne BJ. Dynamical fluctuating charge force fields: application to liquid water. *J Chem Phys.* 1994;101:6141–6156.
100. Stern HA, Rittner F, Berne BJ, Friesner RA. Combined fluctuating charge and polarizable dipole models: application to a five-site water potential function. *J Chem Phys.* 2001;115:2237–2251.
101. Patel S, MacKerell Jr AD, Brooks III CL. CHARMM fluctuating charge force field for proteins: II protein/solvent properties from molecular dynamics simulations using a nonadditive electrostatic model. *J Comput Chem.* 2004;25:1504–1514.
102. Morita A. Water polarizability in condensed phase: *Ab initio* evaluation by cluster approach. *J Comput Chem.* 2002;23(15):1466–1471.
103. Lamoureux G, MacKerell Jr AD, Roux B. A simple polarizable model of water based on classical Drude oscillators. *J Chem Phys.* 2003;119(10):5185–5197.
104. Kaminski GA, Stern HA, Berne BJ, et al. Development of a polarizable force field for proteins via *Ab Initio* quantum chemistry: first generation model and gas phase tests. *J Comput Chem.* 2002;23(16):1515–1531.
105. Ponomarev SY, Kaminski GA. Polarizable simulations with second-order interaction model (POSSIM) force field: developing parameters for alanine dipeptide and protein backbone. *J Chem Theory Comput.* 2011;7(5):1415–1427.
106. Kaminski GA, Ponomarev SY, Liu AB. Polarizable simulations with second-order interaction model-force field and software for fast polarizable calculations: parameters for small model systems and free energy calculations. *J Chem Theory Comput.* 2009;5(11):2935–2943.
107. Ponomarev SY, Sa Q, Kaminski GA. Effects of lysine substitution on stability of polyalanine α helix. *J Chem Theory Comput.* 2012;8(11):4691–4706.
108. Li X, Ponomarev SY, Sigalovsky DL, Cvitkovic JP, Kaminski GA. POSSIM: parameterizing complete second-order polarizable force field for proteins. *J Chem Theory Comput.* 2014;10(11):4896–4910.
109. Sharma I, Kaminski GA. Using polarizable POSSIM force field and fuzzy-border continuum solvent model to calculate pK_a shifts of protein residues. *J Comput Chem.* 2017;38(2):65–80.
110. Ponder JW, Wu C, Ren P, et al. Current status of the AMOEBA polarizable force field. *J Phys Chem B.* 2010;114(8):2549–2564.
111. Shi Y, Xia Z, Zhang J, et al. Polarizable atomic multipole-based AMOEBA force field for proteins. *J Chem Theory Comput.* 2013;9:4046–4063.
112. Ren P, Ponder JW. Polarizable atomic multipole water model for molecular mechanics simulation. *J Phys Chem B.* 2003;107(24):5933–5947.
113. Grossfield A, Ren P, Ponder JW. Ion solvation thermodynamics from simulation with a polarizable force field. *J Am Chem Soc.* 2003;125(50):15671–15682.
114. Halgren TA. The representation of van der Waals (vdW) interactions in molecular mechanics force fields: potential form, combination rules, and vdW parameters. *J Am Chem Soc.* 1992;114(20):7827–7843.
115. Thole BT. Molecular polarizabilities calculated with a modified dipole interaction. *Chem Phys.* 1981;59(3):341–350.
116. Célerse F, Lagardère L, Derat E, Piquemal J-P. Massively parallel implementation of steered molecular dynamics in tinker-HP: comparisons of polarizable and non-polarizable simulations of realistic systems. *J Chem Theory Comput.* 2019;15(6):3694–3709.
117. Zhang C, Lu C, Jing Z, et al. AMOEBA polarizable atomic multipole force field for nucleic acids. *J Chem Theory Comput.* 2018;14(4):2084–2108.
118. Ahmed MC, Papaleo E, Lindorff-Larsen K. How well do force fields capture the strength of salt bridges in proteins? *PeerJ.* 2018;6:e4967.

119. Drude P, Millikan RA, Mann RC. *The Theory of Optics*. New York: Longmans, Green, and Co.; 1902.
120. Lopes PEM, Roux B, MacKerell Jr AD. Molecular modeling and dynamics studies with explicit inclusion of electronic polarization: theory and applications. *Theor Chem Acc*. 2009;124:11–28.
121. Lemkul JA, Huang J, Roux B, MacKerell Jr AD. An empirical polarizable force field based on the classical Drude oscillator model: development history and recent applications. *Chem Rev*. 2016;116(9):4983–5013.
122. Lopes PEM, Huang J, Shim J, et al. Polarizable force field for peptides and proteins based on the classical Drude oscillator. *J Chem Theory Comput*. 2013;9(12):5430–5449.
123. Lemkul JA, MacKerell Jr AD. Polarizable force field for DNA based on the classical Drude oscillator: I. Refinement using quantum mechanical base stacking and conformational energetics. *J Chem Theory Comput*. 2017;13(5):2053–2071.
124. Lemkul JA, MacKerell Jr AD. Polarizable force field for DNA based on the classical Drude oscillator: II. Microsecond molecular dynamics simulations of duplex DNA. *J Chem Theory Comput*. 2017;13(5):2072–2085.
125. Lemkul JA, MacKerell Jr AD. Polarizable force field for RNA based on the classical Drude oscillator. *J Comput Chem*. 2018;39(32):2624–2646.
126. Miller KJ. Additivity methods in molecular polarizability. *J Am Chem Soc*. 1976;112(23):8533–8542.
127. Harder E, Anisimov VM, Whitfield T, MacKerell Jr AD, Roux B. Understanding the dielectric properties of liquid amides from a polarizable force field. *J Phys Chem B*. 2008;112(11):3509–3521.
128. Harder E, Anisimov VM, Vorobyov IV, et al. Atomic level anisotropy in the electrostatic modeling of lone pairs for a polarizable force field based on the classical Drude oscillator. *J Chem Theory Comput*. 2006;2(6):1587–1597.
129. Lamoureux G, Harder E, Vorobyov IV, Roux B, MacKerell Jr AD. A polarizable model of water for molecular dynamics simulations of biomolecules. *Chem Phys Lett*. 2006;418:245–249.
130. Lemkul JA, MacKerell Jr AD. Balancing the interactions of Mg^{2+} in aqueous solution and with nucleic acid moieties for a polarizable force field based on the classical Drude oscillator model. *J Phys Chem B*. 2016;120(44):11436–11448.
131. Ngo VA, Fanning JK, Noskov SY. Comparative analysis of protein hydration from MD simulations with additive and polarizable force fields. *Adv Theory Simul*. 2018;2(2): 1800106.
132. Lin F-Y, Lopes PEM, Harder E, Roux B, MacKerell Jr AD. Polarizable force field for molecular ions based on the classical Drude oscillator. *J Chem Inf Model*. 2018;58(5):993–1004.
133. Huang J, Lopes PEM, Roux B, MacKerell Jr AD. Recent advances in polarizable force fields for macromolecules: microsecond simulations of proteins using the classical Drude oscillator model. *J Phys Chem Lett*. 2014;5(18):3144–3150.
134. Vorobyov IV, Anisimov VM, MacKerell Jr AD. Polarizable empirical force field for alkanes based on the classical Drude oscillator model. *J Phys Chem B*. 2005; 109(40):18988–18999.
135. Huang J, MacKerell Jr AD. Induction of peptide bond dipoles drives cooperative helix formation in the $(AAQAA)_3$ peptide. *Biophys J*. 2014;107(4):991–997.
136. Shalongo W, Dugad L, Stellwagen E. Distribution of helicity within the model peptide acetyl$(AAQAA)_3$amide. *J Am Chem Soc*. 1994;116(18):8288–8293.
137. Hazel AJ, Walters ET, Rowley CN, Gumbart JC. Folding free energy landscapes of β-sheets with non-polarizable and polarizable CHARMM force fields. *J Chem Phys*. 2018;149:072317.

138. Lemkul JA, Huang J, MacKerell Jr AD. Induced dipole-dipole interactions influence the unfolding pathways of wild-type and mutant amyloid β-peptides. *J Phys Chem B*. 2015;119(51):15574–15582.
139. Davidson DS, Brown AM, Lemkul JA. Insights into stabilizing forces in amyloid fibrils of differing sizes from polarizable molecular dynamics simulations. *J Mol Biol*. 2018;430(20):3819–3834.
140. Coles M, Bicknell W, Watson AA, Fairlie DP, Craik DJ. Solution structure of amyloid β-peptide(1-40) in a water-micelle environment. is the membrane-spanning domain where we think it is? *Biochemistry*. 1998;37(31):11064–11077.
141. Lindert S, Bucher D, Eastman P, Pande V, McCammon JA. Accelerated molecular dynamics simulations with the AMOEBA polarizable force field on graphics processing units. *J Chem Theory Comput*. 2013;9(11):4684–4691.
142. Huang J, Lemkul JA, Eastman PK, MacKerell Jr AD. Molecular dynamics simulations using the Drude polarizable force field on GPUs with OpenMM: implementation, validation, and benchmarks. *J Comput Chem*. 2018;39(21):1682–1689.
143. Lemkul JA, Roux B, van der Spoel D, MacKerell Jr AD. Implementation of extended Lagrangian dynamics in GROMACS for polarizable simulations using the classical Drude oscillator model. *J Comput Chem*. 2015;36(19):1473–1479.
144. Case DA, Cheatham III TE, Darden T, et al. The Amber biomolecular simulation programs. *J Comput Chem*. 2005;26(16):1668–1688.
145. Salomon-Ferrer R, Case DA, Walker RC. An overview of the Amber biomolecular simulation package. *WIREs Comput Mol Sci*. 2013;3(2):198–210.
146. Rackers JA, Wang Z, Lu C, et al. Tinker 8: software tools for molecular design. *J Chem Theory Comput*. 2018;14(10):5273–5289.
147. Lagardère L, Jolly L-H, Lipparini F, et al. Tinker-HP: a massively parallel molecular dynamics package for multiscale simulations of large complex systems with advanced point dipole polarizable force fields. *Chem Sci*. 2018;9:956–972.
148. Eastman P, Swails J, Chodera JD, et al. OpenMM 7: rapid development of high performance algorithms for molecular dynamics. *PLoS Comp Biol*. 2017;13(7):e1005659.
149. Brooks BR, Brooks III CL, MacKerell Jr. AD, et al. CHARMM: The biomolecular simulation program. J Comput Chem 2009; 30: 1545–1614.
150. Lamoureux G, Roux B. Modeling induced polarization with classical Drude oscillators: theory and molecular dynamics simulation algorithm. *J Chem Phys*. 2003;119: 3025–3039.
151. Jiang W, Hardy DJ, Phillips JC, MacKerell Jr AD, Schulten K, Roux B. High-performance scalable molecular dynamics simulations of a polarizable force field based on classical Drude oscillators in NAMD. *J Phys Chem Lett*. 2011;2:87–92.
152. Phillips JC, Braun R, Wang W, et al. Scalable molecular dynamics with NAMD. *J Comput Chem*. 2005;26(16):1781–1802.

CHAPTER TWO

Scale-consistent approach to the derivation of coarse-grained force fields for simulating structure, dynamics, and thermodynamics of biopolymers

Adam Liwo[a,b,*], Cezary Czaplewski[a], Adam K. Sieradzan[a,b], Emilia A. Lubecka[c], Agnieszka G. Lipska[a], Łukasz Golon[a], Agnieszka Karczyńska[a], Paweł Krupa[d], Magdalena A. Mozolewska[e], Mariusz Makowski[a], Robert Ganzynkowicz[a], Artur Giełdoń[a], Maciej Maciejczyk[f]

[a]Faculty of Chemistry, University of Gdańsk, Gdańsk, Poland
[b]School of Computational Sciences, Korea Institute for Advanced Study, Seoul, Republic of Korea
[c]Institute of Informatics, Faculty of Mathematics, Physics, and Informatics, University of Gdańsk, Gdańsk, Poland
[d]Institute of Physics, Polish Academy of Sciences, Warsaw, Poland
[e]Institute of Computer Science, Polish Academy of Sciences, Warsaw, Poland
[f]Department of Physics and Biophysics, Faculty of Food Science, University of Warmia and Mazury in Olsztyn, Olsztyn, Poland
*Corresponding author: e-mail address: adam.liwo@ug.edu.pl

Contents

Progress in Molecular Biology and Translational Science, Volume 170
ISSN 1877-1173
https://doi.org/10.1016/bs.pmbts.2019.12.004

Abstract

In this chapter the scale-consistent approach to the derivation of coarse-grained force fields developed in our laboratory is presented, in which the effective energy function originates from the potential of mean force of the system under consideration and embeds atomistically detailed interactions in the resulting energy terms through use of Kubo's cluster-cumulant expansion, appropriate selection of the major degrees of freedom to be averaged out in the derivation of analytical approximations to the energy terms, and appropriate expression of the interaction energies at the all-atom level in these degrees of freedom. Our approach enables the developers to find correct functional forms of the effective coarse-grained energy terms, without having to import them from all-atom force fields or deriving them on a heuristic basis. In particular, the energy terms derived in such a way exhibit correct dependence on coarse-grained geometry, in particular on site orientation. Moreover, analytical formulas for the multibody (correlation) terms, which appear to be crucial for coarse-grained modeling of many of the regular structures such as, e.g., protein α-helices and β-sheets, can be derived in a systematic way. Implementation of the developed theory to the UNIfied COarse-gRaiNed (UNICORN) model of biological macromolecules, which consists of the UNRES (for proteins), NARES-2P (for nucleic acids), and SUGRES-1P (for polysaccharides) components, and is being developed in our laboratory is described. Successful applications of UNICORN to the prediction of protein structure, simulating the folding and stability of proteins and nucleic acids, and solving biological problems are discussed.

1. Introduction

Coarse-grained models of macromolecules and large molecular assemblies, especially those involving biological (macro)molecules, are often used to simulate large molecular systems.[1–15] In these models, a single interaction site encompasses several atoms, a whole small molecule or polymer unit, or even large structural parts of a macromolecule[16,17] (e.g., protein α-helices[18,19]). The development of coarse-grained model was initiated 43 years ago by the work by Levitt and Warshel who introduced a heavily reduced physics-based model of polypeptide chains.[20,21]

Regardless of the degree of coarse graining, the respective force field is needed to compute the energy and forces. In this regard, the coarse-grained force fields can be divided into four principal kinds: (i) structure-based, which are biased toward the target (native) structure; (ii) the force fields for modeling the basic features of the systems studied (e.g., the HP model of proteins developed by Dill and coworkers[22]), (iii) the statistical potentials,[1,10,12,23,24] and (iv) the physics-based potentials.[2–5,7,9,15] The statistical potentials are derived by applying the Boltzmann inversion to the distribution and correlation functions derived from structural databases.[1,12,24] Although these potentials have been criticized because of their vague connection to the physics of interactions,[25,26] they are quite successful in protein- and RNA-structure modeling, even in database- and data-unassisted mode.[12,24] The physics-based potentials are meant to reproduce the energetics of the interactions at coarse-grained level correctly. Two approaches, often combined together, to their derivation are applied: the bottom-up and the top-down approach, the first one starting from atomistically detailed interactions and merging them into the interactions between the extended groups, the second one being based on the parameterization of the force field to reproduce experimental quantities such as, e.g., free energies of solvation and association.[12] Some of the force fields used to simulate biological macromolecules are summarized in Table 1, while more details can be found in the review articles and book chapters cited[1,5,6,12,14] as well as in the books dedicated to the subject.[4,13]

The physics-based coarse-grained force fields are much more difficult to design than the all-atom ones, because the patterns of interactions between extended interaction sites are much more complex than those at the atomistically detailed level. Dependence on site orientation and not only their distances should be considered to account for the interactions properly. Moreover, harmonic approximation often becomes irrelevant to handle virtual-bond and virtual-bond-angle deformation, because the respective potentials are often multimodal.[31–34] Most importantly, proper modeling of regular structures, such as the secondary structures of proteins, requires the consideration of the interplay between local and backbone hydrogen bonding interactions.[23,35,36] Finding the appropriate functional forms to account for the angular and correlation terms is difficult and, therefore, most of the coarse-grained force fields include long-range interactions in the spherical form and no correlation contribution.[2,9]

About 20 years ago, we started to develop a general method of physics-based coarse graining.[36] We have defined the prototype of the coarse-grained

Table 1 Examples of coarse-grained force fields used in simulations of proteins and other biological macromolecules.

Name	Type	Applications	Webpage	Refs.
AWSEM	Physics-based	Protein-structure prediction, prediction of protein–protein complexes, dimerization interfaces, and aggregation	https://github.com/adavtyan/awsemmd	7
CABS	Statistical	Protein-structure prediction, simulations of protein dynamics, protein–protein and protein–ligand docking, proteins in lipid environment	http://biocomp.chem.uw.edu.pl/tools	23
MARTINI	Physics-based	Dynamics and properties of systems comprising proteins, lipids, nucleic acids, and polysaccharides	http://cgmartini.nl	27, 28
OPEP	Physics-based	Dynamics and properties of systems comprising proteins, nucleic acids and polysaccharides	https://bioserv.rpbs.univ-paris-diderot.fr/services/PEP-FOLD3	3
PRIMO	Statistical	Folding small peptides, protein-structure prediction	—	10
SimRNA	Statistical	RNA structure prediction	http://genesilico.pl/SimRNAweb	24
SIRAH	Physics-based	Simulations of proteins and protein complexes, extension to nucleic acids and glycans	http://www.sirahff.com	15
UNICORN (UNRES)	Physics-based	Protein-structure prediction, dynamics and properties of systems comprising proteins and nucleic acids in water and membrane environments, extension to polysaccharides	http://www.unres.pl	29, 30

effective energy function of a system as a restricted free energy (RFE) or potential of mean force (PMF), in which the degrees of freedom omitted from the coarse-grained representation (the *secondary degrees of freedom*) are integrated out.[36] To get a usable and transferable force field, we express the PMF in terms of Kubo cluster cumulants[37] (factors), corresponding to effective energy terms, keeping only a minimum number of factors necessary to describe the energetics of the system at a desired accuracy level. Approximate functional forms are obtained as the first nonvanishing terms in the cluster-cumulant expansion of the factors.[36] A unique feature of this approach is the natural inclusion of the dependence of the effective energy terms on temperature, because they absorb the motions of the secondary degrees of freedom.[38] This feature causes the highest-order factors, corresponding to concerted interactions (the correlation terms), which are responsible for ordered structures, to vanish fast with increasing temperature, thereby causing the structure to become more disordered as temperature increases.[38]

Recently,[34] we further developed the methodology outlined above, by expressing the interaction energies at the atomistically detailed level in terms of the squares of interatomic distances. These distances and, in turn, the energy, are further expressed in terms of the distances between site centers, site-orientation parameters and collective rotation angles of the atoms about the virtual-bond axes of the sites. Averaging becomes then integration over the powers of sines and cosines. The approximate functional expressions derived in such a way are scale-consistent, because they encode the information of the distribution of interaction between the atoms of the interacting sites. Consequently, they provide proper dependence of the effective energy terms on site orientation and local geometry and, moreover, enable us to derive physics-based expressions for the correlation terms. In this chapter, we describe the methodology outlined above, along with its implementation in the development of the UNified COarse-gRaiNed (UNICORN) model of biological molecules,[29] which is being developed in our laboratory, and the applications of this model in predicting the structure and investigating the dynamics and thermodynamics of proteins and nucleic acids and in solving biological problems.

2. Theoretical background

2.1 Potential of mean force of a coarse-grained system as a prototype of the effective energy function

In our approach to coarse-graining,[34,36] we assume that the effective energy function of a system is derived from its potential of mean force (PMF), with all degrees of freedom that are not present in the coarse-grained representation averaged out. Thus, the variables describing the geometry of the polypeptide-water system are divided into two sets: the *primary* variables ($\mathbf{X}$), which describe the coarse-grained degrees of freedom, and the *"less important"* or *secondary* variables (**Y**) that are averaged over; the secondary variables usually include the solvent degrees of freedom. The degrees of freedom contained in **X** and **Y** are those of atomic nuclei. Consequently, the PMF [$F(\mathbf{X};\ T)$] is expressed by Eq. (1); it should be noted that it depends on temperature.

$$F(\mathbf{X};T) = -\frac{1}{\beta}\ln\left\{\frac{1}{V_{\mathbf{Y}}}\int\limits_{\Omega_{\mathbf{Y}}}\exp\left[-\beta E(\mathbf{X};\mathbf{Y})\right]dV_{\mathbf{Y}}\right\} \tag{1}$$

where $V_{\mathbf{Y}} = \int_{\Omega_{\mathbf{Y}}} dV_{\mathbf{Y}}$, $E(\mathbf{X};\ \mathbf{Y})$ is the original (all-atom) energy function, equivalent to the Born–Oppenheimer energy surface, in which the electron degrees of freedom are averaged out, $\Omega_{\mathbf{Y}}$ is the region of the **Y** subspace of variables over which the integration is carried out, $V_{\mathbf{Y}}$ is the volume of this region, and $\beta = 1/RT$, R being the universal gas constant and T the absolute temperature.

The choice of the PMF as a prototype of coarse-grained energy function is motivated by the fact that the probability density, $P(\mathbf{X};\ T)$ of a given coarse-grained structure is directly related to it (Eq. 2) and also any ensemble and thermodynamic averages can be computed using the PMF or PMF and its temperature derivatives.[34]

$$P(\mathbf{X};T) = \frac{\int\limits_{\Omega_{\mathbf{Y}}}\exp\left[-\beta E(\mathbf{X};\mathbf{Y})\right]dV_{\mathbf{Y}}}{\int\limits_{\Omega_{\mathbf{X}}}\int\limits_{\Omega_{\mathbf{Y}}}\exp\left[-\beta E(\mathbf{X};\mathbf{Y})\right]dV_{\mathbf{Y}}dV_{\mathbf{X}}} = \frac{\exp\left[-\beta F(\mathbf{X};\mathbf{Y})\right]}{\int\limits_{\Omega_{\mathbf{X}}}\exp\left[-\beta F(\mathbf{X};\mathbf{Y})\right]dV_{\mathbf{X}}} \tag{2}$$

It can be noted at this point that the PMF (Eq. 1) defines the prototype of coarse-grained energy functions in the same way as the Born–Oppenheimer potential energy defines all-atom energy functions; in the second case averaging is done over electron degrees of freedom.

2.2 Steps toward a scale-consistent coarse-grained energy function

Similar as the Born–Oppenheimer energy surface, which is specific to the system it was calculated for and, therefore, needs to be approximated by a sum of contributions (energy terms) characteristic of smaller fragments and parameterized to give a usable all-atom force field, it is not possible to use the PMF [$F(\mathbf{X};\ T)$] directly because such a force field would be not transferable in the first place. It would also not be usable, because obtaining it for a real-size system would require much greater effort than doing actual all-atom calculations aimed at solving a specific problem. Therefore, the components of a coarse-grained force field must be derived and parameterized based on PMFs of smaller systems. This is a procedure similar that of the construction of all-atom force fields.[39] However, care must be taken to derive the effective energy terms in a scale-consistent manner, i.e., keeping track of the atomistically detailed structure of interactions, which lead to the emergence of nonradial and correlation terms. This issue is particularly important when extensive coarse graining is applied. The procedure for the construction of a physics-based scale-consistent force field developed in our laboratory[29,30,34,36,38,40–42] is summarized in Fig. 1 and its subsequent stages are described in Sections 2.3-2.6.

2.3 Factorization of the PMF into Kubo cluster-cumulant functions

When designing a coarse-grained model, the first step is to determine the extended interaction sites and assign atoms to them (Fig. 1). Usually the atoms are assigned to a given site once for all but some, e.g., the solvent atoms, are volatile and we can only approximately assign an average number of solvent molecules of the solvation sphere to a given site but the molecules can migrate between sites. It should also be noted that models exist in which the sites overlap and, consequently, an atom can be assigned to more than one site (e.g., in the coarse-grained SURPASS model of proteins developed recently by Gront and coworkers[43]).

With the partition into extended sites and assignment of atoms to sites accomplished, we express the all-atom energy $E(\mathbf{X};\mathbf{Y})$ as a sum of *component energies*, each of which is either the sum of energies between the atoms that belong to the same unit or to different units, as given by Eq. (3). Following this division, the PMF (Eq. 1) can be decomposed into *factors*, each of which is a *Kubo's cluster-cumulant function*,[37] as expressed by Eq. (4) and illustrated in Fig. 2.

$$E(\mathbf{X};\mathbf{Y}) = \sum_{i=1}^{n} \epsilon_i(\mathbf{X};\mathbf{z}_i) \tag{3}$$

where $\epsilon_i(\mathbf{X};\mathbf{z}_i)$ is the ith component energy, $\mathbf{z}_i$ contains the secondary degrees of freedom on which ϵ_i depends, and n is the number of energy

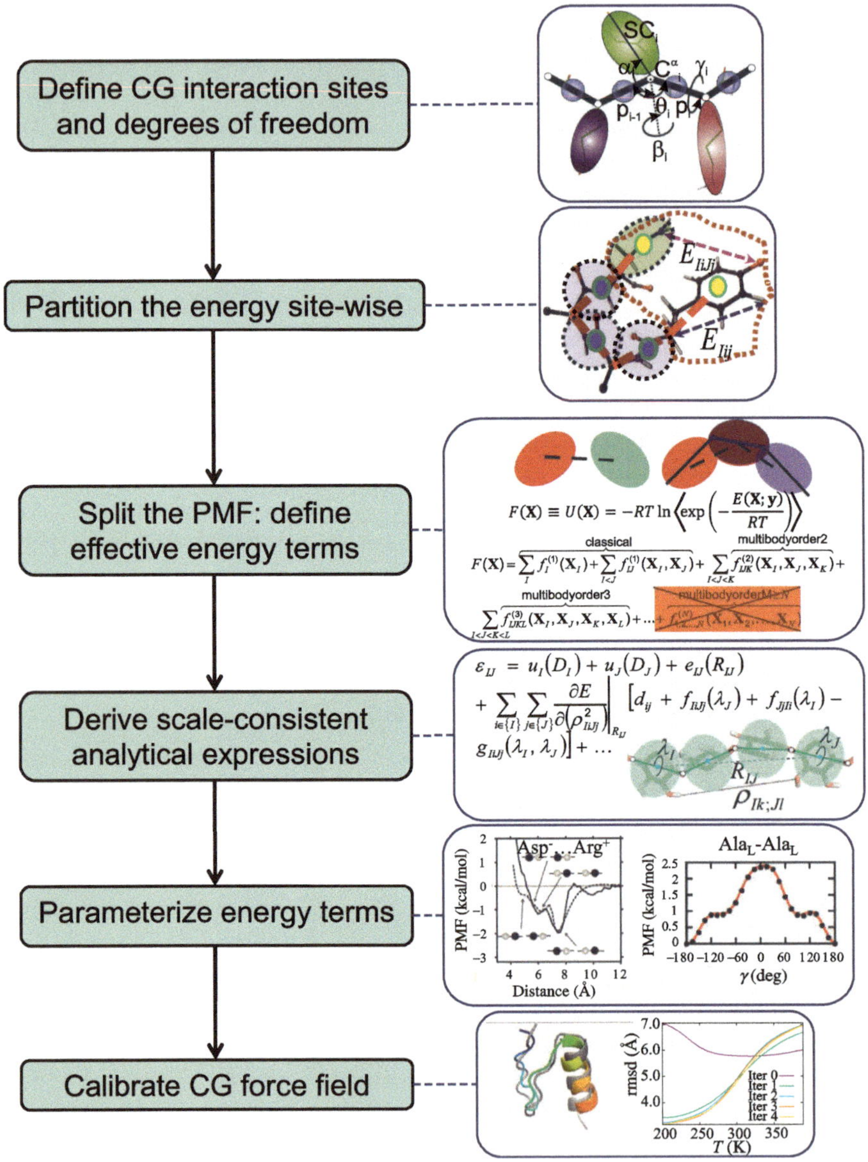

Fig. 1 Scheme of the development of a physics-based scale-consistent model of coarse-grained force fields.

components. It should be noted that, because we do not consider electrons explicitly (their coordinates already averaged out), the component energies can, in principle, contain multibody terms, although these are usually not considered in all-atom force fields (Fig. 1).

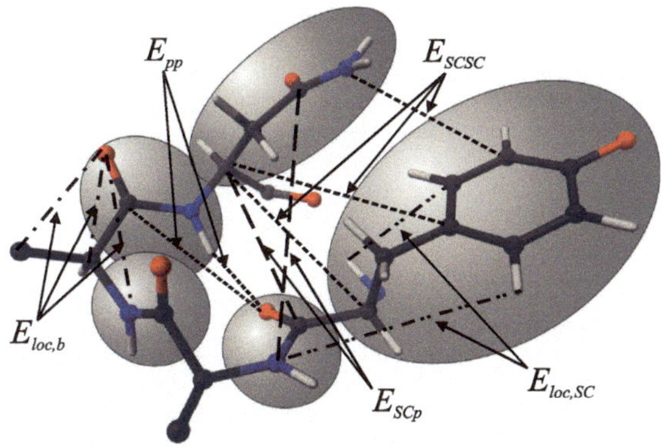

Fig. 2 Splitting of the energy of interactions into component energies illustrated with a fragment of a polypeptide chain. SC denotes a side chain and *p* denotes a peptide group. The atoms of two side chains and three peptide groups of the portion of the polypeptide chain shown in the picture are embedded in *shaded ellipsoids*. For clarity only some of the interactions are shown and the water molecules are not included. E_{AB} denote the energies of interactions between the atoms of site A and B, while $E_{loc,b}$ and $E_{loc,SC}$ denote the energies that contribute to local-interaction energies within the backbone and the side-chain part of a given residue, respectively. *Reproduced with permission from Liwo A, Czaplewski C, Ołdziej S, et al. Simulation of protein structure and dynamics with the coarse-grained UNRES force field. In: Voth G, ed. Coarse-Graining of Condensed Phase and Biomolecular Systems Boca Raton, Florida:CRC Press & Taylor & Francis Group; 2008:1391–1411 [chapter 8]. Copyright 2008, CRC Press, Taylor & Francis Group, LLC.*

We can now write the PMF as a sum of Kubo's cluster-cumulant functions,[37] termed *factors*, of order 1, 2, …, *n*, a factor of order *k* containing integrals with up to *k* ϵs (Fig. 1).

$$F(\mathbf{X};T)=\sum_i f_i^{(1)}(\mathbf{X})+\sum_{i<j} f_{ij}^{(2)}(\mathbf{X})+\sum_{i<j<k} f_{ijk}^{(3)}(\mathbf{X})+\cdots+\sum_{i_1<i_2\ldots<i_n} f_{i_1 i_2\ldots i_n}^{(n)}(\mathbf{X}) \tag{4}$$

The factors are expressed by Eq. (5).

$$\begin{aligned} f_{i_1 i_2\ldots i_k}^{(k)}(\mathbf{X}) &= \langle\langle \epsilon_{i_1}\epsilon_{i_2}\ldots\epsilon_{i_k}\rangle\rangle_f = \sum_{l=1}^{k}\sum_{\substack{i_{m_1}<i_{m_2}<\cdots<i_{m_l}\\ m_i\in[1\ldots k]}} (-1)^{k-l} F_{i_{m_1} i_{m_2}\ldots i_{m_l}}^{(l)}(\mathbf{X}) \\ &= \sum_{l=1}^{k}\sum_{\substack{i_{m_1}<i_{m_2}<\cdots<i_{m_l}\\ m_i\in[1\ldots k]}} (-1)^{k-l}\langle\langle \epsilon_{i_{m_1}}\epsilon_{i_{m_2}}\ldots\epsilon_{i_{m_l}}\rangle\rangle \end{aligned} \tag{5}$$

where

$$F^{(k)}_{i_1,i_2,\ldots i_k}(\mathbf{X}) \equiv \langle\langle \epsilon_{i_1}\epsilon_{i_2}\ldots\epsilon_{i_k}\rangle\rangle = -\frac{1}{\beta}\ln\left\{\frac{1}{V_{\mathbf{y}_\mathcal{I}}}\int\limits_{\Omega_\mathcal{I}}\exp\left[-\beta\sum_{l=1}^{k}\epsilon_{i_l}(\mathbf{X};\mathbf{z}_{i_l})\right]dV_{\mathbf{y}_\mathcal{I}}\right\} \tag{6}$$

(with $V_{\mathbf{y}_\mathcal{I}}$ being the volume of the subspace spanned by variables $\mathbf{y}_{i_1}, \mathbf{y}_{i_2}, \ldots, \mathbf{y}_{i_k}$) is the PMF containing only a subset of component interactions.

The factors of the first order, $f^{(1)}$, correspond to the PMFs containing single component energies, i.e., those pertaining to interactions within single units (such as, e.g., isolated amino acid residues), within pairs of units (such as, e.g., pairs of interacting side chains), including those pertaining to interactions within higher clusters of units if the all-atom potential energy surface (PES) contains substantial multibody terms. Examples of the first-order factors that account for the local and long-range interactions are shown in Fig. 3A. The factors of order 2 and higher correspond to generic coarse-grained *multibody* or *correlation* terms and they contain integrals that have $2, 3, \ldots, n$ component energies in the Boltzmann factor. In a broad sense, the multibody factors in coarse-grained force field have a similar origin as those in all-atom force fields, namely, they emerge because of averaging over some of the degrees of freedom. The degrees of freedom averaged over when constructing all-atom force fields are the electron degrees of freedom, while those averaged over when deriving coarse-grained force fields are atomic-detailed degrees of freedom (cf. Eq. 1). An important difference between the role of these terms in the two kinds of force fields is that they are not necessary for the all-atom fields to work, while coarse-grained force fields without them usually cannot produce regular structures of biopolymers in free simulations.[23,34–36] It should be noted that $f^{(k)}$ is a generic kth order factor, because all contributions of order lower than k are subtracted from it (cf. Eq. 5). The contribution of a factor to the PMF heavily depends on whether the secondary degrees of freedom are shared in the expressions for the component energies (ϵs) that occur in the expression for this factor; if at least one of the ϵs does not share the secondary degrees of freedom with the other ϵs, the corresponding factor is zero. An example of the second-order factor that corresponds to the torsional potential is shown in Fig. 3B. All factors depend on temperature, this dependence increasing with increasing the order of the factor because of the increasing order of the first term in generalized cumulant expansion of this factor.[36,38] In our approach this temperature dependence is explicitly accounted for.[38]

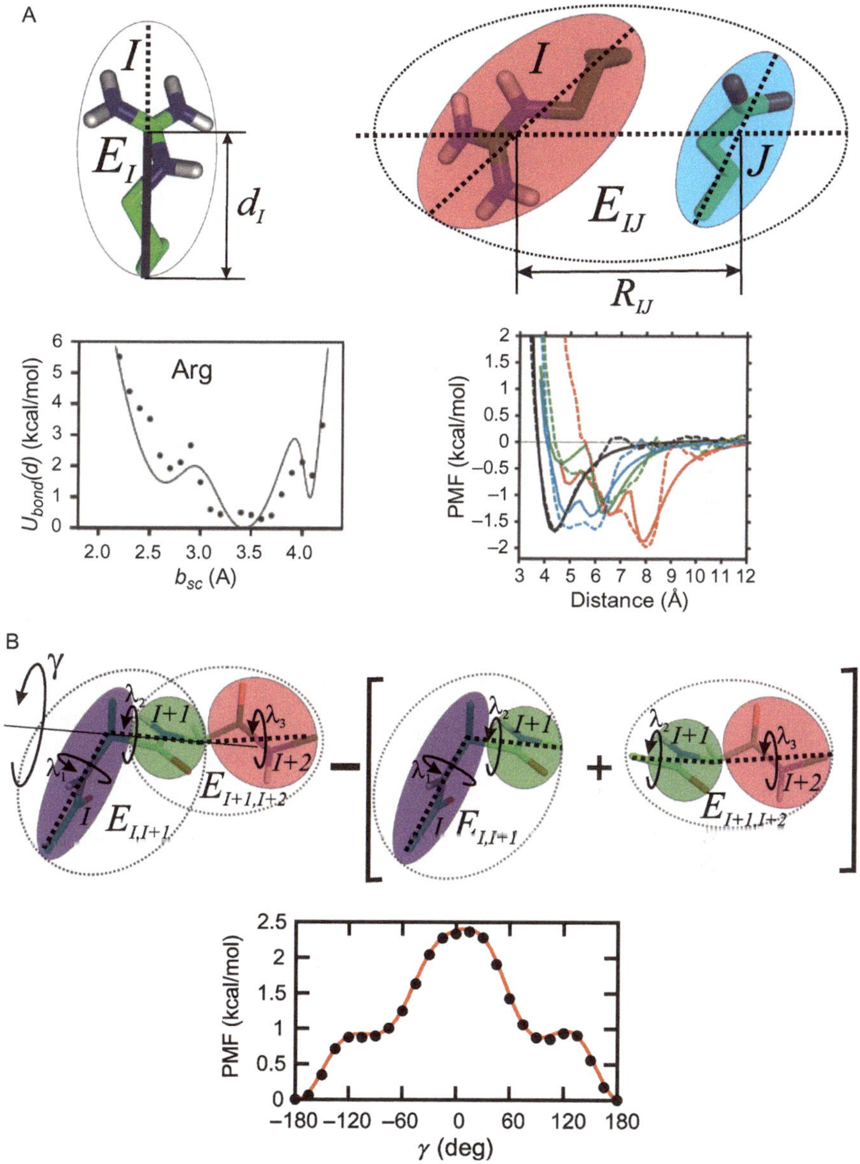

Fig. 3 Examples of the first-order (A) and second-order PMF factors (B). The factor illustrated in the left part of panel (A) corresponds to the virtual-bond potential of a united arginine side chain. The only coarse-grained degree of freedom is the length of the virtual bond (which is shown as a thick *gray line*, the side-chain long axis being shown as the thinner *black dashed line*) from C^{α} atom to the side-chain center, all other degrees of freedom averaged out. The energy consists of the energies of interactions between the side-chain atoms (within the *dashed oval*). Because of side-chain flexibility, the resulting potential is multimodal.[31] The factor illustrated in the right part of panel (A) corresponds to the mean-field potential of interactions of coarse-grained arginine and glutamine side chains in water, the secondary degrees of freedom including the rotation about

(Continued)

The factor expansion is truncated to achieve a compromise between the complexity of the force field and its ability to reproduce the structure and dynamics of the system. In the UNICORN model, we now use only up to third-order factors, which seem sufficient for reproducing regular secondary structure.[30,45] In the neo-classical force fields, such as MARTINI,[27,28] all long-range interactions are approximated by factors of order 1 involving only pairs of sites as in typical all-atom force fields, factors of order 2 are present only as the torsional potentials; these factors account for the coupling between the conformational states of the consecutive polymer units.[34,36] Nevertheless,

Fig. 3—Cont'd the virtual-bond axes. The side-chain long axes and the line linking side-chain centers are shown as *black dashed lines*. The component energy consists of the energies of interactions between the atoms of the arginine and those of the glutamine side chains plus those involving the surrounding solvent but does not include the energies of interactions between the atoms of the same side chain. The resulting PMF profile depends not only on the distance between site centers but also on the orientation of virtual-bond axes.[40] The factor illustrated in panel (B) corresponds to the virtual-bond-torsional potential of a terminally blocked alanine dipeptide. The atoms that belong to the consecutive coarse-grained units (I, $I+1$, and $I+2$) are embedded in *colored ovals*. The factor is obtained by subtracting the PMFs of terminally blocked alanine residues (composed of units I and $I+1$ and units $I+2$ and $I+2$, respectively) from that of the whole dipeptide. The component energies consist of the energies of interactions between the neighboring units; they do not include the energies of interactions within units or between units I and $I+2$. The primary degrees of freedom to be averaged out are the angles λ_I, λ_{I+1}, and λ_{I+2} for the rotation of the atoms of the respective unit about the $C^\alpha \ldots C^\alpha$ virtual-bond axis. The coupling results from sharing the angle λ_{I+1} on which the component energies of interaction $E_{I,I+1}$ and $E_{I+1,I+2}$ depend, this resulting in the dependence of the factor profile on the virtual-bond angles θ and the virtual-bond-dihedral angle γ; only the dependence on γ is shown in the bottom graph for clarity. *The effective potential profiles shown in the bottom sections of the panels were adapted with permission from Kozlowska U, Maisuradze GG, Liwo A, Scheraga HA. Determination of side-chain-rotamer and side-chain and backbone virtual-bond-stretching potentials of mean force from AM1 energy surfaces of terminally-blocked amino-acid residues, for coarse-grained simulations of protein structure and folding. 2. Results, comparison with statistical potentials, and implementation in the UNRES force field. J Comput Chem. 2010;31:1154–1167, Copyright 2010 Wiley; Makowski M, Liwo A, Scheraga HA. Simple physics-based analytical formulas for the potentials of mean force of the interaction of amino-acid side chains in water. VI. Oppositely charged side chains. J Phys Chem B. 2011;115(19):6130–6137, Copyright 2011 American Chemical Society; and Sieradzan AK, Hansmann UHE, Scheraga HA, Liwo A. Extension of UNRES force field to treat polypeptide chains with* d*-amino acid residues. J Chem Theory Comput. 2012;8(11):4746–4757, Copyright 2012 American Chemical Society. The chemical structures shown in the picture were prepared with PyMOL.*[44]

several coarse-grained force fields contain the multibody terms. The CABS statistical force field[23,35] contains the multibody terms describing hydrogen-bond cooperativity in a form of the energy gain added when two pairs of consecutive residues form hydrogen bonds. These terms were introduced based on the observed regularity of the hydrogen-bond patterns in regular secondary structures. The physics-based OPEP force field[3] also has similar multibody terms pertaining to cooperativity in hydrogen bonding, which were introduced based on the cooperativity of electrostatic interactions. Multibody terms have also been introduced in statistical potentials designed for structure recognition and quality assessment of protein models.[46] The multiscale coarse-graining method developed by the Voth group[2] incorporates the multibody terms implicitly in the effective pairwise potentials.

The components of the truncated expansion comprise only few interaction sites and are, thus, transferable to all systems composed of the units they contain, as the all-atom-force field components are.

2.4 Analytical scale-consistent approximations to the coarse-grained energy terms

Although the coarse-grained energy terms are defined by the factor expansion to the PMF (Eqs. 4–6), putting them into analytical expressions is much more difficult than in the case of the all-atom energy terms. The reason is that long-range atom–atom interactions can safely be considered to have spherical symmetry (even if the atoms are in the context of the other atoms bonded to them) and there are usually no multibody terms; if present, they usually pertain to electron polarization the handling of which is computationally expensive but straightforward. Neither of this applies to coarse-grained force fields because (i) the interaction sites are not spherical and (ii) multibody terms enter through PMF factors of order 2 and higher even if the parent all-atom energy function is pairwise. The coarse-grained force field, if meant to model the systems under study correctly, must take into account both factors. We have termed such an approach to the derivation of the coarse-grained force field the *scale-consistent approach*[30,34] to reflect the fact that, although the atomistic details of geometry and interactions are not present in coarse-grained models, they are accounted for (and not arbitrarily omitted) in the coarse-grained energy functions derived using this approach.

In the first step, we use Kubo's generalized cumulant expansion[37] to obtain the approximations to the formulas for the PMF factors (coarse-grained energy terms), as given by Eq. (7) (Refs. 36 and 34).

$$f^{(k)}_{i_1 i_2 \ldots i_k}(\mathbf{X};T) \equiv \langle\langle \epsilon_{i_1}\epsilon_{i_2}\ldots\epsilon_{i_k}\rangle\rangle_f = -\frac{1}{\beta}\sum_{l=k}^{\infty}\sum_{\substack{m_1,m_2,\ldots m_k>0\\ m_1+m_2+\cdots+m_k=l}}(-1)^l\left\langle \epsilon_{i_1}^{m_1}\epsilon_{i_2}^{m_2}\ldots\epsilon_{i_k}^{m_k}\right\rangle_c \tag{7}$$

where $\left\langle \epsilon_{i_1}^{m_1}\epsilon_{i_2}^{m_2}\ldots\epsilon_{i_k}^{m_k}\right\rangle_c$ is the generalized cumulant corresponding to factor $f^{(k)}_{i_1 i_2 \ldots i_k}$. The generalized cumulants are, in turn, expressed by mixed energy moments, in which the products of powers of component energies are integrated over the secondary degrees of freedom. For example, for the lowest-order of the cumulants in the expansion of the first three factors we have

$$\langle \epsilon_i\rangle_c = \langle \epsilon_i\rangle \tag{8}$$

$$\langle \epsilon_i\epsilon_j\rangle_c = \langle \epsilon_i\epsilon_j\rangle - \langle \epsilon_i\rangle\langle \epsilon_j\rangle \tag{9}$$

$$\langle \epsilon_i\epsilon_j\epsilon_k\rangle_c = \langle \epsilon_i\epsilon_j\epsilon_k\rangle - [\langle \epsilon_i\rangle\langle \epsilon_j\epsilon_k\rangle + \langle \epsilon_j\rangle\langle \epsilon_i\epsilon_k\rangle + \langle \epsilon_k\rangle\langle \epsilon_i\epsilon_j\rangle] + 2\langle \epsilon_i\rangle\langle \epsilon_j\rangle\langle \epsilon_k\rangle \tag{10}$$

The regular and mixed moments of the component energies that occur in Eqs. (7)–(10) are obtained by integrating out the secondary degrees of freedom. For polymer chains, the most important secondary degrees of freedom are the angles for collective rotation of all atoms of a unit about the respective virtual-bond axis; we term these angles λ. On the other hand, in the absence of external fields, the energy of a molecular system at the all-atom level can be expressed in terms of the distances between the atoms, as given by Eq. (11).

$$E(\boldsymbol{\rho}_1,\boldsymbol{\rho}_2,\ldots,\boldsymbol{\rho}_N) = E(\{\rho_{ij}\}) \tag{11}$$

where $\boldsymbol{\rho}_i$ is the vector of Cartesian coordinates of atom i and $\{\rho_{ij}\}$ denotes the set of all interatomic distances. Based on these two observations, we recently developed a general method for the derivation of the expressions for the energy moments and, thereby, for the scale-consistent coarse-grained energy terms for the polymer chains,[34] by expressing the squares of interatomic distances in terms of the distances between the centers of the coarse-grained sites they belong to, the variables defining the orientation of the respective virtual-bond axes, and the angles λ, as given by Eqs. (12)–(16) for the distance between atom i of site I and atom j of site J.[34]

$$\rho^2_{Ii;Jj} = R^2_{IJ} + d_{IiJj} + f_{IiJj}(\lambda_J) + f_{JjIi}(\lambda_I) - g_{IiJj}(\lambda_I,\lambda_J) \tag{12}$$

with

$$
\begin{aligned}
d_{IiJj} = {} & \delta_{Ii}^2 + \delta_{Jj}^2 + \varepsilon_{Ii}^2 + \varepsilon_{Jj}^2 - 2R_{IJ}\left[\delta_{Ii}\cos\theta_{IJ}^{(1)} - \delta_{Jj}\cos\theta_{IJ}^{(2)}\right] \\
& - 2\delta_{Ii}\delta_{Jj}\cos\theta_{IJ}^{(12)}
\end{aligned}
\tag{13}
$$

$$
\begin{aligned}
f_{IiJj}(\lambda_J) = {} & 2\varepsilon_{Jj}[R_{IJ}\sin\theta_{IJ}^{(2)}\cos(\lambda_J + \varphi_{Jj} - \psi_{IJ}) \\
& -\delta_{Ii}\sin\theta_{IJ}^{(12)}\cos(\lambda_J + \varphi_{Jj} - \Psi_{IJ})]
\end{aligned}
\tag{14}
$$

$$
\begin{aligned}
f_{JjIi}(\lambda_I) &= 2\varepsilon_{Ii}[R_{JI}\sin\theta_{JI}^{(2)}\cos(\lambda_I + \varphi_{Ii} - \psi_{JI}) - \delta_{Jj}\sin\theta_{JI}^{(12)}\cos(\lambda_I + \varphi_{Ii} - \Psi_{JI})] \\
&= -2\varepsilon_{Ii}[R_{IJ}\sin\theta_{IJ}^{(1)}\cos(\lambda_I + \varphi_{Ii} - \psi_{JI}) \\
&\quad + \delta_{Jj}\sin\theta_{IJ}^{(12)}\cos(\lambda_I + \varphi_{Ii} - \Psi_{JI})]
\end{aligned}
\tag{15}
$$

$$
\begin{aligned}
g_{IiJj}(\lambda_I,\lambda_J) = {} & \varepsilon_{Ii}\varepsilon_{Jj}\Big\{\left(1 - \cos\theta_{IJ}^{(12)}\right)\cos\left[(\lambda_I + \varphi_{Ii} - \Psi_{JI}) + (\lambda_J + \varphi_{Jj} - \Psi_{IJ})\right] \\
& -\left(1 + \cos\theta_{IJ}^{(12)}\right)\cos\left[(\lambda_I + \varphi_{Ii} - \Psi_{JI}) - (\lambda_J + \varphi_{Jj} - \Psi_{IJ})\right]\Big\}
\end{aligned}
\tag{16}
$$

and the geometric variables are defined in Fig. 4 and described in detail in ref 34. After expanding the energy in a Taylor series about R_{IJ}^2 (the square of the distance between the coarse-grained-site centers), all energy moments are computed as integrals over the powers of different functions f and g defined by Eqs (14)–(16) and, thereby, of the products of the powers of sines and cosines, as given by Eq. (17) (Ref. 34).

$$
\begin{aligned}
\left\langle \in_{k_1}^{i_1} \in_{k_2}^{i_2} \ldots \in_{k_m}^{i_m} \right\rangle = {} & \frac{1}{V_{y'}} \int\limits_{y'_{I_1}} \cdots \int\limits_{y'_{I_M}} \Big[\varepsilon(R_{I_1J_1}\ldots, R_{I_nJ_n}) \\
& + \sum_{\mu=1}^{\infty}\frac{1}{\mu!} \sum_{\substack{\nu_1,\nu_2,\ldots,\nu_n \\ \nu_1+\ldots\nu_n=\mu}} \sum_{\substack{k_1\ldots k_n \\ l_1\ldots l_n}} \frac{\partial^{\mu}\varepsilon}{\partial\left(\rho_{I_1i_{k_1};J_1j_{l_1}}^2\right)^{\nu_1}\ldots\partial\left(\rho_{I_ni_{k_n};J_nj_{l_n}}^2\right)^{\nu_n}}\Bigg|_{R_{I_1J_1}\ldots R_{I_nJ_n}} \\
& \times \frac{1}{(2\pi)^M}\int\limits_{-\pi}^{\pi}\cdots\int\limits_{-\pi}^{\pi}\prod_{m=1}^{n}\left[d_{I_m i_{k_m} J_m j_{l_m}} + f_{I_m i_{k_m} J_m j_{l_m}}\left(\lambda_{J_m}\right) + f_{J_m j_{l_m} I_m i_{k_m}}\left(\lambda_{I_m}\right)\right. \\
& \left. - g_{I_m i_{k_m} J_m j_{l_m}}\left(\lambda_{I_m}, \lambda_{J_m}\right)\right]^{\nu_m} d\lambda_{I_1}\ldots d\lambda_{I_M}\ldots d\lambda_{J_M}\Big]dV_{y'}
\end{aligned}
\tag{17}
$$

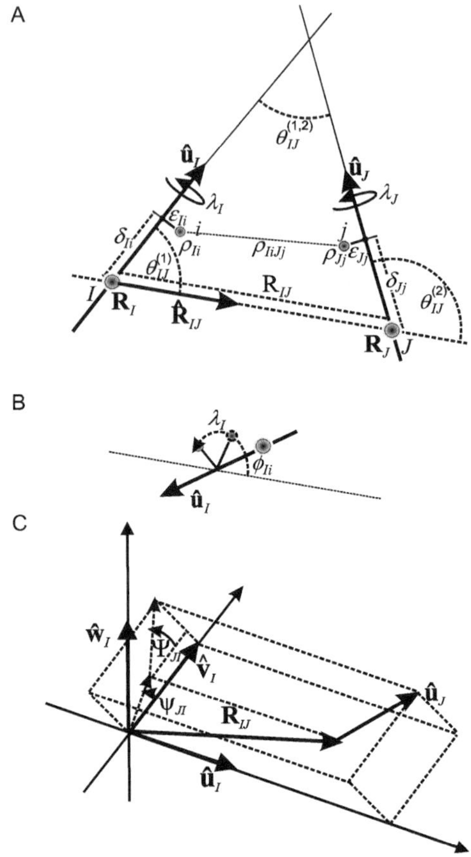

Fig. 4 Illustration of the definitions of the geometric parameters of Eqs. (13)–(16) to express the distance ρ_{IiJj} between atoms i (with Cartesian coordinates $\boldsymbol{\rho}_{Ii}$) and j (with Cartesian coordinates $\boldsymbol{\rho}_{Jj}$) of coarse-grained sites I (with Cartesian coordinates $\mathbf{R}_I$) and J (with the Cartesian coordinates $\mathbf{R}_J$), respectively (indicated in panel A). The centers of the sites are shown as larger *shaded spheres* and the atoms are shown as smaller *shaded spheres* in panels A and B. (A) Location of the sites in space. R_{IJ} is the distance between the centers of sites I and J, $\hat{\mathbf{u}}_I$ and $\hat{\mathbf{u}}_J$ are the unit vectors of the virtual-bond axes of the sites, and $\hat{\mathbf{R}}_{IJ}$ is the unit vector pointing from site I to site J. The angles $\theta_{IJ}^{(1)}$, $\theta_{IJ}^{(2)}$, and $\theta_{IJ}^{(12)}$ are the angles between vectors $\hat{\mathbf{u}}_I$ and $\hat{\mathbf{R}}_{IJ}$, between $\hat{\mathbf{u}}_J$ and $\hat{\mathbf{R}}_{IJ}$, and between $\hat{\mathbf{u}}_I$ and $\hat{\mathbf{u}}_J$, respectively. The projection of the position of atom i on the virtual-bond axis of site I is δ_{Ii} and can be positive or negative depending on whether the atom is to the left or to the right of the center, and that of atom i is δ_{Jj}. The distance of atom i from the virtual-bond axis is ε_{Ii} and that of atom j is ε_{Jj}. The angles λ_I and λ_J are the angles for counterclockwise rotation about the respective virtual-bond axes. (B) Illustration of the phase angle φ_{Ii} of atom i of site I; this angle defines the counterclockwise rotation of this atom about the virtual-bond axis for $\lambda_I = 0$. (C) Illustration of the angles Ψ_{JI} and ψ_{JI}. $\hat{\mathbf{u}}_I$ (the unit vector of the virtual-bond axis of site I), $\hat{\mathbf{v}}_I$, and $\hat{\mathbf{w}}_I$ are the unit vectors of the axes of the right-handed local Cartesian-coordinate system of site I. Ψ_{JI} and ψ_{JI} are the angles of counterclockwise rotation of the projections of $\hat{\mathbf{u}}_J$ and $\hat{\mathbf{R}}_{IJ}$, respectively, onto the $\hat{\mathbf{v}}_I$, $\hat{\mathbf{w}}_I$ plane from the $\hat{\mathbf{v}}_I$ axis. *Reproduced from Sieradzan AK, Makowski M, Augustynowicz A, Liwo A. A general method for the derivation of the functional forms of the effective energy terms in coarse-grained energy functions of polymers. I. Backbone potentials of coarse-grained polypeptide chains. J Chem Phys. 2017;146: 124106, with the permission of AIP Publishing.*

where $\mathbf{y}'_{I_k}$ denotes all fine-grained degrees of freedom of site I_k except for λ_{I_k}, $\mathbf{y}' = (\mathbf{y}'_{I_1}, \ldots, \mathbf{y}'_{I_M})$, $\mathcal{E}$ is the shorthand for $\epsilon_{i_1}^{k_1}\epsilon_{i_2}^{k_2}\ldots\epsilon_{i_m}^{k_m}$ and $i_{k_1}, i_{k_2}, \ldots, i_{k_n}$ and $j_{l_1}, j_{l_2}, \ldots, j_{l_n}$ denote the indices of the atoms. Only those energy moments will be nonzero, which contain even-power cosines of the angles λ. In particular, for mixed moments containing only the first powers of component energies, the terms with $f_{IpJq}f_{JrKs}$ or $f_{IpJq}g_{JrKs}f_{KtLu}$ will contribute to the expansion of the PMF. If $J = I + 1$, $K = I + 2$, and $L = I + 3$, these expressions contribute to the torsional and double-torsional potentials, respectively.[34]

The energy moments (Eq. 17) are evaluated by using the Euler formula to express the cosines. This operation reduces evaluation of the integrals into an algebraic problem because, after doing all the multiplications, only the integrals over the terms without imaginary exponentials in λs are not zero.

It should be noted that the way of derivation of the effective energy expressions does not cover the virtual-bond potentials, in which case averaging has to be carried out over the relative displacements of the atoms constituting the virtual bonds. However, these potentials depend on virtual-bond lengths only and, consequently, it is not difficult to find a reasonable analytical expression (such as a multimodal rational function as implemented in our earlier work to parameterize the side-chain virtual-bond potentials[31]). Also, averaging over the solvent degrees of freedom cannot be handled this way and to do this we used separate models to derive side chain–side chain hydrophobic/hydrophilic association potentials,[40–42,47,48] as well as the potentials for the screening of peptide groups from the solvent.[49]

2.5 Parameterization of the effective energy expressions

To parameterize the expressions for effective coarse-grained energy, the surfaces corresponding to PMF factors need to be computed first. A variety of methods can be used here. For example, to compute the PMFs corresponding to long-range solvent-mediated interactions such as, e.g., the effective interactions between protein side chains in water, all-atom umbrella-sampling molecular dynamics simulations are appropriate.[40–42] For less computationally demanding interactions such as, e.g., interaction of two peptide groups or nucleic-acid bases in nonpolar environment (inside a protein or DNA double helix), straightforward numerical integration with the potential energy calculated with empirical force fields[50,51] or ab initio methods[52] can be used. Finally, numerical integration of the all-atom energy surfaces calculated with ab initio or semiempirical methods of molecular quantum mechanics should be used to handle the terms involving local interactions, i.e., the virtual-bond-angle, virtual-bond-torsional potentials and the correlation energy terms.[30–32,34,52-54] Once the PMFs are calculated, nonlinear least-squares fitting can be used to determine the parameters of the potentials.[30–32,34,40–42,52–54]

2.6 Force field calibration

Following the folding-funnel theory,[55] a force field with folding properties should locate the native structure as the lowest in energy. Many pertinent algorithms were designed following this principle that optimize the energy gap between the native and alternative structure[56–58] or optimize the Z-score.[59,60] We developed the hierarchical optimization method[38,61,62] in which the free-energy gaps between gradually folded conformational subensembles, the order of which follows the folding pathway, are optimized so that the free energy decreases with increased native-likeness below the folding-transition temperature and increases above the folding-transition temperature. In all these methods, an iterative procedure is applied which consists of cycles in each of which the decoys are generated with the current force field parameters and then the force field parameters are optimized given the generated decoy set(s). However, all these methods suffer from poor convergence behavior, because of favoring few rather accidentally generated conformations, which happen to meet the native-likeliness criteria best. Recently[63] we developed an alternative method, which is based on the maximum-likelihood principle. Optimization is carried out to match the conformational ensembles of the selected training molecules(s) to the corresponding experimental ensembles, ideally determined at temperatures bracketing the folding-transition temperature. Because whole ensembles are fitted, the bulk of the generated ensembles rather than a few exceptionally good conformations need to match the experimental ensemble, which results in quick convergence and stable behavior of the iterative process.[63]

The maximum-likelihood term is not the only component of the target function to be optimized; the terms with free energies of folding/hybridization, heat-capacity profiles, etc., can be entered in the form of least-squares (χ^2) components.[64] Recently, we developed an approach to introduce the force-matching terms, in order to make the forces generated by coarse-grained force fields compatible with those computed with all-atom force fields.[65]

3. Implementation

3.1 The UNIfied COarse-gRaiNed (UNICORN) model of biological macromolecules

Based on the formalism presented in Sections 2.1-2.6, we developed the UNited RESidue (UNRES) model of polypeptide chains,[66] which later evolved into the UNIfied COarse-gRaiNed (UNICORN) model of biological macromolecules, encompassing proteins (UNRES), nucleic acids (NARES-2P), and polysaccharides (SUGRES-1P),[29,67,68] including the interactions between these biopolymers. The model is depicted in Fig. 5. As can be seen, the model is heavily coarse grained, with two sites per residue (a united peptide group and a united side chain) in UNRES, two (a united phosphate group and a united sugar-base group) in NARES-2P and one (a united sugar ring) in SUGRES-1P (which has, so far, been developed for linear polysaccharides only). A characteristic feature of all three UNICORN components is that one of the kinds of extended sites (peptide groups in UNRES, nucleic-acid bases in NARES-2P and sugar rings in SUGRES-1) always contains acidic protons and proton acceptors which are, thus, capable of the formation of regular hydrogen-bonded structures. By using the formalism presented in Sections 2.1-2.3, the hydrogen bonding interactions can be represented by average dipole–dipole interactions,[29,34,36,50,51] averaged either freely or in the context of the local interactions with the adjacent groups. The effective energies are expressed by (Eqs. 18–20), respectively.

$$
\begin{aligned}
U_{prot}^{UNRES} = {} & w_{SCSC}\sum_{i<j} U_{SC_iSC_j} + w_{SCp}\sum_{i\neq j} U_{SC_ip_j} + w_{pp}^{VDW}\sum_{i<j-1} U_{p_ip_j}^{VDW} \\
& + f_2(T)w_{pp}^{el}\sum_{i<j-1} U_{p_ip_j}^{el} + w_{ssbond}\sum_{\substack{disulfide\\ bonds}} U_{ssbond}\left(d_i^{ssbond}\right) \\
& + w_{bond}\sum_i U_{bond}(d_i) + w_b\sum_i U_b(\theta_i) + w_{rot}\sum_i U_{rot}(\theta_i, \alpha_{SC_i}, \beta_{SC_i}) \\
& + f_2(T)w_{tor}\sum_i U_{tor}(\gamma_i, \theta_i, \theta_{i+1}) + f_3(T)w_{corr}^{(3)}\sum_{i<j-2} U_{corr;i}^{(3)} \\
& + f_3(T)w_{turn}^{(3)}\sum_i U_{turn;p_ip_{i+2}}^{(3)}
\end{aligned}
\qquad (18)
$$

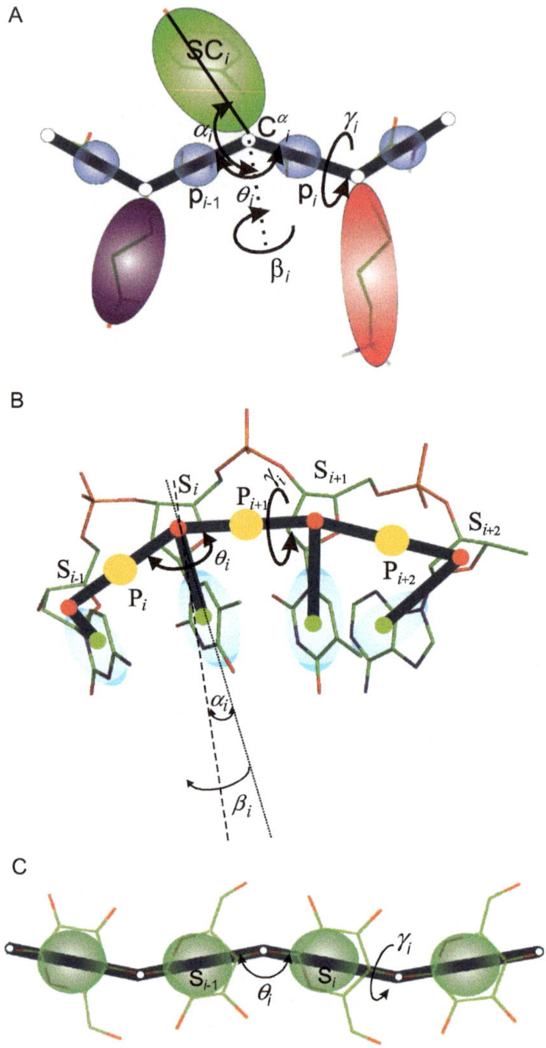

Fig. 5 The components of the UNIfied COarse-gRaiNed (UNICORN) model of biological macromolecules. (A) UNRES (proteins), (B) NARES-2P (nucleic acids), (C) SUGRES-1P (polysaccharides). The virtual bonds are shown as *thick black lines*. All-atom chains are superposed on coarse-grained representations for better illustration. In the UNRES model of polypeptide chains (A), the interaction sites are united side chains, represented as ellipsoids of revolution of different size (SC), attached to the corresponding α-carbon atoms (represented by *small open circles*), and united peptide groups (p), each centered in the middle between the consecutive C$^\alpha$ atoms. The C$^\alpha$s are not interaction sites but serve to define backbone geometry. In the NARES-2P model of the nucleotide chain, the interaction sites are united phosphate groups (P), represented by *yellow circles*, and the sites composed of the nucleic-acid bases fused with sugar rings, represented

(Continued)

$$U_{nucl}^{NARES} = w_{BB}^{GB} \sum_{i<j} U_{B_iB_j}^{GB} + w_{BB}^{el} f_2(T) \sum_{i<j} U_{B_iB_j}^{el} + w_{PP} \sum_{i<i-1} U_{P_iP_j} + w_{PB} \sum_{i \neq j} U_{P_iB_j}$$
$$+ w_{bond} \sum_i U_{bond}(d_i) + w_b \sum_i U_b(\theta_i) + w_{rot} \sum_i U_{rot}(\theta_i, \alpha_i, \beta_i)$$
$$+ w_{tor} f_2(T) \sum_i U_{tor}(\gamma_i, \theta_i, \theta_{i+1}) \tag{19}$$

$$U_{sugar}^{SUGRES} = w_{SS}^{GB} \sum_{i<j-1} U_{S_iS_j}^{GB} + w_{SS}^{el} f_2(T) \sum_{i<j-1} U_{S_iS_j}^{el}$$
$$+ w_{bond} \sum_i U_{bond}(d_i) + w_b \sum_i U_b(\theta_i) + w_{tor} f_2(T) \sum_i U_{tor}(\gamma_i)$$
$$+ w_{corr}^{(3)} f_3(T) \sum_i \sum_{j<i-2} U_{corr;S_iS_j}^{(3)} + w_{turn}^{(3)} f_3(T) \sum_i U_{turn;S_iS_{i+2}}^{(3)} \tag{20}$$

The ws in Eqs. (18)–(20) are the energy-term weights that are determined by force field calibration. The multipliers $f_m(T)$, defined by Eq. (21) (Ref. 38) account for the temperature dependence of the leading terms in the cumulant expansions of the factors corresponding to the consecutive effective energy terms (cf. Eq. 7).

$$f_n(T) = \frac{\ln[\exp(1) + \exp(-1)]}{\ln\left\{\exp\left[\left(\frac{T}{T_\circ}\right)^{n-1}\right] + \exp\left[-\left(\frac{T}{T_\circ}\right)^{n-1}\right]\right\}} \tag{21}$$

where $T_\circ = 300$ K.

Fig. 5—Cont'd by ellipsoids, with their geometric centers at the B's (*green circles*). The P's are located halfway between two consecutive sugar centers. *Small red circles* represent the sugar-ring centers (S), which serve to define backbone geometry. In the SUGRES-1P model of polysaccharide chains, the interaction sites are united sugar rings (S), represented by *blue transparent spheres* (S), while *white circles* represent the glycosidic oxygen atoms, serving to define chain geometry (O). The backbone virtual-bond angles, θ, the virtual-bond dihedral angles, γ, and the angles α and β that define the orientation of amino-acid side-chain (UNRES) or nucleic-acid-base center (NARES-2P) with respect to the backbone are also shown. *Adapted from Liwo A, Baranowski M, Czaplewski C, et al. A unified coarse-grained model of biological macromolecules based on mean-field multipolemultipole interactions. J Mol Model. 2014;20:2306, under the Creative Commons Attribution License.*

We also have already developed the energy function for the peptide–nucleic acid interactions, as given by Eq. (22) (Refs. 69 and 70).

$$\begin{aligned} U_{prot-nucl} &= w_{SC-B}\sum_i\sum_j U_{SC_iB_j} + w_{p-P}\sum_i\sum_j U_{p_iP_j} \\ &+ w_{pB}f_2(T)\sum_i\sum_j U_{p_iB_j} + w_{SC-P}\sum_i\sum_j U_{SC_iP_j} \end{aligned} \tag{22}$$

There are three kinds of terms in Eqs. (18)–(22). The terms of the first kind correspond to long-range two-body site–site interactions, which are factors of order 1 and, except for $U^{el}_{p_ip_j}$, $U^{el}_{B_iB_j}$, $U^{el}_{S_iS_j}$, and $U^{el}_{p_iB_j}$ do not include the temperature multipliers [$f_n(T)$], because the leading cumulant is of order 1. The three terms mentioned above can be treated as average dipole–dipole interactions, in which case the first-order cumulant is zero and, consequently, the leading cumulant is of the second order.[34,71,72] Most of the long-range terms are not radial but have axial symmetry, the axes being the virtual-bond axes, which accounts for scale-consistency. Apart from the average electrostatic-interaction terms mentioned above, these are all terms with the "GB" superscript, where GB stands for the Gay-Berne potential introduced to coarse-grain liquid-crystal molecules,[73] and the protein side-chain interaction terms ($U_{SC_iSC_j}$) which, depending on the version of the force field, are either statistical potentials of the Gay-Berne form[74] or the physics-based potentials derived recently.[40–42] In UNRES, the last of the long-range terms (U_{ssbond}) is that responsible for the dynamic formation/breaking of disulfide bonds that has one minimum corresponding to disulfide-bonded and another one to nonbonded cysteines[75]; it has later been supplemented with a three-body term that prevents from the involvement of a cysteine residue in more than one disulfide bonds.[76]

The terms of the second kind are local-interaction terms describing virtual-bond deformation (U_{bond}), backbone-virtual-angle bending (U_b), rotamer states of amino acid side chains or nucleic bases (U_{rot}), and torsional potentials (U_{tor}). Except for U_{tor}, these terms are first-order factors, while the U_{tor} potentials are second-order factors that describe the coupling between the local conformational states of the neighboring amino acid residues.[34,36] It should be noted that the torsional terms obtained by using the scale-consistent approach depend not only on the virtual-bond-dihedral angles γ but also on the virtual-bond angles θ adjacent to the respective γ (Ref. 34).

The terms of the third kind, $U^{(3)}_{corr}$ and $U^{(3)}_{turn}$, are the correlation terms that arise from the coupling between the backbone-local and backbone-electrostatic interactions between the peptide groups. They are essential in the reproduction of regular (α-helical and β-sheet) structures in the protein structures modeled with UNRES[34,36] and in the polysaccharide structures modeled with SUGRES-1P[29] but do not occur in the NARES-2P force field for nucleic acids because there is no coupling between the backbone and nucleic-acid-base secondary degrees of freedom.

The UNRES force field with the scale-consistent torsional and correlation potentials was recently calibrated with 9 proteins with different structural types.[30] UNRES and NARES-2P are available from http://www.unres.pl and a server version of UNRES has also been created,[77] which is available at http://unres-server.chem.ug.edu.pl.

3.2 Features of selected energy terms of the scale-consistent UNICORN model

In this section the importance of using the scale-consistent derivation of the coarse-grained energy terms is illustrated with the terms that arise from average electrostatic interactions between peptide-group ($U^{el}_{p_ip_j}$) and nucleic-acid bases ($U^{el}_{B_iB_j}$), the torsional terms, and the correlation terms. The first of those provide correct dependence of site–site interaction energy on their orientations and can be obtained by using a Clausius-Mossotti-like approach presented in our early work on UNRES.[50,71] The correct form of torsional potentials cannot be obtained without using the scale-consistent methodology and the correlation potentials can only be introduced on a heuristic basis without the scale-consistent methodology.

3.2.1 Long-range $U^{el}_{p_ip_j}$ and $U^{el}_{B_iB_j}$ terms

These are the terms corresponding to average electrostatic interactions between peptide-group or nucleic-acid-base dipoles. Application of the scale-consistent methodology gives the general formula for the respective average energy (Eq. 23).[34,71]

$$\begin{aligned} f^{el}_{IJ} \approx \left\langle E^{dip}_{IJ}\right\rangle - \frac{\beta}{2}\left\langle \left(E^{dip}_{IJ}\right)^2\right\rangle_c = \frac{p^{\|}_I p^{\|}_J}{R^3_{IJ}} W_{IJ} \\ - \frac{\beta}{2R^6_{IJ}}\left\{ (p^{\|}_I p^{\perp}_J)^2 \left(Y^{(1)}_{IJ}\right)^2 + (p^{\perp}_I p^{\|}_J)^2 \left(Y^{(2)}_{IJ}\right)^2 \right. \\ \left. + (p^{\perp}_I p^{\perp}_J)^2 \left[\left(Z^{(1)}_{IJ}\right)^2 + \left(Z^{(2)}_{IJ}\right)^2 \right] \right\} \\ = U^{(el)\|}_{IJ} + U^{(el)|\perp}_{IJ} + U^{(el)\perp|}_{IJ} + U^{(el)\perp\perp}_{IJ} \end{aligned} \tag{23}$$

with

$$W_{IJ} = \cos\theta_{IJ}^{(12)} - 3\cos\theta_{IJ}^{(1)}\cos\theta_{IJ}^{(2)} \tag{24}$$

$$Y_{IJ}^{(1)} = \sqrt{1 + 3\left(\cos\theta_{IJ}^{(2)}\right)^2 - W_{IJ}^2} \tag{25}$$

$$Y_{IJ}^{(2)} = \sqrt{1 + 3\left(\cos\theta_{IJ}^{(1)}\right)^2 - W_{IJ}^2} \tag{26}$$

$$Z_{IJ}^{(1)} = \frac{1}{2}\sqrt{4\left(1 + \cos\theta_{IJ}^{(12)}\right) + W_{IJ}^2 - \left(\cos\theta_{IJ}^{(1)} + \cos\theta_{IJ}^{(2)}\right)^2} \tag{27}$$

$$Z_{IJ}^{(2)} = \frac{1}{2}\sqrt{4\left(1 - \cos\theta_{IJ}^{(12)}\right) + W_{IJ}^2 - \left(\cos\theta_{IJ}^{(1)} - \cos\theta_{IJ}^{(2)}\right)^2} \tag{28}$$

where R_{IJ} is the distance between the site centers,

$$\theta_{IJ}^{(1)} = \arccos\left(\hat{\boldsymbol{u}}_I \circ \hat{\mathbf{R}}_{IJ}\right) \tag{29}$$

$$\theta_{IJ}^{(2)} = \arccos\left(\hat{\boldsymbol{u}}_J \circ \hat{\mathbf{R}}_{IJ}\right) \tag{30}$$

$$\theta_{IJ}^{(12)} = \arccos\left(\hat{\boldsymbol{u}}_I \circ \hat{\boldsymbol{u}}_J\right) \tag{31}$$

are the angles between the virtual-bond axis of site I ($\hat{\mathbf{u}}_I$ being its unit vector) and the line linking axis centers, with that of virtual-bond axis J ($\hat{\mathbf{u}}_J$ being its unit vector) and the line linking axis centers and the angle between the virtual-bond axes of site I and J ($\hat{\mathbf{R}}_{IJ}$ being its unit vector), respectively (Fig. 4), and $p^{\|}$ and $p^{\perp}$ are the components of the dipole moments of the sites parallel and perpendicular to the group, respectively. For the interacting peptide groups, the perpendicular dipole moment components are dominant and, consequently, the term with $Z_{IJ}^{(1)}$ and $Z_{IJ}^{(2)}$ is dominant.[71] These terms favor nearly parallel/antiparallel orientation of virtual-bond axes as occurs in α-helices or β-sheets and enabled us to model the structures of α-helical proteins,[50,78] as shown in Fig. 6A. Conversely, the terms with $Y_{IJ}^{(1)}$ and $Y_{IJ}^{(2)}$ dominate in the interactions between the nucleic-acid-base pair, one base in a pair being purine with dominant parallel and one pyrimidine with dominant perpendicular component; these terms favor the virtual-bond axes making an angle, as occurs in the Watson-Crick or Hoogsteen base pairs, as shown in Fig. 6B. These relatively simple potentials not only enabled us to model DNA and RNA double helices but also to reproduce the double helix harmonic dynamics and pre-melting transitions in AT-pair-rich DNA sequences.[51] Consequently, even a heavily coarse-grained model can capture

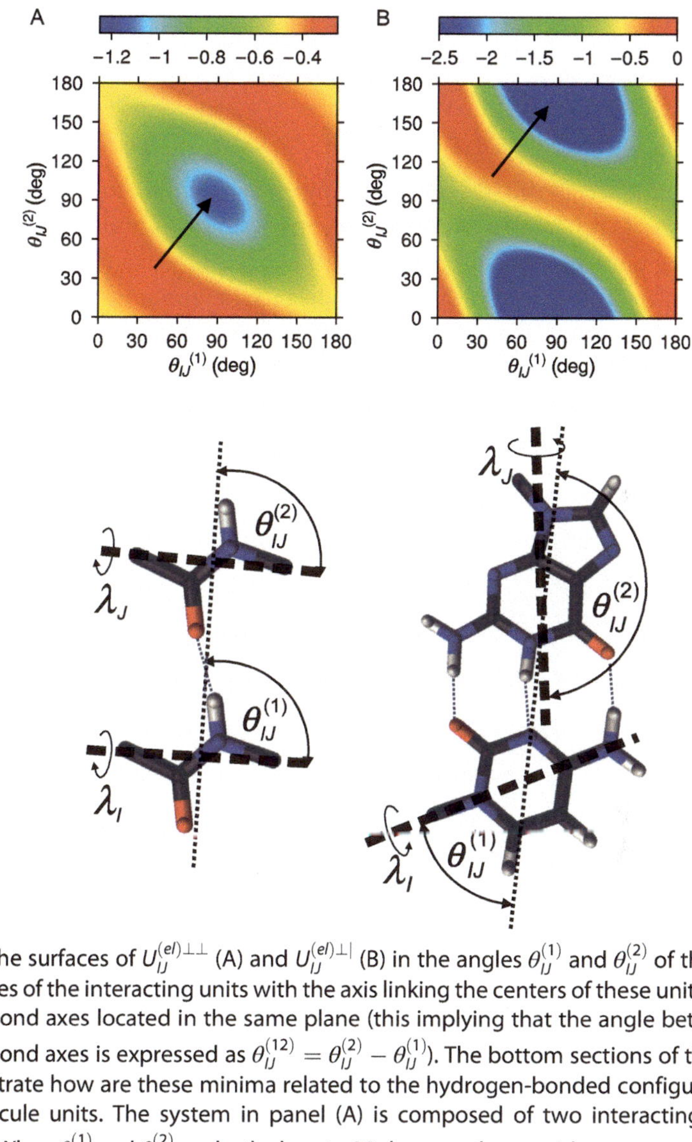

Fig. 6 The surfaces of $U_{IJ}^{(el)\perp\perp}$ (A) and $U_{IJ}^{(el)\perp|}$ (B) in the angles $\theta_{IJ}^{(1)}$ and $\theta_{IJ}^{(2)}$ of the virtual-bond axes of the interacting units with the axis linking the centers of these units for both virtual-bond axes located in the same plane (this implying that the angle between the virtual-bond axes is expressed as $\theta_{IJ}^{(12)} = \theta_{IJ}^{(2)} - \theta_{IJ}^{(1)}$). The bottom sections of the panels demonstrate how are these minima related to the hydrogen-bonded configurations of biomolecule units. The system in panel (A) is composed of two interacting peptide groups. When $\theta_{IJ}^{(1)}$ and $\theta_{IJ}^{(2)}$ are both close to 90 degrees, the peptide groups are stacking over each other, as in the hydrogen-bonded configurations that occur in β-sheets and α-helices; in the second case the virtual-bond axes are not in plane but this results in only minor variation of the $U_{IJ}^{(el)\perp\perp}$ term.[52] The example in the bottom part of the panel corresponds to parallel peptide groups but the surface is the same for antiparallel configuration, because is corresponds to the first non-zero term in the cumulant expansion of the energy of interactions of the components of peptide-group dipoles perpendicular to the peptide-group virtual-bond axes. The system in panel (B) is a cytosine-guanine pair. Here two minima are present: one for $\theta_{IJ}^{(1)} = 90$ degrees and

(Continued)

the orientation dependence of the interaction sites, provided that the expressions for the effective energy are derived with taking care of scale-consistency.

3.2.2 Torsional terms

These terms correspond to the coupling between the local interactions between sites I and $I+1$ with those between sites $I+1$ and $I+2$, the coupling arising because the collective rotation angle λ_{I+1} of site $I+1$ about its virtual-bond axis is shared between the unit that consists of sites I and $I+1$ and that that consists of sites $I+1$ and $I+2$, even though no interactions between units I and $I+2$ are considered.[34,36] The appropriate scale-consistent formula, up to the second order, is expressed by Eq. (32) (Ref. 34) (for clarity sake the index I is omitted).

$$
\begin{aligned}
U_{tor}(\gamma,\theta_1,\theta_2) &\approx -\frac{\beta}{2}\langle e_1 e_2\rangle_c \\
&\approx \frac{\beta}{2}\sin\theta_1\sin\theta_2\left\{\left[\sum_{k=0}^{2}\sum_{l=0}^{2}\left(b_{X21}^{(k)}b_{Y11}^{(l)}+b_{X22}^{(k)}b_{Y12}^{(l)}\right)(\cos\theta_1)^k(\cos\theta_2)^l\right]\cos\gamma\right.\\
&\left.+\left[\sum_{k=0}^{2}\sum_{l=0}^{2}\left(b_{X22}^{(k)}b_{Y11}^{(l)}+b_{X21}^{(k)}b_{Y12}^{(l)}\right)(\cos\theta_1)^k(\cos\theta_2)^l\right]\sin\gamma\right\}\\
&-\frac{\beta}{4}(\sin\theta_1\sin\theta_2)^2\left\{\left[\sum_{k=0}^{1}\sum_{l=0}^{1}\left(c_{X11}^{(k)}d_{Y11}^{(l)}-c_{X12}^{(k)}d_{Y12}^{(l)}\right)(\cos\theta_1)^k(\cos\theta_2)^l\right]\cos 2\gamma\right.\\
&\left.+\left[\sum_{k=0}^{1}\sum_{l=0}^{1}\left(c_{X12}^{(k)}d_{Y11}^{(l)}+c_{X11}^{(k)}d_{Y12}^{(l)}\right)(\cos\theta_1)^k(\cos\theta_2)^l\right]\sin 2\gamma\right\}
\end{aligned}
\tag{32}
$$

where the coefficients are those of the expansion of the local-interaction energy surfaces in the angles λ for collective rotation of sites' atoms about

Fig. 6—Cont'd $\theta_{IJ}^{(2)} = 0$ degrees, a geometry close to the Hoogsteen base pair, and the other one for $\theta_{IJ}^{(1)} = 90$ degrees and $\theta_{IJ}^{(2)} = 180$ degrees, the geometry being close to that of the canonical Watson–Crick base pair; this second instance is illustrated in the bottom section of panel (B). The surface is that of the first nonzero term of the cumulant expansion of the interaction of the component of the pyrimidine-base dipole perpendicular to the virtual-bond axis with the purine-base dipole parallel to the virtual-bond axis. The minima corresponding to the configurations shown in the bottom sections of the panels are pointed to in maps in the top sections by *arrows*. The angles of rotation λ_I and λ_J, which are the degrees of freedom to average out, are indicated in the bottom sections of the panels. The GRI (http://gri.sourceforge.net/index.php) and MOLMOL[79] software were used to prepare the picture.

the virtual-bond axes (Eqs. 12–16)[30]; these coefficients are adjustable parameters. Remarkably, each of these parameters pertain to single-residue type and not to the types of both residues that are on the virtual-bond axis. Moreover, for chiral units, these coefficients are good for both chiral forms, because change of chirality is accomplished by changing the sign of some of the coefficients.[30,67] As follows from Eq. (32), the torsional potentials depend on both virtual-bond-dihedral angles and the virtual-bond angles adjacent to it. In particular, each trigonometric term in γ, which has multiplicity n, is multiplied by the sines of the two adjacent θ angles, each raised to the nth power. In contrast to the commonly used torsional potentials that depend only on γ, this multiplication prevents the potential from becoming indefinite when the virtual bonds become aligned (a situation that can happen while using coarse-grained models) and γ becomes, thus indefinite. The torsional potentials derived based on the scale-consistent methodology tend to 0 in such a case. The strong dependence of the virtual-bond-dihedral potentials, both calculated from the PMF and derived from PDB statistics, on the virtual-bond angles is illustrated in Fig. 7.

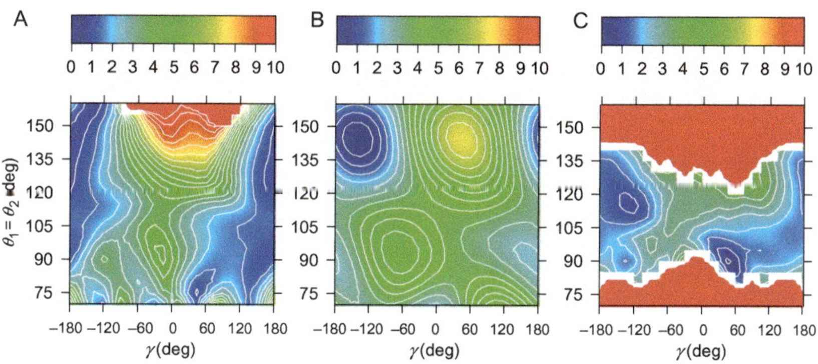

Fig. 7 Coarse-grained torsional-potential profiles for the terminally blocked L-Ala–L-Ala (where L-Ala denotes L-alanine) dipeptide obtained by (A) calculation of the corresponding second-order factor (Kubo cluster-cumulant function) by numerical integration with the ab initio PES of the terminally blocked L-alanine,[30] (B) by fitting the analytical expression (Eq. 32) to the PMF (and, simultaneously, to the PMFs of other dipeptides containing L- and D-Ala, Gly, and L- and D-Pro), (C) the corresponding statistical torsional potentials obtained by analysis of protein structures in the PDB, where all residue types except for Gly and Pro were considered alanine type.[30] *Reproduced from Liwo A, Sieradzan AK, Lipska AG, et al. A general method for the derivation of the functional forms of the effective energy terms in coarse-grained energy functions of polymers. III. Determination of scale-consistent backbone-local and correlation potentials in the UNRES force field and force-field calibration and validation. J Chem Phys. 2019;150:155104, with the permission of AIP Publishing.*

3.2.3 Backbone-local-correlation ($U^{(3)}_{corr}$) terms

These terms pertain to the coupling between the long-range interactions, the local interactions between sites *I* and *J*, and the local interactions involving site *I* or *J* and the sites adjacent to them. Consequently, there are four respective cumulants, pertaining to sites $I \pm 1$, *I*, *J*, and $J \pm 1$ (Refs. 34 and 36). When the long-range interactions are electrostatic interactions between electrostatically neutral groups (e.g., the peptide groups), summing the four cumulant terms results in Eq. (33).[34,36]

$$\begin{aligned} U^{(3)}_{loc-el;IJ} &\approx \frac{\beta^2}{6}\left[\langle e_I E_{IJ} e_J\rangle_c + \langle e_I E_{IJ} e_{J+1}\rangle_c + \langle e_{I+1} E_{IJ} e_J\rangle_c + \langle e_{I+1} E_{IJ} e_{J+1}\rangle_c\right] \\ &= \frac{\beta^2}{6R_{IJ}^3}\left[\boldsymbol{\mu}_I \circ \boldsymbol{\mu}_J - 3(\boldsymbol{\mu}_I \circ \hat{\mathbf{R}}_{IJ})(\boldsymbol{\mu}_J \circ \hat{\mathbf{R}}_{IJ})\right] \end{aligned} \tag{33}$$

where E_{IJ} is the electrostatic-interaction energy between units *I* and *J*, and

$$\begin{aligned} \boldsymbol{\mu}_K &= \boldsymbol{\mu}_K(\gamma_K, \theta_K, \theta_{K+1}) \\ &= \begin{pmatrix} \cos\gamma_K & -\sin\gamma_K \\ \sin\gamma_K & \cos\gamma_K \end{pmatrix} \boldsymbol{\mu}_{2K}(\theta_K) + \boldsymbol{\mu}_{1,K+1}(\theta_{K+1}), \quad K = I, J \end{aligned} \tag{34}$$

with

$$\boldsymbol{\mu}_{1K}(\theta_{K+1}) = \mathbf{b}_{1,K+1}(\theta_{K+1}), \quad \boldsymbol{\mu}_{2K}(\theta_K) = \begin{pmatrix} -1 & 0 \\ 0 & 1 \end{pmatrix} \mathbf{b}_{2K}(\theta_K) \tag{35}$$

and the components of the vectors $\mathbf{b}_{1K}$ and $\mathbf{b}_{2K}$, $K \in \{I, I+1, J, J+1\}$ are those of the second-order expansion of the local-interaction-energy surface of a unit composed of sites *K* and $K+1$ or $K-1$ and *K*, e.g., a terminally-blocked amino acid residue.[34] Thus, for polypeptide chains, the correlation interactions can be represented as the interactions of two fictitious dipoles located on the $C^\alpha \cdots C^\alpha$ virtual-bond axes; the corresponding dipole moments are vector sums of components from the residues that share a virtual-bond axis and, thus, each of them depends on the respective virtual-bond-dihedral angles γ. The components of the dipole moments, in turn, depend on the Ramachandran maps of the residues involved. The $U^{(3)}_{corr}$ terms are illustrated in Fig. 8. The $U^{(3)}_{turn}$ terms are similar; however, they possess an additional components that arises from the connectivity of the interacting peptide groups.[34,36]

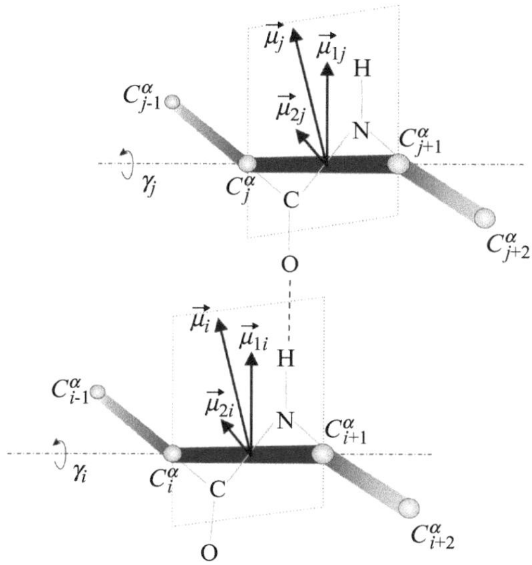

Fig. 8 Illustration of the $U^{(3)}_{loc-el}$ correlation terms. The *arrows* represent the components of the dipole moments (which represent the peptide groups between the C^{α}_i and C^{α}_{i+1} (C^{α}_j and C^{α}_{j+1}) atoms, respectively; these peptide groups are represented by *light-gray letters* and *lines*) from the cumulant-based expression (Eq. 33) for the third-order correlation contribution to the PMF of two interacting peptide groups p_i and p_j considered together with the local interactions within the adjacent amino-acid residues and the vector components of these dipole moments defined by Eq. (34). The positions of atoms C^{α}_i, C^{α}_{i+1}, and C^{α}_{i+2} (C^{α}_j, C^{α}_{j+1}, and C^{α}_{j+2}) are fixed and so are the vector components $\boldsymbol{\mu}_{1i}$ and $\boldsymbol{\mu}_{1j}$ of the dipole moments, while the virtual bonds $C^{\alpha}_{i-1}\cdots C^{\alpha}_i$ ($C^{\alpha}_{j-1}\cdots C^{\alpha}_j$) and the components of $\boldsymbol{\mu}_{2i}$ ($\boldsymbol{\mu}_{2j}$) can rotate about the $C^{\alpha}_i\cdots C^{\alpha}_{i+1}$ ($C^{\alpha}_j\cdots C^{\alpha}_{j+1}$) axis by the angle γ_i (γ_j). *Adapted with permission from Liwo A, Ołdziej S, Czaplewski C, Kozłowska U, Scheraga HA. Parameterization of backbone-electrostatic and multibody contributions to the UNRES force field for protein-structure prediction from ab initio energy surfaces of model systems. J Phys Chem B. 2004;108:9421–9438. Copyright 2004 American Chemical Society.*

This representation of correlation interactions in the UNRES model agrees with the concept of the interplay between local and hydrogen bonding interactions in proteins, because the Ramachandran maps[80] determine the allowed conformations of polypeptide units (that can be represented by terminally-blocked residues), thus restricting the number of ways in which the backbone hydrogen bonds can be formed, exactly as the coefficients of the expansion of the local-interaction energy determine the components of the dipole moments in Eq. (35). The $U^{(3)}_{corr}$ terms are also of key importance for reproducing both the α- and β-structures in proteins.

With U_{pp}^{el} supplemented with the now obsolete fourth-order correlation terms alone, only the α-helical structure could be modeled.[36] The reason for this was that, although U_{pp}^{el} has the global minimum for the stacking orientation of two peptide groups, which is close to that in β-sheets and farther from that in α-helices (Fig. 6), the number of peptide-group contacts and, thereby, the strength of freely (i.e., without the context of local interaction) averaged electrostatic interactions in an α-helix is greater than in a β-sheet of the same chain length.[36] Only adding the context of local interactions, which are more favorable for extended structures, enabled us to model both α-helices and β-sheets with the heavily coarse-grained UNRES model.[36]

As can be seen from Eq. (35), each of the components of the fictitious dipole moment depends on the respective virtual-bond angle θ; this dependence can only be obtained after full implementation of the scale-consistent methodology of coarse graining[34] and was not present in the earlier versions of UNRES.[36] The dependence of the components of the fictitious dipole moments on the backbone-virtual-bond angles θ makes these components align to produce the largest dipole (and, thereby, to maximize the correlation interactions) for $\gamma < 60$ degrees when $\theta \leq 90$ degrees (α-helix geometry) and for $\gamma > 110$ degrees otherwise (β-sheet geometry), this agreeing with the trivariate distribution of the γ and adjacent θ angles in the PDB.[34] Neglecting this dependence in the previous versions of UNRES resulted in overcompressed β-sheets due to the tendency of maximizing the average electrostatic interactions.

3.3 Molecular dynamics and other conformational-search engine implementation with UNICORN

The scale-consistency of the UNICORN models and the implied axial symmetry of the interaction sites also imply that the masses of the coarse-grained sites are not centered in a point, as opposed to most of the molecular dynamics (MD) implementations to coarse-grained models, but, instead, are distributed on the virtual-bond axes, which can be treated as stretchable rods. We applied this model to implement MD in UNRES[81,82] and, later, in NARES-2P[51] and SUGRES-1P.[29] We used the Lagrange formalism to derive the equations of motion[81] and the Langevin approach to account for the friction and random forces of the solvent.[82] We chose the virtual-bond vectors as the generalized variables; given the virtual-bond vectors and the position of the first anchor point (the first C^{α} atom in UNRES, the sugar-ring center in NARES-2P, and glycosidic oxygen atom in SUGRES-1P; cf. Fig. 5), the Cartesian coordinates of all the sites can be calculated. The equations of motion are expressed by Eq. (36).

$$\mathbf{G}\ddot{\mathbf{q}} = -\nabla_{\mathbf{q}} U(\mathbf{q}) - \mathbf{A}^{T}\boldsymbol{\Gamma}\dot{\mathbf{q}} + \mathbf{f}^{rand} \quad (36)$$

where $\mathbf{q}$, $\dot{\mathbf{q}}$, and $\mathbf{q}$ are the generalized coordinates (the components of all virtual-bond vectors plus the Cartesian coordinates of the first anchor point), velocities, and accelerations, respectively, $\mathbf{G} = \mathbf{A}^{T}\mathbf{M}\mathbf{A} + \mathbf{H}$ is the inertia matrix expressed in the virtual-bond vectors, $\mathbf{A}$ is the constant matrix that relates the coordinates of the virtual-bond vectors to those of the interaction sites, $\mathbf{M}$ is the diagonal matrix of masses of the interacting sites, $\mathbf{H}$ is the matrix of the moments of inertia of the virtual bonds, $\boldsymbol{\Gamma}$ is the diagonal matrix of the friction coefficients of the interaction sites, and $\mathbf{f}^{rand}$ is the vector of the random forces.[81,82] Although the inertia matrix $\mathbf{G}$ is not a diagonal matrix, it is a constant matrix, which means that it can be inverted once and then its inverse used to compute accelerations. We implemented the Velocity Verlet (VV) algorithm[83] to integrate the equations of motion and developed the Adaptive Multiple Time Step (A-MTS) algorithm to assure its stability.[84] Because of averaging out the secondary degrees of freedom (including the solvent degrees of freedom), the coarse-grained MD timescale is 1000–10,000 times wider than the all-atom timescale,[82] this enabling us to run simulations at the timescale inaccessible to all-atom approaches.

To enhance the search of the conformational space, we implemented[85] the Replica Exchange Molecular Dynamics (REMD)[86] and Multiplexed Replica Exchange Molecular Dynamics (MREMD)[87] in UNRES. We also implemented[88] the Conformational Space Annealing (CSA) genetic algorithm,[89] which is aimed at finding the lowest-energy conformations, in UNRES[90] and NARES-2P.[91]

4. Applications of UNICORN

From the early stages of its development, UNRES was applied in protein structure prediction within the framework of the Community Wide Experiments on the Critical Assessment of Techniques for protein structure prediction (CASP), achieving success several times in the ab initio[51,78,88,92] and, recently, data-assisted prediction category[93]; it also was one of the component of the WeFold coopetition within CASP.[94,95] UNRES was also applied in studying protein-folding and assembly pathways and kinetics,[96–103] including those involving the formation of disulfide bonds,[76] free-energy landscapes,[104–107] and in a variety of biology-related studies, including amyloid formation,[108–111] signaling,[112] Hsp70 chaperone cycle,[113] iron-sulfur cluster biogenesis,[114] stability of proteins from *H. pylori*,[115] and structure and dynamics of natively unfolded proteins.[116] NARES-2P was used in modeling DNA and RNA structure,[91] determining the origin of telomere stability[117] and investigating mechanical stability of DNA.[118]

In Sections 4.1–4.4, selected examples of application of UNRES and NARES-2P are described.

4.1 Protein structure prediction

Following Anfinsen's thermodynamic hypothesis,[119] physics-based prediction of protein structure essentially means finding those structures which are favored by energy. This general guideline used to be oversimplified to state that the global minimum of the potential-energy function should be sought.[120–122] We initially used this method of structure prediction and, consequently, implemented[90] the conformational space annealing (CSA)[89] global optimization method with UNRES.[90] Although we scored some success in the CASP exercises by applying this methodology,[88,92] for systems with many degrees of freedom the most probable structure at a finite temperature is not that with the lowest potential energy but it should be understood as an ensemble of structures that belong to the basin with the lowest free energy.[38] Consequently, we designed a new procedure for protein structure prediction, in which extensive conformational search is carried out with MREMD,[85] the ensemble of conformations thus obtained dissected by cluster analysis, and representative of each cluster selected, with the free energy of the cluster it belongs to used in ranking the models.[38,123] We recently enriched this procedure with the tools that enable us to run prediction simulations also with bioinformatics-derived restraints such as the template[124] and contact restraints,[125] as well as low-resolution-experiment restraints.[126]

In the CASP10 exercise, the UNRES-based methodology resulted in the best models of two Free Modeling (FM) targets: T0663 and T0740-D1.[127] T0663 is a two-domain protein, each domain consisting of an α-helix packed

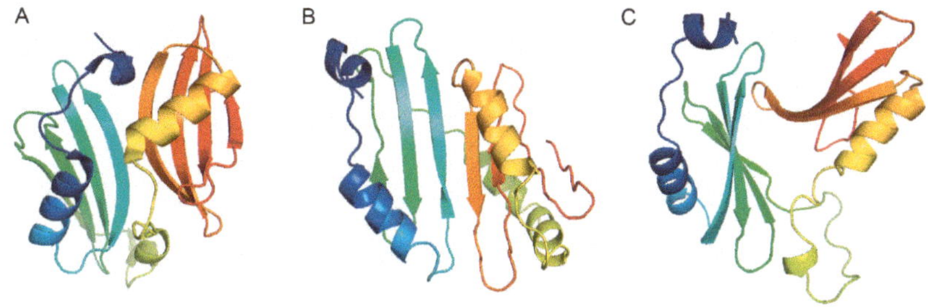

Fig. 9 The experimental 4EXR structure of the CASP10 target T0663 (A), model 1 (best) from our (Cornell-Gdansk) group that used UNRES,[127] which was the best prediction for this target in CASP10 (B) and model 1 from the LEEcon group, which was the best template-based model of this target (C). It can be seen that, though the individual domains in our model 1 are of coarse resolution, the β-sheets being too flat, the two domains are oriented as in the experimental structure. Conversely, each of the domains is modeled well in the template-based model but their packing is not native (C). *Adapted from He Y, Mozolewska MA, Krupa P, et al. Lessons from application of the UNRES force field to predictions of structures of CASP10 targets. Proc Nat Acad Sci USA. 2013;110: 14936–14941. Copyright (2013) National Academy of Sciences.*

against a four-stranded antiparallel β-sheet (Fig. 9A). Each domain is a Template Based Modeling (TMB) target but the domain packing is not trivial, which makes the whole target an FM target. UNRES was able to find the correct domain packing (Fig. 9B), as opposed to the knowledge-based methods, which wrongly predicted nearly orthogonal and not antiparallel packing of the C-terminal part of the β-sheet section of the N-terminal domain to the N-terminal part of the β-sheet section of the C-terminal domain (Fig. 9C). The second of the targets is an α-helical protein and UNRES produced the best model with the aid of contact information.[94]

The protein structures modeled with UNRES usually have correct topology at least in part of the structure but remarkably lower resolution compared to that of the knowledge-based models, which resulted in their lower ranking in CASP experiments compared to the models from the groups that used knowledge-based methods, except the situations when the knowledge-based methods failed to find the correct topology, as demonstrated with the example of T0663. Apparently one reason for this was our using, until very recently, not fully scale-consistent expressions for the torsional and correlation terms; as mentioned, the dependence on virtual-bond angles was not considered. We tested the scale-consistent UNRES in the CASP13 exercise and, although none of our models was featured this time, there was a major increase of the quality of the models in terms of Global Distance Test Score (GDT_TS)[128] of UNRES-modeled structures in

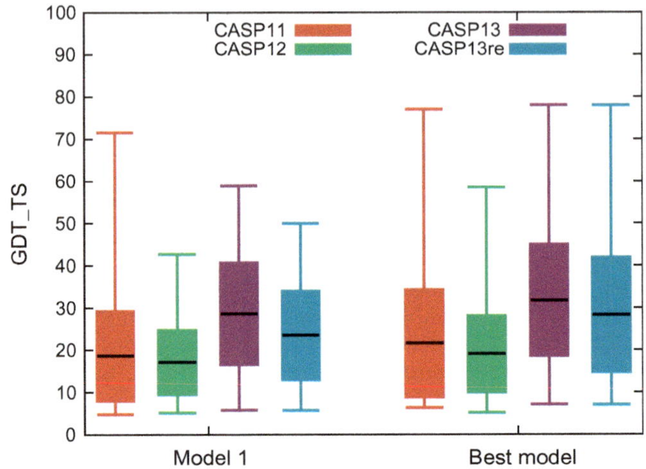

Fig. 10 Candlestick plots of the averages (*black horizontal segments*), standard deviations (filled bars centered at the averages), minimum (*bottom horizontal segments*), and maximum (*top horizontal segments*) of the GDT_TS calculated from the Cornell-Gdansk group models produced in CASP11 (*red*), UNRES group models produced in CASP12 (*green*), all UNRES group models produced in CASP13 (*purple*), and the UNRES group models produced in CASP13 for which simulations were started from random or extended conformations and not from server models (*blue*). *Reproduced with permission from Lubecka EA, Karczyńska AS, Lipska AG, et al. Evaluation of the scale consistent UNRES force field in template-free prediction of protein structures in the CASP13 experiment. J Mol Graph Model. 2019;92:154–166. Copyright (2019) Elsevier.*

CASP13 with respect to those in CASP11 and CASP12 (Fig. 10).[129] We expect that further improvement of resolution will be achieved by splitting the torsional and correlation potentials into the backbone and backbone-side chain part and by lowering the symmetry of the potentials involving flat side chains (especially phenylalanine, tyrosine, and tryptophan) to reproduce the stacking of these side chains better.

4.2 Effect of hydrodynamic interactions on folding kinetics

Hydrodynamic interactions (HIs) are solvent-mediated interactions that are manifested as the apparent drag of two objects moving through a liquid.[130–132] Intermolecular HIs are thought to be important in protein folding and in such biological phenomena as, e.g., formation of lipid membranes,[133] protein–protein association,[134] and the sliding motion of protein along DNA.[135] With implicit solvent representations, these interactions can be accounted for by introducing a nondiagonal matrix of friction coefficients, known as the Rotne–Prager (RP) tensor[136] or its variants.

Alternative approaches to handle hydrodynamic interactions have also been developed such as, e.g., the multiparticle collision dynamics methodology[137] and lattice Boltzmann methodology[138] and were subsequently applied to simulate the effect of hydrodynamic interactions on the structure and dynamics of biological systems.[139,140]

Tanaka[130] suggested that HIs could have a major effect on protein-folding pathways and on the final state. Several researchers studied the effect of HIs on protein folding with the structure-based (Gō-like) models.[131,132] It was found that the HIs accelerated the folding because of faster initial collapse that they cause. However, the Gō-like models are biased toward native interactions and, while nonnative interactions, even leading to the formation of intermediates with nonnative secondary structure occur during folding (e.-g., for lactalbumin A[141]). Therefore, we undertook a study of the effect of hydrodynamic interactions on the folding kinetics with UNRES[103] using two proteins: the N-terminal part of the B-domain of staphylococcal protein A (a 46-residue three-α-helix bundle; PDB: 1BDD) and the Formin-binding protein 28 WW domain (a 37-residue three-stranded antiparallel β-sheet, PDB: 1E0L).

For both proteins, we ran two series of 200 independent canonical MD trajectories, one with plain-Langevin dynamics and another one with Langevin dynamics including hydrodynamic interactions, at $T = 300$ K and $T = 335$ K for 1BDD and 1E0L, respectively, which are below the respective transition midpoints equal to 320 K and 339 K for 1BDD and 1E0L, respectively. The progress of folding was monitored as the variation of the fraction of native-like conformations and that of the conformation of the respective intermediate with time, computed from all trajectories at a given snapshot.[103]

We found that introducing hydrodynamic interactions remarkably slowed down the overall kinetics for both proteins (Fig. 11), which was caused by the presence of nonnative intermediates (Fig. 12), which acted as kinetics traps. For 1BDD, the intermediate was a mirror image of the native three-helix bundle and for 1E0L it was a partially helical conformation. The intermediates persisted for a much longer time when hydrodynamic interactions were included and, for 1E0L, even a steady state is reached (Fig. 11B). The kinetics was found to be biexponential, with the fast route corresponding to direct folding and the slow one involving an intermediate.[103] It can, therefore, be concluded that solvent drag slows down the folding, unless the system is strongly biased toward native interactions.

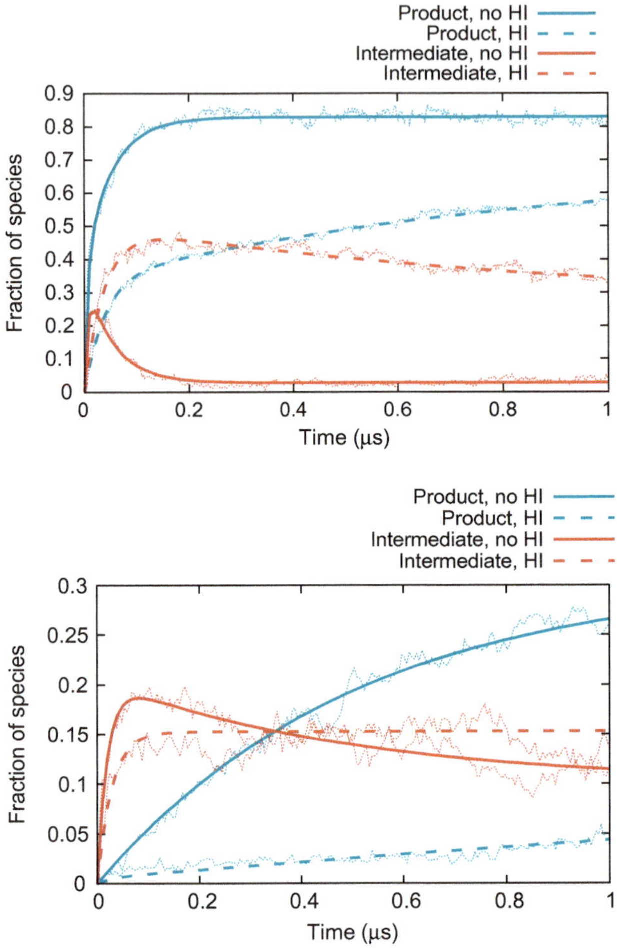

Fig. 11 Plots of the fractions of native (*blue lines*) and intermediate (*red lines*) structures as a function of time obtained in canonical MD simulations without (*solid lines*) and with including HIs (*dashed lines*) for 1BDD (A) and 1E0L (B). *Thin rugged lines* and *thick smooth lines* correspond to averages obtained from simulations and fitting multiexponential functions, respectively. *The figures have been drawn based on the data from Lipska AG, Seidman SR, Sieradzan AK, Giełdoń A, Liwo A, Scheraga HA. Molecular dynamics of protein a and a WW domain with a united residue model including hydrodynamic interaction. J Chem Phys. 2016;144:184110. The graphs were prepared with GNUPLOT (http://www.gnuplot.info).*

4.3 Investigation of Hsp70 chaperone cycle

The 70-kDa heat-shock (Hsp70) proteins are the major chaperones involved in assisting the correct folding or refolding and preventing aggregation of proteins in the cytosol.[142,143] They are composed of an N-terminal

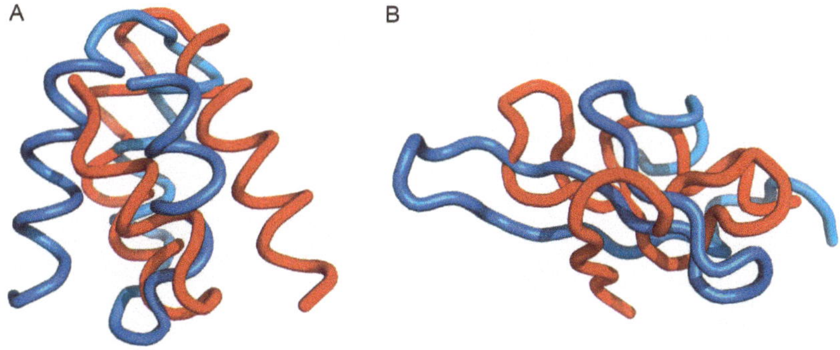

Fig. 12 Superpositions of the experimental (*blue*) and intermediate (*red*) structures of (A) 1BDD and (B) 1E0L proteins. *Reproduced from Lipska AG, Seidman SR, Sieradzan AK, Giełdoń A, Liwo A, Scheraga HA. Molecular dynamics of protein a and a WW domain with a united residue model including hydrodynamic interaction. J Chem Phys. 2016;144:184110, with the permission of AIP Publishing.*

nucleotide-binding (ATPase) domain (NBD), which catalyzes ATP hydrolysis and a C-terminal substrate-binding domain (SBD), connected by a short linker to the NBD.[144] The NBD is further split into NBD-I and NBD-II and the SBD is divided into SBD-α and SBD-β sections respectively. The Hsp70 chaperones perform their functions through cycles of binding and release of unfolded polypeptide chains,[142,143] this activity being controlled by ATP binding (substrate release) and hydrolysis to ADP (substrate binding). When ATP is bound, SBD-α binds to NBD-I, while SBD-β binds to NBD-II, respectively, this resulting in opening the SBD and low affinity toward the substrate. Conversely, when ATP is hydrolyzed, the two parts of the SBD dissociate from the NBD and bind together, closing tightly on the chunk of the substrate protein.[142]

We investigated the dynamics of the opening-closing cycle by using UNRES with canonical MD and MREMD.[113] At that time, only the structure of the SBD-closed (ADP-bound) form was known experimentally (PDB: 2KHO[144]), while only the structures of variants of the ATP-bound NBD domain alone were known. These are characterized by a reduced dihedral angle between the NBD-I and NBD-II compared to that in the SBD-closed form. Because we could not model ATP/ADP binding with UNRES, we used this information to impose distance restraints on the relative geometry of NBD-I and NBD-II. Some of the trajectories started from the 2KHO structures, particularly those started from conformations with the reduced dihedral angle between NBD-I and NBD-II, ended up in the putative open-form structure[113]; an example of such trajectory is shown in Fig. 13. The free-energy landscape determined by MREMD simulations

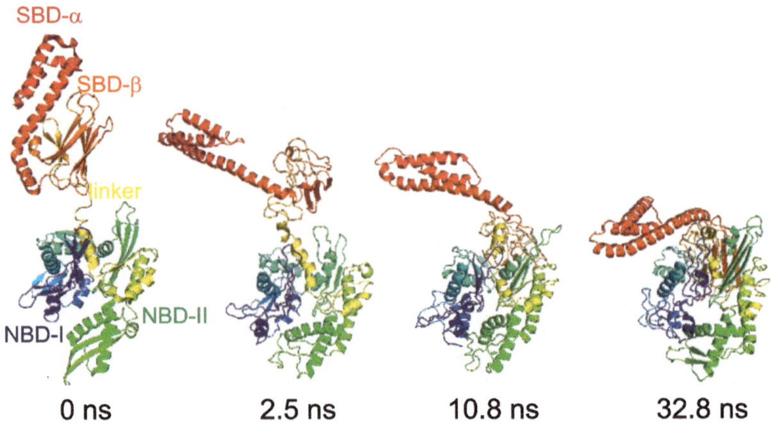

Fig. 13 Snapshots from a sample trajectory of the opening of yeast Hsp70 chaperone. The simulations were started from the experimental 2KHO structure[144] (the leftmost snapshot), in which the sections of the domains are labeled. The substrate-binding domain (SBD) gradually opens following the binding of the linker to the top section of the nucleotide-binding domain (NBD), its α and β sections binding to NBD-I and NBD-II, respectively. The time corresponds to the coarse-grained scale; the real time is by about four orders of magnitude bigger due to preaveraging over the fast degrees of freedom. *The structures have been drawn based on the data from Golas EI, Maisuradze GG, Senet P, et al. Simulation of the opening and closing of Hsp70 chaperones by coarse-grained molecular dynamics. J Chem Theor Comput. 2012;8:1334–1343. The drawings were prepared with PyMOL.*[44]

confirmed that this structure belongs to a stable free-energy basin.[113] After our study was published, the first ever solved structure of the SBD-open form of Hsp70 appeared (PDB: 4B9Q),[145] which was close to the open structures determined in our simulations (Fig. 14).

4.4 Investigation of telomere stability

Telomeres are repetitive nucleotide sequences that occur at linear DNA termini and play a central role in cell biology.[146] Shortening telomeres plays a role of a "cellclock", keeping track of the previous cell divisions and, thereby, controlling the ageing process.[147] The telomeres also adjust the cellular response to stress and growth stimulation[148] and their loop-like or lasso-like structures[149] play an important role in chromosome stability.

The telomere structures are unusually stable given the fact that their sequences contain numerous A and T bases, which usually weaken the double helix, making it prone to pre-melting (bubbling) transitions. In order to find out the origin of telomere stability, we performed a series of steered molecular

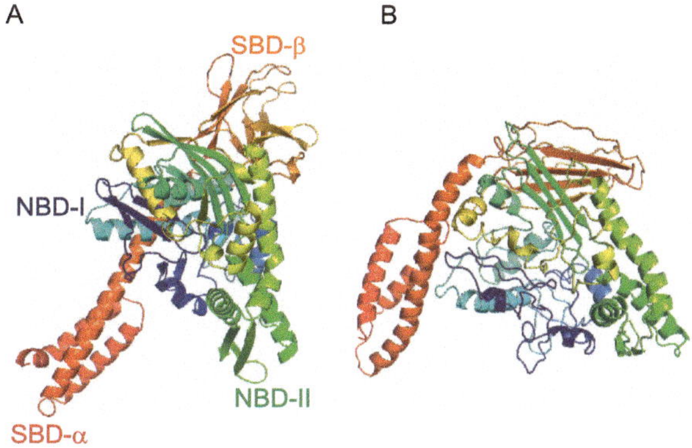

Fig. 14 Comparison of the crystal structure of the open Hsp70 chaperone form (PDB: 4B9Q)[145] (A) with that of the last snapshot of the UNRES/MD trajectory started from the closed form Ref. 113 (B). The sections of the domains are labeled in panel (A). The drawings were prepared with PyMOL.[44]

dynamics (SMD)[150] simulations, with the all-atom AMBER14[151,152] force field and with NARES-2P, of human-like (TTAGGG), plant (TTTAGG), insect (TTAGG), and *Candida guilermondii* (GGTGTAC) telomere repeats with various length, and other nontelomere repeats: AT, CG, and TTTTTCCCC for comparison.[91] The double helices were pulled either longitudinally, each strand from the other end of the helix, or pulled apart vertically, both strands at the same end of the double helix.

The force strength profiles in the extent of stretch exhibited a peak whose position depended on sequence length. With the all-atom AMBER14 force field, fast pulling-force rate had to be applied, this resulting in the formation of a flat double ladder, the required force strength being largely independent of sequence. Conversely, the faster dynamics with NARES-2P enabled us to apply a slower pulling-force rate resulting in sequence dependence of stability. For the CG repeats, which form the strongest Watson-Crick pairs, the force peak was the highest and for the AT repeats the lowest (Fig. 15A).

For the telomere sequences as well as for the TTTTCCCC repeat, the force peaks were almost as high as that for the CG repeat; moreover, the profiles decayed slower with stretch extent than that for the CG repeat and had fine structure (Fig. 15A). By analyzing the telomere structures on the pulling pathways, we found that this fine structure corresponds to cycles of dissociation of parts of one strand from another one and subsequent "regrabbing" them, as shown in Fig. 15. This "regrabbing" seems to be the major reason of preventing chain separation and seems to be a characteristic feature of telomere (and TTTTTCCCC) repeats, because it does not

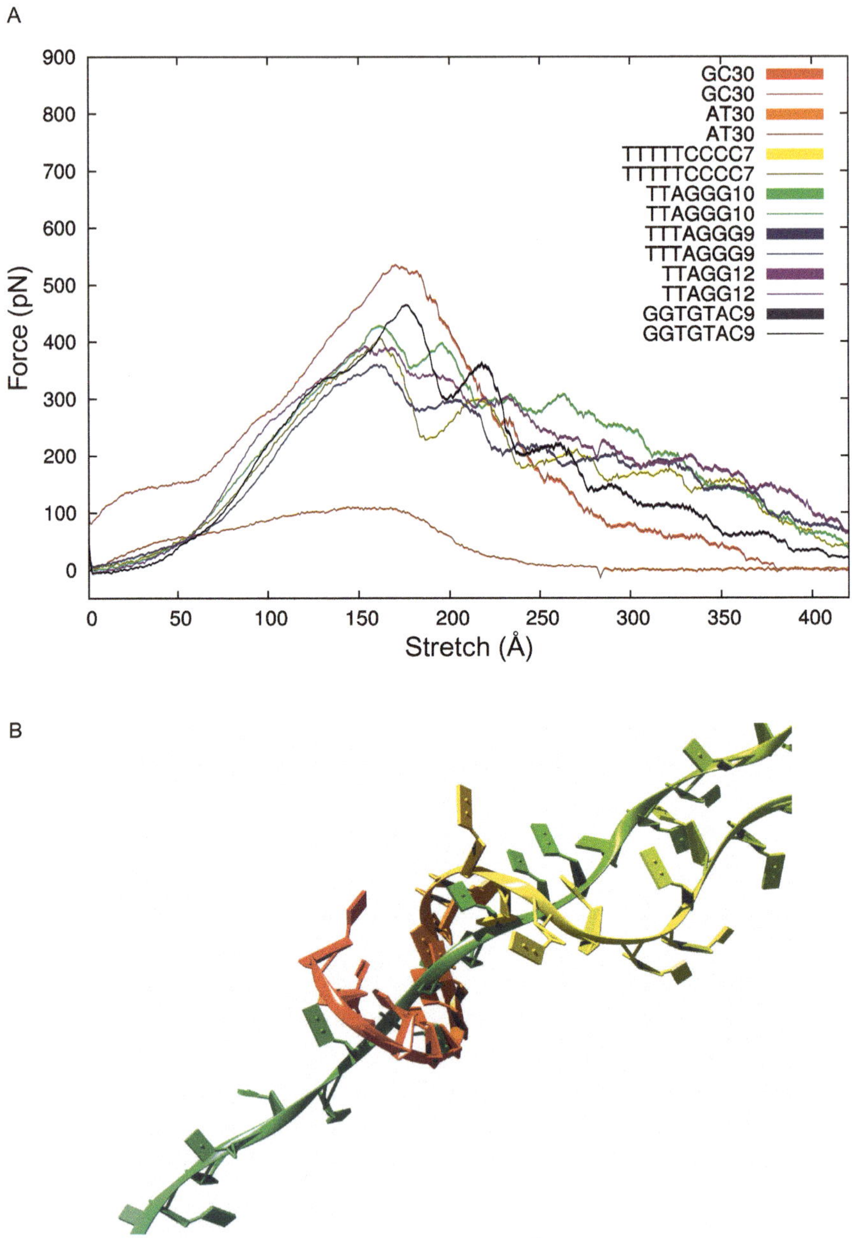

Fig. 15 (A) Plot of the force needed to stretch the DNA perpendicular to the hydrogen-bonds between bases, as a function of sequence. *Thin lines* indicate averages over 64 trajectories and the *colored areas* indicate one standard deviation. (B) Structure of telomere after regrabbing. The free end (*red*) wraps around the second DNA chain. *Reproduced with permission from Sieradzan AK, Krupa P, Wales DJ. What makes telomeres unique? J Phys Chem B. 2017;121:2207–2219, Copyright 2015 American Chemical Society.*

occur for CG repeat sequences. Consequently, it can be the reason of the ability of telomeres to resist damage and, thus, being able to protect the chromatin structure. We also found that the pulling-force peak increases greatly with telomere-repeat length, which explains why shortening telomere sequences upon cell ageing gradually results in reduced protection of chromatin DNA and, consequently, marks a cell for apoptosis.

5. Conclusions

The scale-consistent theory of coarse-grained force fields presented in this chapter and verified by its successful implementation in the UNICORN model of biological macromolecules provides a systematic methodology for the development of coarse-grained models of polymers and the pertinent force fields. This enables us to embed the atomistic details in the effective potential-energy function in a similar way that proper decomposition of the Born–Oppenheimer PES enables us to embed the electron degrees of freedom (that are averaged out) in properly designed effective energy terms, such as torsional potentials. As the all-atom force fields originate from the Born–Oppenheimer PESs, in our approach, the effective coarse-grained free-energy function originates from the potential of mean force of the system under study with secondary degrees of freedom averaged out, which is directly related to probability and can be used to compute ensemble averages and thermodynamic quantities.[34,36,38]

The key points of our methodology are as follows (Fig. 1): (i) decomposition of the PMF into factors, which are Kubo cluster-cumulant functions[37] and can be identified with effective energy terms, (ii) use of the lowest-order nonzero terms of the expansion of the cluster-cumulant functions into generalized Kubo cluster cumulants[37] as the basis of the analytical expressions for the effective energy terms, and (iii) expressing the interatomic distances and, consequently, the potential energy (at the all-atom level) in terms of site–site distances, the orientation of virtual-bond-axes, and the angles λ for collective rotations of polymer units about the respective virtual-bond axes, which converts the calculation of cumulants into an algebraic problem.[34]

From point (i) it follows that, in our approach, the force field is a free-energy and not a potential-energy function and its terms, except for those which correspond to cumulants of order 1, naturally depend on temperature. This is important, because the terms corresponding to higher-order PMF factors are responsible for the formation of ordered structures and should, therefore, decay with increasing temperature.

Embedding the atomic details in the effective energy terms provides not only appropriate dependence of the site–site long-range potentials on site orientation but also leads to correct functional forms of the local and correlation terms on geometric variables. For example, we have found that the torsional terms depend on both virtual-bond-dihedral and the adjacent virtual-bond angles.[34] Adhering to scale-consistency remarkably improves the quality of the modeled structures, as demonstrated by comparing the results obtained with the previous (not fully scale-consistent) and the latest (scale-consistent) versions of UNRES in regular tests[30] and in the CASP13 blind protein structure prediction experiment.[129] Because one of the key points of our methodology is averaging out the collective rotation angle about the virtual-bond axis, it can also be applied to coarse grain other systems with axial symmetry such as, e.g., lipid and liquid-crystal molecules.

Acknowledgments

This work was supported by grants UMO-2017/25/B/ST4/01026, UMO-2017/27/B/ST4/00926, and UMO-2017/26/M/ST4/00044 from the National Science Center of Poland (Narodowe Centrum Nauki). Calculations were carried out using the computational resources provided by (a) the supercomputer resources at the Centre of Informatics - Tricity Academic Supercomputer Center in Gdańsk (CI TASK), (b) the supercomputer resources at the Interdisciplinary Center of Mathematical and Computer Modeling (ICM), University of Warsaw (grant GA76-11), (c) the Polish Grid Infrastructure (PL-GRID), and (d) our Beowulf cluster at the Faculty of Chemistry, University of Gdańsk.

References

1. Kolinski A, Skolnick J. Reduced models of proteins and their applications. *Polymer.* 2004;45:511–524.
2. Ayton GS, Noid WG, Voth GA. Multiscale modeling of biomolecular systems: in serial and in parallel. *Curr Opinion Struct Biol.* 2007;17:192–198.
3. Maupetit J, Tuffery P, Derreumaux P. A coarse-grained protein force field for folding and structure prediction. *Proteins.* 2007;69:394–408.
4. Voth G. *Coarse-graining of Condensed Phase and Biomolecular Systems.* 1st ed. Boca Raton, FL: CRC Press, Taylor & Francis Group; 2008.
5. Czaplewski C, Liwo A, Makowski M, Ołdziej S, Scheraga HA. Coarse-grained models of proteins: theory and applications. In: Koliński A, ed. *Multiscale Approaches to Protein Modeling.* New York, NY: Springer; 2010:35–83. [chapter 3].
6. Tozzini V. Minimalist models for proteins: a comparative analysis. *Quart Rev Biophys.* 2010;43:333–371.
7. Davtyan A, Schafer NP, Zheng W, Clementi C, Wolynes PG, Papoian GA. AWSEM-MD: protein structure prediction using coarse-grained physical potentials and bioinformatically based local structure biasing. *J Phys Chem B.* 2012;116:8494–8503.
8. Sauders MG, Voth GA. Coarse-graining methods for computational biology. *Annu Rev Biophys.* 2013;42:73–93.

9. Ingolfsson HI, Uusitalo JJ, de Jong DH, Gopal S, Periole X, Marrink SJ. The power of coarse-graining in biomolecular simulations. *WIREs Comput Mol Sci*. 2014;4:225–248.
10. Kar P, Gopal SM, Cheng YM, Predeus A, Feig M. PRIMO: a transferable coarse-grained force field for proteins. *J Chem Theory Comput*. 2013;9:3769–3788.
11. Kar P, Feig M. Recent advances in transferable coarse-grained modeling of proteins. *Adv Protein Chem Struct Biol*. 2014;96:143–180.
12. Kmiecik S, Gront D, Koliński M, Wieteska L, Dawid A, Koliński A. Coarse-grained protein models and their applications. *Chem Rev*. 2016;116:7898–7936.
13. Papoian GA. *Coarse-Grained Modeling of Biomolecules*. CRC Press; 2017.
14. Singh N, Li W. Recent advances in coarse-grained models for biomolecules and their applications. *Int J Mol Sci*. 2019;20:3774.
15. Machado MR, Barrera EE, Klein F, Sóñora MS, Silva S, Pantano S. The SIRAH 2.0 force field: Altius, Fortius, Citius. *J Chem Theory Comput*. 2019;15:2719–2733.
16. Dama JF, Sinitskiy AV, McCullagh M, et al. The theory of ultra-coarse-graining. 1. General principles. *J Chem Theory Comput*. 2013;9:2466–2480.
17. Davtyan A, Dama JF, Sinitskiy AV, Voth GA. The theory of ultra-coarse-graining. 2. Numerical implementation. *J Chem Theory Comput*. 2014;10:5265–5275.
18. Nanias M, Chinchio M, Pillardy J, Ripoll DR, Scheraga HA. Packing helices in proteins by global optimization of a potential energy function. *Proc Nat Acad Sci USA*. 2003;100(4):1706–1710.
19. Li X, Jacobson MP, Friesner RA. High-resolution prediction of protein helix positions and orientations. *Proteins*. 2004;55:368–382.
20. Levitt M, Warshell A. Computer simulation of protein folding. *Nature*. 1975;253:694–698.
21. Levitt M. A simplified representation of protein conformations for rapid simulation of protein folding. *J Mol Biol*. 1976;104:59–107.
22. Chan HS, Dill KA. Protein folding in the landscape perspective: Chevron plots and non-Arrhenius kinetics. *Proteins*. 1998;30:2–33.
23. Kolinski A. Protein modeling and structure prediction with a reduced representation. *Acta Biochim Pol*. 2004;51:349–371.
24. Boniecki MJ, Lach G, Dawson WK, et al. SimRNA: a coarse-grained method for RNA folding simulations and 3D structure prediction. *Nucleic Acids Res*. 2016;44:e63.
25. Thomas PD, Dill KA. Statistical potentials extracted from protein structures: how accurate are they? *J Mol Biol*. 1996;257(2):457–469.
26. Ben Naim A. Statistical potentials extracted from protein structures: are these meaningful potentials? *J Chem Phys*. 1997;107:3698–3706.
27. Marrink SJ, Risselada HJ, Yefimov S, Tieleman DP, de Vries AH. The MARTINI force field: coarse grained model for biomolecular simulations. *J Phys Chem B*. 2007;111:7812–7824.
28. Marrink SJ, Tieleman DP. Perspective on the martini model. *Chem Soc Rev*. 2013;42:6801–6822.
29. Liwo A, Baranowski M, Czaplewski C, et al. A unified coarse-grained model of biological macromolecules based on mean-field multipole-multipole interactions. *J Mol Model*. 2014;20:2306.
30. Liwo A, Sieradzan AK, Lipska AG, et al. A general method for the derivation of the functional forms of the effective energy terms in coarse-grained energy functions of polymers. III. Determination of scale-consistent backbone-local and correlation potentials in the UNRES force field and force-field calibration and validation. *J Chem Phys*. 2019;150:155104.
31. Kozłowska U, Maisuradze GG, Liwo A, Scheraga HA. Determination of side-chain-rotamer and side-chain and backbone virtual-bond-stretching potentials of mean force from AM1 energy surfaces of terminally-blocked amino-acid residues, for coarse-

grained simulations of protein structure and folding. 2. Results, comparison with statistical potentials, and implementation in the UNRES force field. *J Comput Chem.* 2010;31:1154–1167.
32. Sieradzan AK, Scheraga HA, Liwo A. Determination of effective potentials for the stretching of $C^{\alpha}\cdots C^{\alpha}$ virtual bonds in polypeptide chains for coarse-grained simulations of proteins from ab initio energy surfaces of N-methylacetamide and N-acetylpyrrolidine. *J Chem Theor Comput.* 2012;8:1334–1343.
33. Sieradzan A, Niadzvedtski A, Scheraga H, Liwo A. Revised backbone-virtual-bond-angle potentials to treat the L- and D-amino acid residues in the coarse-grained united residue (UNRES) force field. *J Chem Theory Comput.* 2014;10:2194–2203. https://doi.org/10.1021/ct500119.
34. Sieradzan AK, Makowski M, Augustynowicz A, Liwo A. A general method for the derivation of the functional forms of the effective energy terms in coarse-grained energy functions of polymers. I. Backbone potentials of coarse-grained polypeptide chains. *J Chem Phys.* 2017;146:124106.
35. Kolinski A, Godzik A, Skolnick J. A general method for the prediction of the three-dimensional structure and folding pathway of globular proteins: application to designed helical proteins. *J Chem Phys.* 1993;98:7420–7433.
36. Liwo A, Czaplewski C, Pillardy J, Scheraga HA. Cumulant-based expressions for the multibody terms for the correlation between local and electrostatic interactions in the united-residue force field. *J Chem Phys.* 2001;115:2323–2347.
37. Kubo R. Generalized cumulant expansion method. *J Phys Soc Japan.* 1962;17:1100–1120.
38. Liwo A, Khalili M, Czaplewski C, et al. Modification and optimization of the united-residue (UNRES) potential energy function for canonical simulations. I. Temperature dependence of the effective energy function and tests of the optimization method with single training proteins. *J Phys Chem B.* 2007;111:260–285.
39. Mackerell Jr AD. Empirical force fields for biological macromolecules: overview and issues. *J Comput Chem.* 2004;25:1584–1604.
40. Makowski M, Liwo A, Scheraga HA. Simple physics-based analytical formulas for the potentials of mean force of the interaction of amino-acid side chains in water. VI. Oppositely charged side chains. *J Phys Chem B.* 2011;115(19):6130–6137.
41. Makowski M, Liwo A, Scheraga HA. Simple physics-based analytical formulas for the potentials of mean force of the interaction of amino acid side chains in water. VII. Charged-hydrophobic/polar and polar-hydrophobic/polar side chains. *J Phys Chem B.* 2017;121:379–390.
42. Makowski M, Liwo A. Physics-based modeling of side chain-side chain interactions in the UNRES force field. *Computational Methods to Study the Structure and Dynamics of Biomolecules and Biomolecular Processes. From Bioinformatics to Molecular Quantum Mechanics.* Cham, Switzerland: Springer Nature Switzerland AG; 2018:89–115.
43. Dawid AE, Gront D, Kolinski A. SURPASS low-resolution coarse-grained protein modeling. *J Chem Theory Comput.* 2017;13:5766–5779.
44. Schrödinger LLC. The PyMOL Molecular Graphics System, Version 1.3r1; 2010.
45. Krupa P, Hałabis A, Żmudzińska W, Ołdziej S, Scheraga HA, Liwo A. Maximum likelihood calibration of the UNRES force field for simulation of protein structure and dynamics. *J Chem Inf Model.* 2017;57:2364–2377.
46. Gniewek P, Leelananda SP, Koliński A, Jernigan RL, Kloczkowski A. Multibody coarse-grained potentials for native structure recognition and quality assessment of protein models. *Proteins.* 2011;79:1923–1929.
47. Makowski M, Liwo A, Scheraga HA. Simple physics-based analytical formulas for the potentials of mean force for the interaction of amino acid side chains in water. 1. Approximate expression for the free energy of hydrophobic association based on a Gaussian-overlap model. *J Phys Chem B.* 2007;111:2910–2916.

48. Makowski M, Liwo A, Maksimiak K, Makowska J, Scheraga HA. Simple physics-based analytical formulas for the potentials of mean force for the interaction of amino acid side chains in water. 2. Tests with simple spherical systems. *J Phys Chem B*. 2007; 111:2917–2924.
49. Sieradzan AK, Lipska AG, Lubecka EA. Shielding effect in protein folding. *J Mol Graph Model*. 2018;79:118–132.
50. Liwo A, Pincus MR, Wawak RJ, Rackovsky S, Scheraga HA. Prediction of protein conformation on the basis of a search for compact structures; test on avian pancreatic polypeptide. *Protein Sci*. 1993;2:1715–1731.
51. He Y, Maciejczyk M, Ołdziej S, Scheraga HA, Liwo A. Mean-field interactions between nucleic-acid-base dipoles can drive the formation of a double helix. *Phys Rev Lett*. 2013;110:098101.
52. Liwo A, Ołdziej S, Czaplewski C, Kozłowska U, Scheraga HA. Parameterization of backbone-electrostatic and multibody contributions to the UNRES force field for protein-structure prediction from *ab initio* energy surfaces of model systems. *J Phys Chem B*. 2004;108:9421–9438.
53. Sieradzan AK, Hansmann UHE, Scheraga HA, Liwo A. Extension of UNRES force field to treat polypeptide chains with D-amino acid residues. *J Chem Theory Comput*. 2012;8(11):4746–4757.
54. Sieradzan AK, Krupa P, Scheraga HA, Liwo A, Czaplewski C. Physics-based potentials for the coupling between backbone- and side-chain-local conformational states in the united residue (UNRES) force field for protein simulations. *J Chem Theory Comput*. 2015;11:817–831.
55. Hardin C, Eastwood MP, Prentiss M, Luthey-Schulten Z, Wolynes PG. Folding funnels: the key to robust protein structure prediction. *J Comput Chem*. 2002;23:138–146.
56. Seetharamulu P, Crippen GM. A potential function for protein folding. *J Math Chem*. 1991;6:91–110.
57. Lee J, Ripoll DR, Czaplewski C, Pillardy J, Wedemeyer WJ, Scheraga HA. Optimization of parameters in macromolecular potential energy functions by conformational space annealing. *J Phys Chem B*. 2001;105:7291–7298.
58. Lee J, Park K, Lee J. Full optimization of linear parameters of a united residue protein potential. *J Phys Chem B*. 2002;106:11647–11657.
59. Eastwood MP, Hardin C, Luthey-Schulten Z, Wolynes PG. Statistical mechanical refinement of protein structure prediction schemes: cumulant expansion approach. *J Chem Phys*. 2002;117:4602–4615.
60. Eastwood MP, Hardin C, Luthey-Schulten Z, Wolynes PG. Statistical mechanical refinement of protein structure prediction schemes. II. Mayer cluster expansion approach. *J Chem Phys*. 2003;118:8500–8512.
61. Liwo A, Arłukowicz P, Czaplewski C, Ołdziej S, Pillardy J, Scheraga HA. A method for optimizing potential-energy functions by a hierarchical design of the potential-energy landscape: application to the UNRES force field. *Proc Natl Acad Sci USA*. 2002;99:1937–1942.
62. Ołdziej S, Łągiewka J, Liwo A, et al. Optimization of the UNRES force field by hierarchical design of the potential-energy landscape. 3. Use of many proteins in optimization. *J Phys Chem B*. 2004;108:16950–16959.
63. Zaborowski B, Jagieła D, Czaplewski C, et al. A maximum-likelihood approach to force-field calibration. *J Chem Inf Model*. 2015;55:2050–2070.
64. He Y, Liwo A, Scheraga HA. Optimization of a nucleic acids united-RESidue 2-point model (NARES-2P) with a maximum-likelihood approach. *J Chem Phys*. 2015; 143:243111.
65. Liwo A, Czaplewski C. Extension of the force matching method to anisotropic coarse-grained transferable force fields: application to the UNRES model of proteins. *Biophys J*. 2019;116:145a.

66. Liwo A, Czaplewski C, Ołdziej S, et al. Simulation of protein structure and dynamics with the coarse-grained UNRES force field. In: Voth G, ed. *Coarse-Graining of Condensed Phase and Biomolecular Systems*. Boca Raton, Florida: CRC Press & Taylor & Francis Group; 2008:1391–1411. [chapter 8].
67. Lubecka EA, Liwo A. A general method for the derivation of the functional forms of the effective energy terms in coarse-grained energy functions of polymers. II. Backbone-local potentials of coarse-grained O1→4-bonded polyglucose chains. *J Chem Phys*. 2017;146:124106.
68. Samsonov SA, Lubecka EA, Bojarski KK, Ganzynkowicz R, Liwo A. Local and long range potentials for heparin-protein systems for coarse-grained simulations. *Biopolymers*. 2019;110:e23269.
69. Yin Y, Sieradzan AK, Liwo A, He Y, Scheraga HA. Physics-based potentials for coarse-grained modeling of protein-DNA interactions. *J Chem Theory Comput*. 2015;11(4):1792–1808. https://doi.org/10.1021/ct5009558.
70. Sieradzan AK, Giełdoń A, Yin Y, He Y, Scheraga HA, Liwo A. A new protein nucleic-acid coarse-grained force field based on the UNRES and NARES-2P force fields. *J Comput Chem*. 2018;39:2360–2370.
71. Liwo A, Pincus MR, Wawak RJ, Rackovsky S, Scheraga HA. Calculation of protein backbone geometry from α-carbon coordinates based on peptide-group dipole alignment. *Protein Sci*. 1993;2:1697–1714.
72. Liwo A, Kaźmierkiewicz R, Czaplewski C, et al. United-residue force field for off-lattice protein-structure simulations; III. Origin of backbone hydrogen-bonding cooperativity in united-residue potentials. *J Comput Chem*. 1998;19:259–276.
73. Gay JG, Berne BJ. Modification of the overlap potential to mimic a linear site-site potential. *J Chem Phys*. 1981;74:3316–3319.
74. Liwo A, Ołdziej S, Pincus MR, Wawak RJ, Rackovsky S, Scheraga HA. A united-residue force field for off-lattice protein-structure simulations. I. functional forms and parameters of long-range side-chain interaction potentials from protein crystal data. *J Comput Chem*. 1997;18:849–873.
75. Chinchio M, Czaplewski C, Liwo A, Ołdziej S, Scheraga HA. Dynamic formation and breaking of disulfide bonds in molecular dynamics simulations with the UNRES force field. *J Chem Theory Comput*. 2007;3:1236–1248.
76. Krupa P, Sieradzan AK, Mozolewska MA, Li H, Liwo A, Scheraga HA. Dynamics of disulfide-bond disruption and formation in the thermal unfolding of ribonuclease a. *J Chem Theory Comput*. 2017;13:5721–5730.
77. Czaplewski C, Karczyńska A, Sieradzan AK, Liwo A. UNRES server for physics-based coarse-grained simulations and prediction of protein structure, dynamics and thermodynamics. *Nucleic Acids Res*. 2018;46:W304–W309.
78. Liwo A, Lee J, Ripoll DR, Pillardy J, Scheraga HA. Protein structure prediction by global optimization of a potential energy function. *Proc Natl Acad Sci USA*. 1999;96:5482–5485.
79. Koradi R, Billeter M, Wüthrich K. MOLMOL: a program for display and analysis of macromolecular structures. *J Mol Graph*. 1996;14:51–55.
80. Ramachandran GN, Sasisekharan V. Conformation of polypeptides and proteins. *Adv Protein Chem*. 1968;23:283–437.
81. Khalili M, Liwo A, Rakowski F, Grochowski P, Scheraga HA. Molecular dynamics with the united-residue model of polypeptide chains. I. Lagrange equations of motion and tests of numerical stability in the microcanonical mode. *J Phys Chem B*. 2005; 109:13785–13797.
82. Khalili M, Liwo A, Jagielska A, Scheraga HA. Molecular dynamics with the united-residue model of polypeptide chains. II. Langevin and Berendsen-bath dynamics and tests on model α-helical systems. *J Phys Chem B*. 2005;109:13798–13810.

83. Swope WC, Anderson HC, Berens PH, Wilson KR. A computer simulation method for the calculation of equilibrium constants for the formation of physical clusters of molecules: application to small water clusters. *J Chem Phys*. 1982;76: 637–649.
84. Rakowski F, Grochowski P, Lesyng B, Liwo A, Scheraga HA. Implementation of a symplectic multiple-time-step molecular dynamics algorithm, based on the united-residue mesoscopic potential energy function. *J Chem Phys*. 2006;125:204107.
85. Czaplewski C, Kalinowski S, Liwo A, Scheraga HA. Application of multiplexing replica exchange molecular dynamics method to the UNRES force field: tests with α and $\alpha + \beta$ proteins. *J Chem Theor Comput*. 2009;5:627–640.
86. Hansmann UHE, Okamoto Y, Eisenmenger F. Molecular dynamics, Langevin and hybrid Monte Carlo simulations in multicanonical ensemble. *Chem Phys Lett*. 1996;259:321–330.
87. Rhee YM, Pande VS. Multiplexed-replica exchange molecular dynamics method for protein folding simulation. *Biophys J*. 2003;84:775–786.
88. Lee J, Liwo A, Ripoll DR, Pillardy J, Scheraga HA. Calculation of protein conformation by global optimization of a potential energy function. *Proteins*. 1999;37(Suppl 3): 204–408.
89. Lee J, Scheraga HA, Rackovsky S. Conformational analysis of the 20-residue membrane-bound portion of melittin by conformational space annealing. *Biopolymers*. 1998;46: 103–115.
90. Lee J, Liwo A, Scheraga HA. Energy-based *de novo* protein folding by conformational space annealing and an off-lattice united-residue force field: application to the 10–55 fragment of staphylococcal protein A and to *apo*-calbindin D9K. *Proc Natl Acad Sci USA*. 1999;96:2025–2030.
91. Sieradzan AK, Golon L, Liwo A. Prediction of DNA and RNA structure with the NARES-2P force field and conformational space annealing. *Phys Chem Chem Phys*. 2018;20:19656–19663.
92. Ołdziej S, Czaplewski C, Liwo A, et al. Physics-based protein-structure prediction using a hierarchical protocol based on the UNRES force field: assessment in two blind tests. *Proc Natl Acad Sci USA*. 2005;102:7547–7552.
93. Karczyńska A, Mozolewska MA, Krupa P, et al. Use of the UNRES force field in template-based prediction of protein structures and the refinement of server models: test with CASP12 targets. *J Mol Graph Model*. 2018;83:92–99.
94. Khoury GA, Liwo A, Khatib F, et al. Wefold: a coopetition for protein structure prediction. *Proteins*. 2014;82:1850–1868.
95. Keasar C, McGuffin LJ, Wallner EB, et al. An analysis and evaluation of the WeFold collaborative for protein structure prediction and its pipelines in CASP11 and CASP12. *Sci Rep*. 2018;8:9939.
96. Liwo A, Khalili M, Scheraga HA. Ab initio simulations of protein-folding pathways by molecular dynamics with the united-residue model of polypeptide chains. *Proc Natl Acad Sci USA*. 2005;102:2362–2367.
97. Khalili M, Liwo A, Scheraga HA. Kinetic studies of folding of the B-domain of staphylococcal protein A with molecular dynamics and a united-residue (UNRES) model of polypeptide chains. *J Mol Biol*. 2006;355:536–547.
98. He Y, Chen C, Xiao Y. United-residue (UNRES) Langevin dynamics simulations of trpzip2 folding. *J. Comput. Biol*. 2009;16:1719–1730.
99. Tuszynska I, Bujnicki JM. Predicting atomic details of the unfolding pathway for YibK, a knotted protein from the SPOUT superfamily. *J Biomol Struct Dyn*. 2010;27:511–520.
100. Sieradzan A, Liwo A, Hansmann UHE. Folding and self-assembly of a small protein complex. *J Chem Theory Comput*. 2012;8:3416–3422.

101. Gao K, Yang M. Molecular dynamics simulations of helix bundle proteins using UNRES force field and all-atom force field. *J Theor Comput Chem*. 2012;11: 1201–1215.
102. Zhou R, Maisuradze GG, Sunol D, et al. Folding kinetics of WW domains with the united residue force field for bridging microscopic motions and experimental measurements. *Proc Natl Acad Sci USA*. 2014;111:18243–18248.
103. Lipska AG, Seidman SR, Sieradzan AK, Giełdoń A, Liwo A, Scheraga HA. Molecular dynamics of protein a and a WW domain with a united-residue model including hydrodynamic interaction. *J Chem Phys*. 2016;144:184110.
104. Maisuradze GG, Liwo A, Scheraga HA. Principal component analysis for protein folding dynamics. *J Mol Biol*. 2009;385:312–329.
105. Maisuradze GG, Liwo A, Ołdziej S, Scheraga HA. Evidence, from simulations, of a single state with residual native structure at the thermal denaturation midpoint of a small globular protein. *J Am Chem Soc*. 2010;132:9444–9452.
106. Maisuradze GG, Zhou R, Liwo A, Xiao Y, Scheraga HA. Effects of mutation, truncation, and temperature on the folding kinetics of a WW domain. *J Mol Biol*. 2012;420:350–365.
107. Yin Y, Maisuradze GG, A L. Scheraga HA. Hidden protein folding pathways in free-energy landscapes uncovered by network analysis. *J Chem Theory Comput*. 2012; 8:1176–1189.
108. Rojas A, Liwo A, Browne D, Scheraga HA. Mechanism of fiber assembly; treatment of Aβ-peptide aggregation with a coarse-grained united-residue force field. *J Mol Biol*. 2010;404:537–552.
109. Rojas A, Liwo A, Scheraga HA. A study of the α-helical intermediate preceding the aggregation of the amino-terminal fragment of the Aβ-amyloid peptide (1–28). *J Phys Chem B*. 2011;115:12978–12983.
110. Rojas A, Maisuradze N, Kachlishvili K, Scheraga HA, Maisuradze GG. Elucidating important sites and the mechanism for amyloid fibril formation by coarse-grained molecular dynamics. *ACS Chem Neurosci*. 2017;8:201–209.
111. Rojas AV, Maisuradze GG, Scheraga HA. Dependence of the formation of Tau and A beta peptide mixed aggregates on the secondary structure of the N-terminal region of A beta. *J Phys Chem B*. 2018;122:7049–7056.
112. He Y, Liwo A, Weinstein H, Scheraga HA. PDZ binding to the BAR domain of PICK1 is elucidated by coarse-grained molecular dynamics. *J Mol Biol*. 2011;405: 298–314.
113. Golas EI, Maisuradze GG, Senet P, et al. Simulation of the opening and closing of Hsp70 chaperones by coarse-grained molecular dynamics. *J Chem Theor Comput*. 2012;8:1334–1343.
114. Mozolewska M, Krupa P, Scheraga HA, Liwo A. Molecular modeling of the binding modes of the iron-sulfur protein to the Jac1 co-chaperone from *Saccharomyces cerevisiae* by all-atom and coarse-grained approaches. *Proteins*. 2015;83:1414–1426.
115. Grzeszczuk MJ, Ba¸k A, Banaś AM, et al. Impact of selected amino acids of HP0377 (Helicobacter pylori thiol oxidoreductase) on its functioning as a CcmG (cytochrome c maturation) protein and Dsb (disulfide bond) isomerase. *PLoS ONE*. 2018;13: e0195358.
116. Vallet SD, Miele AE, Uciechowska-Kaczmarzyk U, et al. Insights into the structure and dynamics of lysyl oxidase propeptide, a flexible protein with numerous partners. *Sci Rep*. 2018;8:11768.
117. Sieradzan AK, Krupa P, Wales DJ. What makes telomeres unique? *J Phys Chem B*. 2017;121:2207–2219.
118. Krupa P, Wales DJ, Sieradzan AK. Computational studies of the mechanical stability for single-strand break DNA. *J Phys Chem B*. 2018;122:8166–8173.

119. Anfinsen CB. Principles that govern the folding of protein chains. *Science*. 1973; 181:223–230.
120. Androulakis IP, Maranas CD, Floudas CA. Prediction of oligopeptide conformations via deterministic global optimization. *J Glob Opt*. 1997;11:1–34.
121. Wales DJ, Scheraga HA. Global optimization of clusters, crystals, and biomolecules. *Science*. 1999;285:1368–1372.
122. Scheraga HA, Pillardy J, Liwo A, et al. Evolution of physics-based methodology for exploring the conformational energy landscape of proteins. *J Comput Chem*. 2002; 23:28–34.
123. Krupa P, Mozolewska MA, Wiśniewska M, et al. Performance of protein-structure predictions with the physics-based UNRES force field in CASP11. *Bioinformatics*. 2016;32(21):3270–3278.
124. Mozolewska MA, Krupa P, Zaborowski B, et al. Use of restraints from consensus fragments of multiple server models to enhance protein-structure prediction capability of the UNRES force field. *J Chem Inf Model*. 2016;56:2263–2279.
125. Lubecka EA, Liwo A. Introduction of a bounded penalty function in contact-assisted simulations of protein structures to omit false restraints. *J Comput Chem*. 2019;40:2164–2178.
126. Karczyńska AS, Mozolewska MA, Krupa P, Giełdoń A, Liwo A, Czaplewski C. Prediction of protein structure with the coarse-grained UNRES force field assisted by small X-ray scattering data and knowledge-based information. *Proteins*. 2018; 86(S1):228–239.
127. He Y, Mozolewska MA, Krupa P, et al. Lessons from application of the UNRES force field to predictions of structures of CASP10 targets. *Proc Nat Acad Sci USA*. 2013;110:14936–14941.
128. Zemla A. LGA: a method for finding 3D similarities in protein structures. *Nucleic Acids Res*. 2003;13:3370–3374.
129. Lubecka EA, Karczyńska AS, Lipska AG, et al. Evaluation of the scale-consistent UNRES force field in template-free prediction of protein structures in the CASP13 experiment. *J Mol Graph Model*. 2019;92:154–166.
130. Tanaka H. Roles of hydrodynamic interactions in structure formation of soft matter: protein folding as an example. *J Phys Condens Matter*. 2005;17:S2795–S2803.
131. Cieplak M, Niewieczerzał S. Hydrodynamic interactions in protein folding. *J Chem Phys*. 2009;130:124906. https://doi.org/10.1063/1.3050103.
132. Frembgen-Kesner T, Elcock AH. Striking effects of hydrodynamic interactions on the simulated diffusion and folding of proteins. *J Chem Theory Comput*. 2009;5:242–256. https://doi.org/10.1021/ct800499p.
133. Ando T, Skolnick J. On the importance of hydrodynamic interactions in lipid membrane formation. *Biophys J*. 2013;104:96–105.
134. Frembgen-Kesner T, Elcock AH. Absolute protein-protein association rate constants from flexible, coarse-grained Brownian dynamics simulations: the role of intermolecular hydrodynamic interactions in barnase-barstar association. *Biophys J*. 2010;99:L75-L77.
135. Ando T, Skolnick J. Sliding of proteins non-specifically bound to DNA: Brownian dynamics studies with coarse-grained protein and DNA models. *PLoS Comput Biol*. 2014;10:e1003990.
136. Rotne J, Prager S. Variational treatment of hydrodynamic interaction in polymers. *J Chem Phys*. 1969;50(11):4831–4837. https://doi.org/10.1063/1.1670977.
137. Malevanets A, Kapral R. Mesoscopic model for solvent dynamics. *J Chem Phys*. 1999;110:8605–8613.
138. Giupponi G, De Fabrilis G. Hybrid method coupling fluctuating hydrodynamics and molecular dynamics for the simulation of macromolecules. *J Chem Phys*. 2007;126: 154903.

139. Chen JX, Kapral R. Mesoscopic dynamics of diffusion-influenced enzyme kinetics. *J Chem Phys*. 2011;134:044503.
140. Sterpone F, Derreumaux P, Melchionna S. Protein simulations in fluids: coupling the OPEP coarse-grained force field with hydrodynamics. *J Chem Theory Comput*. 2015;11:1843–1853.
141. Kuwajima K. The molten globule state of α-lactalbumin A. *FASEB J*. 1996;10:102–109.
142. Mayer MP, Bukau P. Hsp70 chaperones: cellular functions and molecular mechanism. *Cell Mol Life Sci*. 2005;62:670–684.
143. Kampinga HH, Craig EA. The Hsp70 chaperone machinery: J proteins as drivers of functional specificity. *Nat Rev Mol Cell Biol*. 2010;11:589–591.
144. Bertelsen EB, Chang L, Gestwick JE, Zuiderweg ERP. Solution conformation of wild-type E. coli Hsp70 (DnaK) chaperone complexed with ADP and substrate. *Proc Natl Acad Sci USA*. 2009;106:8471–8476.
145. Kityk R, Koop J, Sinning I, Mayer MP. Structure and dynamics of the ATP-bound open conformation of Hsp70 chaperones. *Mol Cell*. 2012;48:863–874.
146. Aubert G, Lansdorp PM. Telomeres and aging. *Physiol Rev*. 2008;88:557–579.
147. Wright WE, Piatyszek MA, Rainey WE, Byrd W, Shay JW. Telomerase activity in human germline and embryonic tissues and cells. *Dev Genet*. 1996;18:173–179.
148. von Zglinicki T, Saretzki G, Döcke W, Lotze C. Mild hyperoxia shortens telomeres and inhibits proliferation of fibroblasts: a model for senescence? *Exp Cell Res*. 1995;220:186–193.
149. Galati A, Micheli E, Cacchione S. Chromatin structure in telomere dynamics. *Front Oncol*. 2013;3:46.
150. Grubmüller H, Heymann B, Tavan P. Ligand binding: molecular mechanics calculation of the streptavidin-biotin rupture force. *Science*. 1996;271:997–999.
151. Pearlman DA, Case DA, Caldwell JW, et al. AMBER, a package of computer programs for applying molecular mechanics, normal mode analysis, molecular dynamics and free energy calculations to simulate the structural and energetic properties of molecules. *Comp Phys Commun*. 1995;91:1–41.
152. Galindo-Murillo R, Robertson JC, Zgarbová M, et al. Assessing the current state of Amber force field modifications for DNA. *J Chem Theory Comput*. 2016;12:4114–4127.

CHAPTER THREE

How to learn from inconsistencies: Integrating molecular simulations with experimental data

Simone Orioli[†,a,b], **Andreas Haahr Larsen**[†,‡,a,b], **Sandro Bottaro**[a,c], **Kresten Lindorff-Larsen***[,a]

[a]Structural Biology and NMR Laboratory & Linderstrøm-Lang Centre for Protein Science, Department of Biology, University of Copenhagen, Copenhagen, Denmark

[b]Structural Biophysics, Niels Bohr Institute, Faculty of Science, University of Copenhagen, Copenhagen, Denmark

[c]Atomistic Simulations Laboratory, Istituto Italiano di Tecnologia, Genova, Italy

*Corresponding author: e-mail address: lindorff@bio.ku.dk

Contents

† These authors contributed equally to this work.

‡ Present address: Structural Bioinformatics and Computational Biochemistry Unit, Department of Biochemistry, University of Oxford, Oxford, United Kingdom

Progress in Molecular Biology and Translational Science, Volume 170
ISSN 1877-1173
https://doi.org/10.1016/bs.pmbts.2019.12.006

Abstract

Molecular simulations and biophysical experiments can be used to provide independent and complementary insights into the molecular origin of biological processes. A particularly useful strategy is to use molecular simulations as a modeling tool to interpret experimental measurements, and to use experimental data to refine our biophysical models. Thus, explicit integration and synergy between molecular simulations and experiments is fundamental for furthering our understanding of biological processes. This is especially true in the case where discrepancies between measured and simulated observables emerge. In this chapter, we provide an overview of some of the core ideas behind methods that were developed to improve the consistency between experimental information and numerical predictions. We distinguish between situations where experiments are used to refine our understanding and models of specific systems, and situations where experiments are used more generally to refine transferable models. We discuss different philosophies and attempt to unify them in a single framework. Until now, such integration between experiments and simulations have mostly been applied to equilibrium data, and we discuss more recent developments aimed to analyze time-dependent or time-resolved data.

1. Introduction

Molecular mechanics simulations of biological systems have matured over the last decades, to a state where reliable predictions and interpretations of biological phenomena can be achieved.[1,2] Simulations can readily be used either before experiments to suggest hypotheses and design experiments, or after experiments to analyze and interpret the data. With simulations it is possible to probe timescales and spatial details that are, yet, impossible to access experimentally, and thus they provide a unique tool to study, e.g., conformational transitions,[3–7] signaling events,[8] ion transport,[8,9] and many other phenomena. Since most experimental observables are averaged over time and a large number of molecules, simulations can help resolve the underlying structural and dynamical distribution with atomistic spatial accuracy and femtosecond time resolution.

As simulations are in essence a theoretical framework, experimental verification is pivotal and there are still, in many cases, significant differences between experimental data and the corresponding observables calculated from state-of-the art simulations.[10] These discrepancies are usually due to imperfect force fields,[11–14] insufficient sampling,[15] or inaccurate forward models.[16] Even with accurate forward models and robust sampling, however, the quality of the simulation results are bound by the accuracy of the underlying force field.[13] Indeed, molecular mechanics is intrinsically limited by the fact that it employs classical approximations for phenomena that are known to happen on a quantum scale. Force fields have been developed to approximate these quantum interactions within a classical framework in a way to balance between computational simplicity and accuracy: thus, an exact match is not to be expected and synergy between simulations and experiments remains of fundamental importance. Substantial efforts in the biophysical community have been therefore directed toward the development of methods that correct the results of simulations by accounting for experimental information, either in a system-specific or generalizable fashion. In this chapter we aim to provide a self-contained overview of the principles and strategies that have been applied to infer conformational ensembles, i.e., collections of structures reflecting the dynamism and plasticity of a molecule, as well as to improve force fields using experimental data. In other words, we will make a clear cut distinction between system-specific and general force field corrections: in the former case, simulations are performed according to experimental conditions, and if inconsistencies emerge, the resulting ensemble is corrected according to the available experimental information; in the latter case, instead, discrepancies between simulations and experimental observables are used to actively improve the physical potential employed to describe the system. We note that the goal of the chapter is not to be comprehensive of all the literature that has been previously published: rather, it focuses on some cornerstone principles and methods that summarize the efforts of the community. For this reason we also focus mostly on the methods and theory, and only provide few examples of applications.

This chapter is meant for readers experienced in molecular simulations and comfortable with the fundamentals of statistics and statistical mechanics. It is particularly suited for scientists that approach the combination of experiments with simulations for the first time or readers interested in a nontechnical overview of the field.

The remainder of the chapter is comprised of six sections, each one concerned with a different aspect of the literature on the subject. Section 2 is dedicated to the description of some of the main strategies to obtain consistency between simulation and data by manipulating the ensemble *after* simulations have been performed. Differently, in Section 3 we will discuss how consistency can be achieved by introducing a system-specific empirical energy term in the force field. In this case, the refinement step occurs *before* the actual simulation, as the experimental bias will guide the sampling toward only the most relevant regions of conformational space. In Section 4 we discuss how to employ experimental data to refine the physical description of macromolecules, i.e., the force field, instead of the conformational ensemble of a particular system. Like the biased force field approaches described in Section 3, these methods also adjust the force field *before* the production simulation. In Section 5 we discuss how to combine time-dependent and time-resolved data with simulations. This is particularly relevant as computational power[17–20] and enhanced sampling techniques[3,21] have pushed the accessible timescales to the ones resolved by some experimental techniques. In Section 6 we discuss some of the most important overall challenges arising from combining experimental data and simulations, and, finally, in Section 7 we conclude by summarizing similarities and differences between the various approaches. We thus aim to give an overview of the present state of the field as well as pointing out key aspects where further collective effort is needed to improve the predictive power of combined computational and experimental methods for structural studies of biological systems.

2. Reweighting strategies

Consistency between simulations and experimental data can in many cases be achieved by reweighting a trajectory (or a set of trajectories) obtained with a given force field (Fig. 1). Let us assume we have performed a molecular dynamics (MD) or Monte Carlo (MC) simulation and have saved N snapshots, $X_i, \ldots, X_N$, from the trajectory for analysis. Each frame in the simulation is given an initial weight that reflects the population of that structure as predicted by the force field. We denote the initial set of weights, $\boldsymbol{w^0} = w_1^0, w_2^0, \ldots, w_N^0$, the *reference distribution* and we assume them to be non-negative and normalized as probabilities, $\sum_i w_i^0 = 1$. When performing a standard MD/MC simulation and under the assumption that enough sampling has been performed, the initial weights are constant, $w_i^0 = 1/N$, as such simulations generate conformations distributed according to the

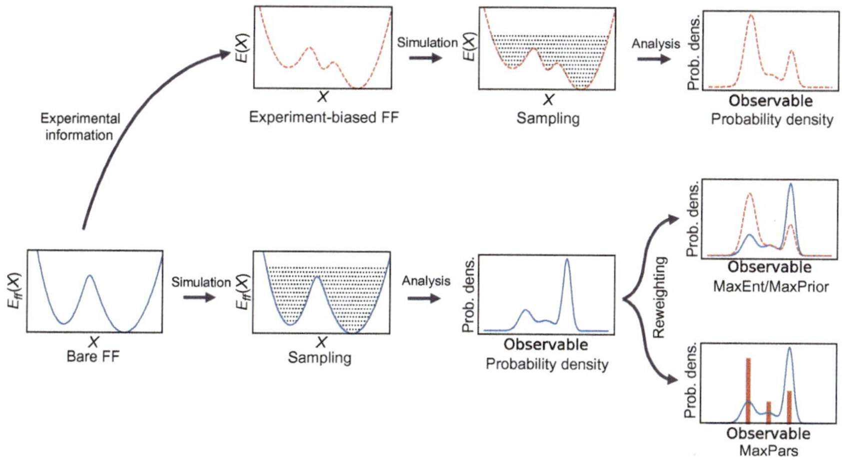

Fig. 1 Overview of methods to achieve consistency between simulation and experimental data for the distribution of a generic static observable. All methods start with an initial force field $E_{ff}(x)$ ('bare' FF, *blue line*) which is either employed as is or modified in accordance to some experimentally driven bias (experiment-biased FF, *red dashed line*). In the upper case, the biased force field is sampled to directly generate an observable distribution which is coherent with experimental data. In the lower case, the observable distribution is subsequently reweighted against experimental data using either the MaxEnt/MaxPrior or MaxPars principles to yield some reweighted distributions (*red dashed line and bars*).

Boltzmann distribution. When using enhanced sampling techniques that do not directly sample the Boltzmann distribution or, in the case of shorter off-equilibrium simulations, the initial weights are in general nonuniform.[22] Given the N snapshots and the reference distribution, a static observable can, in the simplest case, be calculated as a weighted ensemble average:

$$\langle O^{\text{calc}} \rangle (\boldsymbol{w}) = \sum_{i=1}^{N} w_i O^{\text{calc}}(X_i). \tag{1}$$

Some experimental quantities (e.g., NMR relaxation rates and other inherently time-dependent observables) cannot generally be expressed as linear ensemble averages over individual structures, i.e., employing Eq. (1). Therefore, throughout the chapter the validity of Eq. (1) will be taken as a working assumption until Section 5, where we will discuss the case of time-dependent observables.

In a reweighting process, the weights are modified until the calculated observables become consistent with corresponding experimental observables, O_j^{exp}, where $j = 1, \ldots, M$. We denote the reweighted set of weights,

$\boldsymbol{w} = w_1, w_2, \ldots, w_N$, the *optimal distribution*.[23] Unfortunately, many possible sets of weights can lead to the same averages, and thus to consistency with data. That is, it is an ill-posed problem, and strategies are therefore needed to regularize the problem, so the optimal among these distributions can be selected. There exist several strategies to regularize and solve the resulting optimization problem,[24] and we here focus on three of them: the principle of maximum entropy (MaxEnt, Section 2.1), the principle of maximum parsimony (MaxPars, Section 2.2) and Bayesian inference (MaxPrior, Section 2.3). Each method encodes a different philosophy for what it means for a set of weights to be *optimal*, which translates into different optimal distributions after reweighting.

2.1 Maximum entropy

The principle of maximum entropy (MaxEnt)[25] states that the optimal distribution is the one adding the least amount of information given some imposed *constraints*, e.g., $\langle O^{\text{calc}} \rangle = O^{\text{exp}}$. In other words, the solution with highest entropy which is also consistent with the data is considered optimal. In this context, entropy is quantified via the *relative entropy* which we define as the negative of the Kullback–Leibler divergence[26]:

$$S(\boldsymbol{w}) = -\sum_{i=1}^{N} w_i \log\left(\frac{w_i}{w_i^0}\right), \tag{2}$$

where w_i^0 and w_i are the weights of frame X_i before and after reweighting. Other expressions for the entropy are possible,[27,28] but Eq. (2) is usually preferred.[29–32] As the distributions of initial and refined weights diverges, $S(\boldsymbol{w})$ becomes smaller, and vice versa. In the simplest case, the set of weights that maximizes the entropy is chosen by the only constraint that the weights are normalized as probabilities, i.e., we look for weights for which the function $g_0(\boldsymbol{w}) = 1 - \sum_i w_i$ is zero. This problem can be solved by constructing a Lagrangian:

$$\mathcal{L}(\boldsymbol{w}, \lambda_0) = S(\boldsymbol{w}) - \lambda_0 g_0(\boldsymbol{w}), \tag{3}$$

where λ_0 is a Lagrange multiplier. The solution of the set of equations

$$\nabla \mathcal{L} = 0 \tag{4}$$

gives the highest entropy weights which fulfill the normalization condition, which are simply $\boldsymbol{w} = \boldsymbol{w}^{\boldsymbol{0}}$. It is clear then, that the problem is interesting only when more constraints are introduced. For example, exact consistency with a given set of experimental data can be imposed via functions of the form $g_j(\boldsymbol{w}) = O_j^{\text{exp}} - \langle O_j^{\text{calc}} \rangle(\boldsymbol{w})$. The Lagrangian in Eq. (3), in the case of M experimental constraints and the normalization constraint, becomes:

$$\mathcal{L}(\boldsymbol{w}, \boldsymbol{\lambda}) = S(\boldsymbol{w}) - \sum_{j=0}^{M} \lambda_j g_j(\boldsymbol{w}), \tag{5}$$

where a Lagrange multiplier λ_j is introduced for each constraint. The solution of Eq. (4) for the Lagrangian in Eq. (5) provides weights that ensure consistency between the reweighted trajectory and the experimental data. The optimal weights are uniquely given in terms of the M Lagrange multipliers[33–37]:

$$w_i(\boldsymbol{\lambda}) = \frac{w_i^0 \exp\left[-\sum_{j=0}^{M} \lambda_j O_j^{\text{calc}}(X_i)\right]}{\sum_{k=0}^{N} w_k^0 \exp\left[-\sum_{j=0}^{M} \lambda_j O_j^{\text{calc}}(X_k)\right]}, \tag{6}$$

where $O_j^{\text{calc}}(X_i)$ is the value of the jth observable, as calculated from the ith frame in the simulation. Note that the denominator just ensures proper normalization of the weights.

Fitting data tightly to gain full consistency by imposed constraints might, however, be unrealistic and lead to overfitting because of errors in the data and in the model.[34] Indeed, experimental observables are only known with some limited certainty and, likewise, forward models used to calculate observables from simulations have an associated uncertainty.[16] Systematic errors in the data lead to further disagreement with the model,[23,38,39] and finally, the finite number of structures N used to compute the theoretical observable (see Eq. 1) introduces additional error to the estimated averages.[39] All these effects add up to an effective uncertainty for each observable, σ_j. Therefore it is generally preferable to apply *restraints* rather than constraints, i.e., to impose consistency with the data only within the estimated uncertainty (restraining) rather than exactly (constraining). This translates into relaxing the equality $\langle O^{\text{calc}} \rangle = O^{\text{exp}}$ to a similarity $\langle O^{\text{calc}} \rangle \sim O^{\text{exp}}$ within a given threshold.

There are two main strategies to introduce restraints in the MaxEnt method using Lagrange multipliers. The first one consists in modifying the constraints in such a way that the calculated and experimental observables are allowed to differ by some small quantity ε, i.e., $g_j^{\sigma}(\boldsymbol{w}) = (O_j^{\text{exp}} + \varepsilon(\lambda_j, \sigma_j)) - \langle O_j^{\text{calc}} \rangle(\boldsymbol{w})$. ε depends on the effective uncertainty, σ_j and the Lagrange multiplier associated to the constraint, λ_j.[36,40,41] The second way to include experimental errors was introduced by Gull and Daniel[42] and, in the assumption of normally distributed errors, describes the discrepancy between data and model via the χ^2 distribution:

$$\chi^2(\boldsymbol{w}) = \sum_{j=1}^{M} \left(\frac{O_j^{\exp} - \langle O_j^{\text{calc}} \rangle(\boldsymbol{w})}{\sigma_j} \right)^2, \tag{7}$$

The expectation value of χ^2 is equal to the number of degrees of freedom ν, so a reduced χ^2 can be defined such that its expectation value is unity,

$$\chi_r^2 = \frac{1}{\nu}\chi^2. \tag{8}$$

Experimental restraints can thus be included by imposing the constraint function $g_{\chi_r^2}(\boldsymbol{w}) = \chi_r^2(\boldsymbol{w}) - 1$ to be zero, with a Lagrange multiplier that we call θ^{-1}, and whose role we discuss in further detail below. This treatment is correct in the assumption of normally distributed errors and for datasets that guarantee $\chi_r^2 = 1$. The latter assumption is generally not safe, as point estimates can deviate from the χ_r^2 expectation value: more details on this are provided in Section 6. To overcome this problem, the optimal distribution can be obtained by minimizing the equation[23,29,32]

$$T_{\text{MaxEnt}}(\boldsymbol{w}) = \chi^2(\boldsymbol{w}) - \theta S(\boldsymbol{w}), \tag{9}$$

where θ is treated as an adjustable parameter rather than a Lagrange multiplier, and controls how tightly the data should be fitted. The solution is then said to be regularized by the entropy, i.e., at a fixed value of θ the optimal solution is the one with the highest entropy $S(\boldsymbol{w})$ among the ones that minimize the discrepancy with the data, $\chi^2(\boldsymbol{w})$.

After reweighting, some of the weights may be close to zero and the corresponding structures will thus effectively be ignored in the calculation of the new averages. MaxEnt ensures that the optimized ensemble preserves as many structures from the reference one as possible. In this sense, the entropy term S can be interpreted via a more intuitive quantity, ϕ_{eff}, which represents the effective fraction of frames used in the reweighted ensemble compared to the initial ensemble[32]:

$$\phi_{\text{eff}}(\boldsymbol{w}) = \exp\left[S(\boldsymbol{w})\right]. \tag{10}$$

If the reference distribution is unaltered, $\phi_{\text{eff}}(\boldsymbol{w} = \boldsymbol{w}_0) = 1$. In the opposite extreme all weight is given to a few frames of the simulation and $\phi_{\text{eff}}(\boldsymbol{w}) \ll 1$, meaning that the force field is in poor agreement with data and/or sampling is poor. We note that other possibilities exist to estimate the effective size of the sample, e.g., the Kish formula[43,44]:

$$\mathcal{K} = \frac{\left(\sum_{i=1}^{N} w_i\right)^2}{\sum_{i=1}^{N} w_i^2}. \tag{11}$$

When a single frame is dominant over all the others, Eq. (11) returns $\mathcal{K} = 1$, while in a situation where no particular frame is preferred one has $\mathcal{K} = N$.

2.2 Maximum parsimony

The principle of maximum parsimony (MaxPars), also sometimes referred to as *Occam's razor*, is another strategy that can help choosing among several models consistent with the data. As we shall discuss later, many different interpretations of Occam's razor exist. In this section, the principle is interpreted as follows: the optimal distribution coherent with a given set of data is the one providing the smallest possible ensemble while still fitting the data. The basic idea is that if a model (ensemble) with few parameters (structures) can explain the data, there is no reason to further complicate the model by introducing more parameters. The above principle can for example be quantified by the Akaike information criterion[45]:

$$\mathrm{AIC} = 2n - 2\log L, \tag{12}$$

where the optimal solution is found by simultaneously maximizing the likelihood, L, that quantifies agreement with experimental data, and minimizing the number of model parameters, n. While the number n is effectively equal to the number of frames in the ensemble, it is customary to associate it to the number of nonzero weights, so $n \ll N$. For normally distributed experimental restraints one has $\log L = -\chi^2(\boldsymbol{w})/2$, so:

$$T_{\mathrm{MaxPars}}(\boldsymbol{w}, n) = \mathrm{AIC} = \chi^2(\boldsymbol{w}) + \theta n, \tag{13}$$

where $\theta = 2$ if Eq. (12) is directly applied. θ was introduced here to emphasize similarity with the MaxEnt methods, i.e., that the difference is in the form of the regularization term.

The Akaike information criterion was directly applied by Bowerman et al.[46] to find a minimal ensemble of tri-ubiquitin, using MD simulations and SAXS data. Bouma et al.[47], on the other hand, used θ as a free adjustable parameter to tune the strength of the MaxPars regularization term. In most methods however, the optimal n is found by increasing it by incremental steps and, for each step, finding the ensemble with lowest $\chi^2(\boldsymbol{w})$. The optimal n is then found when $\chi^2(\boldsymbol{w})$ has converged, either judged by manual assessment[48,49] or by an automatic convergence criterion,[50–52] thus avoiding having to set an explicit value for θ.

2.3 Bayesian inference or MaxPrior

An alternative method to MaxEnt and MaxPars is provided by Bayesian statistics.[23,37,53,54] While MaxEnt regularizes the negative log-likelihood by the entropy, and MaxPars by parsimony, Bayesian inference employs the prior for the same goal. For this reason we will refer to this approach as *MaxPrior.*

In Bayesian inference all available information is expressed by means of probabilities, as captured in Bayes' theorem:

$$P(\boldsymbol{w}|\mathbf{O}^{\mathbf{exp}}, \boldsymbol{\sigma}) = \frac{P(\mathbf{O}^{\mathbf{exp}}|\boldsymbol{w}, \boldsymbol{\sigma})}{P(\mathbf{O}^{\mathbf{exp}}, \boldsymbol{\sigma})} P(\boldsymbol{w}). \tag{14}$$

$P(\boldsymbol{w})$ is called the prior and it quantifies the information known about the system prior to the introduction of experimental data. In case of reweighting, it represents the prior probability associated to the weights, and typically comes from the distribution encoded in an energy function. The term $P(\mathbf{O}^{\mathbf{exp}}|\boldsymbol{w}, \boldsymbol{\sigma})$, known as the likelihood $L(\boldsymbol{w})$, represents the probability of measuring the experimental observables, given the set of weights and uncertainties. Assuming normally distributed errors, $L(\boldsymbol{w})$ is given as:

$$P(\mathbf{O}^{\mathbf{exp}}|\boldsymbol{w}, \boldsymbol{\sigma}) = L(\boldsymbol{w}) \propto \exp\left(-\frac{1}{2}\chi^2(\boldsymbol{w})\right). \tag{15}$$

Note that this is not the only possible expression for the likelihood. More complicated expressions, even nonanalytical ones, can occur if different types of data are combined or when sources of error cannot be assumed to be Gaussian.[55,56] The term on the left-hand side of Eq. (14) is known as the posterior and represents the probability of the weights after experimental data have been considered. Finally, the term in the denominator is treated as a normalization constant in which case Bayes' theorem takes the form of a proportionality relation:

$$P(\boldsymbol{w}|\mathbf{O}^{\mathbf{exp}}, \boldsymbol{\sigma}) \propto P(\mathbf{O}^{\mathbf{exp}}|\boldsymbol{w}, \boldsymbol{\sigma}) P(\boldsymbol{w}). \tag{16}$$

Given Eq. (16), Bayesian inference defines the optimal distribution by following a two-step procedure: first, the information from data is included, i.e., a likelihood function is defined; second, all other forms of data are included, i.e., a prior on the weights is provided. If we assume the employed force field to represent the best estimate of our prior knowledge on the system, then the prior probability $P(\boldsymbol{w})$ must decrease as $\boldsymbol{w}$ deviates from $\boldsymbol{w}^{\mathbf{0}}$. This observation can e.g., be quantified through the relative entropy $S(\boldsymbol{w})$ from Eq. (2),[32,57,58] in which case the prior takes the form:

$$P(\boldsymbol{w}) \propto \exp\left(\frac{\theta}{2} S(\boldsymbol{w})\right). \tag{17}$$

The factor 1/2 is introduced for the sake of simplicity of the final result. Other forms of the prior can be used. For example, Gaussian errors are used as priors in several reweighting methods,[59–62] while other methods assume a Dirichlet distribution for the weights.[63] If the relative entropy term is used as prior, Bayesian inference and MaxEnt approaches result in the same regularized expression. Indeed, by inserting the likelihood (Eq. 15) and prior (Eq. 17) into Bayes theorem (Eq. 16) and taking the negative logarithm, we obtain a regularized functional, $T_{\text{MaxPrior}}(\boldsymbol{w})$, that must be minimized to find the optimal distribution, $\boldsymbol{w}$:

$$-2\log P(\boldsymbol{w}|\mathbf{O}^{\mathbf{exp}}, \boldsymbol{\sigma}) \propto T_{\text{MaxPrior}}(\boldsymbol{w}) = \chi^2(\boldsymbol{w}) - \theta S(\boldsymbol{w}). \tag{18}$$

It is important to note that, differently from the case of MaxEnt where θ was empirically used to tune the strength of the regularization term, in a Bayesian framework the interpretation of θ is clear. Indeed, it accounts for the uncertainty on the weights, $\sigma_{w_i^0}$, and thus effectively on the force field, i.e., $\theta = (\sigma_{w_i^0})^{-1}$. This means that if the uncertainty on the weights, data and model were known accurately, θ could be exactly determined. Unfortunately, the uncertainty on the force field parameters, and thus on the weights, is generally unknown and θ must therefore be treated as a hyperparameter of the model. In practice, θ plays a double-role: it expresses our trust in the force field and at the same time it can be used to compensate for over- or underestimated experimental errors, as well as errors in the forward model.[23,64] See more on the determination of θ in Section 6.

The usefulness of Bayes theorem in refining structural ensembles from simulations and data is clear from the amount of studies on the subject, all including the word *Bayesian* in their title[23,31, 38, 39, 50, 57, 59–67] (see also the overview in Ref. 68). The methodology has become so widespread in the field, that one could argue that researchers should start stating in the title when Bayesian methods are not used, rather than the opposite.

We note also that different methods may differ substantially in how and the extent to which they apply the Bayesian formalism including whether priors are defined over all parameters and whether these are integrated out during the procedures.

Several methods combine MaxPars, MaxEnt, and MaxPrior approaches (see an overview in Table 1 in the review by Bonomi et al.[68]). Bayesian priors can, e.g., be constructed to prefer minimal ensembles,[63,69] thus combining MaxPrior with MaxPars. Additional information can also be used together with prior information from the simulation, e.g., secondary structure restraints, like in the case where MD simulated data are fitted into low-resolution cryo-electron microscopy density maps.[70] Therefore, the Bayesian approach should be seen more as a toolbox than an actual principle for selection among ensembles.

The Bayesian framework also provides an interpretation of the principle of parsimony, which is different from the Akaike information criterion, where the model with fewer parameters is considered the simplest one. In a Bayesian setting, the principle of maximum parsimony is a matter of *surprise*: the further a posterior result is from the prior, the more *surprising* it is. This leads to a Bayesian Occam's term: a measure of the distance between the reference distribution (prior) and the optimal distribution (posterior).[71,72] The principle is implicitly built into Bayes theorem. In that sense, the $S(\boldsymbol{w})$ term in the Bayesian inference and MaxEnt methods can be interpreted as an Occam's term, and both these methods could claim to fulfill the principle of maximum parsimony. However, to keep notation clear, we will use the Akaike interpretation of Occam's razor.

2.4 Comparing MaxEnt, MaxPars, and MaxPrior reweighting

Different interpretations of the concept of optimal distribution can lead to dramatically different conformational ensembles. Indeed, MaxEnt includes as many configurations of the initial ensemble as possible, while MaxPars tries to include only the ones that are strictly necessary to model the data. The behavior of MaxPrior, on the other hand, depends on the prior employed. In most approaches, however, a term similar to MaxEnt is used, preferring solutions that are consistent with the reference distribution,[68] so that MaxPrior will lead to solutions with as many frames as possible. Therefore, we will only distinguish between MaxEnt and MaxPars in the following, when discussing which method is more appropriate under what circumstances. To our knowledge, no systematic and direct comparison of the different reweighting methods have been made. However, the methods have some clear and important differences that will be discussed below.

2.4.1 Interpretation of the results

A relevant point of comparison between the methods is how easily the results are interpreted. The results of the MaxPars method are easier to interpret and visualize, as the optimal ensemble usually contains only 2–5 representative structures,[65,73] whereas the MaxEnt method may result in an optimal ensemble with thousands of structures, whereof many are similar.

2.4.2 General applicability

As argued by Bonomi et al.[68] and Ravera et al.[24], both methods can be used when the system's free energy landscape shows a few distinct and well-defined minima. In that case, the ensemble is represented well by a few structures, one for each free energy minimum and weighted by its corresponding depth (see Fig. 1). On the other hand, if the energy landscape representing the ensemble is flat or high-dimensional, i.e., in case of high-entropy systems, a few structures provide a poor description of the ensemble, even if they may fit the data. So in the case of high-entropy systems, the MaxEnt method should be preferred over MaxPars. A simplistic example of a high-entropy system is a regular die.[24,33,35] For an unbiased die, all six faces, which represent the states of the system, are equally likely. Let us consider the experimental observation that the average result of throwing the die is 3.5. The application of MaxPars to this problem would lead to the conclusion that the die is well described by two states only, e.g., states *3* and *4* with weights $w_3 = w_4 = 0.5$, or states *2* and *4* with weights $w_2 = 0.25$ and $w_4 = 0.75$, or any other pair of states consistent with the observation. This result is ambiguous, as many combinations are equally *optimal*, and it is also a poor description of the regular die. MaxEnt, on the other hand, would lead to a uniform distribution for all outcomes. More relevant examples of high-entropy systems are multidomain proteins with flexible linkers,[74] intrinsically disordered proteins,[75] and unfolded proteins,[76] which are all described poorly by a few distinct states. Thus, the MaxEnt method is more versatile, in the sense that it gives reliable results also for high-entropy systems.[24,68] The dimensionality of the system also plays an important role. For reasonably small systems, a description in terms of low-dimensional free energy landscapes might be possible and relevant. That is, a few structures can adequately represent the system. However, for high-dimensional biological systems the assumption that a few coordinates, or states, can properly describe the collective dynamics of the molecule might be ventured. Recent studies have highlighted how also the dynamics of some fast-folding proteins might be too complex to be represented in fewer than ~10 dimensions.[77] Therefore, we suggest that MaxEnt, or related methods, should be considered the method of choice when no safe assumptions can be made about the free energy landscape of the ensemble.

2.4.3 Imperfect force fields

The choice of the reweighting method also depends on the quality of the simulations used to generate the reference distribution $\boldsymbol{w}^0$. Indeed, the results of MaxEnt and MaxPrior (with, e.g., Gaussian[59–62] and Dirichlet priors[63]) reweighting will strongly depend on the quality of the simulation, because the optimal distribution is expected not to deviate too much from $\boldsymbol{w}^0$. On the other hand, MaxPars reweighting depends less on the quality of the initial distribution, as it freely picks the simulated structures that best represent the data. This decoupling from the initial simulation is both the strength and the weakness of the MaxPars method. It implies that, in principle, a realistic reference distribution is not necessarily preferred over a poor one, as long as some good representative structures have been sampled (Fig. 2). Consequently, approximate but efficient simulation

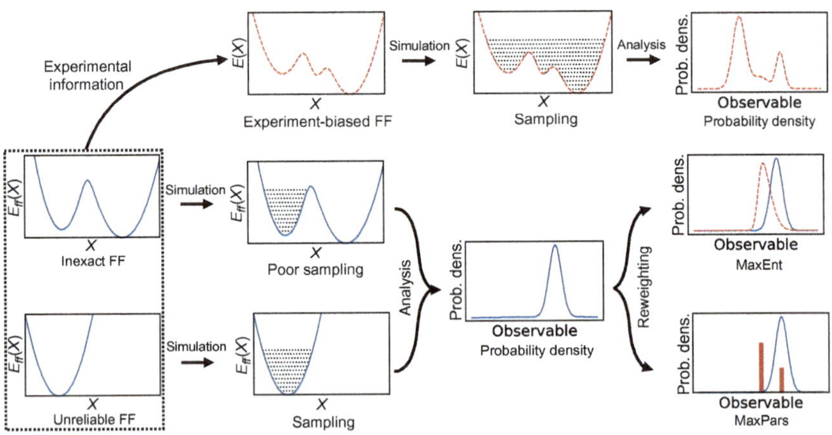

Fig. 2 *Lower two rows*: The two cases of poor sampling of an inexact force field and the one of full sampling of an unreliable force field can both lead to a result where the distribution of an observable of interest is poorly estimated. The performance of subsequent MaxEnt/MaxPrior and MaxPars reweighting differ, as a poor prior inevitably leads to a poor reweighted distribution in the case of MaxEnt/MaxPrior, while MaxPars can still select some relevant conformations from the initial unreliable pool. *Top row*: In both cases, experiment-biased simulations can help alleviate the problem.

methods, e.g., Monte Carlo sampling with implicit water, can be used in conjunction with the MaxPars method (e.g., Rosetta[78]), allowing for a fast exploration of the conformational space and better overall sampling without the need for enhanced sampling methods. Such approximate simulation methods are evidently less suitable for the MaxEnt and MaxPrior reweighting and therefore, more costly explicit solvent MD simulations should be employed. On the other hand, MaxPars does not benefit much from the qualified prior information of a good force field. We want to stress here that one of the main advantages of using accurate simulations as a prior is when the provided experimental dataset is sparse. In that case, the refined ensembles obtained from MaxEnt/MaxPrior are generally more reliable than MaxPars, as they are regularized by the simulated model.

2.5 Numerical challenges

Reweighting methods, as we presented them here, are essentially optimization problems. As such, they are cursed by the problem of dimensionality[79] or, in other words, the higher the dimension of the parameter space the more computationally challenging it becomes to find the true minimum of the functional of interest. Several strategies have been employed to reduce the computational burden of reweighting.

In MaxEnt, for example, the principal numerical challenge is to find the set of weights that minimize Eq. (5). This can become highly nontrivial when N is large, e.g., for tens of thousands of structures. The computational complexity can nonetheless be alleviated by optimizing the values of the M Lagrange multipliers (Eq. (6)) rather than the N weights, as usually $M \ll N$.[23,29,32] We note however that, depending on the implementation, the optimization of the N weights can be faster than the one of the Lagrange multipliers.[37] Moreover, the functional in Eq. (18) is strictly convex for finite θ,[80] so the corresponding optimization can be carried out in a straightforward fashion by means of highly efficient routines such as limited memory Broyden–Fletcher–Goldfarb–Shanno (L-BFGS) algorithm.[81]

In MaxPars, the numerically challenging part is provided by the test of the $N!/(n!(N-n)!)$ combinations of n-sized ensembles out of N structures. A naive approach where all the possible combinations are tested is intractable for realistic situations as, for example, in the case of $n = 5$ and $N = 1000$ the number of combinations becomes $\sim 10^{13}$. Therefore, the best fitting combinations of n structures has to be determined with some other strategies, e.g., via Monte Carlo minimization[82] or more complex algorithms such as the one implemented in ASTEROID.[83] Other strategies limit the search to the subset of structures that best match the experimental data individually,[52] which, however, violates the principle that the goodness of fit should only be assessed only against the ensemble average (Eq. 1), and not against the fit to single structures.

As a final remark, we note that both MaxPars and MaxEnt methods can benefit from clustering, where structures are grouped according to their structural resemblance, to reduce N before reweighting.[29,60,61]

3. Experiment-biased simulations

In Section 2 we focused on three possible principles, MaxEnt, MaxPars, and MaxPrior, to reweight the results of simulations a posteriori. However, as we argued in Section 1, this is not the only possible approach. Instead of changing the simulated ensemble after the simulation has been performed, the simulation can be guided by adding a bias to the underlying force field, $E_{ff}(X)$, employed in the simulation to ensure consistency with the experimental data. In this section we provide an overview of the methods that have been applied with this approach, which are either based on the MaxEnt principle (Section 3.1), on the addition of empirical energy terms (Section 3.2), or on the MaxPrior principle (Section 3.3). Finally, in Section 3.4 we compare and discuss the differences between experiment-biased simulations and a posteriori reweighting strategies, discussing their possible advantages and shortcomings.

3.1 Maximum entropy

In Section 2.1, we discussed how maximum entropy can be used to obtain consistency between data and simulations by optimizing a set of Lagrange multipliers (see Eq. (6) and Ref. 36). The mathematical formulation of the principle makes it immediate to extend it to the problem of experiment-biased simulations. Starting from an unbiased force field $E_{ff}(X)$ (Fig. 1), the idea is to perturb it as little as possible under the constraint/restraint of consistency with some given experimental data. Maximum entropy principle can be expressed in terms of the probability of a conformation X given the new force field $P(X) \propto \exp(-\beta E(X))$, where $\beta = (k_B T)^{-1}$ with k_B being the Boltzmann constant and T the temperature of the system, and the probability of the same configuration obtained from the unbiased force field $P_{ff}(X) \propto \exp(-\beta E_{ff}(X))$[34,41]:

$$S[P(X)] = -\int dXP(X)\log\left(\frac{P(X)}{P_{\text{ff}}(X)}\right). \tag{19}$$

Practical examples of this approach are provided by experiment directed simulations (EDS)[84] and experiment directed metadynamics (EDM).[85] The effective error from experiment and forward model can also be accounted for in the Lagrangian framework, thus only restraining the solution within the given uncertainty.[40,41]

3.2 Empirical energy terms

Rather than approaching with MaxEnt, experimental restraints can be directly included as empirical penalty terms[86]:

$$\begin{aligned} E(X) &= E_{\text{ff}}(X)+E_{\text{exp}}(\langle \mathbf{O}^{\text{calc}}\rangle, \mathbf{O}^{\text{exp}}, \boldsymbol{\sigma}) \\ &= E_{\text{ff}}(X)+\sum_{j=1}^{M}\theta_j h_j(\langle O_j^{\text{calc}}\rangle, O_j^{\text{exp}}, \sigma_j), \end{aligned} \tag{20}$$

where the sum is carried out over M restraints. θ_j are the force constants associated to each restraint, while the specific choice for the functional form of each h_j depends on the distribution of the corresponding observable. The constraints are usually expressed as squared residuals[39,86] or, when errors are taken into account, as χ^2 terms[68] (Eq. 7).

For concreteness and simplicity, we assume here normally distributed experimental observables. We also assume that the same force constant can be used for all observables. This is a good assumption, e.g., when the observables come from SAXS data,[38] but cannot generally be assumed when more than one experimental techniques are combined. With this assumption, the energy function reduces to:

$$E(X) = E_{\text{ff}}(X)+\theta\chi^2(\langle \mathbf{O}^{\text{calc}}\rangle, \mathbf{O}^{\text{exp}}, \boldsymbol{\sigma}). \tag{21}$$

Just as in reweighting approaches, parameter θ in Eq. (21) takes into account unknown uncertainties of the force field parameters as well as unknown or imperfectly determined errors in the data and the forward model. Therefore, θ is generally not known and determining it is one of the key challenges of the method.[23,37]

Another issue in the practical implementation of this approach is that the average $\langle \mathbf{O}^{\text{calc}}\rangle$ can only be determined *after* the simulation is over. Thus the restraints need to be applied iteratively.[35] As an alternative approach to obtain ensemble averages at each point in the simulation, replica methods have been introduced,[23], [87–89] where N independent replicas of the same system are simulated and observables are calculated as ensemble averages from these. Interestingly, replica experiment-biased methods mathematically converge to a maximum entropy *constrained* solution as $N \to \infty$ and $\theta \to \infty$, as discussed by Pitera and Chodera,[90] Roux and Weare,[34] and Cavalli et al.[91], and reviewed by Boomsma et al.[35]. Very recently, Köfinger et al.[37] combined reweighting and experiment-biased methods to simultaneously ensure that a large ensemble (as in MaxEnt) is considered and that all relevant states are visited by adding a biasing energy term to the force field.

3.3 Bayesian inference

Just as in the case of MaxEnt, Bayesian inference, or the MaxPrior principle (see Section 2.3), can be also employed to bias a priori molecular simulations rather than just reweighting them a posteriori. The method known as Metainference[39] is an implementation of this principle: it employs a Bayesian approach to quantify how much the prior is modified by the introduction of noisy and heterogeneous sources of data. In its essence, the strategy works by running N replicas of the system and guiding the sampling by means of a log-posterior scoring function:

$$s(X, \boldsymbol{\sigma}) = -\sum_{i=1}^{N} \log P(X_i, \sigma_i) + \Delta^2(X) \sum_{i=1}^{N} \frac{1}{2\sigma_i^2}, \tag{22}$$

where $\boldsymbol{\sigma}$ takes into account all the sources of errors, $P(X_i, \sigma_i)$ is the prior and the factor $\Delta^2(X)$ estimates the deviation between the predicted observables and the experimental ones. It can be shown that in the single-replica limit, $N=1$, Eq. (22) reduces to Eq. (18). Therefore, the log-prior plays the role of an effective entropy term, while the second component of $s(X, \boldsymbol{\sigma})$ is a χ^2 term computed over the replicas. It is interesting to notice here that the equivalence between Eq. (22) and Eq. (18) is valid only in the $\theta = 1$ case: this comes from the fact that the method proposes a model to take into account all the sources of error,[39] which means that θ is not expected to be a free parameter anymore. We notice that Eq. (22) is only valid for Gaussian error sources, and more complicated and complete expressions can be obtained in the general case (see the discussion in section Materials and Methods in Refs. 39 and 92).

Metainference has been effectively combined with Metadynamics[93–95] in its parallel bias formulation[92,96]: this synergy enables one to explore the configuration space in an efficient way while simultaneously sampling conformations that are coherent with experimental data.

3.4 Comparing reweighting with experiment-biased methods

3.4.1 Adaptability

Reweighting methods can be used with many different types of simulations and force fields, as the reweighting process is independent from the simulation and sampling (Fig. 1). This makes it a rather adaptable module-like tool, with the input being the trajectory and the experimental data only.[32,36,44] Also, new experimental data can easily be incorporated in the reweighting process without having to rerun the conformational sampling. In the experiment-biased simulations, the implementation is more specific, as the empirical energy term is an integrated part of the simulation.[38] Decisions about types of experiments and force field constants have therefore to be taken before the simulation. It is still possible, however, to reweight experiment-biased simulations a posteriori to remove or add sets of experimental data, though the procedure is technically somewhat more challenging than reweighting MD trajectories; indeed, the applied experimental bias has to be estimated and subtracted from the simulation before further reweighting.[23,44] For the same reason, experiment-biased simulations are often not carried out together with enhanced sampling techniques. A notable exception is provided by metadynamics,[97] which has been combined with metainference[39,92] to increase its efficiency, and in principle metadynamics (and other enhanced sampling methods) could be used in other methods for experiment-biased simulations.

3.4.2 Forward models

When a trajectory is reweighted a posteriori, the forward model is only evaluated on the frames that are to be reweighted, which are typically only a small fraction of the frames generated during the simulation. Also, as long as the observables are calculated according to Eq. (1), the calculations are done independently of one another and may thus be easily parallelized. For these reasons, the forward model can be of high complexity (e.g., quantum calculations can be carried out on the ensemble structures[98]). This is not the case, however, of experiment-biased simulations: when implemented within a molecular simulation, forward models are evaluated and differentiated at (almost) every step, so they have to be sufficiently simple to assure computational efficiency[44] and their gradients have to be known analytically. Therefore, for complex forward models reweighting might be more applicable than a priori experimental bias. In some cases, such as for NMR chemical shifts, it is possible to employ a fast forward model[99] that is almost as accurate as more refined and complex models,[100] which would be too complex to be computed at each step in a simulation. In other cases, it might not be possible to derive sufficiently computationally efficient and accurate forward models. We suggest that a possibility would be to use simpler and less accurate models to bias the simulations and then reweight a posteriori the simulation with the more realistic models.

3.4.3 Imperfect force fields

Reweighting methods and experiment-biased simulation methods may in practice perform differently in cases where the force field provides a relatively poor description of the system's conformational space. With a poor force field, some relevant states may be rarely or never visited when running an unbiased simulation (Fig. 2). Consequently, reweighting methods may fail in predicting the correct average of an observable. An indication that this is happening is usually provided by the fraction of effective frames ϕ_{eff} becoming close to 0 and by the reweighted distributions of relevant observables being skewed toward the experimental average (Fig. 2). In principle, this can be overcome by sufficient sampling, but it might be computationally very expensive and practically unrealistic. Experiment-biased force fields, on the other hand, can better provide reasonable results even when a rather poor unperturbed force field is used as basis, as the empirical energy term will alter the energy landscape such that even relevant states with high energies in the unbiased force field become reachable (Fig. 2). The effect of the added empirical energy term can be monitored by comparison with a control simulation without the additional energy term ($\theta = 0$ in Eq. (21)).

4. Force field optimization

In Sections 2 and 3 we described techniques to refine simulations in a system-specific manner by including experimental data. From a different perspective, substantial progress has been made when using experimental data to improve the force field as a general and transferable predictive tool. In this section we focus on this point and review some of the fundamental advancements in the subject.

4.1 Background on force field parametrization

As widely discussed in literature and assessed by empirical knowledge, the quality of the physical description provided by force fields is fundamental for the accuracy of biophysical simulations. The reliability of these simulations indeed depends critically on the ability of the underlying physical description to effectively model all the relevant interatomic interactions. After decades of force fields development,[101–104] MD simulations have reached a high level of reliability and the evergrowing amount of experimental observations calls for a systematic and detailed comparison of theoretical predictions, coming from MD simulations, with the available data.

Unfortunately, force fields do not always provide results that are in perfect agreement with experimental findings. To understand the reasons and the sources of these emerging discrepancies, we shall first recall how a force field is usually designed. For a comprehensive introduction on the subject we refer the reader to chapter "Pairwise-additive and polarizable atomistic force fields for molecular dynamics simulations of proteins" by Lemkul of this book, and here we instead focus on how experimental data may be used in force field parameterization.

The typical force field is composed by two fundamental elements: (i) A functional form $E(X)$; (ii) A set of parameters $\boldsymbol{\xi} = (r_0, \theta_0, \phi_0, k_b, k_\theta, k_\phi, \ldots)$. The first element embeds our physical understanding of molecular processes by providing a classical parametrization of the Born–Oppenheimer energy surface.[105] The typical functional form of the force fields used for biomolecular simulations (with minor readjustments among the different interpretations) is given by[106]:

$$\begin{aligned} E(X) &= E_{\text{bonded}}(X) + E_{\text{non-bonded}}(X) \\ &= \sum_{\text{bonds}} k_b (r - r_0)^2 + \sum_{\text{angles}} k_\theta (\theta - \theta_0)^2 + \sum_{\text{dihedrals}} k_\phi [1 + \cos(n\phi - \phi_0)] \\ &+ \sum_{i,j \in \text{non-bonded}} \left[\frac{A_{ij}}{r_{ij}^{12}} - \frac{B_{ij}}{r_{ij}^{6}} + \frac{q_i q_j}{r_{ij}} \right], \end{aligned} \tag{23}$$

where X is a configuration of the system and r, θ, and ϕ are functions of the atomic coordinates and q are atomic charges. k_b, k_θ, and k_ϕ denote the different strengths of the interactions and are usually tensors, as they depend on the specific group of atoms involved in the interaction. The functional expression of the force field depends on the choice of the set of parameters $\boldsymbol{\xi}$ which set, for example, the strength of interactions, equilibrium distances, and angles. To determine their values, it is necessary to fit them against known experimental and quantum mechanical (i.e., ab initio) properties. The specific choice of these properties depends on the philosophy underlying the force field development.

Historically, force field development has always been a daunting task because of its technical complexity and the amount of time, experimental data, and simulations required.[107–110] To give an example of this, let us focus on the AMBER class of force fields. In its first version,[104] bonded angles were fit to the vibrational frequencies of single amino acids or small molecules, in order to reproduce experimental frequencies; fixed charges were fit to reproduce the results of quantum calculations.[111–113] Lennard-Jones parameters, instead, were set in such a way to reproduces enthalpies of vaporization and densities in organic liquids.[101] Finally, dihedral and torsional angles were fit to reproduce quantum calculations of single amino acids or experimental barrier heights of small molecules. With the increasing quality of experimental and ab initio data, modern and widely used versions of the AMBER force fields[114] have reached a high level of complexity. Nonetheless, the underlying general philosophy remained the same: fit the force field parameters to ab initio and experimental data of single amino acids or small molecules and compare the results against data available for larger systems. Force field ff19SB, obtained by only employing ab initio simulations, constitutes a notable exception.[115] Despite its historical significance and great success, this procedure shows some important limitations: (i) it is difficult, within this approach, to improve the force field parameters by making use of discrepancies with respect to experimental data on large biomolecules; (ii) errors induced by the forward models used to calculate experimental observables are not consistently taken into account, and errors on the force field parameters are not estimated and/or not used. In the last decade, many new strategies for force field parametrization have been developed that less strictly follow the traditional approaches and instead embrace a more Bayesian approach.[116–124] These efforts, combined with the growing interest in automated force field parametrization (as exemplified by the development of the ForceBalance framework[121,122,124]), partially solve the issues reported in point (ii). A deeper discussion on point (ii) will be left for Section 6. In this section we focus on point (i) or, more precisely, describe and summarize the strategies available in the literature to optimize force field parameters using discrepancies between MD simulations and experimental observations on longer peptides or even entire proteins.

As we will see, these methods are tightly connected to common reweighting algorithms (see Section 2) but, when applied to a wide set of molecules, provide an answer of more general purpose than the sole ensemble refinement.

4.2 Refining protein and RNA force fields

NMR data have been widely and successfully used in combination with ensemble reweighting strategies of proteins[30,32,48,57,58,61,62,64,83,125–135]; other approaches employed various sources of experimental data, e.g., for small molecules. This fact opens to the appealing possibility to use NMR data on full proteins to directly optimize force fields. Such an approach is more general than ensemble reweighting, in the sense that experimental NMR data can be collected for a larger number of proteins and used to increase the quality of the force field rather than optimizing the description of a single system of interest. To understand how this could be possible in practice, we need first to realize that the general scheme of ensemble reweighting methods basically comprises of three steps: (i) generation of an ensemble of structures with a given force field; (ii) calculation of experimental observables from the ensemble; (iii) determination of weights, associated to each structure, that better describe the experimental observables.

If this procedure is repeated for a large set of proteins, the information carried by the optimized weights can be in principle used to increase the quality of the underlying physical model, i.e., the values of the force field fit parameters, rather than just the single molecule ensemble (Fig. 3). We note, however, that this approach is less flexible than, e.g., MaxEnt where the weights of all the simulation frames can be fine tuned. When optimizing a force field, one is inevitably constrained by its functional form and only limited adjustments can be made.

In the following we review the different strategies that have been developed to compute and use the weights in force field refinement for proteins (Section 4.2.1) and RNA (Section 4.2.2).

4.2.1 Proteins

Let us assume we can access a given set of N snapshots X_i obtained from a sampling strategy of choice (MD or Monte Carlo) and generated using a given force field E_{ff}, defined by a set of parameters $\boldsymbol{\xi}_0$. These snapshots can be used to calculate the ensemble average of some observables of interest, $\langle O_j^{\mathrm{calc}} \rangle$, for which we possess experimental information O_j^{exp}. The goal is to determine a set of force field parameters $\boldsymbol{\xi}$ that decreases the discrepancy between experimental and simulated averages. This discrepancy can be

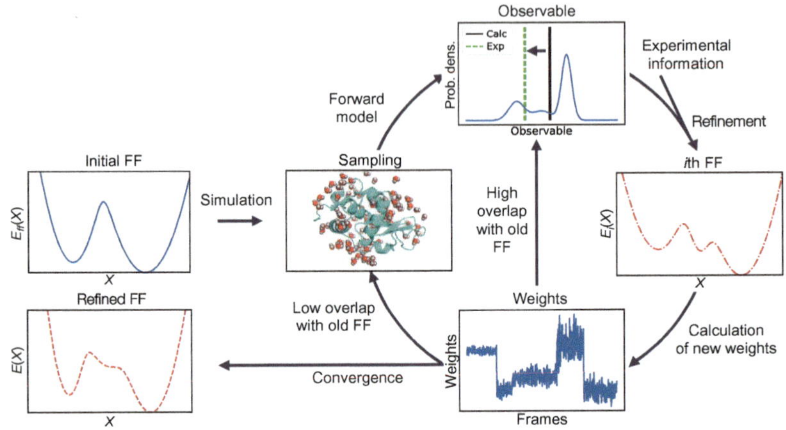

Fig. 3 Schematic representation of the force field optimization process. First, an initial force field $E_{ff}(X)$ is employed to sample an ensemble of configurations of some systems of interest. A forward model is then applied to the computed frames, to reproduce the known experimental average of a set of relevant observables. If the computed and the experimental observables disagree, further experimental information is used to refine the current force field parameters and defining a new force field $E_i(X)$. A weight is associated to each frame obtained from the previous sampling using Eq. (25). If the overlap between the old and the new force field is high (i.e., most of the weights do not change substantially) the new weights are used to refine the estimation of the observable's average, which in turn is used to further refine the force field parameters and to estimate new weights. If the overlap is instead small, a further round of conformational sampling is carried out with the new force field $E_i(X)$. The process is repeated until some convergence criterion is satisfied.

estimated via the χ^2 (Eq. 7), whose minimization with respect to the force field parameters would maximize the compatibility between simulations and the experiments of choice. The search for the χ^2 minimum can in principle be done in a brute force fashion, by infinitesimally perturbing the set of force field parameters by a quantity $\delta\xi_i$ and resimulating the system of interest. However, this strategy is particularly inefficient, as it corresponds to a search in a high-dimensional parameter space, where the evaluation of each new trial force field requires a full resampling of the system's conformational ensemble. To avoid this, it is possible to use some ideas coming from statistical mechanics. In principle, if we knew the *exact* force field E_{exact} describing our system, we could reweight each configuration using the Boltzmann relationship:

$$w_i^{\text{exact}} = w_i^0 e^{-\beta(E_{\text{exact}}(X_i) - E_{\text{ff}}(X_i))}. \tag{24}$$

Eq. (24) is the same idea behind the method of free energy perturbation[136] devised by Zwanzing to determine the changes in the free energy of a system when a perturbing potential is introduced. In this case, force field E_{ff} plays the role of the unperturbed potential of the system, while the exact force field E_{exact} is the contribution to the total energy provided by the perturbation. Even though E_{exact} can never be known, Eq. (24) can be used as the basis of an efficient optimization strategy.[116,137–139] In this method, a force field E_{old} is iteratively optimized to obtain a new one, E_{new}, that is used to recompute the weights of the parent MD simulation via:

$$w_i^{new} = w_i^{old} e^{-\beta(E_{new}(X_i) - E_{old}(X_i))}. \tag{25}$$

The new weights w_i^{new} are used to refine the estimation of the simulated average by means of Eq. (1). E_{new} force field is obtained by minimizing the χ^2, where the specific strategy for minimization can change (e.g., Levenberg–Marquardt procedure,[116,140] simplex minimization[137] or simulated annealing followed by simplex minimization[138]). The strength of the reweighting strategy in Eq. (25) is that it does not require one to sample new conformations every time the force field parameters are updated. Nonetheless, it only works under the assumption that minimal perturbations of the force field are introduced by the χ^2 minimization and therefore that the snapshots X_i could have been reasonably sampled also by means of the E_{new} force field. For this reason, a sensible overlap between E_{old} and E_{new} force fields is expected to exist, and particular care is dedicated to the assessment of this overlap,[138,141] which can be quantified, e.g., through ϕ_{eff} (Eq. (10)). A resampling of the full conformational ensemble needs to be repeated every time the overlap between the two force fields becomes too small: when $\phi_{eff} < \varepsilon$, with $\varepsilon \ll 1$ being a given threshold, one can argue that the overlap between force fields is too small to allow a further round of optimization. As described, this strategy is based on a method designed to compute free energy differences between different energy function. Thus, for future applications other and more general methods could be employed including Bennett Acceptance Ratio[142] and its multistate extension.[143] These methods may likely also be fruitfully combined with the use of (adaptive) surrogate models,[120] though work remains to be done to make such models accurate and efficient for high-dimensional problems.

Norgaard et al.[116] introduced the method discussed above to optimize a force field for coarse-grained simulations of unfolded proteins. Reference data were collected from paramagnetic relaxation enhancement NMR, useful to determine long-range effects in unfolded proteins.[144–147] In the first round of optimization, all the interaction parameters of the coarse-grained force field were set to zero for the sake of simplicity, while conformational ensembles were generated using Metropolis Monte Carlo. The method was applied to the Δ131Δ fragment of staphylococcal nuclease[144,148] to obtain reproducible and consistent results.

Li and Brüschweiler[137] applied the strategy from Norgaard et al.[116] to all atom simulations to derive the Amber ff99SBnmr1 force field starting from backbone dihedral angle potential of Amber ff99SB.[149] This was done by employing as observables the time-averaged chemical shifts of C_α, C_β, and C' carbon atoms of 4 trial proteins and subsequently benchmarking it against a test set of 18 proteins of different topologies. The obtained results show an average improvement of the comparison with the experimental data for the proteins in the test set, and was later refined further.[138]

More recently Chen et al.[139] extended the method from Norgaard et al.[116] to use Markov State Models (refer to Section 5.1 for more details on Markov State Models) as an intermediate step to calculate observables from simulations and parameterize the conformational landscape. Despite the different framework, the general philosophy of the method, named ODEM (observable-driven design of effective molecular models), remains unaltered. ODEM has been applied to design a $C_\alpha - C_\beta$ coarse-grained model of protein FIP35 which is able to reproduce relevant pair-distance distributions measured by FRET.

4.2.2 RNA

Molecular simulations may also be used to study the conformational landscapes and dynamics of RNA molecules. Unfortunately, the precision of RNA force fields is still limited and not comparable to the one reached by force fields designed for proteins.[150–153] The approach of integrative structural biology therefore becomes a powerful tool to enhance the comprehension of fundamental processed governing RNA dynamics and indeed the effectiveness of a posteriori reweighting on RNA simulations has been established in several studies.[32,154–157] The lack of a common strategy to increase the reliability of RNA force fields from first principles,[158–160] however, makes it particularly appealing to resort to approaches that exploit discrepancies between MD simulations and experimental data to refine force fields. Here we review a recent study[134] that tackles this problem by proposing a likelihood minimization scheme. Let us assume the system of interest is described by a force field $E_{ff}(X)$ and a corresponding Boltzmann probability distribution $P_0(X) \propto e^{-\beta E_{ff}(X)}$. For the same system, M experimental data have also been collected. In this work, the authors seek for an optimized probability density

$$P(X, \boldsymbol{\mu}) \propto P_0(X) e^{-\beta \sum_{i=0}^{N} \mu_i f_i(X)}, \tag{26}$$

where now the force field is expressed by an expansion on a basis $f_i(X)$, where each function is associated to a weight μ_i. We stress that weights $\boldsymbol{\mu}$ should not be interpreted as the previously introduced weights $\boldsymbol{w}$, as the former are arbitrarily normalized and not meant to be interpreted as probability densities. Each of the N terms helps in reproducing the M experimental constraints, but, differently from MaxEnt methods, $f_i(X)$ do not represent the forward model connecting configuration X to an experimental measurements. Rather, $f_i(X)$ can be generic functions and N is sought in such a way that $N \ll M$. While the analytic form of the basis functions is enforced at the beginning of the optimization process, the value of the weights μ_i needs to be determined through the minimization of a function describing the discrepancy between the simulated and the experimental observables. We note that in this strategy the functional form of the force field is actively modified by introducing basis functions $f_i(X)$ that are not explicitly included in $E_{ff}(X)$. Alternatively, this can be seen as a way to associate nonzero weights to terms in the force field that have an effective null coupling constant. If functions $f_i(X)$ are already part of the potential, instead, the corresponding weights μ_i amount to a refinement of the interaction strength. Supposing we are interested in M observables O_j^{calc}, the target for the minimization takes the form of a regularized error function

$$T(\boldsymbol{\mu}) = T(\langle O_1^{\text{calc}} \rangle(\boldsymbol{\mu}), \ldots, \langle O_M^{\text{calc}} \rangle(\boldsymbol{\mu})) + \theta |\boldsymbol{\mu}|^2 \qquad \theta \geq 0, \tag{27}$$

where the averages $\langle O_j \rangle(\boldsymbol{\mu})$ are computed in the refined ensemble

$$\langle O_j \rangle(\boldsymbol{\mu}) = \frac{1}{Z} \int dX \; O_j^{\text{calc}}(X) P(X, \boldsymbol{\mu}). \tag{28}$$

The error function is designed to enforce both equalities and inequalities, i.e., $\langle O_j^{\text{calc}} \rangle(\boldsymbol{\mu}) = O_j^{\text{exp}}$ and $\langle O_j^{\text{calc}} \rangle(\boldsymbol{\mu}) < O_j^{\text{exp}}$ and the optimal set of weights $\boldsymbol{\mu}$ is obtained as the one minimizing T. The strength θ of the regularization term is key to the strategy: for $\theta \rightarrow \infty$ the error function does not feel the contribution of the experimental constraints and thus the potential resulting from the optimization is just the original one $E_{ff}(X)$. Instead, for $\theta \rightarrow 0$ the deviations from the original potential are not restrained, and the prior information from the simulations is effectively ignored. Therefore, an optimal value of θ has to be determined or alternative statistical tools have to be applied to integrate out the variable. In this work the authors employ cross-validation to determine the optimal θ, but this is a general issue common to many ensemble optimization methods. We refer therefore to Section 6 and references therein for further details on the choice of the optimal prior strength.

The RNA systems chosen for this application were four tetranucleotides and two tetraloops, and NOEs (nuclear overhauser effect) as well as scalar coupling NMR data were used in the force field refinement procedure. The basis functions were chosen to be sines and cosines acting on torsional angles, while the employed guess force field was Amber ff99bsc0 + χ_{OL3} with the OPC water model.[104,161–163]

5. Matching time-dependent and time-resolved data

When dealing with systems at equilibrium, the average of an observable $\langle O \rangle$ is consistent with many possible distributions of the same observable $p(O)$. As we discussed in Section 2, MaxEnt, MaxPars, and MaxPrior principles can help discriminate, among the several possibilities, which one is the optimal distribution. Suppose, however, we can access a good description of some relevant conformations of a molecule (by experiments or previous sampling) and we are now interested in understanding the dynamics of interconversion between them. From an MD perspective, one can hope to see interconversions happening by running several trajectories starting from the relevant configurations. If extensive sampling is achieved and enough state transitions are captured, one can, for example, use the trajectories to build a Markov State Model for the system, and obtain the desired information about the interstate dynamics.[164–166] However, full sampling of the conformational dynamics might be challenging to achieve, especially when the desired motions happen in the μs timescale or above. Moreover, as discussed in detail in Section 4, limitations in the force field employed in the simulations might lead to conclusions which disagree with key experimental data, even in the case of excellent sampling. This may be the case, for example, for NMR spin relaxation experiments: spin relaxation rates are sensitive to both short- and long-range dynamics of a protein, occurring from the picosecond to the nanosecond timescale.[167] For dynamical observables like NMR spin relaxation rates, methods developed to increase the consistency between computed and experimental values of static observables might be of only limited help and different strategies are needed. In this section we describe some recent efforts in this direction and how it is possible to use simulations to match experimental knowledge on *time-dependent* observables (i.e., observables that depend on time because of fast processes happening in the system and cannot therefore be expressed by means of Eq. (1), e.g., NOEs with spin diffusion[135]) and *time-resolved* ones (i.e., observables that are at equilibrium locally in time, but have been monitored for long timescales and can therefore be expressed as in Eq. (1) by including a time label, e.g., time-resolved SAXS). In the following we will focus on applications concerning maximum entropy and Likelihood estimation applied to Markov State Models[168,169] (Section 5.1), the principle of Maximum Caliber[170] (Section 5.2) and Average Block Selection[171] (Section 5.3) (Fig. 4).

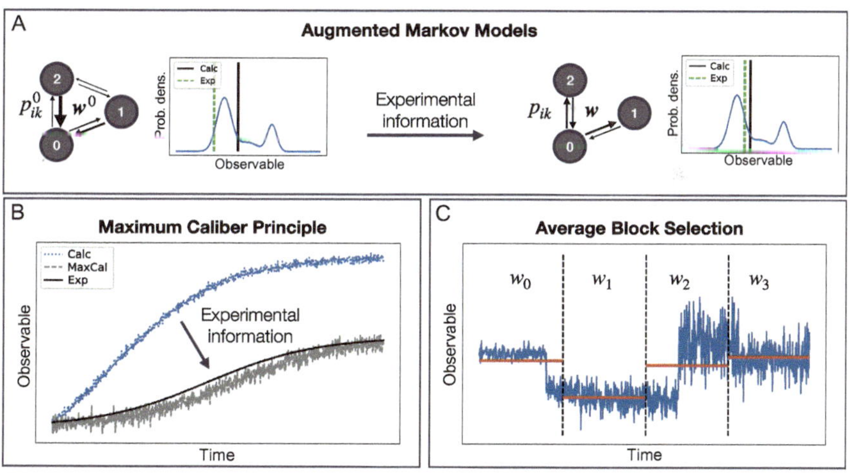

Fig. 4 Graphical summary of methods implemented to recover time-dependent and time-resolved data. (A) Augmented Markov Models employ experimental information to refine a guess Markov State Model and consequently better estimate kinetic observables of interests; (B) Maximum Caliber, in its experiment-biased formulation, is employed to bias the sampling to match some time-resolved experimental quantities; (C) Average Block Selection is used to assign weights to subtrajectories and better estimate time-dependent experimental data.

5.1 Maximum entropy and likelihood in dynamical systems

In a kinetic model, the equilibrium distribution $\boldsymbol{w}$ can be determined from a transition probability matrix $T(\tau)$, i.e., a matrix whose elements p_{ik} provide the probabilities for the system to be found in state k at time $t+\tau$, given it was in the state i at time t. $T(\tau)$, together with the structures of the states among which transitions happen, is usually known as a *Markov State Model* (MSM).[165,168] In this section, we will discuss a method to refine a MSM by adding static (not time-dependent or time-resolved) experimental information. This framework, that balances data coming from simulations and experimental averages, is called Augmented Markov Model (AMM).[172]

The construction of an AMM starts from the definition of an initial MSM, $T^0(\tau)$, with transition probabilities p^0_{ik}. The MSM is typically built by following a multistep procedure[165,168]: (i) relevant features (e.g., dihedral angles, contact maps) of the system are selected. The dynamics of these features needs to play a key role in the conformational transitions under consideration, so their identification might not be trivial. This task can become easier with the help of automatic selection tools[173,174]; (ii) the features are used to generate a lower-dimensional representation of the system's dynamics[175,176]; (iii) the lower dimensionality[177] allows one to cluster similar configurations in the dimensionally reduced space. The N clusters will define the states of the model; (iv) finally, statistical tools[168] are employed to estimate the transition matrix $T^0(\tau)$. Among other possibilities, the transition probability matrix can be obtained by maximization of the likelihood[165,168]:

$$L(T^0) \propto \prod_{i,k} (p^0_{ik})^{c_{ik}}, \tag{29}$$

where c_{ik} is the number of transitions occurring between state i and k, which are obtained by explicitly counting the transitions between states in the simulation.[178,179] The resulting transition matrix has to satisfy further criteria, e.g., it has to be row- or column-stochastic, and detailed balance can be enforced explicitly. Maximization of Eq. (29) is equivalent to minimizing the negative log-likelihood:

$$S(T^0) = -\sum_{i,k}^{N} c_{ik} \log p^0_{ik}, \tag{30}$$

We note that while Eq. (30) resembles an entropy term, does not represent a proper one, as the counts c_{ik} enter Eq. (29) unnormalized. However, $S(T^0)$ can be used in a similar way as a regularizing prior when including experimental information, as we describe further below. The equilibrium distribution of the system, usually referred to as the stationary probability distribution of the MSM, is obtained from the transition probability matrix by solving the eigenvalue equation $T^0(\tau)\boldsymbol{w}^0 = \boldsymbol{w}^0$. For more information about Markov State Models, we refer the reader to some excellent reviews on the subject.[164–166,168,180]

Given the initial Markov State Model $T^0(\tau)$, the framework of Augmented Markov Models employs the MaxEnt principle and likelihood maximization to build an optimized MSM, $T(\tau)$. Let us assume we can access both the expectation values of M experimental observables and an MSM built on simulation data, with the reference stationary distribution $\boldsymbol{w}_0$. To obtain the optimal distribution $\boldsymbol{w}$, one can apply the MaxEnt principle, as reported in Eq. (6), which bridges the model distribution with the experimental one by means of the provided experimental averages. Rather than estimating the weights corresponding to the optimized distribution, $\boldsymbol{w}$, the MaxEnt formulation allows one to optimize the set of Lagrange multipliers $\boldsymbol{\lambda}$ (Eq. (6)). As usual, the Lagrange multipliers are obtained by enforcing constraints on the experimental averages. In order to account for the statistical errors in both sampling and experiments, one can introduce the so-called *Augmented Markov Model likelihood*, assuming Gaussian errors:

$$L(\boldsymbol{w}, T) \propto \left(\prod_{i,k} (p_{ik})^{c_{ik}} \right) e^{-\frac{1}{2}\chi^2(\boldsymbol{w})}. \tag{31}$$

The term in the parentheses is the MSM likelihood, Eq. (29), while the second term incorporates the Gaussian error model proposed for the observables. The corresponding negative log-likelihood is thus proportional to:

$$T_{\mathrm{AMM}}(\boldsymbol{w}, T) = \chi^2(\boldsymbol{w}) - \theta S(T), \tag{32}$$

where $S(T)$ is the entropy-like term introduced in Eq. (30). Note that $\boldsymbol{w}$ is uniquely determined by T by solving the eigenvalue equation $T(\tau)\boldsymbol{w} = \boldsymbol{w}$, so Eq. (32) is effectively a minimization problem for the matrix elements p_{ik} alone. We stress that we have introduced Eq. (32) only to show the similarity with the reweighting methods for equilibrium ensembles, Eqs. (9) and (13), and that it does not directly represent the way AMMs are practically implemented. Rather, one obtains an AMM that optimally balances between the information from simulation and the experimental data (Fig. 4A) by employing fixed-point iteration algorithm to maximize

Eq. (31) with respect to the unknowns p_{ik} and the weights $\boldsymbol{w}$ (see the supplementary information of Ref. 172 for further details). Olsson et al.[172] applied this method to two 1 ms simulations of ubiquitin. They built AMMs using NMR scalar and residual dipolar couplings, i.e., static data, and their results were compared against NMR relaxation dispersion data. The results of such experiments can either be calculated directly from a long molecular dynamics simulation[18,181] or indirectly from a MSM[182]. The authors found that the AMM that had been optimized against the experimental data was in overall better agreement this independent data compared to the raw MSM. This is a strong indication of a nontrivial fact: reweighting an MSM with respect to experimental equilibrium observables helps increase the reliability in the prediction of time-dependent data.

5.2 Maximum Caliber

The principle of Maximum Caliber (MaxCal) can be regarded as a generalization of MaxEnt and enables one to compute the probabilities associated to dynamical pathways, rather than probabilities (weights) associated with equilibrium states. To infer pathway probabilities, the MaxCal principle maximizes a *path entropy*[183] defined over all the possible pathways, constrained to reproduce a given dynamical observable. The path entropy is defined as:

$$S[p_0(\gamma)] = -\sum_{\gamma} p_0(\gamma) \log p_0(\gamma), \tag{33}$$

where $p_0(\gamma)$ is the probability that the system follows the structural path γ. Each path is considered as a collection of configurations, $\gamma = \{X_0^\gamma, \ldots, X_T^\gamma\}$, labeled by a discrete time index $t = 0, \ldots, T$. Maximization of Eq. (33), together with M constraints of the form

$$g^j[p_0(\gamma)] = 0 \quad j = 1, \ldots, M \tag{34}$$

yields the optimal distribution $p(\gamma)$ over pathways. Just like in the MaxEnt case, constraints are usually enforced by Lagrange multipliers, and typically they concern the calculated dynamical observables $O^{\mathrm{calc}}(X_t^\gamma)$ and corresponding experimental values at the same time point t, O_t^{exp}:

$$\sum_{\gamma} p_0(\gamma) O^{\mathrm{calc}}(X_t^\gamma) - O_t^{\mathrm{exp}} = 0. \tag{35}$$

The normalization of the probability distribution of pathways is also enforced:

$$\sum_{\gamma} p_0(\gamma) - 1 = 0. \tag{36}$$

If it is possible to introduce an estimate for the time-dependent error σ_t, then the notion of χ^2 can be extended to a time-dependent form as well, $\chi^2 = \chi_t^2$. In this way, by analogy with MaxEnt, the MaxCal principle can also be expressed in a regularized fashion as

$$T_{\text{MaxCal}}[p_0(\gamma)] = \chi_t^2[p_0(\gamma)] - \theta_t S[p_0(\gamma)]. \tag{37}$$

where the parameter θ of Eq. (9) acquires a dependence on time. All the terms in Eq. (37) are functionals of the probability density in the space of pathways: as such, the minimization of $T_{\text{MaxCal}}[p_0(\gamma)]$ would require a search in path space, which is knowingly a hard task.[184–187] To our knowledge, indeed, the MaxCal principle has never been used in the acceptation of Eq. (37). Instead, the principle of Maximum Caliber can be employed to run restrained MD simulations, where restraints are provided by time-resolved experimental data (Fig. 4B).

Note, when comparing simulations with time-resolved experiments, a single simulation cannot be used to generate an ensemble as in the MaxEnt method (Section 2), due to the nonequilibrium conditions. Therefore, replicas are needed to describe ensemble development over time (see also Section 3). Capelli et al.[188] introduced such replica-averaging implementation of MaxCal, so ensembles can be determined at each time t. In this formulation, Eqs. (35) and (36) are employed to constrain the minimization of the path entropy, together with the path probability density normalization and two more equations. The first one constrains the system's diffusion constant D:

$$\frac{1}{2\Delta t}\sum_{\gamma} p_0(\gamma)\left[X_{t+1}^{\gamma} - X_t^{\gamma}\right]^2 - D = 0, \tag{38}$$

while the other constrains the standard deviation σ_t of the observable of interest at time t, obtained by averaging it over all the N replicas

$$\sum_{\gamma^i} p_0(\gamma^i)\left(\langle O^{\text{calc}}(X_t^{\gamma,i})\rangle - O_t^{\text{exp}}\right)^2 - \sigma_t^2 = \sum_{\gamma^i} p_0(\gamma^i)\xi_t^2 - \sigma_t^2 = 0, \tag{39}$$

where $\sum_{\gamma^i}$ is used to specify that the sum is carried out over all the possible pathways of all replicas and the average $\langle\cdot\rangle$ is computed over replicas. The expression of the path entropy, together with all the constraints (in this particular case, Eq. (35), (36), (38) and (39)) imposed via Lagrange multipliers is called *caliber*. The result of caliber maximization provides the optimal distribution of pathways:

$$p(\gamma) = \frac{1}{Z}\exp\left[-\sum_{t,i}(\nu_t^i\left[X_{t+1}^{\gamma,i} - X_t^{\gamma,i}\right]^2 + \lambda_t^i O(X_t^{\gamma,i}) + \mu_t^i \xi_t^2)\right], \tag{40}$$

where Z is a normalization factor, ν_t^i, λ_t^i, and μ_t^i are replica- and time-dependent Lagrange multipliers. Despite the complicated expression of the optimal path probability distribution, it is possible to prove that, in the assumption where the system's dynamics is Brownian and by analogy with MaxEnt,[188–190] the MaxCal distribution in Eq. (40) can be sampled by adding a time-dependent, harmonic bias potential to the underlying force field:

$$E_{\text{exp}}(\gamma^i, t) = Nk\left(\langle O^{\text{calc}}(X_t^\gamma)\rangle - O_t^{\text{exp}}\right)^2, \tag{41}$$

where the constant k is used to tune the strength of the interaction. This approach was applied to the case of the second hairpin of protein G B1 domain by employing synthetically generated time-resolved SAXS data.[188]

Rather than using MaxCal principle as an experiment-biasing technique, it is also possible to employ it for reweighting, in particular for the case of MSMs.[191–193] Suppose we built a MSM from a set of trajectories and we want to predict how a given nonequilibrium observable O_{ik} changes in the transition between state i and state k. A possible example would be the change in a spectroscopic signal (e.g., Förster resonance energy transfer or circular dichroism) when passing from a partially unfolded state i to a helical state k of a protein. The average, computed over many transitions, is provided by:

$$\langle O^{\text{calc}}\rangle(\boldsymbol{w}) = \sum_{i,k} w_i^0 p_{ik} O_{ik}^{\text{calc}}, \tag{42}$$

where $\boldsymbol{w}^0$ is the stationary probability distribution of the MSM. Let us suppose we also possess experimental information of such observable, O^{exp} and that the computed average does not match the expected result. In this case, we can use the MaxCal principle to select among the models with the correct average.[192] To do so, first the entropy is built as:

$$S(T, \boldsymbol{w}) = -\sum_{i,k} w_i p_{ik} \log\left(\frac{p_{ik}}{p_{ik}^0}\right), \tag{43}$$

where $\boldsymbol{w}$ is the optimal stationary distribution and p_{ik} are the refined transition probabilities. Note that Eq. (43) is a version of Eq. (33) in the case of discrete pathways. The entropy has to be maximized together with the constraint that the function

$$g(\boldsymbol{w}) = \langle O^{\text{calc}}\rangle(\boldsymbol{w}) - O^{\text{exp}} \tag{44}$$

is zero, where $\langle O^{\text{calc}}\rangle(\boldsymbol{w})$ is the average over transitions introduced in Eq. (42). Moreover, w_i and p_{ik} are interdependent because of probability conservation, so three more constraints naturally emerge:

$$\sum_i w_i p_{ik} - w_k = 0 \quad \sum_k w_i p_{ik} - w_i = 0 \quad \sum_{i,k} w_i p_{ik} - 1 = 0. \tag{45}$$

The minimization of the entropy in Eq. (43), with the constraints in Eq. (44) and (45), can be carried out with the method of Lagrange multipliers. It yields:

$$p_{ik} = \eta \phi_k \phi_i^{-1} M_{ik}(\lambda, p_{ik}^0, O_{ik}^{\text{calc}}), \tag{46}$$

where $M_{ik}(\lambda, p_{ik}^0, O_{ik}^{\text{calc}})$ is a non-Hermitian matrix, ϕ is its only right-eigenvector having only positive elements (whose existence and uniqueness is guaranteed by the Perron–Frobenius theorem), η the corresponding eigenvalue and λ is the Lagrange multiplier controlling the constraint in Eq. (44). The positivity of the elements of vector ϕ is necessary to guarantee the positivity of the transition probabilities. Eq. (46) provides a new set of transition probabilities defining an optimal transition probability matrix $T(\tau)$. Effectively, similarly to what happened in the case of AMMs, the introduction of experimental information actively modifies the transition probabilities between states, and interconversions can become more or less favorable depending on the cases. Finally, the optimal stationary distribution is obtained as:

$$w_i = \psi_i \phi_i, \tag{47}$$

where ψ is the left-eigenvector of matrix $M_{ik}(\lambda, p_{ik}^0, O_{ik}^{\text{calc}})$ corresponding to the eigenvalue η. Matrix $M_{ik}(\lambda, p_{ik}^0, O_{ik}^{\text{calc}})$ only depends on the Lagrange multiplier λ, which is used as a parameter, and on quantities that can be readily computed from the initial MSM, i.e., the transition probabilities and the observables of interest. Therefore, the optimal transition probabilities and equilibrium probability distribution can be obtained by single value decomposition of $M_{ik}(\lambda, p_{ik}^{0}, O_{ik}^{\mathrm{calc}})$ for different values of λ: the optimal Lagrange multiplier will be the one for which the constraint in Eq. (44), estimated a posteriori from the updated MSM, holds true.

In the original work, the method was tested on a toy model for a growth factor activation pathway, showing how the incorporation of experimental information can help correcting the transition probabilities in a guess MSM. Despite the method, as illustrated here, does not incorporate experimental errors, it is possible to take them into account in a Bayesian fashion.[192]

5.3 Average Block Selection

Time-dependent data can also be applied directly in the ensemble refinement by means of time-dependent Average Block Selection. Suppose we have N MD trajectories, started from different conformations, and a set of M time-dependent experimental observables O_j^{exp}. As anticipated at the beginning of this section, by time-dependent we here mean observables that cannot be calculated using Eq. (1) because the values depend on the underlying dynamics as well as on the stationary distribution. To increase the accuracy of the MD predictions, given the experimental data, and preserve information about dynamical processes, Salvi et al.[171] proposed to divide each trajectory into B blocks representing subsequent time-windows. The total number of blocks is then $N \times B$, each one associated to a weight w_b. The block-weights are then optimized by minimizing the residual sum:

$$\mathcal{R}^2(\boldsymbol{w}) = \sum_{j=1}^{M} \left(O_j^{\mathrm{exp}} - \sum_{b=1}^{N \cdot B} w_b O_{bj}^{\mathrm{calc}} \right)^2, \tag{48}$$

where O_{bj}^{calc} is the jth observable computed from the bth block. In later studies, the authors included errors in the expression by the usual $\chi^2(\boldsymbol{w})$ expression.[194] Division in blocks allows one to determine how much each block contributes to the time-dependent experimental signal. For example, blocks with null weights are excluded and can be interpreted as nonphysical artifacts. Despite the similarity with other approaches used for equilibrium observables, this method shows two important differences: on the one hand, weights are not associated to a single structure of an ensemble, but rather to subtrajectories (Fig. 4C). It is evident, then, that the method can be effective only in the case where the experimental timescales of interest can be sampled within a single trajectory. On the other hand the method as it stands is unregularized, i.e., it assumes that the optimal solution is obtained by minimizing the sum of residuals without adding any restraint. This is a crucial difference with respect to MaxEnt, MaxPars, and MaxPrior strategies: one would expect nonetheless the method to benefit from regularization terms. The addition of such terms would be advisable in further applications, but it is not clear what kind of prior would be needed in this case. Because of the fact that Eq. (48) associates weights to pathways and not conformations, it might be possible to explore a connection with a posteriori reweighting using MaxCal principle.

In its original formulation, the method (called ABSURD—Average Block Selection Using Relaxation Data)[171] was designed to deal in particular with NMR spin relaxation rates, but its application can be extended to any experimental source sharing similar timescales in a straightforward way. ABSURD was applied to the C-terminal domain of the nucleoprotein of Sandai virus. A cumulative time of approximately 6.5 μs of MD was sampled, employing two different water models. While the reproduction of spin relaxation rates by MD simulations alone was imperfect, the ABSURD-optimized trajectories were found in much better agreement with relaxation data on a wide spectrum of timescales, even in the simplest case where a single experimental rate was employed in the optimization.

6. Challenges

In Sections 2–5 we focused on major efforts that have been done to merge information coming from experiments and simulations to refine conformational ensembles or improve existing force fields. Despite many important steps forward in the last decade, some major challenges remain. In this section we will point out some of these and try to draw a possible paths for solutions. In Section 6.1 we will focus on the problem of setting the parameter θ in a robust way. In Section 6.2 we will discuss the option of combining force field corrections and reweighting for a given system. In Section 6.3 we discuss possibilities and obstacles to employ kinetic data to reweight equilibrium ensembles. Finally, Section 6.4 is dedicated to a discussion on the next generation of force fields.

6.1 Balance between simulations and experimental data

In the MaxEnt and MaxPars reweighting methods, and equivalent methods biasing on the fly, the prior knowledge coming from simulations is balanced against experimental data by tuning the parameter θ (Eqs. (9), (13), and (21)). In principle, the balance could be exactly known if the effective uncertainties on the weights, force field, sampling, and on the both the calculated and experimental observables were known (as also discussed in Section 3.3). However, the uncertainties on the weights are usually unknown and the experimental uncertainty stems from different sources, only some of which are typically known or easy to estimate. We can subdivide the sources to the experimental uncertainty in the following categories: (i) Statistical errors on experimental data; (ii) statistical errors in the calculation of average observables from a limited amount of structures (Eq. 1); (iii) systematic errors on experimental data; (iv) inaccuracy of the forward model. For normally distributed errors, these add up to a total variance.[23] In the following we will shortly discuss each of them.

The statistical error on experimental data is usually estimated by repetitive measurements and counting statistics, and it is included in most approaches, e.g., as standard deviation in the χ^2. Over- or underestimated errors may be detected by visual good fits having χ_r^2 much lower or higher than unity.

The statistical error on the mean of the observables calculated from the ensemble can easily be taken into account.[68] This is important when the ensemble is small, which is typically the case for MaxPars methods and experiment-biased approaches with replicas.

Systematic errors in experimental data are generally difficult to take into account, as their magnitude and nature are usually unknown and may be system specific (e.g., incorrect buffer subtraction in SAXS[38]). Shevchuk and Hub[195] treated the systematic errors with Bayesian statistics as a *nuisance* parameter, i.e., an unknown (and uninteresting) parameter that should be determined together with the model parameters. Bonomi et al.[39] discussed the more general case of outliers, and showed that an approach that combines reweighting and experiment-biased force fields is more robust against outliers than other related methods. Similar methods are also discussed by Köfinger et al.[37]

Inaccuracies in the forward model are likewise highly nontrivial to estimate. While negligible in some cases, they are the dominant error source in others, e.g., NMR chemical shifts, where they can be orders of magnitude greater than the statistical error. SAXS forward models share similar problems.[16] The source of the model inaccuracy might also come from neglecting an intrinsic time-dependency of the observable (e.g., spin diffusion or dynamic effects in the estimation of NOEs[135]).

In summary, as long as errors in the force field, systematic errors in the experimental data, and errors on the forward model continue to be challenging, if not impossible, to estimate accurately, the parameter θ remains necessary as an effective scaling of the total error σ. A key challenge is therefore to determine it. A simple way is to tune θ until χ_r^2 reaches unity (Eq. 8). This is, however, not generally a valid approach.[196] First, the number of degrees of freedom ν is ill-defined when reweighting is concerned. Conventionally, ν is estimated as $M - k$, where M is the number of data points and k is the number of fitted parameters. In MaxEnt, for example, k is given by the number of Lagrange multipliers, $k = M$, therefore ν is effectively zero and χ_r^2 has consequently no meaning. Second, although the expectation value of χ_r^2 is one, point estimates are typically different from unity and χ_r^2-distributed. Therefore, this method may provide incorrect balance between data and simulation. Another strategy to determine θ is to plot $S(\boldsymbol{w})$ against $\chi^2(\boldsymbol{w})$ and look for an *elbow* in the curve.[32,37] Indeed, if plotted with double-logarithmic axis, $\chi^2(S)$ often becomes an L-shaped curve, and the optimal θ can be estimated by finding the kink of the curve.[197] A third strategy is cross-validation, i.e., fitting the optimal ensemble to a separate set of data that has not been used in the analysis.[61,86,134,198,199] The reweighting or experiment-biased simulation may then be done for several values of θ to monitor when the goodness of fit to the unused data starts to decrease. While intuitively appealing, the practical implementation may be difficult. First, the data needs to be divided into independent subsets, which may sometimes be difficult for highly correlated or interdependent data. Second, as different sources of data may report on very different aspects of the system, they may in practice not be useful for cross-validation.[199] A fourth strategy for determining θ is strictly Bayesian, and θ is in that context treated as a nuisance parameter. The posterior probabilities at each value of θ are calculated and used to find the most probable value of θ,[72] or to integrate out θ completely.[23]

6.2 Interplay between reweighting and force field corrections

In Section 3.4 we mentioned that reweighting might fail in reproducing the experimental averages when the prior (force field) used is too inaccurate, because relevant states are either poorly sampled or not sampled at all. A pragmatic, and potentially transferable, solution to this problem would be to identify the parameters in the force field responsible for the incorrect behavior and modify them slightly in order to move closer to the expected averages. Such rescaling approaches have proved successful for specific systems and force fields, e.g., for adjusting the protein–protein interaction strength in the coarse-grained Martini force field[200,201] or protein–water interactions for simulations of disordered proteins.[202] A subsequent reweighting of the obtained trajectories might correct for inconsistencies that are not related to the identified and rescaled parameter, and could therefore lead to better consistency with the experimental data.

Applying such approaches, however, raises some relevant questions. For example, to what extent should one correct the force field? Considerable modifications to a force field would typically require the need to iteratively resample the system under consideration, an exercise that might become computationally expensive. Moreover, after any substantial reparametrization the force field should be benchmarked against other data and ab initio calculations, as done in the original parametrization of the force field. Therefore, should one perhaps modify the force field just enough to allow for reweighting? And, more generally, should force field corrections and reweighting be employed together at all? Standard force field reparameterization effectively corresponds to reweighting a conformational ensemble, but the extent is limited by the functional form of the energy function, and the simultaneous consideration of data on other systems. A concurrent application of reweighting and force field optimization could violate the Bayesian principle of having well-defined prior and likelihood in the reweighting process, as the rescaled prior used for reweighting has already been adjusted against experimental data. The data are, so to say, used twice in such protocol. We do not have an answer to the aforementioned questions, nonetheless we believe they provide important points of discussion for future applications of reweighting, and that Bayesian methods for force field parameterization would make it easier to merge these two different strategies.

6.3 Using kinetic data to reweight equilibrium ensembles

As discussed in Section 5.1, Augmented Markov Models are a framework based on maximum entropy and maximum likelihood to include experimental information in a Markov State Model. Olsson et al.[172] used

AMMs to predict NMR relaxation dispersion data, by employing only static experimental data to reweight the model. This fact leads to intriguing questions which, to our knowledge, have been largely unanswered in literature. As a matter of principle, it should be possible to directly employ kinetic data for ensemble reweighting, just like it is done with the ABSURD method.[171] What would happen then to equilibrium observables? Would the amount of provided information be enough to increase the accuracy of simulated averages with respect to experimental equilibrium quantities? What would be the optimal framework to test this hypothesis? While the first question has no clear answer yet, concerning the latter we believe that a suitable framework would be the one of Markov State Models, because of their intrinsic ability to encode kinetic information.[172,191] We note, however, that MSMs, by construction, ignore the fastest dynamical timescales of the system, which can, however, be important for some types of experimental measurements. Therefore, a careful choice of kinetic data is recommended.

6.4 A new generation of force fields

Force field parametrization has been historically guided by a combination of chemical and biological intuition, ab initio quantum calculation of small molecules and trial and error approaches.[108] This approach has been proven to be extremely successful[203] but, as the ever growing amount of experimental information calls for force fields which are easily improvable once new data are collected, the complexity and the amount of expertise required in this procedure have made it somewhat impractical. We discussed some advancements in Section 4.2, but none of the applications we reviewed provided an ultimate solution. Indeed, suggested improvements are usually small adjustments (e.g., modifications in the backbone dihedral angles terms[137,138]), though in some cases more extensive changes have been introduced using such fitting to experiments.[13,121,122,124] Therefore, in this section we want to discuss some principles on which a new generation of force fields could be built.

One of the main challenges in force field development lies in how to include new data, which is potentially conflicting with old information, and use it to improve the force field after it has already been parametrized. Automatic reparametrization would make it extremely easy to modify and update force field once such new experimental data become available.

The development of a Bayesian formalism[116] and later the more systematic ForceBalance[121,122,124] framework, for example, goes exactly in this direction and shows that it is possible in principle to approach to the problem in an automated fashion. It would be also important to assign some level of *trust* to force field parameters, such that it can be assessed to what degree new data should alter their values. At the same time, a fully Bayesian approach would involve distribution of force fields, rather than point estimates, and thus parallel simulations could be used to integrate out force field uncertainty. We also stress that such developments would ideally be carried concurrently with the construction of worldwide accessible and curated databases of experimental and simulation data. In order to include new data in a more automatic fashion, there have to be data quality checks and consensus on experimental and forward model errors, as this all affects how much the new data should be able to alter the existing parametrization (in case of inconsistency). At the same time, it is also important to keep in mind that such models should ideally capture well-understood physical effects, and that lack of agreement with experiments might indicate important effects that are missing from the functional form or parameter combining rules.[110]

The molecular mechanics force fields were developed as a classical parametrization of the Born–Oppenheimer energy surface that balances computational speed and precision and the choices of functional forms of the force field terms has always been guided by chemical intuition. This has led to families of force fields with differences in the parametrization[106] and to the proliferation of big sets of parameters needed to accommodate empirical choices (see, for example, the discussion on atom types in Refs. 204 and 205). We expect the next generation of force fields to be more flexible with respect to specific choices of parametrization: more specifically, tools are required to define not only the force field parameters, but also the functional shapes of each term in a data-driven fashion. These developments go hand in hand with a robust estimation of both statistical and systematic errors: the two sources of errors need to be fully decoupled and it should be clear when poor estimates are due to poor training datasets.

Some of these advancements have already been applied to the SMIRNOFF99Frosst force field and the Open Force Field Toolkit,[204] developed by the Open Force Field Initiative.[206] For now, SMIRNOFF99Frosst has been tested only on a wide set of pharmaceutically relevant small molecules, but it represents nonetheless an important step toward a new generation of force fields.

7. Conclusions

Much research in the field of structural biology and molecular biophysics requires one to integrate several heterogeneous sources of data, coming both from diverse experimental techniques and simulations. Experiments are employed to characterize thermodynamic and kinetic quantities of the system under study and simulations aid the interpretation of or complement these results thanks to their high spatial and time resolutions. However, technological[17–20] and theoretical[3,21] advancements in the field of molecular mechanics have highlighted the existence of discrepancies between simulations and experiments.[11–16] In this review, we have approached this issue through a specific perspective: inconsistencies between computational and experimental results carry information that can be systematically extracted and exploited to improve our understanding of biochemical entities and more general biophysical models. To pursue this perspective, we focused on the cornerstone ideas that have guided the optimization of system-specific conformational ensembles (Sections 2 and 3) or general purpose force fields (Section 4) against experimental information. We provided a summary of alternative strategies and discussed their strengths and shortcomings depending on the level of trust one places in the force field used to carry out the simulations. For each method, we critically assessed how much a realistic scenario deviates from the ideal version of the approach, examining the challenges that arise when the systems grow in size and complexity. We stressed that many sources of data can be employed and different strategies are needed depending on whether observables can or cannot be represented as a linear combination of a given forward model applied to the single structures in the ensemble (see Eq. (1) and Section 5). Despite the differences in the technical implementation, however, we argued that most of the approaches presented here share a common framework, rooted in the minimization of the functional

$$T = \chi^2 - \theta R, \tag{49}$$

where χ^2 is the negative log-likelihood in the case of normally distributed errors, θ is a hyperparameter that balances between simulations and experimental data and R is a regularization term. Depending on the method, R assumes different functional shapes: in MaxEnt, it becomes a cross-entropy term, Eq. (9); in MaxPars, it reduces a parsimony term, i.e., the negative number of conformations with nonzero associated weight, Eq. (13); in

MaxPrior, the regularization term is provided by the log-prior, Eq. (18); in AMMs R represents the logarithm of the MSM likelihood, Eq. (32); in MaxCal the regularization term takes the form of a path entropy, Eq. (37); finally, Average Block Selection is unregularized and so $R = 0$, Eq. (48). Therefore, the different philosophies that distinguish among the several approaches are encoded in the choice of the regularization term.

As a final remark, we devoted Section 6 to the challenges that the community is facing and we anticipate some open questions that we believe will capture experts' attention in the incoming years.

Acknowledgments

We acknowledge support by a grant from the Lundbeck Foundation to the BRAINSTRUC structural biology initiative, the NordForsk Nordic Neutron Science Programme, the Carlsberg Foundation, a grant from the Velux Foundations and a Hallas-Møller Stipend from the Novo Nordisk Foundation. We thank members of the Linderstrøm-Lang Centre for Protein Science for numerous discussions on these topics, and thank Giovanni Bussi, Ramon Crehuet, Clemens Kauffmann and Simon Olsson for insightful comments on the chapter.

References

1. Lopes PEM, Guvench O, MacKerell AD. Current status of protein force fields for molecular dynamics simulations. In: *Molecular Modeling of Proteins*. New York, NY: Springer New York; 2015:47–71. https://doi.org/10.1007/978-1-4939-1465-4_3.
2. Bottaro S, Lindorff-Larsen K. Biophysical experiments and biomolecular simulations: a perfect match? *Science*. 2018;361(6400):355–360.
3. Maximova T, Moffatt R, Ma B, et al. Principles and overview of sampling methods for modeling macromolecular structure and dynamics. *PLoS Comput Biol*. 2016;12(4): e1004619.
4. Adcock SA, McCammon JA. Molecular dynamics: survey of methods for simulating the activity of proteins. *Chem Rev*. 2006;106(5):1589–1615.
5. Klepeis JL, Lindorff-Larsen K, Dror RO, et al. Long-timescale molecular dynamics simulations of protein structure and function. *Curr Opin Struct Biol*. 2009;19(2): 120–127.
6. Abrams C, Bussi G. Enhanced sampling in molecular dynamics using metadynamics, replica-exchange, and temperature-acceleration. *Entropy*. 2013;16(1):163–199.
7. Elber R. Perspective: computer simulations of long time dynamics. *J Chem Phys*. 2016;144(6):060901.
8. Hollingsworth SA, Dror RO. Molecular dynamics simulation for all. *Neuron*. 2018;99(6):1129–1143.
9. Maffeo C, Bhattacharya S, Yoo J, et al. Modeling and simulation of ion channels. *Chem Rev*. 2012;112(12):6250–6284.
10. van Gunsteren WF, Daura X, Hansen N, et al. Validation of molecular simulation: an overview of issues. *Angew Chem Int Ed*. 2018;57(4):884–902.

11. Rauscher S, Gapsys V, Gajda MJ, et al. Structural ensembles of intrinsically disordered proteins depend strongly on force field: a comparison to experiment. *J Chem Theory Comput.* 2015;11(11):5513–5524.
12. Henriques J, Cragnell C, Skepö M. Molecular dynamics simulations of intrinsically disordered proteins: force field evaluation and comparison with experiment. *J Chem Theory Comput.* 2015;11(7):3420–3431.
13. Robustelli P, Piana S, Shaw DE. Developing a molecular dynamics force field for both folded and disordered protein states. *Proc Natl Acad Sci.* 2018;115(21):E4758–E4766.
14. Nerenberg PS, Head-Gordon T. New developments in force fields for biomolecular simulations. *Curr Opin Struct Biol.* 2018;49:129–138.
15. Bernardi RC, Melo MCR, Schulten K. Enhanced sampling techniques in molecular dynamics simulations of biological systems. *Biochim Biophys Acta Gen Subj.* 2015; 1850(5):872–877.
16. Cordeiro TN, Chen PC, De Biasio A, et al. Disentangling polydispersity in the PCNA-p15PAF complex, a disordered, transient and multivalent macromolecular assembly. *Nucleic Acids Res.* 2017;45(3):1501–1515. https://doi.org/10.1093/nar/gkw1183.
17. Piana S, Lindorff-Larsen K, Shaw DE. Atomic-level description of ubiquitin folding. *Proc Natl Acad Sci.* 2013;110(15):5915–5920.
18. Lindorff-Larsen K, Maragakis P, Piana S, et al. Picosecond to millisecond structural dynamics in human ubiquitin. *J Phys Chem B.* 2016;120(33):8313–8320.
19. Voelz VA, Jäger M, Yao S, et al. Slow unfolded-state structuring in Acyl-CoA binding protein folding revealed by simulation and experiment. *J Am Chem Soc.* 2012; 134(30):12565–12577.
20. Bowman GR, Voelz VA, Pande VS. Atomistic folding simulations of the five-helix bundle protein λ6-85. *J Am Chem Soc.* 2010;133(4):664–667.
21. Camilloni C, Pietrucci F. Advanced simulation techniques for the thermodynamic and kinetic characterization of biological systems. *Adv Phys X.* 2018;3(1):1477531.
22. Wu H, Nüske F, Paul F, et al. Variational Koopman models: slow collective variables and molecular kinetics from short off-equilibrium simulations. *J Chem Phys.* 2017;146(15):154104.
23. Hummer G, Köfinger J. Bayesian ensemble refinement by replica simulations and reweighting. *J Chem Phys.* 2015;143:243150. https://doi.org/10.1063/1.4937786.
24. Ravera E, Sgheri L, Luchinat C. A critical assessment of methods to recover information from averaged data. *Phys Chem Chem Phys.* 2016;18:5686–5701. https://doi.org/10.1039/C5CP04077A.
25. Jaynes ET. Information theory and statistical mechanics. *Phys Rev.* 1957; 106(4):620–630.
26. Kullback S, Leibler RA. On information and sufficiency. *Ann Math Stat.* 1951; 22(1):79–86.
27. Hansen S, Pedersen JS. A comparison of three different methods for analysing small-angle scattering data. *J Appl Cryst.* 1991;24:541–548. URL: http://scripts.iucr.org/cgi-bin/paper?gk0508.
28. Skilling J. *Maximum Entropy and Bayesian Methods.* Springer Netherlands; 1989:45–52. http://weekly.cnbnews.com/news/article.html?no=124000.
29. Różycki B, Kim YC, Hummer G. SAXS ensemble refinement of ESCRT-III CHMP3 conformational transitions. *Structure.* 2011;19(1):109–116. https://doi.org/10.1016/j.str.2010.10.006.
30. Leung HTA, Bignucolo O, Aregger R, et al. A rigorous and efficient method to reweight very large conformational ensembles using average experimental data and to determine their relative information content. *J Chem Theory Comput.* 2016; 12(1):383–394. https://doi.org/10.1021/acs.jctc.5b00759.

31. Reichel K, Stelzl LS, Köfinger J, et al. Precision DEER distances from spin-label ensemble refinement. *J Phys Chem Lett.* 2018;9(19):5748–5752. https://doi.org/10.1021/acs.jpclett.8b02439.
32. Bottaro S, Bussi G, Kennedy SD, et al. Conformational ensembles of RNA oligonucleotides from integrating NMR and molecular simulations. *Sci Adv.* 2018;4(5): eaar8521.
33. Jaynes ET. Where do we stand on maximum entropy? In: Rosenkrantz R.D. (Ed.), Papers on Probability, Statistics and Statistical Physics. Dordrecht: Springer; 15–118, https://doi.org/10.1007/978-94-009-6581-2_10.
34. Roux B, Weare J. On the statistical equivalence of restrained-ensemble simulations with the maximum entropy method. *J Chem Phys.* 2013;138(8):084107. https://doi.org/10.1063/1.4792208.
35. Boomsma W, Ferkinghoff-Borg J, Lindorff-Larsen K. Combining experiments and simulations using the maximum entropy principle. *PLoS Comput Biol.* 2014;10(2): e1003406. https://doi.org/10.1371/journal.pcbi.1003406.
36. Cesari A, Reißer S, Bussi G. Using the maximum entropy principle to combine simulations and solution experiments. *Computation.* 2018;6(1):15. https://doi.org/10.3390/computation6010015.
37. Köfinger J, Stelzl LS, Reuter K, et al. Efficient ensemble refinement by reweighting. *J Chem Theory Comput.* 2019;15(5):3390–3401.
38. Shevchuk R, Hub JS. Bayesian refinement of protein structures and ensembles against SAXS data using molecular dynamics. *PLoS Comput Biol.* 2017;13(10):1–27. https://doi.org/10.1371/journal.pcbi.1005800.
39. Cavalli A, Bonomi M, Camilloni C, et al. Metainference: a Bayesian inference method for heterogeneous systems. *Sci Adv.* 2016;2(1):e1501177. https://doi.org/10.1126/sciadv.1501177.
40. Cesari A, Gil-Ley A, Bussi G. Combining simulations and solution experiments as a paradigm for RNA force field refinement. *J Chem Theory Comput.* 2016;12(12): 6192–6200. https://doi.org/10.1021/acs.jctc.6b00944.
41. Amirkulova DB, White AD. Recent advances in maximum entropy biasing techniques for molecular dynamics. *Mol Simul.* 2019;1–10.
42. Gull SF, Daniell GJ. Image reconstruction from incomplete and noisy data. *Nature.* 1978;272:686–690.
43. Wiegand H. Kish, L.: Survey sampling. John Wiley & Sons, Inc., New York, London 1965, IX + 643 S., 31 Abb., 56 Tab., Preis 83 s. *Biochem Z.* 1968;10(1):88–89. https://doi.org/10.1002/bimj.19680100122.
44. Rangan R, Bonomi M, Heller GT, et al. Determination of structural ensembles of proteins: restraining vs reweighting. *J Chem Theory Comput.* 2018;14(12): 6632–6641.
45. Akaike HAI. A new look at the statistical model identification. *IEEE Trans Autom Control.* 1974;19(6):716–723.
46. Bowerman S, Rana ASJB, Rice A, et al. Determining atomistic SAXS models of tri-ubiquitin chains from Bayesian analysis of accelerated molecular dynamics simulations. *J Chem.* 2017;13(6):2418–2429. https://doi.org/10.1097/CCM.0b013e31823da96d. Hydrogen.
47. Boura E, RóŻycki B, Herrick DZ, et al. Solution structure of the ESCRT-I complex by small-angle X-ray scattering, EPR, and FRET spectroscopy. *Proc Natl Acad Sci.* 2011;108(23):9437–9442.
48. Chen Y, Campbell SL, Dokholyan NV. Deciphering protein dynamics from NMR data using explicit structure sampling and selection. *Biophys J.* 2007;93(7): 2300–2306. https://doi.org/10.1529/biophysj.107.104174.

49. Francis DM, Rä B, Koveal D, et al. Structural basis of p38 regulation by hematopoietic tyrosine phosphatase. *Nat Chem Biol.* 2011;7(12):916–924. https://doi.org/10.1038/nchembio.707.
50. Cossio P, Hummer G. Bayesian analysis of individual electron microscopy images: towards structures of dynamic and heterogeneous biomolecular assemblies. *J Struct Biol.* 2013;184(3):427–437. https://doi.org/10.1016/j.jsb.2013.10.006.
51. Berlin K, Castañeda CA, Schneidman-Duhovny D, et al. Recovering a representative conformational ensemble from underdetermined macromolecular structural data. *J Am Chem Soc.* 2013;135(44):16595–16609. https://doi.org/10.1021/ja4083717.
52. Schneidman-Duhovny D, Hammel M, Tainer JA, et al. FoXS, FoXSDock and MultiFoXS: single-state and multi-state structural modeling of proteins and their complexes based on SAXS profiles. *Nucleic Acids Res.* 2016;44(Web server issue):424–429. https://doi.org/10.1093/nar/gkw389.
53. Rieping W. Inferential structure determination. *Science.* 2005;309(5732):303–306. https://doi.org/10.1126/science.1110428.
54. Olsson S, Frellsen J, Boomsma W, et al. Inference of structure ensembles of flexible biomolecules from sparse, averaged data. *PLoS ONE.* 2013;8(11):1–7. https://doi.org/10.1371/journal.pone.0079439.
55. Dutta R, Brotzakis ZF, Mira A. Bayesian calibration of force-fields from experimental data: TIP4P water. *J Chem Phys.* 2018;149(15):154110.
56. Pernot P, Cailliez F. A critical review of statistical calibration/prediction models handling data inconsistency and model inadequacy. *AIChE J.* 2017;63(10):4642–4665.
57. Beauchamp KA, Pande VS, Das R. Bayesian energy landscape tilting: towards concordant models of molecular ensembles. *Biophys J.* 2014;106(6):1381–1390. https://doi.org/10.1016/j.bpj.2014.02.009.
58. Brookes DH, Head-Gordon T. Experimental inferential structure determination of ensembles for intrinsically disordered proteins. *J Am Chem Soc.* 2016;138(13):4530–4538.
59. Fisher CK, Huang A, Stultz CM. Modeling intrinsically disordered proteins with Bayesian statistics. *J Am Chem Soc.* 2010;132(42):14919–14927. https://doi.org/10.1021/ja105832g.
60. Sethi A, Anunciado D, Tian J, et al. Deducing conformational variability of intrinsically disordered proteins from infrared spectroscopy with Bayesian statistics. *Chem Phys.* 2013;422:143–155. https://doi.org/10.1016/j.chemphys.2013.05.005.
61. Xiao X, Kallenbach N, Zhang Y. Peptide conformation analysis using an integrated Bayesian approach. *J Chem Theory Comput.* 2014;10(9):4152–4159. https://doi.org/10.1021/ct500433d.
62. Ge Y, Voelz VA. Model selection using BICePs: a Bayesian approach for force field validation and parameterization. *J Phys Chem B.* 2018;122(21):5610–5622. https://doi.org/10.1021/acs.jpcb.7b11871.
63. Potrzebowski W, Trewhella J, Andre I. Bayesian inference of protein conformational ensembles from limited structural data. *PLoS Comput Biol.* 2018;14(12):e1006641.
64. Bottaro S, Bengtsen T, Lindorff-Larsen K. Integrating molecular simulation and experimental data: a Bayesian/maximum entropy reweighting approach. *bioRxiv.* 2018, 457952.
65. Molnar KS, Bonomi M, Pellarin R, et al. Cys-Scanning disulfide crosslinking and Bayesian modeling probe the transmembrane signaling mechanism of the histidine kinase, PhoQ. *Structure.* 2014;22(9):1239–1251. https://doi.org/10.1016/j.str.2014.04.019.
66. Mechelke M, Habeck M. Bayesian weighting of statistical potentials in NMR structure calculation. *PLoS ONE.* 2014;9(6):e100197.

67. Antonov LD, Olsson S, Boomsma W, et al. Bayesian inference of protein ensembles from SAXS data. *Phys Chem Chem Phys.* 2016;18(8):5832–5838. https://doi.org/10.1039/c5cp04886a.
68. Bonomi M, Heller GT, Camilloni C, et al. Principles of protein structural ensemble determination. *Curr Opin Struct Biol.* 2017;42:106–116. https://doi.org/10.1016/j.sbi.2016.12.004.
69. Fisher CK, Ullman O, Stultz CM. Efficient construction of disordered protein ensembles in a Bayesian framework with optimal selection of conformations. *Pac Symp Biocomput.* 2012;82–93. https://doi.org/10.1038/cdd.2010.172.MicroRNAs.
70. Kirmizialtin S, Loerke J, Behrmann E, et al. Using molecular simulation to model high-resolution cryo-EM reconstructions. In: Woodson SA, Allain FHT, eds. *Methods In Enzymology.* Academic Press; 2015; vol. 558:497–514.
71. MacKay DJC. Bayesian model comparison and backprop nets. In: Moody JE, Hanson SJ, Lippmann RP, eds. *Advances in Neural Information Processing Systems 4.* Morgan-Kaufmann; 1992:839–846. http://papers.nips.cc/paper/488-bayesian-model-comparison-and-backprop-nets.pdf.
72. Larsen AH, Arleth L, Hansen S. Analysis of small-angle scattering data using model fitting and Bayesian regularization. *J Appl Cryst.* 2018;51(4):1151–1161. https://doi.org/10.1107/S1600576718008956. http://scripts.iucr.org/cgi-bin/paper?S1600576718008956.
73. Pelikan M, Hura GL, Hammel M. Structure and flexibility within proteins as identified through small angle X-ray scattering. *Gen Physiol Biophys.* 2009;28(2):174.
74. Vogel C, Bashton M, Kerrison ND, et al. Structure, function and evolution of multidomain proteins. *Curr Opin Struct Biol.* 2004;14(2):208–216. https://doi.org/10.1016/j.sbi.2004.03.011.
75. Dyson HJ, Wright PE. Intrinsically unstructured proteins and their functions. *Nat Rev Mol Cell Biol.* 2005;6(3):197–208. https://doi.org/10.1038/nrm1589.
76. Lindorff-Larsen K, Trbovic N, Maragakis P, et al. Structure and dynamics of an unfolded protein examined by molecular dynamics simulation. *J Am Chem Soc.* 2012;134(8):3787–3791.
77. Rodriguez A, d'Errico M, Facco E, et al. Computing the free energy without collective variables. *J Chem Theory Comput.* 2018;14(3):1206–1215.
78. Alford RF, Leaver-Fay A, Jeliazkov JR, et al. The rosetta all-atom energy function for macromolecular modeling and design. *J Chem Theory Comput.* 2017;13:3031–3048. https://doi.org/10.1021/acs.jctc.7b00125.
79. Bellman RE. *Adaptive Control Processes: A Guided Tour,* vol. 2045. Princeton University Press; 2015.
80. Agmon N, Alhassid Y, Levine RD. An algorithm for finding the distribution of maximal entropy. *J Comput Phys.* 1979;30(2):250–258.
81. Malouf R. A comparison of algorithms for maximum entropy parameter estimation. In: Proceedings of the 6th Conference on Natural Language Learning, vol. 20. Association for Computational Linguistics; 1–7.
82. Bernadó P, Mylonas E, Petoukhov MV, et al. Structural characterization of flexible proteins using small-angle X-ray scattering. *J Am Chem Soc.* 2007;129(17):5656–5664. https://doi.org/10.1021/ja069124n.
83. Nodet G, Salmon L, Ozenne V, et al. Quantitative description of backbone conformational sampling of unfolded proteins at amino acid resolution from NMR residual dipolar couplings. *J Am Chem Soc.* 2009;131(49):17908–17918. https://doi.org/10.1021/ja9069024.
84. White AD, Voth GA. Efficient and minimal method to bias molecular simulations with experimental data. *J Chem Theory Comput.* 2014;10(8):3023–3030. https://doi.org/10.1021/ct500320c.

85. White AD, Dama JF, Voth GA. Designing free energy surfaces that match experimental data with metadynamics. *J Chem Theory Comput*. 2015;11(6):2451–2460. https://doi.org/10.1021/acs.jctc.5b00178.
86. Lindorff-Larsen K, Best RB, DePristo MA, et al. Simultaneous determination of protein structure and dynamics. *Nature*. 2005;433(7022):128–132. https://doi.org/10.1038/nature03199.
87. Vendruscolo M. Determination of conformationally heterogeneous states of proteins. *Curr Opin Struct Biol*. 2007;17(1):15–20. https://doi.org/10.1016/j.sbi.2007.01.002.
88. Burnley BT, Afonine PV, Adams PD, et al. Modelling dynamics in protein crystal structures by ensemble refinement. *eLife*. 2012;1:e00311. https://doi.org/10.7554/eLife.00311.
89. Levin EJ, Kondrashov DA, Wesenberg GE, et al. Ensemble refinement of protein crystal structures: validation and application. *Structure*. 2007;15(9):1040–1052.
90. Pitera JW, Chodera JD. On the use of experimental observations to bias simulated ensembles. *J Chem Theory Comput*. 2012;8(10):3445–3451. https://doi.org/10.1021/ct300112v.
91. Cavalli A, Camilloni C, Vendruscolo M. Molecular dynamics simulations with replica-averaged structural restraints generate structural ensembles according to the maximum entropy principle. *J Chem Phys*. 2013;138(9):094112. https://doi.org/10.1063/1.4793625.
92. Löhr T, Camilloni C, Bonomi M, Vendruscolo M. A practical guide to the simultaneous determination of protein structure and dynamics using metainference.. In: Bonomi M, Camilloni C, eds. *Biomolecular Simulations*. New York, NY: Humana; 2019:313–340.
93. Laio A, Gervasio FL. Metadynamics: a method to simulate rare events and reconstruct the free energy in biophysics, chemistry and material science. *Rep Prog Phys*. 2008;71(12):126601.
94. Barducci A, Bonomi M, Parrinello M. Metadynamics. *Wiley Interdiscip Rev Comput Mol Sci*. 2011;1(5):826–843.
95. Sutto L, Marsili S, Gervasio FL. New advances in metadynamics. *Wiley Interdiscip Rev Comput Mol Sci*. 2012;2(5):771–779.
96. Pfaendtner J, Bonomi M. Efficient sampling of high-dimensional free-energy landscapes with parallel bias metadynamics. *J Chem Theory Comput*. 2015;11(11):5062–5067.
97. Bonomi M, Camilloni C, Vendruscolo M. Metadynamic metainference: enhanced sampling of the metainference ensemble using metadynamics. *Sci Rep*. 2016;6:31232.
98. Ianeselli A, Orioli S, Spagnolli G, et al. Atomic detail of protein folding revealed by an ab initio reappraisal of circular dichroism. *J Am Chem Soc*. 2018;140(10):3674–3682.
99. Kohlhoff KJ, Robustelli P, Cavalli A, et al. Fast and accurate predictions of protein NMR chemical shifts from interatomic distances. *J Am Chem Soc*. 2009;131(39):13894–13895.
100. Shen Y, Bax A. SPARTA+: a modest improvement in empirical NMR chemical shift prediction by means of an artificial neural network. *J Biomol NMR*. 2010;48(1):13–22.
101. Jorgensen WL, Maxwell DS, Tirado-Rives J. Development and testing of the OPLS all-atom force field on conformational energetics and properties of organic liquids. *J Am Chem Soc*. 1996;118(45):11225–11236.
102. Oostenbrink C, Villa A, Mark AE, et al. A biomolecular force field based on the free enthalpy of hydration and solvation: the GROMOS force-field parameter sets 53A5 and 53A6. *J Comb Chem*. 2004;25(13):1656–1676.

103. MacKerell Jr AD, Bashford D, Bellott MLDR, et al. All-atom empirical potential for molecular modeling and dynamics studies of proteins. *J Phys Chem B*. 1998;102(18): 3586–3616.
104. Cornell WD, Cieplak P, Bayly CI, et al. A second generation force field for the simulation of proteins, nucleic acids, and organic molecules. *J Am Chem Soc*. 1995;117(19):5179–5197.
105. Tuckerman M. *Statistical Mechanics: Theory and Molecular Simulation*. Oxford University Press; 2010.
106. Guvench O, MacKerell AD. Comparison of protein force fields for molecular dynamics simulations. In: Kukol A, ed. *Molecular Modeling of Proteins*. Totowa, NJ: Humana Press; 2008:63–88. https://doi.org/10.1007/978-1-59745-177-2_4.
107. Ponder JW, Case DA. Force fields for protein simulations. *Adv Protein Chem*. 2003;66:27–85.
108. Zhu X, Lopes PEM, MacKerell Jr AD. Recent developments and applications of the CHARMM force fields. *Wiley Interdiscip Rev Comput Mol Sci*. 2012;2(1): 167–185.
109. Dauber-Osguthorpe P, Hagler AT. Biomolecular force fields: where have we been, where are we now, where do we need to go and how do we get there? *J Comput Aided Mol Des*. 2019;33(2):133–203.
110. Hagler AT. Force field development phase II: relaxation of physics-based criteria…or inclusion of more rigorous physics into the representation of molecular energetics. *J Comput Aided Mol Des*. 2019;33(2):205–264.
111. Bayly CI, Cieplak P, Cornell W, et al. A well-behaved electrostatic potential based method using charge restraints for deriving atomic charges: the RESP model. *J Phys Chem*. 1993;97(40):10269–10280.
112. Jakalian A, Bush BL, Jack DB, et al. Fast, efficient generation of high-quality atomic charges. AM1-BCC model: I. Method. *J Comput Chem*. 2000;21(2):132–146.
113. Jakalian A, Jack DB, Bayly CI. Fast, efficient generation of high-quality atomic charges. AM1-BCC model: II. Parameterization and validation. *J Comput Chem*. 2002; 23(16):1623–1641.
114. Lindorff-Larsen K, Piana S, Palmo K, et al. Improved side-chain torsion potentials for the Amber ff99SB protein force field. *Proteins*. 2010;78(8):1950–1958.
115. Tian C, Kasavajhala K, Belfon K, et al. ff19SB: amino-acid-specific protein backbone parameters trained against quantum mechanics energy surfaces in solution. *J Chem Theory Comput*. 2020;16:528–552.
116. Norgaard AB, Ferkinghoff-Borg J, Lindorff-Larsen K. Experimental parameterization of an energy function for the simulation of unfolded proteins. *Biophys J*. 2008; 94(1):182–192.
117. Best RB, Hummer G. Optimized molecular dynamics force fields applied to the helix-coil transition of polypeptides. *J Phys Chem B*. 2009;113(26):9004–9015.
118. Cailliez F, Pernot P. Statistical approaches to forcefield calibration and prediction uncertainty in molecular simulation. *J Chem Phys*. 2011;134(5):054124.
119. Rizzi F, Najm HN, Debusschere BJ, et al. Uncertainty quantification in MD simulations. Part II: Bayesian inference of force-field parameters. *Multiscale Model Simul*. 2012;10(4):1460–1492.
120. Angelikopoulos P, Papadimitriou C, Koumoutsakos P. Bayesian uncertainty quantification and propagation in molecular dynamics simulations: a high performance computing framework. *J Chem Phys*. 2012;137(14):144103.
121. Wang LP, Chen J, Van Voorhis T. Systematic parametrization of polarizable force fields from quantum chemistry data. *J Chem Theory Comput*. 2012;9(1):452–460.
122. Wang LP, Martinez TJ, Pande VS. Building force fields: an automatic, systematic, and reproducible approach. *J Phys Chem Lett*. 2014;5(11):1885–1891.

123. Wu S, Angelikopoulos P, Papadimitriou C, et al. A hierarchical Bayesian framework for force field selection in molecular dynamics simulations. *Philos Trans R Soc A Math Phys Eng Sci*. 2016;374(2060):20150032.
124. Wang LP, McKiernan KA, Gomes J, et al. Building a more predictive protein force field: a systematic and reproducible route to AMBER-FB15. *J Phys Chem B*. 2017;121(16):4023–4039.
125. Bonvin AMJJ, Boelens R, Kaptein R. Time-and ensemble-averaged direct NOE restraints. *J Biomol NMR*. 1994;4(1):143–149.
126. Jr Huang, Grzesiek S. Ensemble calculations of unstructured proteins constrained by RDC and PRE data: a case study of urea-denatured ubiquitin. *J Am Chem Soc*. 2009;132(2):694–705.
127. Lange OF, Lakomek NA, Farès C, et al. Recognition dynamics up to microseconds revealed from an RDC-derived ubiquitin ensemble in solution. *Science*. 2008; 320(5882):1471–1475.
128. Olsson S, Vögeli BR, Cavalli A, et al. Probabilistic determination of native state ensembles of proteins. *J Chem Theory Comput*. 2014;10(8):3484–3491.
129. Esteban-Martín S, Fenwick RB, Salvatella X. Refinement of ensembles describing unstructured proteins using NMR residual dipolar couplings. *J Am Chem Soc*. 2010;132(13):4626–4632.
130. Scheek RM, Torda AE, Kemmink J, et al. Structure determination by NMR: the modeling of NMR parameters as ensemble averages. In: Hoch JC, Poulsen FM, Redfield C, eds. *Computational Aspects of the Study of Biological Macromolecules by Nuclear Magnetic Resonance Spectroscopy*. Springer; 1991:209–217.
131. Mantsyzov AB, Maltsev AS, Ying J, et al. A maximum entropy approach to the study of residue-specific backbone angle distributions in α-synuclein, an intrinsically disordered protein. *Protein Sci*. 2014;23(9):1275–1290.
132. Mantsyzov AB, Shen Y, Lee JH, et al. MERA: a webserver for evaluating backbone torsion angle distributions in dynamic and disordered proteins from NMR data. *J Biomol NMR*. 2015;63(1):85–95.
133. Graf J, Nguyen PH, Stock G, et al. Structure and dynamics of the homologous series of alanine peptides: a joint molecular dynamics/NMR study. *J Am Chem Soc*. 2007;129(5):1179–1189.
134. Cesari A, Bottaro S, Lindorff-Larsen K, et al. Fitting corrections to an RNA force field using experimental data. *J Chem Theory Comput*. 2019;15:3425–3431.
135. Vasile F, Tiana G. Determination of structural ensembles of flexible molecules in solution from NMR data undergoing spin diffusion. *J Chem Inf Model*. 2019;59:2973–2979.
136. Zwanzig RW. High-temperature equation of state by a perturbation method. I. Nonpolar gases. *J Chem Phys*. 1954;22(8):1420–1426.
137. Li DW, Brüschweiler R. NMR-based protein potentials. *Angew Chem Int Ed*. 2010;49(38):6778–6780.
138. Li DW, Brüschweiler R. Iterative optimization of molecular mechanics force fields from NMR data of full-length proteins. *J Chem Theory Comput*. 2011;7(6):1773–1782.
139. Chen J, Chen J, Pinamonti G, et al. Learning effective molecular models from experimental observables. *J Chem Theory Comput*. 2018;14(7):3849–3858.
140. Teukolsky SA, Flannery BP, Press WH, et al. Numerical recipes in C. *SMR*. 1992;693(1):59–70.
141. Brüschweiler R. Collective protein dynamics and nuclear spin relaxation. *J Chem Phys*. 1995;102(8):3396–3403.
142. Bennett CH. Efficient estimation of free energy differences from Monte Carlo data. *J Comput Phys*. 1976;22(2):245–268.
143. Shirts MR, Chodera JD. Statistically optimal analysis of samples from multiple equilibrium states. *J Chem Phys*. 2008;129(12):124105.

144. Gillespie JR, Shortle D. Characterization of long-range structure in the denatured state of staphylococcal nuclease. I. Paramagnetic relaxation enhancement by nitroxide spin labels. *J Mol Biol*. 1997;268(1):158–169.
145. Lietzow MA, Jamin M, Dyson HJ, et al. Mapping long-range contacts in a highly unfolded protein. *Am J Mol Biol*. 2002;322(4):655–662.
146. Yi Q, Scalley-Kim ML, Alm EJ, et al. NMR characterization of residual structure in the denatured state of protein L. *J Mol Biol*. 2000;299(5):1341–1351.
147. Teilum K, Kragelund BB, Poulsen FM. Transient structure formation in unfolded acyl-coenzyme a-binding protein observed by site-directed spin labelling. *J Mol Biol*. 2002;324(2):349–357.
148. Gillespie JR, Shortle D. Characterization of long-range structure in the denatured state of staphylococcal nuclease. II. Distance restraints from paramagnetic relaxation and calculation of an ensemble of structures. *J Mol Biol*. 1997;268(1):170–184.
149. Hornak V, Abel R, Okur A, et al. Comparison of multiple amber force fields and development of improved protein backbone parameters. *Proteins*. 2006;65(3):712–725.
150. Condon DE, Kennedy SD, Mort BC, et al. Stacking in RNA: NMR of four tetramers benchmark molecular dynamics. *J Chem Theory Comput*. 2015;11(6):2729–2742.
151. Bergonzo C, Henriksen NM, Roe DR, et al. Highly sampled tetranucleotide and tetraloop motifs enable evaluation of common RNA force fields. *RNA*. 2015;21(9):1578–1590.
152. Kuhrova P, Best RB, Bottaro S, et al. Computer folding of RNA tetraloops: identification of key force field deficiencies. *J Chem Theory Comput*. 2016;12(9):4534–4548.
153. Bottaro S, Banáš P, Sponer J, et al. Free energy landscape of GAGA and UUCG RNA tetraloops. *J Phys Chem Lett*. 2016;7(20):4032–4038.
154. Borkar AN, Bardaro MF, Camilloni C, et al. Structure of a low-population binding intermediate in protein-RNA recognition. *Proc Natl Acad Sci*. 2016;113(26):7171–7176.
155. Krepl M, Blatter M, Cléry A, et al. Structural study of the Fox-1 RRM protein hydration reveals a role for key water molecules in RRM-RNA recognition. *Nucleic Acids Res*. 2017;45(13):8046–8063.
156. Podbevšek P, Fasolo F, Bon C, et al. Structural determinants of the SINE B2 element embedded in the long non-coding RNA activator of translation AS Uchl1. *Sci Rep*. 2018;8(1):3189.
157. Kooshapur H, Choudhury NR, Simon B, et al. Structural basis for terminal loop recognition and stimulation of pri-miRNA-18a processing by hnRNP A1. *Nat Commun*. 2018;9(1):2479.
158. Aytenfisu AH, Spasic A, Grossfield A, et al. Revised RNA dihedral parameters for the amber force field improve RNA molecular dynamics. *J Chem Theory Comput*. 2017;13(2):900–915.
159. Tan D, Piana S, Dirks RM, et al. RNA force field with accuracy comparable to state-of-the-art protein force fields. *Proc Natl Acad Sci*. 2018;115(7):E1346–E1355.
160. Kuhrova P, Mlynsky V, Zgarbova M, et al. Improving the performance of the amber RNA force field by tuning the hydrogen-bonding interactions. *J Chem Theory Comput*. 2019;15(5):3288–3305.
161. Pérez A, Marchán I, Svozil D, et al. Refinement of the AMBER force field for nucleic acids: improving the description of α/γ conformers. *Biophys J*. 2007;92(11):3817–3829.
162. Zgarbová M, Otyepka M, Šponer J, et al. Refinement of the Cornell et al. nucleic acids force field based on reference quantum chemical calculations of glycosidic torsion profiles. *J Chem Theory Comput*. 2011;7(9):2886–2902.
163. Izadi S, Anandakrishnan R, Onufriev AV. Building water models: a different approach. *J Phys Chem Lett*. 2014;5(21):3863–3871.

164. Pande VS, Beauchamp K, Bowman GR. Everything you wanted to know about Markov State Models but were afraid to ask. *Methods*. 2010;52(1):99–105.
165. Bowman GR, Pande VS, Noé F. *An Introduction to Markov State Models and Their Application to Long Timescale Molecular Simulation*. vol. 797. Springer Science & Business Media; 2013.
166. Husic BE, Pande VS. Markov state models: from an art to a science. *J Am Chem Soc*. 2018;140(7):2386–2396.
167. Palmer IIIAG. NMR characterization of the dynamics of biomacromolecules. *Chem Rev*. 2004;104(8):3623–3640.
168. Prinz JH, Wu H, Sarich M, et al. Markov models of molecular kinetics: generation and validation. *J Chem Phys*. 2011;134(17):174105.
169. Schwantes CR, McGibbon RT, Pande VS. Perspective: Markov models for long-timescale biomolecular dynamics. *J Chem Phys*. 2014;141(9):090901.
170. Jaynes ET. The minimum entropy production principle. *Annu Rev Phys Chem*. 1980;31(1):579–601.
171. Salvi N, Abyzov A, Blackledge M. Multi-timescale dynamics in intrinsically disordered proteins from NMR relaxation and molecular simulation. *J Phys Chem Lett*. 2016;7(13):2483–2489.
172. Olsson S, Wu H, Paul F, et al. Combining experimental and simulation data of molecular processes via augmented Markov models. *Proc Natl Acad Sci*. 2017;114(31): 8265–8270.
173. Bowman GR, Beauchamp KA, Boxer G, et al. Progress and challenges in the automated construction of Markov State Models for full protein systems. *J Chem Phys*. 2009;131(12):124101.
174. Sultan MM, Kiss G, Shukla D, et al. Automatic selection of order parameters in the analysis of large scale molecular dynamics simulations. *J Chem Theory Comput*. 2014;10(12):5217–5223.
175. Pérez-Hernández G, Paul F, Giorgino T, et al. Identification of slow molecular order parameters for Markov model construction. *J Chem Phys*. 2013;139(1):07B604_1.
176. Schwantes CR, Pande VS. Improvements in Markov state model construction reveal many non-native interactions in the folding of NTL9. *J Chem Theory Comput*. 2013;9(4):2000–2009.
177. Noé F, Clementi C. Kinetic distance and kinetic maps from molecular dynamics simulation. *J Chem Theory Comput*. 2015;11(10):5002–5011.
178. Scherer MK, Trendelkamp-Schroer B, Paul F, et al. PyEMMA 2: a software package for estimation, validation, and analysis of Markov models. *J Chem Theory Comput*. 2015;11(11):5525–5542.
179. Harrigan MP, Sultan MM, Hernández CX, et al. Msmbuilder: statistical models for biomolecular dynamics. *Biophys J*. 2017;112(1):10–15.
180. Noé F, Rosta E. Markov models of molecular kinetics. *J Chem Phys*. 2019;151:190401.
181. Xue Y, Ward JM, Yuwen T, et al. Microsecond time-scale conformational exchange in proteins: using long molecular dynamics trajectory to simulate NMR relaxation dispersion data. *J Am Chem Soc*. 2012;134(5):2555–2562.
182. Olsson S, No'e F. Mechanistic models of chemical exchange induced relaxation in protein NMR. *J Am Chem Soc*. 2016;139(1):200–210.
183. Pressé S, Ghosh K, Lee J, et al. Principles of maximum entropy and maximum caliber in statistical physics. *Rev Mod Phys*. 2013;85(3):1115.
184. Bolhuis PG, Chandler D, Dellago C, et al. Transition path sampling: throwing ropes over rough mountain passes, in the dark. *Annu Rev Phys Chem*. 2002;53(1):291–318.
185. Eastman P, Grønbech-Jensen N, Doniach S. Simulation of protein folding by reaction path annealing. *J Chem Phys*. 2001;114(8):3823–3841.
186. Pan AC, Sezer D, Roux B. Finding transition pathways using the string method with swarms of trajectories. *J Phys Chem B*. 2008;112(11):3432–3440.

187. Weinan E, Ren W, Vanden-Eijnden E. String method for the study of rare events. *Phys Rev B*. 2002;66(5):052301.
188. Capelli R, Tiana G, Camilloni C. An implementation of the maximum-caliber principle by replica-averaged time-resolved restrained simulations. *J Chem Phys*. 2018; 148(18):184114.
189. Roux B, Weare J. On the statistical equivalence of restrained-ensemble simulations with the maximum entropy method. *J Chem Phys*. 2013;138(8):02B616.
190. Cavalli A, Camilloni C, Vendruscolo M. Molecular dynamics simulations with replica-averaged structural restraints generate structural ensembles according to the maximum entropy principle. *J Chem Phys*. 2013;138(9):03B603.
191. Dixit PD, Dill KA. Caliber corrected Markov modeling (C2M2): correcting equilibrium Markov models. *J Chem Theory Comput*. 2018;14(2):1111–1119.
192. Dixit PD. Communication: introducing prescribed biases in out-of-equilibrium Markov models. *J Chem Phys*. 2018;148(9):091101.
193. Bause M, Wittenstein T, Kremer K, Bereau T. Microscopic reweighting for non-equilibrium steady-state dynamics. *Phys Rev E*. 2019;100(6): 060103.
194. Salvi N, Abyzov A, Blackledge M. Solvent-dependent segmental dynamics in intrinsically disordered proteins. *Sci Adv*. 2019;5(6):eaax2348. https://doi.org/10.1126/sciadv.aax2348.
195. Shevchuk R, Hub JS. Bayesian refinement of protein structures and ensembles against SAXS data using molecular dynamics. *PLoS Comput Biol*. 2017;13(10):e1005800.
196. Andrae R, Schulze-Hartung T, Melchior P. Dos and don'ts of reduced chi-squared. *arXiv*. 2010;1012.374v1. https://doi.org/10.1088/0004-637X/733/1/66. http://arxiv.org/abs/1012.3754.
197. Hansen PC. The L-curve and its use in the numerical treatment of inverse problems. In: Johnston PR, ed. *Computational Inverse Problems in Electrocardiology*. WIT Press; 2000:119–142.
198. Chen PC, Shevchuk R, Strnad FM, et al. Combined small-angle X-ray and neutron scattering restraints in molecular dynamics simulations. *J Chem Theory Comput*. 2019;15:4687–4698. https://doi.org/10.1021/acs.jctc.9b00292.
199. Crehuet R, Buigues PJ, Salvatella X, Lindorff-Larsen K. Bayesian-maximum-entropy reweighting of IDP ensembles based on NMR chemical shifts. *Entropy*. 2019; 21(9):898.
200. Stark AC, Andrews CT, Elcock AH. Toward optimized potential functions for protein-protein interactions in aqueous solutions: osmotic second virial coefficient calculations using the MARTINI coarse-grained force field. *J Chem Theory Comput*. 2013;9(9):4176–4185. https://doi.org/10.1021/ct400008p.
201. Javanainen M, Martinez-Seara H, Vattulainen I. Excessive aggregation of membrane proteins in the martini model. *PLoS ONE*. 2017;12(11):1–20. https://doi.org/10.1371/journal.pone.0187936.
202. Best RB, Zheng W, Mittal J. Balanced protein-water interactions improve properties of disordered proteins and non-specific protein association. *J Chem Theory Comput*. 2014;10(11):5113–5124.
203. Hospital A, Goñi JR, Orozco M, et al. Molecular dynamics simulations: advances and applications. *Adv Appl Bioinform Chem*. 2015;8:37.
204. Mobley DL, Bannan CC, Rizzi A, et al. Escaping atom types in force fields using direct chemical perception. *J Chem Theory Comput*. 2018;14(11):6076–6092.
205. Zanette C, Bannan CC, Bayly CI, et al. Toward learned chemical perception of force field typing rules. *J Chem Theory Comput*. 2018;15(1):402–423.
206. Open Force Field Initiative. *Open Force Field Initiative*; 2019. https://openforcefield.org/. Accessed 28.07.19.

CHAPTER FOUR

Enhanced sampling and free energy calculations for protein simulations

Qinghua Liao*
Science for Life Laboratory, Department of Chemistry-BMC, Uppsala University, Uppsala, Sweden
*Corresponding author: e-mail address: qinghua.liao@kemi.uu.se

Contents

Abstract

Molecular dynamics simulation is a powerful computational technique to study biomolecular systems, which complements experiments by providing insights into the structural dynamics relevant to biological functions at atomic scale. It can also be used to calculate the free energy landscapes of the conformational transitions to better understand the functions of the biomolecules. However, the sampling of biomolecular configurations is limited by the free energy barriers that need to be overcome, leading to considerable gaps between the timescales reached by MD simulation and those governing biological processes. To address this issue, many enhanced sampling methodologies have been developed to increase the sampling efficiency of molecular dynamics simulations and free energy calculations. Usually, enhanced sampling algorithms can be classified into methods based on collective variables (CV-based) and approaches which do not require predefined CVs (CV-free). In this chapter, the

Progress in Molecular Biology and Translational Science, Volume 170
ISSN 1877-1173
https://doi.org/10.1016/bs.pmbts.2020.01.006

theoretical basis of free energy estimation is briefly reviewed first, followed by the reviews of the most common CV-based and CV-free methods including the presentation of some examples and recent developments. Finally, the combination of different enhanced sampling methods is discussed.

1. Introduction

Thanks to the tremendous progress in both computing power and methodology, molecular dynamics (MD) simulation has been widely applied in computational modeling of biomolecules ranging from proteins to membrane lipids and nucleic acids since the first MD simulation of a protein, bovine pancreatic trypsin inhibitor in 1977.[1] In an MD simulation, the trajectory of the interacting atoms of a biological system is determined by solving Newton's equations of motion numerically, where the forces on atoms and the potential energies are usually calculated using molecular mechanics force fields (see chapters "Pairwise-additive and polarizable atomistic force fields for molecular dynamics simulations of proteins" by Justin Lemkul and "Scale-consistent approach to the derivation of coarse-grained force fields for simulating structure, dynamics, and thermodynamics of biopolymers" by Adam Liwo of this book for more information on force fields). Structural and dynamical information of the biomolecules can be extracted from MD trajectories, which can be compared to and used to interpret experimental results determined from, for example, neutron scattering,[2] nuclear magnetic resonance (NMR),[3–5] small-angle X-ray scattering (SAXS)[6] and cryogenic electron microscopy (cryo-EM).[7] Methods and protocols were also developed for using MD simulations to refine structures obtained from NMR spectroscopy,[8] cryo-EM,[9,10] X-ray crystallography,[11] and SAXS.[12]

Currently, MD simulations are limited to tens of microseconds for normal-size systems, and even shorter timescales for larger systems. However, the results obtained from MD simulations are reliable only if the simulations are long enough for the systems to visit all the relevant configurations in phase space. Unfortunately, MD trajectories are mostly not ergodic as many relevant regions in configurational space are left unexplored in MD simulations. In most cases the reason for this is that relevant configurations are separated by high free energy barriers, thus it would take milliseconds or even longer for configurational transitions across these energy barriers to occur. There is often a gap between the timescales than can be covered with MD simulations and those of biological events. This results in insufficient sampling and thus unconverged simulations of biomolecular systems, which also causes the free energy calculations to be unreliable.

To solve the problem of insufficient sampling, many enhanced sampling approaches have been developed during the last decades.[13–15] Usually, these methods can be divided into two classes: one class is based on adding bias potentials along predefined collective variables (CVs) of the system (CV-based), while the methods of other class do not need predefined CVs (CV-free). The CV-based sampling methods greatly improve the convergence of free energy calculations when the CVs are well selected. Umbrella sampling (US),[16] metadynamics (MetaD),[17] steered molecular dynamics (SMD),[18] adaptive biasing force (ABF),[19] conformational flooding,[20] string method,[21] and temperature accelerated MD (TAMD)[22] are widely applied methods belonging to this class of enhanced sampling methods. However, the selection of the proper CVs is a nontrivial task for many systems. In such cases, CV-free sampling methods are advantageous as they allow sampling biomolecular configurations without requiring a priori knowledge of the system. They are useful for exploring possible biomolecular structural transition pathways and determining possible unknown intermediate states of the systems. Replica exchange (RE),[23] or parallel tempering (PT), self-guided molecular Langevin dynamics (SGLD),[24,25] and accelerated molecular dynamics (aMD)[26] are well-known methods of this second class of sampling methods. For larger or more complex systems, a combination of enhanced sampling methods is a powerful approach for efficient sampling of conformational transitions, such as replica exchange umbrella sampling (REUS)[27,28] or bias-exchange metadynamics (BE-MetaD).[29]

In this chapter, three CV-based (US, MetaD, SMD) and two CV-free (REMD, aMD) enhanced sampling methods for biomolecular simulations are reviewed by providing introductions to their theoretical basis, presenting illustrative applications, and discussing recent developments. To better understand these methods, various methods for the calculation of free energies from simulations are introduced first. The end of the chapter is focused on methods that combine different enhanced sampling algorithms, followed by an overview of the availability of the different methods in the common biomolecular MD simulation engines.

2. Collective variable and free energy

For MD simulations, collective variables (CVs) instead of atomic coordinates are often used to quantify certain properties of the simulated system, thereby reducing the highly dimensional problem. A CV is a function of atomic coordinates which can describe a certain motion or transition of interest. For example, a distance between two atoms is a simple CV which can be used to describe bond forming and breaking, while a dihedral angle is usually used to characterize an isomerization process as a CV. For a proton transfer process, the distances between the proton and each of the proton donor and acceptor could be two reasonable CVs. In short, a CV is coarse-grained description of the system.

With a defined CV s, the probability of observing the system at a given CV point s_i is the weight of all configurations q mapping to s_i:

$$P(s_i) = \langle \delta[s(q) - s_i] \rangle \tag{1}$$

where the Dirac δ function picks only s_i from all possible $s(q)$, and the angular bracket $\langle \cdot \rangle$ denotes an ensemble average. A free energy can be obtained from the probability by:

$$G(s_i) = -k_B T \ln \langle \delta[s(q) - s_i] \rangle \tag{2}$$

where k_B is Boltzmann constant and T is the temperature. Based on the free energy landscape $G(s)$, the energetic cost of a maximally efficient transition from one region of the CV space to another one can be measured. Fig. 1 shows an illustrative free energy landscape along a CV involving a conformational transition between states A and B. State A is the global minimum, and state B is a metastable state. States A and B are connected by an intermediate region with high energy, transition state C. The transition rate between A and B is heavily dependent on the height of the free energy barrier compared with the thermal energy $k_B T$, according to the Arrhenius-like equation:

$$\nu_{A \to B} = \nu_0 \exp\left(-\frac{\Delta G^{\ddagger}}{k_B T}\right) \tag{3}$$

where ν_0 is a prefactor, and $\Delta G^{\ddagger}$ is the free energy difference between state A and the transition state C:

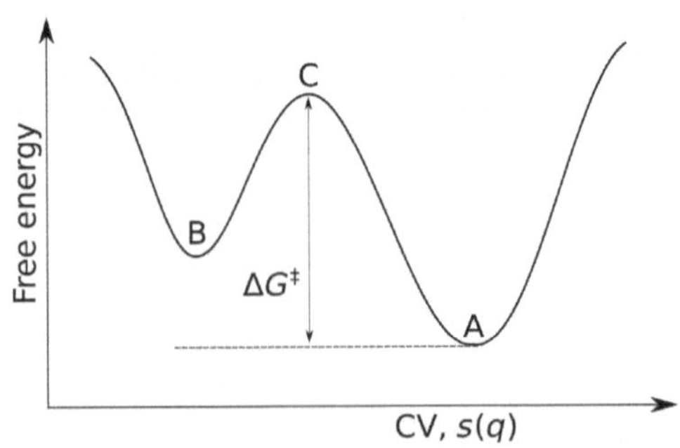

Fig. 1 A double-well potential model. The free energy is plotted against a CV. The stable state is termed as A, the metastable one is termed as B, while the transition state is C.

$$\Delta G^{\ddagger} = G_C - G_A = -k_B T \ln \frac{\langle \delta[s(q) - s_C] \rangle}{\langle \delta[s(q) - s_A] \rangle} \tag{4}$$

If an MD simulation is perfectly ergodic (long enough), all possible configurations in the CV space would be sampled. In the case of the double-well potential (Fig. 1) this would imply that enough transitions between states A and B crossing C were observed. This would allow one to determine how often each point s_i in the CV space is sampled, and the free energy can be computed by normalizing this histogram into a probability $P(s_i)$ via Eqs. (1) and (2). Histograms are concerned with computing CV distributions and averages. They are related to the fluctuations in the ensemble of interest, and further enable us to calculate free energies and entropies. There are many approaches developed to estimate free energy differences between discrete states (histograms).

The Zwanzig equation is the most well-known method historically for calculating free energies.[30] The classical free energy between two states characterized by Hamiltonians $H_i(q)$ and $H_j(q)$ can be estimated as:

$$\begin{aligned} \Delta G_{i \to j} &= -k_B T \ln \left\langle \exp\left[-\frac{H_j(q) - H_i(q)}{k_B T} \right] \right\rangle_i \\ &= -k_B T \ln \left\langle \exp\left[-\frac{\Delta H(q)}{k_B T} \right] \right\rangle_i \end{aligned} \tag{5}$$

This method is denoted as EXP, for exponential averaging. However, when the phase space overlap of the two states is too low, EXP becomes less accurate. The statistical uncertainty of EXP grows exponentially when the overlap drops. When the overlap between two states is poor, introducing a number of intermediate states can efficiently improve the computation of free energy differences.

A significantly more robust estimator was originally proposed by Bennett[31] in 1976, and further improved by Shirts et al.[32] The resulting Bennett acceptance ratio (BAR) method accounts for the expectations computed at both states:

$$\Delta G_{ij} = -\beta^{-1} \ln \frac{Q_j}{Q_i} = \beta^{-1} \ln \frac{\langle \alpha(q) \exp[-\beta H_i(q)] \rangle_j}{\langle \alpha(q) \exp[-\beta H_j(q)] \rangle_i} \tag{6}$$

where β equals to $(k_B T)^{-1}$, Q_i is the canonical partition function of state i, and $\alpha(q)$ is a weight function, which can be any everywhere-finite function of the coordinates q. Bennett used variational calculus to obtain the values of

$\alpha(q)$ minimizing the variance of the free energy.[31] The final result is an implicit function of the free energy and the number of samples n_i and n_j at each states have to fulfill the equation:

$$\sum_{i=1}^{n_i} \frac{1}{1+\exp(\ln(n_i/n_j)+\beta\Delta H_{ij}-\beta\Delta G)} - \sum_{j=1}^{n_j} \frac{1}{1+\exp(\ln(n_j/n_i)-\beta\Delta H_{ji}+\beta\Delta G)} = 0 \quad (7)$$

When multiple intermediate states are sampled, the weighted histogram analysis method (WHAM)[33] can be applied to combine information from all intermediates to obtain an optimal estimation of the free energy differences between all pairs of intermediates. WHAM is an improved version of the histogram weighted technique which was first proposed by Ferrenberg and Swendsen.[34] WHAM has been proven to be the lowest uncertainty method to calculate the free energy for samples collected from discrete intermediate states.[35] As the WHAM approach is widely used together with umbrella sampling, more details about WHAM are presented in the Umbrella Sampling section.

The multistate Bennett acceptance ratio (MBAR)[36] is a direct extension of BAR, which allows for accessing the data from all states and predicting the free energy at unsampled regions. MBAR is equivalent to BAR in the limit that only two states are assessed. The equation for the corresponding free energy calculation can be considered as a zero-width bin WHAM:

$$G_i = -\beta^{-1}\ln\sum_{k=1}^{K}\sum_{n=1}^{N_k}\frac{\exp[-\beta H_i(q_{kn})]}{\sum_{k'=1}^{K} N_{k'}\exp[\beta G_{k'} - \beta H_{k'}(q_{kn})]} \quad (8)$$

Here, k and k' are indexes of K, which is the number of states, and N_k and $N_{k'}$ are the number of configurations at state k and k', respectively. G_i is the free energy of state i, and $H_i(q_{kn})$ is the Hamiltonian at state i of configuration n at state k. To be noted, only the free energy difference $\Delta G_{ij} = G_j - G_i$ is meaningful.

In many cases, simulations are trapped in minima (like A and B in Fig. 1)—and often the configurations from where simulations were started are minima—and barely cross the barriers to visit other metastable states. Therefore, normal MD simulations are insufficient for free energy calculations. MD simulations with enhanced sampling are thus necessary to speed up the crossing of free energy barriers with limited computational resources.

3. CV-based sampling

CV-based enhanced sampling is also called constrained enhanced sampling. With CV-based enhanced sampling, a bias potential as a function of chosen collective variables (CVs) is added to the potential energy of the simulated system to improve the sampling efficiency. As a result, the sampled space is largely constrained to the desired CVs space. The selection of CVs is critical to the biased sampling. More details about selecting CVs are provided in previous reviews.[37,38] Bias potentials can be added to the simulated system in a variety of ways to obtain efficient sampling. In this section, three widely used CV-based methods are discussed.

3.1 Umbrella sampling

Umbrella sampling (US) was developed by Torrie and Valleau[16] in 1977. It is one of the most widely used approaches to overcome free energy barriers in free energy calculation. As shown in Fig. 2, umbrella sampling uses a series of independent windows along a selected CV to model a conformational transition. For each of the windows, a simple harmonic potential $\Delta V_i(q)$, whose strength defined by the force constant k is chosen depending on the potential energy at each of the reference points s_i along the CV $s(q)$, is added to the system's Hamiltonian:

$$\Delta V_i(q) = \frac{k}{2} \times (s(q) - s_i)^2 \tag{9}$$

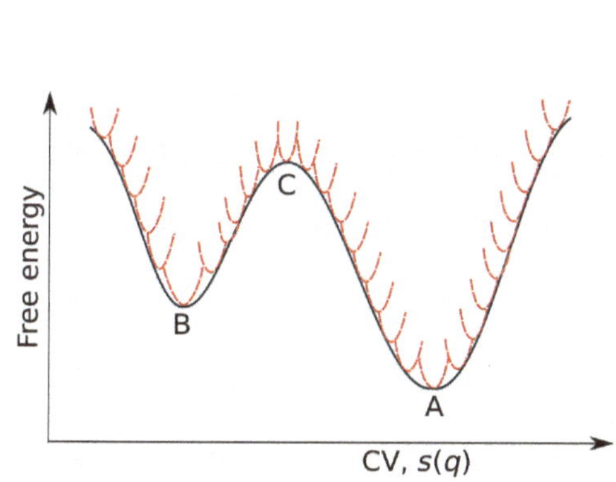

Fig. 2 Schematic illustration of the umbrella sampling method. The *red dash lines* represent the harmonic bias potentials that are added to the system Hamiltonian at different CV points (windows) along the CV space.

and MD simulations are then performed for all windows. When these simulations are finished, WHAM[33] is the most popular method to combine the statistics from all the independent windows. The WHAM equations are expressed as follows:

$$P_u(s) = \frac{\sum_{l=1}^{N} h_l(s)}{\sum_{k=1}^{N} n_k \exp\left([F_k - V(s)]/k_B T\right)} \tag{10}$$

$$F_k = -k_B T \ln \sum_q P_u(s) \exp\left(-V(s)/k_B T\right) \tag{11}$$

where $P_u(s)$ is the unbiased probability distribution along s, N in the number of umbrella windows, l and k are indexes of the umbrella windows, $h(s)$ is the counts at bin s, n_k is the number of data points from window k, and F_k is the factor to be determined for window k. The two equations are coupled and solved iteratively until a self-consistent solution is reached. More details of the WHAM iteration method were previously described by Kumar et al.[33] and Roux.[39] WHAM was shown to be equivalent to BAR when only two states are assessed. Moreover, other WHAM variants were also developed based on maximum-likelihood[40] and Bayesian methods.[41,42] Alternatively, the umbrella integration (UI)[43] method and multistate Bennett acceptance ratio (MBAR)[36] can also be used to recover the unbiased free energy landscape from umbrella sampling. More details about umbrella sampling are discussed by Kästner.[35]

Due to its quick convergence and the fact that the MD simulations for the different windows can be run independently from each other, allowing to add further windows to the system to improve convergence, umbrella sampling has been demonstrated great successes by a tremendous amount of studies that employed umbrella sampling in a wide range of biological systems including protein folding,[44] conformational changes of large proteins, protein–protein interactions,[45] protein–lipid association,[46] ligand–protein binding,[47] DNA conformational change,[48] enzyme catalysis.[49,50] A recent study where lipid-protein binding was studied using US[46] is chosen as illustrative example. The authors of this study applied US to characterize the free energy profile, also called potential of mean force (PMF), for the binding of a cardiolipin (CL) molecule to the mitochondrial transport protein ANT within membranes—modeled by a POPC (1-palmitoyl-2-oleoyl-sn-glycero-3-phosphocholine) membrane—using coarse-grained (CG) MD simulations. The CV was defined as the distance between the CG glycerol particle of CL and the backbone bead of the proline residue in a conserved motif of the protein family (Fig. 3A). MD simulations of 7 μs were performed for 32 windows linearly spaced between 1.6 and 4.5 nm along the chosen CV with a force constant

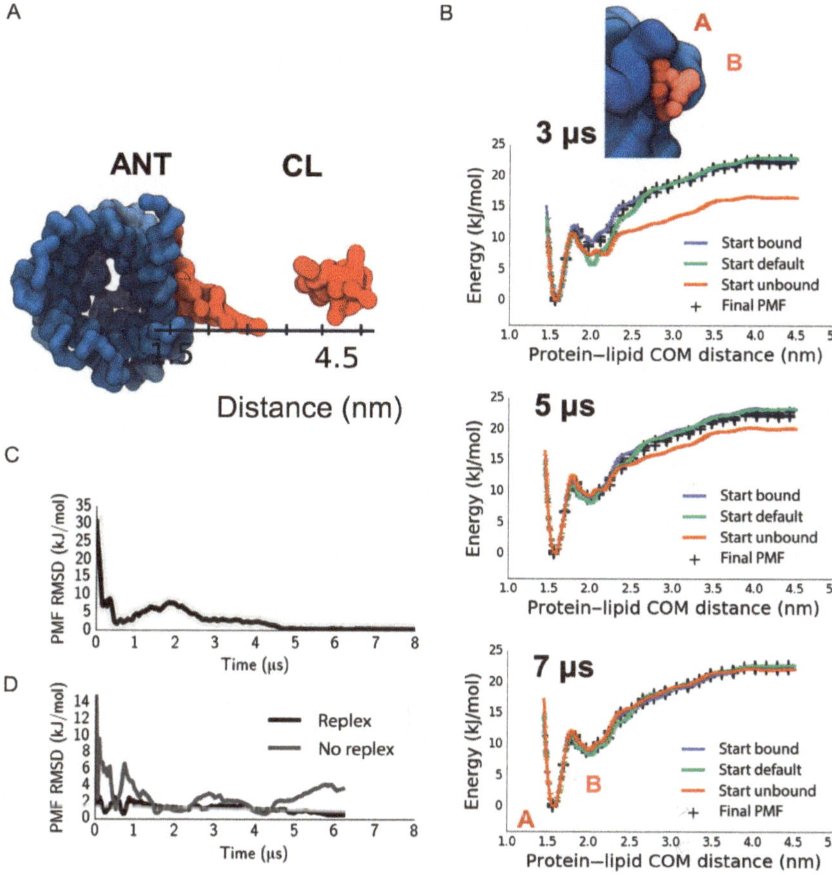

Fig. 3 Cardiolipin binding to ANT, providing an example of calculating a free energy profile for a lipid interacting with a specific binding site on an integral membrane protein. (A) Representative structures of the closest ($d = 1.5$ nm) and furthest ($d = 4.5$ nm) protein–lipid COM distance in the US simulation. (B) PMFs obtained from three different US schemes with different initial configurations (default in *cyan*, unbound in *orange*, bound in *blue*), computed after 3, 5, and 7 μs of the simulation (within each case the first half of the simulation was discarded as equilibration). The final PMF (indicated by *black crosses*) was calculated from the two US replica exchange simulations, corresponding to the starting bound and unbound initial conditions. (C) RMSD between the two PMFs started from all bound and all unbound configurations as a function of time. (D) RMSD between the final PMF and US simulations started from the default initial condition and either allowing for replica exchange or without exchange. *Reprinted with permission from Domański J, Hedger G, Best RB, et al. Convergence and sampling in determining free energy landscapes for membrane protein association. J Phys Chem B. 2017;121:3364–3375, copyright 2017 American Chemistry Society.*

of $k = 1000$ kJ·mol^{-1} ·nm^{-2} (see Eq. 9). Three US schemes were tested, termed as default, bound and unbound. The default scheme corresponds to windows starting from the bias center of each window, which were generated by steered MD simulations (see Section 3.3 for details). The other two schemes were started with all windows being in the bound state (protein–lipid center of mass distance $d = 1.5$ nm) or in the unbound state ($d = 4.5$ nm). The latter two schemes are special as here every 200 ps it was attempted to exchange the configurations between windows according to the Boltzmann criterion, i.e., the same idea as applied in replica exchange MD (see Section 4.1 for details) was used. The unbiased free energy profiles from the three US simulations were determined by WHAM. Two distinct minima were observed, at $d = 1.5$ nm and $d = 2.0$ nm with well-depths of 20 and 10 kJ·mol^{-1}, respectively (Fig. 3B). After 7 μs, the PMF profiles had well converged. The progress of convergence was also monitored by calculating the root-mean-square difference (RMSD) between the PMFs derived from the US simulation started with all windows bound and that started with all windows unbound. This PMF RMSD becomes less than 1 kJ·mol^{-1} after 5 μs (Fig. 3C). Up to this difference, the PMF is thus insensitive to the initial configuration. The authors also demonstrated that exchange between the windows greatly improved the sampling efficiency and speed up the convergence of US (Fig. 3D).

In practice, the harmonic potentials for all the windows need to be tuned manually, so that reasonable positional overlap between the windows along the CVs can be obtained. This implies that more windows are needed, the higher the force constants of the harmonic potentials are, which is especially required to ensure the sampling at high energy regions. The results obtained from US can be very sensitive to the starting configurations as the sampling is restrained to a simple pathway, other important pathways could be neglected while the free energy profiles may appear to be converged. Furthermore, when the number of CVs becomes larger (≥ 3) or the system of interest is complicated, the computation will be demanding, and it will be difficult to determine accurate results. Thus, self-learning adaptive US[51] was proposed. Moreover, the replica exchange algorithm was combined with umbrella sampling (REUS)[27,28] to further improve the sampling efficiency.

3.2 Metadynamics

Metadynamics (MetaD) was originally developed by Parrinello and coworkers to enhance the sampling of rare events.[17] In MetaD simulation, an external

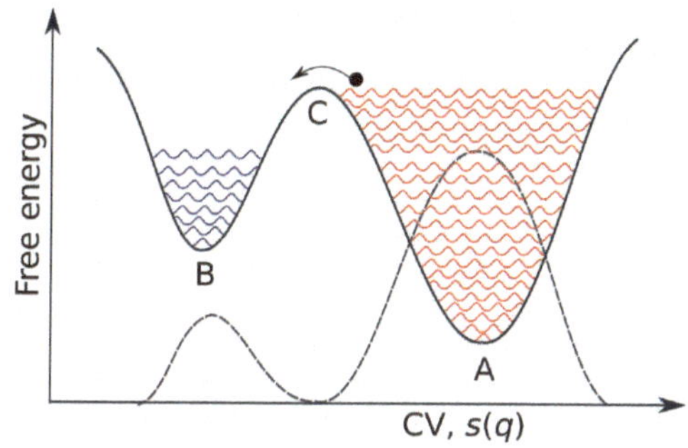

Fig. 4 Schematic illustration of metadynamics simulation. In MetaD, history-dependent Gaussian-type bias potentials (first *red* Gaussians, then *blue* Gaussians) are deposited along the CV space. The deposited potentials encourage the system to explore disfavored configurations. As shown here, when basin A is filled up with Gaussians, the system can explore configurations at with high energies, followed by crossing the transition state C and going to basin B. After basin B is also fully filled up, the free energy landscape along CV space is flattened. The summation of the deposited bias potentials provides an estimation of the negative free energy landscape (*gray dash line*).

history-dependent bias potential is imposed on the system during the simulation, the bias potential which is a function of the CVs can be written as a sum of Gaussians added along the CVs space to encourage the system to visit configurations which have not already been sampled (Fig. 4). The Gaussian-type bias potential $V(s(q), t)$ can be expressed as:

$$V(s(q),t) = \sum_{k\tau<t} W(k\tau)\exp\left(-\sum_{i=1}^{d}\frac{(s_i - s_i(q(k\tau)))^2}{2\sigma_i^2}\right) \tag{12}$$

where τ is the Gaussian deposition rate, σ_i is the width of the Gaussian for the ith CV, and $W(k\tau)$ is the height of the Gaussian at simulation time $k\tau$. The effect of the positive bias potential is to push the system out of any local minimum in the CVs space, to encourage visiting unexplored configurations. In the long time limit, the bias potential will converge to the negative free energy as a function of the CVs, thus the free energy landscapes can be reconstructed by reversing the imposed bias potential:

$$V(s, t\rightarrow\infty) = -G(s)+C \tag{13}$$

where C is a constant. The added bias potential leads to highly efficient sampling in which the energy barriers separating different local minima can be easily crossed. In standard MetaD, the height of the deposited Gaussian-type bias potential is constant during the simulation. As a result, the estimation of the free energy landscape calculated from the bias potential oscillates around the real values, and there is also a risk of driving the system into physically irrelevant regions of phase space. To solve this problem of convergence difficulties, the well-tempered metadynamics (WT-MetaD)[52] was developed, where the height of the added Gaussian potentials decreases with simulation time according to:

$$W(k\tau) = W_0 \exp\left(-\frac{V(s(q(k\tau)), k\tau)}{k_B \Delta T}\right) \tag{14}$$

where W_0 is the initial height and ΔT is a parameter with the dimension of a temperature. This rescaling technique indeed ensures that the bias potential converges more smoothly though it does not compensate the underlying free energy:

$$V(s, t \rightarrow \infty) = -\frac{\Delta T}{T + \Delta T} G(s) + C \tag{15}$$

where T is the temperature of the system. When $\Delta T = 0$, it is identical to standard molecular dynamics, while $\Delta T \rightarrow \infty$ corresponds to standard MetaD. For WT-MetaD as implemented in PLUMED,[53,54] the corresponding input parameter is termed "bias factor" γ:

$$\gamma = \frac{T + \Delta T}{\Delta T} \tag{16}$$

MetaD is robust and easy to use, it is one of the most widely used enhanced sampling methods for free energy calculations, and continues to be applied in simulations by the community. Comparing to methods such as US,[16] a major advantage of MetaD is that no a priori knowledge of the end states is required and that multiple CVs can be used, it is very helpful to explore possible pathways while umbrella sampling is strictly constrained to certain pathways. However, the efficiency of MetaD scales poorly with the number of CVs used and in practice is limited to three CVs. To some extent, the accuracy and convergence of MetaD are also depending on the choices of the parameters of the Gaussians ($W(k\tau)$, σ in Eq. (12) and γ in Eq. (16)), but there is no universal method to choose the parameters, which is one disadvantage of MetaD.

A lot of variants were further developed to address the issues and to further improve sampling efficiency including metadynamics with parallel tempering (PTMetaD),[55,56] metadynamics with multiple walker (MW-MetaD),[57] funnel metadynamics,[58] and bias-exchange metadynamics (BE-MetaD).[29]

More recently, parallel bias metadynamics (PBMetaD)[59] was developed to enhance the crossing of high energy barriers by applying multiple parallel bias potentials in low dimensionality. A newly proposed infrequent metadynamics (InMetaD)[60] can be used to accurately predict the long time scale kinetics, and it was successfully used to estimate ligand unbinding kinetics.[61–64] Wang et al. further developed this scheme termed as frequency-adaptive metadynamics (FaMetaD).[65] More details on the theory and applications of MetaD simulation were presented in previous reviews.[37,38,66]

MetaD has been successfully applied to a wide range of biological problems, including unbinding kinetics of drug molecules[61–64], exploration of conformational changes of biomolecules,[67,68] predictions of ligand binding poses[69] and enzyme catalysis.[70,71] In one of our recent studies,[68] we applied both Hamiltonian replica exchange molecular dynamics (HREX)[72] and bias-exchange metadynamics (BE-MetaD)[29] simulations to study the conformational dynamics of loop 6 of triosephosphate isomerase (TIM), which is critical to the catalytic efficiency of this enzyme (5A and B). TIM catalyzes a simple reversible isomerization of dihydroxyacetone phosphate (DHAP) and (R)-glyceraldehyde-3-phosphate (GAP).[73,74] Loop 6 prefers the closed state when the substrate DHAP is bound to TIM, but the open state when no substrate is bound. For the reaction to occur loop 6 needs to be closed, and it has to open to release the product when the reaction is finished.[73,74] In the HREX simulations, loops 6 (Pro166-Ala176) and 7 (Tyr208-Ser211) were chosen as the hot regions (see Section 4.1 for the theory of HREX). The simulations were carried out using 8 replicas with scaling factors exponentially ranging from 1.0 to 0.6, corresponding to an effective temperature range from 300 to 500 K. Each replica was sampled for 200 ns for apo-TIM and for 300 ns for DHAP-TIM. Our HREX simulations failed to capture the closed state of loop 6. The closed conformation of loop 6 appeared to be briefly sampled in the DHAP-TIM complex (Fig. 5D), due to the fact that the simulations were started from the closed state, and the loop opened within very short simulation time and did not close again. Then we turned to BE-MetaD simulations, in which 7 biased and 1 unbiased replicas were simulated. The distances between the backbone atoms (N, C_α, C, O) of each of the 7 residues Ala169-Ala175 of loop 6 and the backbone atoms of the residues of loop 7 (Tyr208-Ser211) were used to train the native contacts[75] (ranging from 1 to 0) as the 7 CVs in the 7 biased replicas. For the MetaD simulation on each of the 7 replicas, a time-dependent Gaussian-type potential was added to the 7 CVs every 2 ps with an initial height of 0.1 kcal·mol^{-1}, which was gradually decreased with a

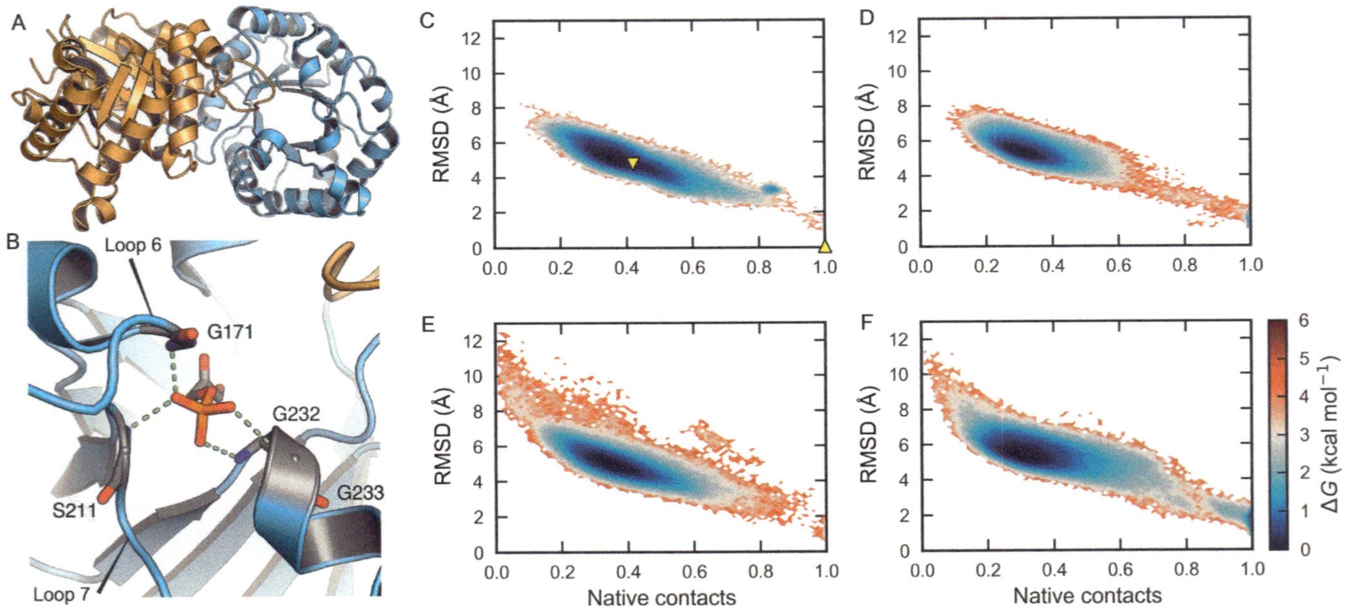

Fig. 5 (A) The overall structure and (B) active site with bound DHAP of TIM (PDB ID: 1NEY[76,77]), (C) Relative free energy surfaces ($T = 300$ K) calculated from (C) and (D) HREX and (E) and (F) BE-MetaD simulations of (C, E) apo-TIM and (D) and (F) DHAP-TIM. The surfaces are defined in terms of the native contacts between loops 6 and 7 and the RMSD of loop 6 relative to its conformation in the crystal structure of DHAP-TIM. In (C) the open and closed states of the loops are projected from chains A of crystal structures 1YPI (▽) and 1NEY (△), respectively. *Reprinted with permission from Liao Q, Kulkarni Y, Sengupta U, et al. Loop motion in triosephosphate isomerase is not a simple open and shut case. J Am Chem Soc. 2018;140:15889–15903, copyright 2018 American Chemistry Society.*

bias factor γ of 10. The width of the Gaussians was set at 0.05. Exchanges between replicas were attempted every 2 ps, and each replica was sampled for 500 ns. Unlike the HREX simulation, the BE-MetaD simulations captured the opening and closing of loop 6 in the presence of DHAP (Fig. 5F). Based on the BE-MetaD sampling, we also explored the influence of the loop 6 conformational dynamics on the catalytic activity of TIM.[68]

3.3 Steered molecular dynamics

Inspired by atomic force microscopy (AFM) experiments, steered molecular dynamics (SMD) was developed.[18,78,79] SMD is often applied to drive the system from one initial state to a final state by pulling in one or multiple directions. In SMD simulations, an external bias potential, which is usually a function of the selected CVs and simulation time, is added to the Hamiltonian of the system to drive it from one state to another:

$$V_\lambda(q,t) = \frac{k(t)}{2}(s(q,t) - \lambda(t))^2 = \frac{k(t)}{2}(s(q,t) - s_0 - \nu t)^2 \tag{17}$$

where $k(t)$ is the force constant of the harmonic potential, ν is the pulling speed, $s(q)$ are the chosen CVs to define the pulling directions, λ is an external parameter that is correlated with $s(q)$. SMD is a nonequilibrium sampling method, but the free energy differences between states i and j driven by SMD can also be calculated by measuring the work $W_{i\to j}$ done on the system as a result of pulling.[78,80,81] To this end, the Jarzynski equation[80] is used:

$$\Delta G = -\beta^{-1} \ln \langle \exp(-\beta W_{i\to j}) \rangle_0 \tag{18}$$

Later on, Crooks proposed a new version[82]:

$$\Delta G = -\beta^{-1} \ln \frac{\langle f(W_{i\to j}) \rangle_i}{\langle f(W_{j\to i}) \rangle_j} \tag{19}$$

where $f(W_{i\to j})$ denotes any finite function of the work. The pulling causes an instantaneous transformation, where the instantaneous work is simply the change in potential energy. Thus, the Jarzynski equation can be considered identical to EXP in Eq. (5), and the Crooks relation can be used to derive the BAR equation (see Eq. (6)) for nonequilibrium work.[32,82]

The SMD approach is widely applied in studies of conformational changes of proteins,[83] unbinding processes of ligands from their binding pockets, exploring possible ligand unbinding pathways,[84,85] and the free energy profiles along dissociation pathways can be evaluated.[86,87] Its role in rational drug design has been reviewed.[88] Furthermore, SMD is powerful in exploring the CVs space in short simulation time, thus it is widely applied together in combination with US and MetaD simulations, for example, to prepare starting conformations for umbrella sampling.[46,89,90] An adaptive SMD method was also proposed,[91] where the idea is that the initial configuration for a given step is obtained from the previous step of the trajectories. To be noted, when the CV is a root-mean-square deviation (RMSD), SMD is then also called targeted MD (TMD),[92] which was initially proposed to search pathways of conformational transitions.

4. CV-free sampling

Although CV-based enhanced sampling algorithms can greatly extend the timescales of MD simulations, predefined CVs are required to precisely describe the actual biological events in these methods, and they usually suffer from hidden energy barriers. It is nontrivial to determine the optimal CVs in MD simulations, especially when the transition processes are not straightforward. The simulations can be very difficult to converge when the CVs are not chosen properly. An example of a hidden barrier is shown in Fig. 6; there is only one basin along CV1, but two basins along CV2. If only CV1 is chosen for biasing sampling, the convergence of CV2 cannot be guaranteed while CV1 can be well sampled. The issue of hidden barriers and thus hidden CVs is explained in more details in other reviews.[14,38] CV-free enhanced sampling techniques, such as replica exchange molecular dynamics and

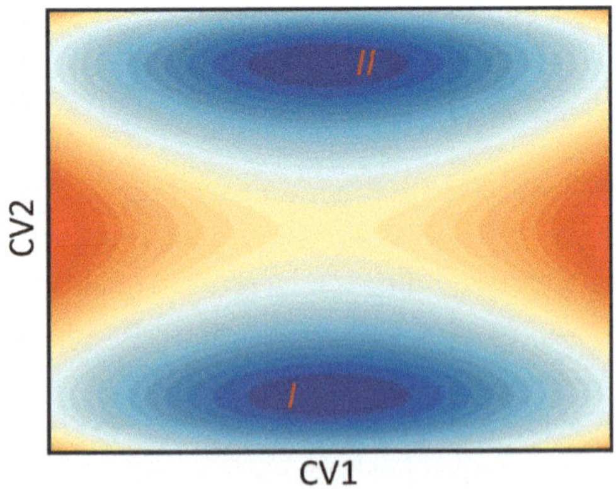

Fig. 6 An illustration of a hidden energy barrier along CV2 between the two basins, I and II.

accelerated molecular dynamics can be used to avoid this problem. Opposite to CV-based enhanced sampling methods, predefined CVs are not required in CV-free enhanced sampling approaches, thus a priori knowledge about the biological process to be studied is not necessarily needed. In Sections 4.1 and 4.2, brief overviews of the just mentioned CV-free enhanced sampling approaches are presented.

4.1 Replica exchange molecular dynamics

Replica exchange molecular dynamics (REMD),[23] also termed as parallel tempering (PT) was developed to enhance conformational exploration. In REMD, as shown in Fig. 7, several independent copies of the target system, which are called replicas, are simulated in parallel at different temperatures. At regular time intervals, exchanges of either the temperatures or the configurations of neighbored replicas are attempted. An exchange between replicas i and j is only accepted when the Metropolis criterion is satisfied:

$$P(q_i \leftrightarrow q_j) = \min\left(1, \exp\left[\left(\frac{1}{k_B T_j} - \frac{1}{k_B T_i}\right)[H(q_j) - H(q_i)]\right]\right) \quad (20)$$

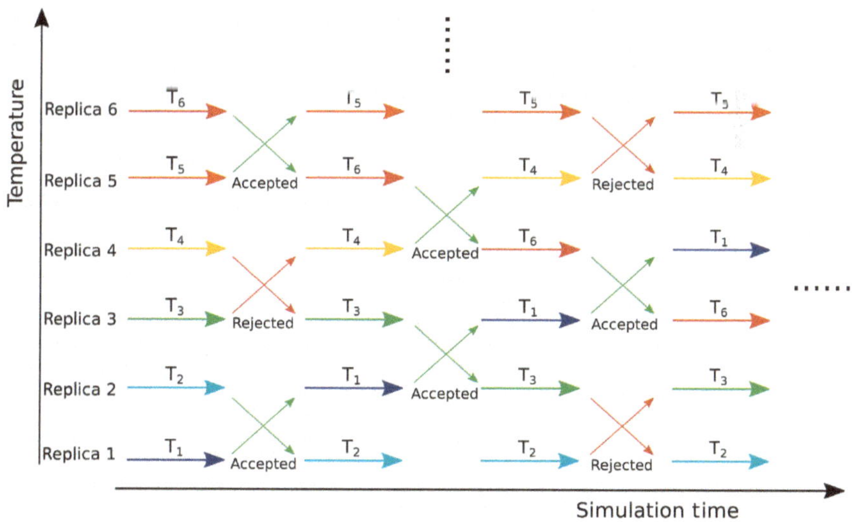

Fig. 7 Schematic of the replica exchange molecular dynamics protocol. In REMD, a number of MD simulations called replicas (six shown here) are performed in parallel and with different temperatures. At recurring intervals, temperatures, or coordinates of neighboring replicas are exchanged with a probability that meets the Metropolis criterion.

where $H(q_i)$ and and $H(q_j)$ are the potential energies for replicas i and j, respectively, and T_i and T_j are the temperatures of replicas i and j, respectively. Random walks of the replicas in the temperature ladder are realized, each replica experiences both high and low temperatures. The sampling of the conformational space is thus enhanced by transferring configurations which are only accessible at high temperatures to the replicas at lower temperatures, which in turn allows one to derive accurate thermodynamics information or free energy landscapes. The main issue of temperature REMD (T-REMD) is that the number of replicas greatly increases as the simulated systems become large, thus it is computationally demanding for complicated biological systems.

In T-REMD, the replicas differ in their temperature at which the MD simulations are performed. However, differences in the replicas are not limited to that. Any control parameter can be changed; for instance, also the Hamiltonian of the simulated system can be changed for the different replicas.[93,94] In a more generalized way, every replica is simulated at a different Hamiltonian and a different temperature (HT-REMD); the acceptance criterion is then expressed as:

$$P(q_i \leftrightarrow q_j) = \min\left[1, \exp\left(\frac{H_i(q_i) - H_i(q_j)}{k_B T_i} + \frac{H_j(q_j) - H_j(q_i)}{k_B T_j}\right)\right] \quad (21)$$

Various ways for modifying the Hamiltonian in replica exchange simulations have been proposed[72,95–99] to address the issues of T-REMD and promote the sampling efficiency. Solute tempering for simulations of solvated systems was an excellent idea among of them. In replica exchange with solute scaling (REST),[96,97] the solute–solute and solute–solvent interactions are scaled by different scaling factors in the different replicas, while solvent–solvent interactions are left unchanged. Thus a relative small number of replicas are sufficient to obtain good exchange probabilities, for which sufficient overlap in the energy contributions of the replicas to be exchanged is required. This algorithm was further improved in REST2,[99] in which all replicas are simulated at the same temperature, while the conformational sampling is enhanced through scaling of the intramolecular potential energy of the solute. Compared to T-REMD, REST2 showed significantly lower computational cost and better sampling efficiency. Later on, a new variant of REST2 with optimized scaling and more flexibility with respect to which parts of the system the scaling is applied was developed by Bussi and termed as Hamiltonian replica exchange molecular dynamics (also known as REST2 or HREX).[72] The Metropolis criterion of HREX for exchange attempts between two replicas i and j is

$$P(q_i \leftrightarrow q_j) = \min\left[1, \exp\left(\frac{H_i(q_i) - H_i(q_j)}{k_B T} + \frac{H_j(q_j) - H_j(q_i)}{k_B T}\right)\right] \tag{22}$$

When all replicas are simulated at the same temperature, Eq. (21) becomes Eq. (22). In HREX, all replicas are simulated at the same temperature, and the Hamiltonian of the solute is scaled by different scaling factors ($0 < \lambda \leq 1$) at different replicas, but the solvent–solvent interactions are not modified. With HREX, one can limit the scaling to a certain part of the simulated system which is considered to be interesting. For instance, in the example introduced in Section 3.2, HREX simulations were also mentioned. In that example only the interactions of loops 6 and 7 of TIM were modified during the HREX simulation.[68]

As REMD enhances unrestrained sampling, this method including its variants are widely applied to problems where the definition of a CV, especially prior sampling, is difficult or even impossible. Such problems include the sampling of peptides and their aggregation,[100,101] protein–protein recognition mechanisms,[102] or the identification of ligand binding sites and poses,[103–105] which could be useful for ligand identification in computational drug discovery. We used HREX to study the dynamics and dimerization of the amyloid peptide $A\beta_{1-42}$ under varying external conditions, such as different pH values, the presence of Cu^{2+} that binds to $A\beta$, and oxidation of $A\beta$ (oxidized Gly25).[106,107] For each dimer system, 24 replicas with scaling factors λ ranging from 1.0 to 0.4 (corresponding to effective temperatures of 300–750 K) were used in the HREX simulations, and each replica was subjected to 500 ns sampling, leading to 12 μs of cumulative simulation time for each dimer system. As shown in Fig. 8, we could characterize the different structural preferences of the four $A\beta_{1-42}$ dimer systems at different conditions. Moreover, transition network analysis[108,109] on the HREX trajectories revealed that the acidic pH considerably increased the β-sheet content, while Cu^{2+} binding increased the exposed hydrophobic surface area. Both of the two factors can contribute to an increased oligomerization and toxicity of $A\beta$.

Replica exchange is also widely combined with other computational methods, such as, constant pH replica exchange,[110] REMD with elastic

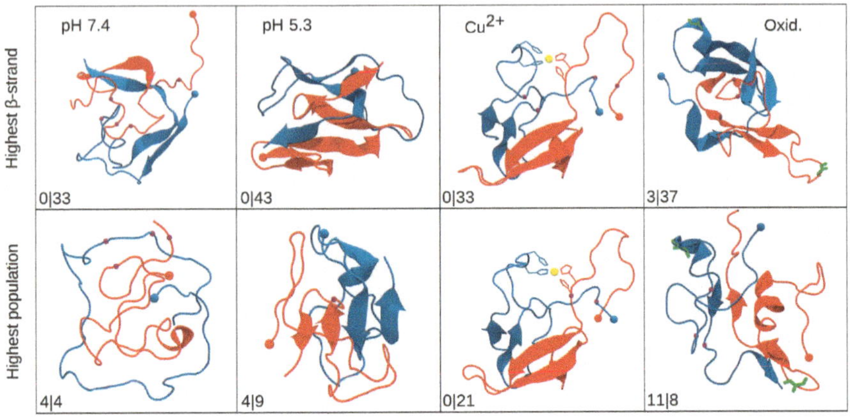

Fig. 8 Representative conformations with highest β-sheet content (*top*) and the most abundant conformations (*bottom*) of the $A\beta_{1-42}$ dimer at different conditions. The peptides are rendered as a cartoon with one peptide of the dimer shown in *red* and the other one in *blue*. The N-terminus of each peptide is indicated by a *sphere* and β-bridges are marked by *violet spheres*. In the Cu^{2+} system, Cu^{2+} is shown as *yellow sphere* and the His13 and His14 residues bound to it as sticks. In the oxidized system, the oxidized Gly25 residues are shown as *green sticks*. In the *lower left corner* of each panel the number of residues in the $\alpha|\beta$ conformation is given. *Reproduced with permission from Liao Q, Owen MC, Bali S, et al. Aβ under stress: the effects of acidosis, Cu^{2+}-binding, and oxidation on amyloid β-peptide dimers. Chem Commun. 2018;54:7766–7769, The Royal Society of Chemistry.*

network analysis,[111] REMD with FEP,[112–115] multiplexed-REMD,[116] replica exchanging self-guided Langevin dynamics[117,118] and quantum mechanics/molecular mechanics (QM/MM)[119] replica exchange.[49] Moreover, multidimensional REMD has also been applied.[93,120,121]

4.2 Accelerated molecular dynamics

In REMD, the sampling is enhanced by exchanging configurations sampled at high temperature replicas with configurations from low temperature replicas. In 2004, the McCammon group developed another promising CV-free enhanced sampling method, accelerated molecular dynamics (aMD).[26] In aMD, as shown in Fig. 9, a boost potential is added to the potential energy surface of the studied system to facilitate the crossing of energy barriers between different conformational states. The boost potential $\Delta V(q)$ is activated when the potential energy of the system is below a threshold energy E:

$$\begin{aligned} V^*(q) &= V(q) & V(q) &\geq E \\ V^*(q) &= V(q) + \Delta V(q) & V(q) &< E \end{aligned} \tag{23}$$

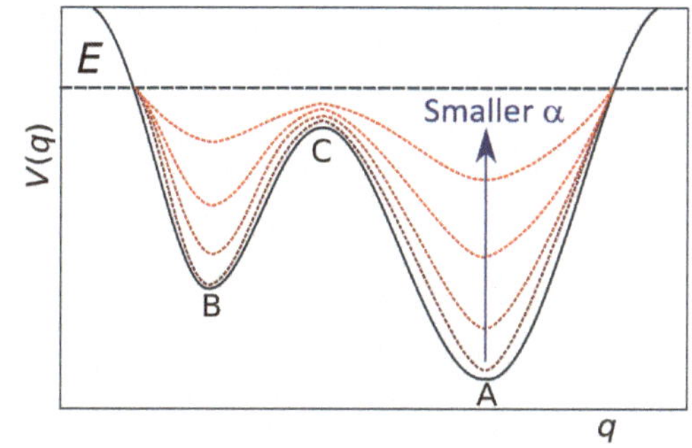

Fig. 9 An illustration of aMD. A nonnegative boost potential is imposed on the system potential energy surface when it is lower than a threshold energy *E*. As the acceleration factor α gets smaller, the potential energy surface becomes flatter and the number of transitions between states A and B increases. To be noted, q is for coordinates of the simulated system, but not a collective variable.

where $V(q)$ is the original potential energy of the system at configuration q, E is the threshold energy, and $V^*(q)$ is the modified potential. $\Delta V(q)$ is the boost potential, which is given by:

$$\Delta V(q) = \frac{(E - V(q))^2}{\alpha + E - V(q)} \tag{24}$$

where α is a predefined acceleration factor which controls the flatness of the potential energy. The boost potential can be either acting on the torsional potential,[26] the whole potential, or even both of them.[122–124] Similarly to scaling only parts of the solute in HREX, the boost potential can also be added only to regions of the biomolecule to enhance specific conformational transitions.[125]

aMD is a powerful approach for enhancing conformational sampling, and in theory, the free energy landscape can be recovered by a reweighting method termed "exponential average"[126] by applying the Boltzmann factor $e^{\Delta V(q)/k_B T}$ accounting for the boost potential acting on each conformational state. However, in practice, this algorithm is known to result in big statistical noise[126–128] as the Boltzmann reweighting factors are usually dominated by a small number of configurations with high boost potentials. The RaMD[129] (only the rotatable dihedrals are subjected to aMD) and the selective aMD[130] methods were developed to reduce the noise through applying boost potentials only to rotatable dihedrals, or certain regions of the simulated systems. The statistical noise can also be reduced by adapting a different reweighting algorithm. For example, less noise was obtained when the exponential term in the Boltzmann factor was replaced by a summation of the

Maclaurin series expansion of the boost potential,[123] which was further improved by a cumulant expansion to the second order.[131] A new variant called Gaussian accelerated MD (GaMD) was developed[132] to reduce the statistical noise by applying a harmonic boost potential instead to flatten the potential energy surface of the system. In GaMD, the system potential $V(q)$ is altered by adding a boost potential when it goes below a threshold energy E:

$$\Delta V(q) = \frac{k}{2} \times (E - V(q))^2 \tag{25}$$

where k is the force constant of the harmonic potential. The boost potential in GaMD follows a Gaussian distribution, and accurate free energies can be recovered through reweighting using a cumulant expansion to the second order.[132]

The boost potential in aMD simulations can be on the order of tens to hundreds of kcal·mol^{-1}, which is much larger in magnitude and wider in distribution than the boost potentials applied in CV-based biasing simulations, thus millisecond-timescale events can be captured by hundreds of nanoseconds aMD simulations.[123,128] A large range of applications has shown the success of aMD in enhancing the sampling of biomolecular systems, including peptide–protein binding,[133] ligand–protein binding,[134] conformational change of proteins.[135] aMD was also used to explore the most likely binding poses of ligands to their targets.[136] To be noted, GaMD was shown to be more efficient than aMD in characterizing protein folding and ligand binding pathways.[132,137] In a recent study, Palermo[138] used GaMD to study the conformational change of R976 in the CRISPR-Cas9 complex. For GaMD, the boost potential was applied in a dual-boost scheme, in which two acceleration potentials are applied simultaneously to the dihedral terms and entire potential of the system. As shown in Fig. 10, free energy landscapes were obtained by reweighting over the aggregated trajectories of four individual runs (~1.6 μs). The results clearly show the conformational preference of R976 in the RuvC domain (which performs DNA cleavages) of the activated CRISPR-Cas9 complex.

A number of further variants of aMD were developed to improve the configurational sampling efficiency, including the windowed aMD,[139] the integrated aMD,[140] and adaptive aMD.[141] Moreover, aMD was combined with many other computational techniques including replica exchange[98], [142–144] and thermodynamics integration[145] for better sampling. More details about the development and applications of aMD can be found in a previous review.[14]

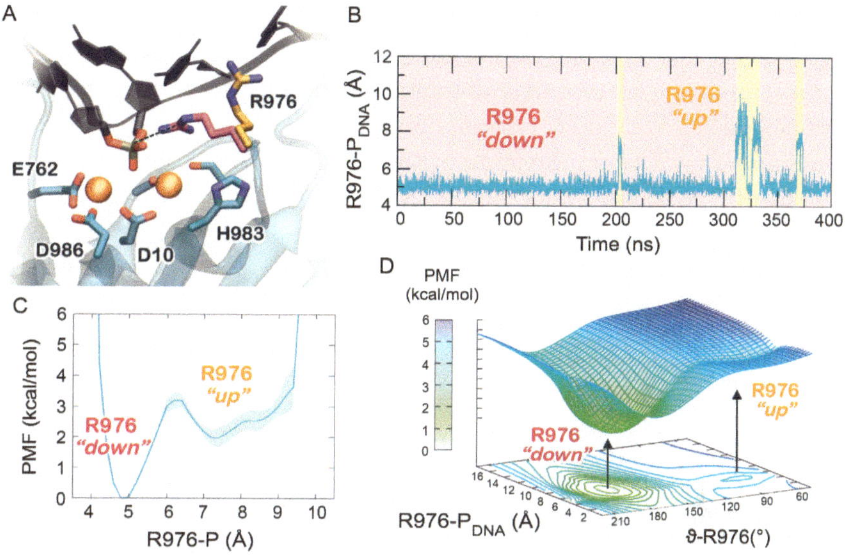

Fig. 10 (A) Catalytic site of the RuvC domain in the CRISPR-Cas9 complex, displaying two conformations of R976. In the activated CRISPR-Cas9 complex, R976 assumes a "down" conformational state (*magenta*), bringing its positive side chain in close contact to the scissile phosphate. In the inactive state of the CRISPR-Cas9 complex, R976 moves away from the scissile phosphate and assumes an "up" conformation (*yellow*). (B) Time evolution of the distance between the terminal carbon of R976 and the scissile phosphate (R976-P_{DNA}) during a ~400 ns GaMD simulation of the activated CRISPR-Cas9 complex. Time windows indicate the "down" (*magenta*) and "up" (*yellow*) conformations of R976. (C) PMF of the R976-P_{DNA} distance computed over the aggregated data from four individual runs for ~1.6 μs of total sampling. The PMF is expressed in kcal·mol^{-1}. (D) Two-dimensional PMF of the R976-P_{DNA} distance in combination with the dihedral angle between the C_α-C_β-C_γ-C_δ atoms of R976 (θ-R976). *Reprinted with permission from Palermo G. Structure and dynamics of the CRISPRCas9 catalytic complex. J Chem Inf Model. 2019;59:2394–2406, copyright 2019 American Chemistry Society.*

5. Combination of enhanced sampling approaches

The features of several popular enhanced sampling approaches have been described above. Due to the methodology differences between CV-based and CV-free approaches, methods from these classes can be easily combined to further optimize the efficiency of conformational space sampling. As mentioned above, any control parameter can be used in Eq. (21) to calculate the exchange probability between replicas in REMD (e.g., bias potentials in REUS, BE-MetaD, etc.). In this way, other enhanced sampling methods can be combined with replica exchange. For example, replica exchange (parallel tempering) can enhance the sampling of all degrees of freedom by simulating the system with multiple replicas at different temperatures simultaneously, while metadynamics is used to enhance the sampling along predefined CVs. The combination of the two approaches leads to the methodology of PTMetaD,[55] however, PTMetaD is still suffering from the poor scaling of computational resources with the system size similar to T-REMD. This issue can be solved by applying the well-tempered ensemble (WTE) algorithm[146] during the PTMetaD simulation, which is then called PTMetaD-WTE scheme.[56] In this ensemble, the average energy value is still close to the canonical average value, but the enhancement of its fluctuations can be tuned to improve the overlap, thus improve the sampling efficiency. PTMetaD-WTE is widely used to study conformational changes of proteins.[147,148]

Inspired by the success of PTMetaD, another combination was also introduced as bias-exchange metadynamics (BE-MetaD) simulations. In BE-MetaD,[29] multiple metadynamics simulations are carried out in parallel, bias potentials are imposed on different CVs in different replicas, configuration exchange between different replicas is attempted at regular time intervals according to the bias potential. It is a powerful method to study protein folding,[149,150] ligand binding[149] and conformational change of disordered peptides.[151–154] As mentioned above, in our study of the loop motion of TIM only BE-MetaD but not standard MD or HREX were able to characterize the full conformational space of loop 6 of TIM with bound substrate.[68]

Further combinations involving replica exchange were developed. For instance, Bussi introduced replica exchange with collective-variable tempering (RECT),[155] which improved the efficiency of exploring the conformational space of RNA. Umbrella sampling was combined with replica exchange (REUS), and it was shown to be powerful in the study of the association and dissociation of a ligand-DNA system by Zeller and Zacharias.[28] Fajer et al.[98] combined replica exchange and aMD, leading to a replica exchange of various degrees of acceleration (REXAMD) to obtain better reweighting statistics while achieving enhanced sampling. This technique was validated and benchmarked on two simple gas-phase model systems. Roe et al.[142] evaluated the advantages of the combination of accelerated molecular dynamics and Hamiltonian replica exchange methods, which was applied to an RNA tetranucleotide system. Huang et al.[143] developed the replica exchange Gaussian aMD, and it was tested on three systems:

alanine dipeptide, chignolin, and HIV protease. Very recently, a combination of replica exchange umbrella sampling with Gaussian accelerated molecular dynamics was proposed.[156]

6. Programs and tutorials

In order to perform enhanced sampling simulations and free energy calculations, one has to be proficient with at least one MD program. Various popular MD programs are available and under active development, such as Amber,[160,161] Gromacs,[162,163] NAMD,[164] OpenMM,[165,166] and LAMMPS.[167] Several enhanced sampling methods are implemented in these programs. For example, Amber Gromacs support US, SMD, REMD, and REUS, while Amber also offers aMD and GaMD. The Colvars module in NAMD is powerful for performing biasing simulations, and very flexible with regard to the definition of specific CVs through Tcl scripts. The plugin PLUMED[53,54] can be interfaced with most biomolecular MD programs allowing to perform enhanced sampling simulation and free energy calculations. It should be noted that it is also a useful tool for analyzing MD simulations. There are a lot of functions for defining CVs available in PLUMED. A more complete overview of the enhanced sampling approaches that are provided by the different MD programs are listed in Table 1. For each program, there is also one, and in some cases even several tutorials provided for the users to learn specific methods. Among them, the hands-on PLUMED tutorials need mentioning as they are well explained and provide details

Table 1 Availability of enhanced sampling and free energy calculation methods in MD programs.

Methods	Bias potential	Free energy estimator	Variants	Programs
US	Eq. (9)	UI, WHAM, MBAR	Adaptive US	Gromacs, Amber, PLUMED[a], Colvars[b]
MetaD	Eq. (12)	Sum up Gaussians	WT-MetaD	PLUMED, Colvars
SMD	Eq. (17)	Jarzynski Eq. (18)	Adaptive SMD	Gromacs, Amber, Colvars
REMD	N/A	WHAM, MBAR	HREX	Gromacs, Amber, NAMD
aMD	Eq. (24)	Reweighting	GaMD	Amber NAMD

Continued

Table 1 Availability of enhanced sampling and free energy calculation methods in MD programs.—cont'd

Methods	Bias potential	Free energy estimator	Variants	Programs
Methods	**Codes**			
UI	https://github.com/ATB-UQ/umbrella_integration[c]			
WHAM	Gromacs (g_wham),[157] PLUMED, PyEMMA[158], wham[159]			
MBAR	PyEMMA,[158] pymbar (https://github.com/choderalab/pymbar)[36]			
Programs	**Web Pages & Tutorials**			
Gromacs	www.gromacs.org & www.mdtutorials.com/gmx			
Amber	https://ambermd.org & https://ambermd.org/tutorials			
NAMD	www.ks.uiuc.edu/Research/namd & www.ks.uiuc.edu/Training/Tutorials			
Plumed	www.plumed.org & www.plumed.org/doc-v2.5/user-doc/html/tutorials.html			

[a]PLUMED is a plugin for analyzing and biasing simulations; it can be interfaced with multiple MD programs, including Gromacs, Amber, NAMD, openMM, and LAMMPS.
[b]Colvars is a module available in NAMD and LAMMPS for biasing simulations.
[c]https://doi.org/10.5281/zenodo.164996.

ranging from theory, how to run simulations to analyzing them. Especially beginners are recommended to go through these tutorials to get familiar with the different enhanced sampling methods.

7. Conclusion and outlook

In the last decades, many methodologies have been proposed to increase the efficiency of sampling in MD simulations and of free energy calculations. They have been found to be highly powerful and successful for exploring the conformational space of biomolecular systems and improving the convergence of free energy calculations. In this chapter, comprehensive reviews of three CV-based and two CV-free enhanced sampling methods were provided, presenting their theories, recent developments and example applications. CV-based methods (e.g., US, MetaD, SMD) can effectively guide the simulations along predefined CVs, while CV-free methods (e.g., REMD, aMD) do not require predefined CVs for the simulations. Normally, there is no absolute standard for the decision which of these methods should be chosen for an enhanced sampling biomolecular simulation. When good CVs exist for describing the biological events of interest, US and MetaD are good choices. When little information is known for the process to be simulated (e.g., protein folding or exploring the dynamics of an intrinsically disordered protein), REMD (or HREX) or aMD should be used. It should be noted that enhanced sampling methods and free energy calculations can also combined with the QM/MM (quantum mechanics/molecular mechanics) method[119] in computational enzymology.[168]

Although enhanced sampling and free energy calculations have demonstrated great efficiency and successes in biomolecular simulations, there are still some challenges. Future developments will be on the following directions to soften the challenges. It is known that CV-based methods suffer from the hidden CV problem, which may lead to inefficient samplings and inaccurate free energy calculations. The same consequence may result from selecting improper CVs that ignore certain dimensions of freedom in biasing simulations of complex or flexible systems. Furthermore, CV-based methods do not scale well with the number of CVs being biased, normally, US is limited to 2 CVs and MetaD is limited to 3 CVs. Although CV-free methods do not have this problem, they may still not assure sufficient sampling along the CVs of interest, as the sampling is enhanced over the whole system. To address these issues, new smart CVs which could describe high dimensional complex processes are proposed continually. For example, principal components (PCs),[169] native contacts,[75] conformational entropy[170] and time-lagged independent component analysis (tICA)[171,172] are very helpful for studies of conformational changes and folding of proteins, while eRMSD[173] was proposed to study RNA folding. Moreover, combinations of CV-based and CV-free methods (e.g., REUS, PTMetaD, BE-MetaD) can enhance the sampling along the interested CVs and other relevant degrees of freedom simultaneously, thus further improve the sampling efficiency and speed up the convergence. To be noted, path collective variables (PCV)[174] and the string method[21] are more advanced approaches to study more complex biological processes. The sampling with PCV or the string method are focused along the minimal free energy paths in high dimensional configuration space, thus promoting the efficiency of sampling. The string method was used to calculate binding free energies of protein–protein systems,[175] which is extremely challenging for other methods.

In general, longer timescales and more massive computational resources are required for more complex systems with millions of atoms, and simulating them with REMD (even HREX) would further increase the computational expense. CV-based methods working well for smaller systems may not be efficient in simulations of huge systems due to the complex and large-scale nature of the relevant CVs. Enhanced sampling methods combined with coarse-grained force fields[176] or multiple time-stepping methods[177] can offer help by greatly accelerating the sampling efficiency for large systems. In all-atom simulation, the use of larger time steps (e.g., 4 fs) is advisable, which can be accomplished by employing virtual sites for hydrogens or hydrogen mass repartition[178] to remove fastest degrees of freedom. Machine learning has become very popular in recent years, more and more powerful deep unsupervised and reinforced learning techniques were developed, which in turn provided powerful tools to develop more intelligent sampling methods.[179–181] Machine learning was also used to train smart CVs.[182,183] It would be interesting to see more developments in this direction, in addition to more applications of machine-learning related methodologies in biomolecular simulations for validation of these methods.

Acknowledgments

The author gratefully acknowledges the help from the editor (BS) in providing detailed feedbacks on the draft of this chapter.

References

1. McCammon JA, Gelin BR, Karplus M. Dynamics of folded proteins. *Nature.* 1977;267:585–590.
2. Smith JC, Kneller GR. Combination of neutron scattering and molecular dynamics to determine internal motions in biomolecules. *Mol Sim.* 1993;10:363–375.
3. Ma P, Schillinger O, Schwarten M, et al. Conformational polymorphism in autophagy-related protein GATE-16. *Biochemistry.* 2015;54:5469–5479.
4. Narayanan C, Bafna K, Roux LD, Agarwal PK, Doucet N. Applications of NMR and computational methodologies to study protein dynamics. *Arch Biochem Biophys.* 2017;628:71–80.
5. Chan-Yao-Chong M, Durand D, Ha-Duong T. Molecular dynamics simulations combined with nuclear magnetic resonance and/or small-angle X-ray scattering data for characterizing intrinsically disordered protein conformational ensembles. *J Chem Inf Model.* 2019;59:1743–1758.
6. Boldon L, Laliberte F, Liu L. Review of the fundamental theories behind small angle X-ray scattering, molecular dynamics simulations, and relevant integrated application. *Nano Rev.* 2015;6:25661.
7. Lundborg M, Narangifard A, Wennberg CL, Lindahl E, Daneholt B, Norlén L. Human skin barrier structure and function analyzed by cryo-EM and molecular dynamics simulation. *J Struct Biol.* 2018;203:149–161.
8. Dolenc J, Missimer JH, Steinmetz MO, van Gunsteren WF. Methods of NMR structure refinement: molecular dynamics simulations improve the agreement with measured NMR data of a C-terminal peptide of GCN4-p1. *J Biomol NMR.* 2010; 47:221–235.

9. Perilla JR, Zhao G, Lu M, et al. CryoEM structure refinement by integrating NMR chemical shifts with molecular dynamics simulations. *J Phys Chem B*. 2017;121: 3853–3863.
10. Igaev M, Kutzner C, Bock LV, Vaiana AC, Grubmüller H. Automated cryo-EM structure refinement using correlation-driven molecular dynamics. *eLife*. 2019;8:e43542.
11. Srivastava A, Nagai T, Srivastava A, Miyashita O, Tama F. Role of computational methods in going beyond X-ray crystallography to explore protein structure and dynamics. *Int J Mol Sci*. 2018;19:3401.
12. Shevchuk R, Hub JS. Bayesian refinement of protein structures and ensembles against SAXS data using molecular dynamics. *PLoS Comput Biol*. 2017;13:1–27.
13. Abrams C, Bussi G. Enhanced sampling in molecular dynamics using metadynamics, replica-exchange, and temperature-acceleration. *Entropy*. 2014;16:163–199.
14. Miao Y, McCammon JA. Unconstrained enhanced sampling for free energy calculations of biomolecules: a review. *Mol Simul*. 2016;42:1046–1055.
15. Yang YI, Shao Q, Zhang J, Yang L, Gao YQ. Enhanced sampling in molecular dynamics. *J Chem Phys*. 2019;151:070902.
16. Torrie GM, Valleau JP. Nonphysical sampling distributions in Monte Carlo free-energy estimation: umbrella sampling. *J Comput Phys*. 1977;23:187–199.
17. Laio A, Parrinello M. Escaping free-energy minima. *Proc Natl Acad Sci USA*. 2002;99:12562–12566.
18. Lu H, Schulten K. Steered molecular dynamics simulations of force-induced protein domain unfolding. *Proteins*. 1999;35:453–463.
19. Darve E, Rodríguez-Gómez D, Pohorille A. Adaptive biasing force method for scalar and vector free energy calculations. *J Chem Phys*. 2008;128:144120.
20. Grubmüller H. Predicting slow structural transitions in macromolecular systems: conformational flooding. *Phys Rev E*. 1995;52:2893–2906.
21. Weinan E, Ren W, Vanden-Eijnden E. String method for the study of rare events. *Phys Rev B*. 2002;66:052301.
22. Maragliano L, Vanden-Eijnden E. A temperature accelerated method for sampling free energy and determining reaction pathways in rare events simulations. *Chem Phys Lett*. 2006;426:168–175.
23. Sugita Y, Okamoto Y. Replica-exchange molecular dynamics method for protein folding. *Chem Phys Lett*. 1999;314:141–151.
24. Wu X, Wang S. Self-guided molecular dynamics simulation for efficient conformational search. *J Phys Chem B*. 1998;102:7238–7250.
25. Wu X, Brooks BR, Vanden-Eijnden E. Self-guided Langevin dynamics via generalized Langevin equation. *J Comput Chem*. 2016;37:595–601.
26. Hamelberg D, Mongan J, McCammon JA. Accelerated molecular dynamics: a promising and efficient simulation method for biomolecules. *J Chem Phys*. 2004;120: 11919–11929.
27. Sabri Dashti D, Roitberg AE. Optimization of umbrella sampling replica exchange molecular dynamics by replica positioning. *J Chem Theory Comput*. 2013;9: 4692–4699.
28. Zeller F, Zacharias M. Adaptive biasing combined with Hamiltonian replica exchange to improve umbrella sampling free energy simulations. *J Chem Theory Comput*. 2014;10:703–710.
29. Piana S, Laio A. A bias-exchange approach to protein folding. *J Phys Chem B*. 2007;111:4553–4559.
30. Zwanzig RW. High-temperature equation of state by a perturbation method. I. Nonpolar gases. *J Chem Phys*. 1954;22:1420–1426.
31. Bennett CH. Efficient estimation of free energy differences from Monte Carlo data. *J Comput Phys*. 1976;22:245–268.

32. Shirts MR, Bair E, Hooker G, Pande VS. Equilibrium free energies from non-equilibrium measurements using maximum-likelihood methods. *Phys Rev Lett.* 2003;91:140601.
33. Kumar S, Rosenberg JM, Bouzida D, Swendsen RH, Kollman PA. The weighted histogram analysis method for free-energy calculations on biomolecules. I. The method. *J Comput Chem.* 1992;13:1011–1021.
34. Ferrenberg AM, Swendsen RH. Optimized Monte Carlo data analysis. *Phys Rev Lett.* 1989;63:1195–1198.
35. Kästner J. Umbrella sampling. *WIREs Comput Mol Sci.* 2011;1:932–942.
36. Shirts MR, Chodera JD. Statistically optimal analysis of samples from multiple equilibrium states. *J Chem Phys.* 2008;129:124105.
37. Laio A, Gervasio FL. Metadynamics: a method to simulate rare events and reconstruct the free energy in biophysics, chemistry, and material science. *Rep Prog Phys.* 2008;71:126601.
38. Barducci A, Bonomi M, Parrinello M. Metadynamics. *WIREs Comput Mol Sci.* 2011;1:826–843.
39. Roux B. The calculation of the potential of mean force using computer simulations. *Comput Phys Commun.* 1995;91:275–282.
40. Bartels C, Karplus M. Multidimensional adaptive umbrella sampling: Applications to main chain and side chain peptide conformations. *J Comput Chem.* 1997;18: 1450–1462.
41. Gallicchio E, Andrec M, Felts AK, Levy RM. Temperature weighted histogram analysis method, replica exchange, and transition paths. *J Phys Chem B.* 2005;109: 6722–6731.
42. Ferguson AL. BayesWHAM: a Bayesian approach for free energy estimation, reweighting, and uncertainty quantification in the weighted histogram analysis method. *J Comput Chem.* 2017;38:1583–1605.
43. Kästner J, Thiel W. Bridging the gap between thermodynamic integration and umbrella sampling provides a novel analysis method: "umbrella integration". *J Chem Phys.* 2005;123:144104.
44. Meshkin H, Zhu F. Thermodynamics of protein folding studied by umbrella sampling along a reaction coordinate of native contacts. *J Chem Theory Comput.* 2017;13: 2086–2097.
45. Lemkul JA, Bevan DR. Assessing the stability of Alzheimer's amyloid protofibrils using molecular dynamics. *J Phys Chem B.* 2010;114:1652–1660.
46. Domański J, Hedger G, Best RB, Stansfeld PJ, Sansom MSP. Convergence and sampling in determining free energy landscapes for membrane protein association. *J Phys Chem B.* 2017;121:3364–3375.
47. Lee MS, Olson MA. Calculation of absolute protein-ligand binding affinity using path and endpoint approaches. *Biophys J.* 2006;90:864–877.
48. Banavali NK, Roux B. Free energy landscape of A-DNA to B-DNA conversion in aqueous solution. *J Am Chem Soc.* 2005;127:6866–6876.
49. Fedorov DG, Sugita Y, Choi CH. Efficient parallel implementations of QM/MM-REMD (quantum mechanical/molecular mechanics-replica-exchange MD) and umbrella sampling: isomerization of H_2O_2 in aqueous solution. *J Phys Chem B.* 2013; 117:7996–8002.
50. Lence E, Van der Kamp MW, González-Bello C, Mulholland AJ. QM/MM simulations identify the determinants of catalytic activity differences between type II dehydroquinase enzymes. *Org Biomol Chem.* 2018;16:4443–4455.
51. Wojtas-Niziurski W, Meng Y, Roux B, Bernèche S. Self-learning adaptive umbrella sampling method for the determination of free energy landscapes in multiple dimensions. *J Chem Theory Comput.* 2013;9:1885–1895.

52. Barducci A, Bussi G, Parrinello M. Well-tempered metadynamics: a smoothly converging and tunable free-energy method. *Phys Rev Lett.* 2008;100:020603.
53. Bonomi M, Branduardi D, Bussi G, et al. PLUMED: a portable plugin for free-energy calculations with molecular dynamics. *Comp Phys Commun.* 2009;180:1961–1972.
54. Tribello GA, Bonomi M, Branduardi D, Camilloni C, Bussi G. PLUMED 2: new feathers for an old bird. *Comp Phys Commun.* 2014;185:604–613.
55. Bussi G, Gervasio FL, Laio A, Parrinello M. Free-energy landscape for β hairpin folding from combined parallel tempering and metadynamics. *J Am Chem Soc.* 2006; 128:13435–13441.
56. Deighan M, Bonomi M, Pfaendtner J. Efficient simulation of explicitly solvated proteins in the well-tempered ensemble. *J Chem Theory Comput.* 2012;8:2189–2192.
57. Raiteri P, Laio A, Gervasio FL, Micheletti C, Parrinello M. Efficient reconstruction of complex free energy landscapes by multiple walkers metadynamics. *J Phys Chem B.* 2006;110:3533–3539.
58. Limongelli V, Bonomi M, Parrinello M. Funnel metadynamics as accurate binding free-energy method. *Proc Natl Acad Sci USA.* 2013;110:6358–6363.
59. Pfaendtner J, Bonomi M. Efficient sampling of high-dimensional free-energy landscapes with parallel bias metadynamics. *J Chem Theory Comput.* 2015;11:5062–5067.
60. Tiwary P, Parrinello M. From metadynamics to dynamics. *Phys Rev Lett.* 2013; 111:230602.
61. Tiwary P, Limongelli V, Salvalaglio M, Parrinello M. Kinetics of protein-ligand unbinding: predicting pathways, rates, and rate-limiting steps. *Proc Natl Acad Sci USA.* 2015;112:E386–E391.
62. Tiwary P, Mondal J, Berne BJ. How and when does an anticancer drug leave its binding site? *Sci Adv.* 2017;3:e1700014.
63. Casasnovas R, Limongelli V, Tiwary P, Carloni P, Parrinello M. Unbinding kinetics of a p38 MAP kinase type II inhibitor from metadynamics simulations. *J Am Chem Soc.* 2017;139:4780–4788.
64. Wang Y, Martins JM, Lindorff-Larsen K. Biomolecular conformational changes and ligand binding: from kinetics to thermodynamics. *Chem Sci.* 2017;8:6466–6473.
65. Wang Y, Valsson O, Tiwary P, Parrinello M, Lindorff-Larsen K. Frequency adaptive metadynamics for the calculation of rare-event kinetics. *J Chem Phys.* 2018;149: 072309.
66. Valsson O, Tiwary P, Parrinello M. Enhancing important fluctuations: rare events and metadynamics from a conceptual viewpoint. *Ann Rev Phys Chem.* 2016;67:159–184.
67. Barrozo A, Liao Q, Esguerra M, et al. Computer simulations of the catalytic mechanism of wild-type and mutant β-phosphoglucomutase. *Org Biomol Chem.* 2018;16: 2060–2073.
68. Liao Q, Kulkarni Y, Sengupta U, et al. Loop motion in triosephosphate isomerase is not a simple open and shut case. *J Am Chem Soc.* 2018;140:15889–15903.
69. Clark AJ, Tiwary P, Borrelli K, et al. Prediction of protein-ligand binding poses via a combination of induced fit docking and metadynamics simulations. *J Chem Theory Comput.* 2016;12:2990–2998.
70. Ensing B, De Vivo M, Liu Z, Moore P, Klein ML. Metadynamics as a tool for exploring free energy landscapes of chemical reactions. *Acc Chem Res.* 2006;39:73–81.
71. Hsu WL, Furuta T, Sakurai M. ATP hydrolysis mechanism in a maltose transporter explored by QM/MM metadynamics simulation. *J Phys Chem B.* 2016;120: 11102–11112.
72. Bussi G. Hamiltonian replica exchange in GROMACS: a flexible implementation. *Mol Phys.* 2014;112:379–384.
73. Knowles JR, Albery WJ. Perfection in enzyme catalysis: the energetics of triosephosphate isomerase. *Acc Chem Res.* 1977;10:105–111.

74. Wierenga RK, Kapetaniou EG, Venkatesan R. Triosephosphate isomerase: a highly evolved biocatalyst. *Cell Mol Life Sci*. 2010;67:3961–3982.
75. Best RB, Hummer G, Eaton WA. Native contacts determine protein folding mechanisms in atomistic simulations. *Proc Natl Acad Sci USA*. 2013;110: 17874–17879.
76. Berman HM, Westbrook J, Feng Z, et al. The protein data bank. *Nucleic Acids Res*. 2000;28:235–242.
77. Jogl G, Rozovsky S, McDermott AE, Tong L. Optimal alignment for enzymatic proton transfer: structure of the Michaelis complex of triosephosphate isomerase at 1.2-Å resolution. *Proc Natl Acad Sci USA*. 2003;100:50–55.
78. Park S, Khalili-Araghi F, Tajkhorshid E, Schulten K. Free energy calculation from steered molecular dynamics simulations using Jarzynski's equality. *J Chem Phys*. 2003;119:3559–3566.
79. Park S, Schulten K. Calculating potentials of mean force from steered molecular dynamics simulations. *J Chem Phys*. 2004;120:5946–5961.
80. Jarzynski C. Nonequilibrium equality for free energy differences. *Phys Rev Lett*. 1997;78:2690–2693.
81. Jarzynski C. Equilibrium free-energy differences from nonequilibrium measurements: a master-equation approach. *Phys Rev E*. 1997;56:5018–5035.
82. Crooks GE. Path-ensemble averages in systems driven far from equilibrium. *Phys Rev E*. 2000;61:2361–2366.
83. Xiao BL, Ning YN, Niu NN, et al. Steered molecular dynamic simulations of conformational lock of Cu, Zn-superoxide dismutase. *Sci Rep*. 2019;9:4353.
84. Capelli AM, Costantino G. Unbinding pathways of VEGFR2 inhibitors revealed by steered molecular dynamics. *J Chem Inf Model*. 2014;54:3124–3136.
85. Jin H, Zhu J, Dong Y, Han W. Exploring the different ligand escape pathways in acylaminoacyl peptidase by random acceleration and steered molecular dynamics simulations. *RSC Adv*. 2016;6:10987–10996.
86. Patel JS, Berteotti A, Ronsisvalle S, Rocchia W, Cavalli A. Steered molecular dynamics simulations for studying protein-ligand interaction in cyclin-dependent kinase 5. *J Chem Inf Model*. 2014;54:470–480.
87. Colizzi F, Perozzo R, Scapozza L, Recanatini M, Cavalli A. Single-molecule pulling simulations can discern active from inactive enzyme inhibitors. *J Am Chem Soc*. 2010;132:7361–7371.
88. Do PC, Lee EH, Le L. Steered molecular dynamics simulation in rational drug design. *J Chem Inf Model*. 2018;58:1473–1482.
89. Marzinek JK, Bond PJ, Lian G, et al. Free energy predictions of ligand binding to an α-helix using steered molecular dynamics and umbrella sampling simulations. *J Chem Inf Model*. 2014;54:2093–2104.
90. Zhou JK, Yang DY, Sheu SY. The molecular mechanism of ligand unbinding from the human telomeric G-quadruplex by steered molecular dynamics and umbrella sampling simulations. *Phys Chem Chem Phys*. 2015;17:12857–12869.
91. Ozer G, Valeev EF, Quirk S, Hernandez R. Adaptive steered molecular dynamics of the long-distance unfolding of neuropeptide Y. *J Chem Theory Comput*. 2010; 6:3026–3038.
92. Schlitter J, Engels M, Krüger P. Targeted molecular dynamics: a new approach for searching pathways of conformational transitions. *J Mol Graph*. 1994;12:84–89.
93. Sugita Y, Kitao A, Okamoto Y. Multidimensional replica-exchange method for free-energy calculations. *J Chem Phys*. 2000;113:6042–6051.
94. Sugita Y, Okamoto Y. Replica-exchange multicanonical algorithm and multicanonical replica-exchange method for simulating systems with rough energy landscape. *Chem Phys Lett*. 2000;329:261–270.

95. Fukunishi H, Watanabe O, Takada S. On the Hamiltonian replica exchange method for efficient sampling of biomolecular systems: application to protein structure prediction. *J Chem Phys*. 2002;116:9058–9067.
96. Liu P, Kim B, Friesner RA, Berne BJ. Replica exchange with solute tempering: a method for sampling biological systems in explicit water. *Proc Natl Acad Sci USA*. 2005;102:13749–13754.
97. Huang X, Hagen M, Kim B, Friesner RA, Zhou R, Berne BJ. Replica exchange with solute tempering: efficiency in large scale systems. *J Phys Chem B*. 2007;111:5405–5410.
98. Fajer M, Hamelberg D, McCammon JA. Replica-exchange accelerated molecular dynamics (REXAMD) applied to thermodynamic integration. *J Chem Theory Comput*. 2008;4:1565–1569.
99. Wang L, Friesner RA, Berne BJ. Replica exchange with solute scaling: a more efficient version of replica exchange with solute tempering (REST2). *J Phys Chem B*. 2011;115:9431–9438.
100. Laghaei R, Mousseau N, Wei G. Effect of the disulfide bond on the monomeric structure of human amylin studied by combined Hamiltonian and temperature replica exchange molecular dynamics simulations. *J Phys Chem B*. 2010;114:7071–7077.
101. Laghaei R, Mousseau N, Wei G. Structure and thermodynamics of amylin dimer studied by Hamiltonian-temperature replica exchange molecular dynamics simulations. *J Phys Chem B*. 2011;115:3146–3154.
102. Leahy CT, Kells A, Hummer G, Buchete NV, Rosta E. Peptide dimerization-dissociation rates from replica exchange molecular dynamics. *J Chem Phys*. 2017; 147:152725.
103. Wang K, Chodera JD, Yang Y, Shirts MR. Identifying ligand binding sites and poses using GPU-accelerated Hamiltonian replica exchange molecular dynamics. *J Comput Aid Mol Des*. 2013;27:989–1007.
104. Luitz MP, Zacharias M. Protein-ligand docking using Hamiltonian replica exchange simulations with soft core potentials. *J Chem Inf Model*. 2014;54:1669–1675.
105. Ostermeir K, Zacharias M. Accelerated flexible protein-ligand docking using Hamiltonian replica exchange with a repulsive biasing potential. *PLoS ONE*. 2017;12:1–17.
106. Liao Q, Owen MC, Olubiyi OO, Barz B, Strodel B. Conformational transitions of the amyloid-β peptide upon copper(II) binding and pH changes. *Isr J Chem*. 2017; 57:771–784.
107. Liao Q, Owen MC, Bali S, Barz B, Strodel B. Aβ under stress: the effects of acidosis, Cu^{2+}-binding, and oxidation on amyloid β-peptide dimers. *Chem Commun*. 2018; 54:7766–7769.
108. Barz B, Wales DJ, Strodel B. A kinetic approach to the sequence-aggregation relationship in disease-related protein assembly. *J Phys Chem B*. 2014;118:1003–1011.
109. Barz B, Olubiyi OO, Strodel B. Early amyloid β-protein aggregation precedes conformational change. *Chem Commun*. 2014;50:5373–5375.
110. Meng Y, Roitberg AE. Constant pH replica exchange molecular dynamics in biomolecules using a discrete protonation model. *J Chem Theory Comput*. 2010;6:1401–1412.
111. Ostermeir K, Zacharias M. Hamiltonian replica exchange combined with elastic network analysis to enhance global domain motions in atomistic molecular dynamics simulations. *Proteins*. 2014;82:3410–3419.
112. Jiang W, Hodoscek M, Roux B. Computation of absolute hydration and binding free energy with free energy perturbation distributed replica-exchange molecular dynamics. *J Chem Theory Comput*. 2009;5:2583–2588.
113. Jiang W, Roux B. Free energy perturbation Hamiltonian replica-exchange molecular dynamics (FEP/H-REMD) for absolute ligand binding free energy calculations. *J Chem Theory Comput*. 2010;6:2559–2565.

114. Jiang W, Luo Y, Maragliano L, Roux B. Calculation of free energy landscape in multi-dimensions with Hamiltonian-exchange umbrella sampling on petascale supercomputer. *J Chem Theory Comput.* 2012;8:4672–4680.
115. Jiang W, Thirman J, Jo S, Roux B. Reduced free energy perturbation/Hamiltonian replica exchange molecular dynamics method with unbiased alchemical thermodynamic axis. *J Phys Chem B.* 2018;122:9435–9442.
116. Rhee YM, Pande VS. Multiplexed-replica exchange molecular dynamics method for protein folding simulation. *Biophys J.* 2003;84:775–786.
117. Lee MS, Olson MA. Protein folding simulations combining self-guided Langevin dynamics and temperature-based replica exchange. *J Chem Theory Comput.* 2010; 6:2477–2487.
118. Wu X, Hodoscek M, Brooks BR. Replica exchanging self-guided Langevin dynamics for efficient and accurate conformational sampling. *J Chem Phys.* 2012;137:044106.
119. Warshel A, Levitt M. Theoretical studies of enzymic reactions: dielectric, electrostatic and steric stabilization of the carbonium ion in the reaction of lysozyme. *J Mol Biol.* 1976;103:227–249.
120. Bergonzo C, Henriksen NM, Roe DR, Swails JM, Roitberg AE, Cheatham TE. Multidimensional replica exchange molecular dynamics yields a converged ensemble of an RNA tetranucleotide. *J Chem Theory Comput.* 2014;10:492–499.
121. Cruzeiro VWD, Roitberg AE. Multidimensional replica exchange simulations for efficient constant pH and redox potential molecular dynamics. *J Chem Theory Comput.* 2019;15:871–881.
122. Hamelberg D, de Oliveira CAF, McCammon JA. Sampling of slow diffusive conformational transitions with accelerated molecular dynamics. *J Chem Phys.* 2007;127:155102.
123. Pierce LCT, Salomon-Ferrer R, de Oliveira CAF, McCammon JA, Walker RC. Routine access to millisecond time scale events with accelerated molecular dynamics. *J Chem Theory Comput.* 2012;8:2997–3002.
124. Miao Y, Feixas F, Eun C, McCammon JA. Accelerated molecular dynamics simulations of protein folding. *J Computat Chem.* 2015;36:1536–1549.
125. Lan P, Tan M, Zhang Y, et al. Structural insight into precursor tRNA processing by yeast ribonuclease P. *Science.* 2018;362:eaat6678.
126. Shen T, Hamelberg D. A statistical analysis of the precision of reweighting-based simulations. *J Chem Phys.* 2008;129:034103.
127. Sinko W, Miao Y, de Oliveira CAF, McCammon JA. Population based reweighting of scaled molecular dynamics. *J Phys Chem B.* 2013;117:12759–12768.
128. Miao Y, Nichols SE, McCammon JA. Free energy landscape of G-protein coupled receptors, explored by accelerated molecular dynamics. *Phys Chem Chem Phys.* 2014;16:6398–6406.
129. Doshi U, Hamelberg D. Improved statistical sampling and accuracy with accelerated molecular dynamics on rotatable torsions. *J Chem Theory Comput.* 2012;8:4004–4012.
130. Wereszczynski J, McCammon JA. Using selectively applied accelerated molecular dynamics to enhance free energy calculations. *J Chem Theory Comput.* 2010;6: 3285–3292.
131. Miao Y, Sinko W, Pierce L, Bucher D, Walker RC, McCammon JA. Improved reweighting of accelerated molecular dynamics simulations for free energy calculation. *J Chem Theory Comput.* 2014;10:2677–2689.
132. Miao Y, Feher VA, McCammon JA. Gaussian accelerated molecular dynamics: unconstrained enhanced sampling and free energy calculation. *J Chem Theory Comput.* 2015;11:3584–3595.
133. Wang J, Alekseenko A, Kozakov D, Miao Y. Improved modeling of peptide-protein binding through global docking and accelerated molecular dynamics simulations. *Front Mol Biosci.* 2019;6:112.

134. Miao Y, Huang YM, Walker RC, McCammon JA, Chang CA. Ligand binding pathways and conformational transitions of the HIV protease. *Biochemistry*. 2018;57: 1533–1541.
135. Miao Y, Feixas F, Eun C, McCammon JA. Accelerated molecular dynamics simulations of protein folding. *J Comput Chem*. 2015;36:1536–1549.
136. Fratev F, Steinbrecher T, Jónsdóttir SO. Prediction of accurate binding modes using combination of classical and accelerated molecular dynamics and free-energy perturbation calculations: an application to toxicity studies. *ACS Omega*. 2018;3:4357–4371.
137. Huang YM, Raymundo MAV, Chen W, Chang CA. Mechanism of the association pathways for a pair of fast and slow binding ligands of HIV-1 protease. *Biochemistry*. 2017;56:1311–1323.
138. Palermo G. Structure and dynamics of the CRISPRCas9 catalytic complex. *J Chem Inf Model*. 2019;59:2394–2406.
139. Sinko W, de Oliveira CAF, Pierce LCT, McCammon JA. Protecting high energy barriers: a new equation to regulate boost energy in accelerated molecular dynamics simulations. *J Chem Theory Comput*. 2012;8:17–23.
140. Peng X, Zhang Y, Li Y, et al. Integrating multiple accelerated molecular dynamics to improve accuracy of free energy calculations. *J Chem Theory Comput*. 2018;14:1216–1227.
141. Markwick PRL, Pierce LCT, Goodin DB, McCammon JA. Adaptive accelerated molecular dynamics (Ad-AMD) revealing the molecular plasticity of P450cam. *J Phys Chem Lett*. 2011;2:158–164.
142. Roe DR, Bergonzo C, Cheatham TE. Evaluation of enhanced sampling provided by accelerated molecular dynamics with Hamiltonian replica exchange methods. *J Phys Chem B*. 2014;118:3543–3552.
143. Huang YM, McCammon JA, Miao Y. Replica exchange Gaussian accelerated molecular dynamics: improved enhanced sampling and free energy calculation. *J Chem Theory Comput*. 2018;14:1853–1864.
144. Fajer M, Swift RV, McCammon JA. Using multistate free energy techniques to improve the efficiency of replica exchange accelerated molecular dynamics. *J Comput Chem*. 2009;30:1719–1725.
145. de Oliveira CAF, Hamelberg D, McCammon JA. Coupling accelerated molecular dynamics methods with thermodynamic integration simulations. *J Chem Theory Comput*. 2008;4:1516–1525.
146. Bonomi M, Parrinello M. Enhanced sampling in the well-tempered ensemble. *Phys Rev Lett*. 2010;104:190601.
147. Sutto L, Gervasio FL. Effects of oncogenic mutations on the conformational free-energy landscape of EGFR kinase. *Proc Natl Acad Sci USA*. 2013;110:10616–10621.
148. Papaleo E, Sutto L, Gervasio FL, Lindorff-Larsen K. Conformational changes and free energies in a proline isomerase. *J Chem Theory Comput*. 2014;10:4169–4174.
149. Baftizadeh F, Cossio P, Pietrucci F, Laio A. Protein folding and ligand-enzyme binding from bias-exchange metadynamics simulations. *Curr Phys Chem*. 2012;2:79–91.
150. Granata D, Camilloni C, Vendruscolo M, Laio A. Characterization of the free-energy landscapes of proteins by NMR-guided metadynamics. *Proc Natl Acad Sci USA*. 2013;110:6817–6822.
151. Baftizadeh F, Pietrucci F, Biarnés X, Laio A. Nucleation process of a fibril precursor in the C-terminal segment of amyloid-β. *Phys Rev Lett*. 2013;110:168103.
152. Granata D, Baftizadeh F, Habchi J, et al. The inverted free energy landscape of an intrinsically disordered peptide by simulations and experiments. *Sci Rep*. 2015;5:15449.
153. Liu N, Duan M, Yang M. Structural properties of human IAPP dimer in membrane environment studied by all-atom molecular dynamics simulations. *Sci Rep*. 2017; 7:7915.
154. Guo AZ, Fluitt AM, de Pablo JJ. Early-stage human islet amyloid polypeptide aggregation: mechanisms behind dimer formation. *J Chem Phys*. 2018;149:025101.

155. Gil-Ley A, Bussi G. Enhanced conformational sampling using replica exchange with collective-variable tempering. *J Chem Theory Comput.* 2015;11:1077–1085.
156. Oshima H, Re S, Sugita Y. Replica-exchange umbrella sampling combined with Gaussian accelerated molecular dynamics for free-energy calculation of biomolecules. *J Chem Theory Comput.* 2019;15:5199–5208.
157. Hub JS, de Groot BL, van der Spoel D. g_wham—a free weighted histogram analysis implementation including robust error and autocorrelation estimates. *J Chem Theory Comput.* 2010;6:3713–3720.
158. Scherer MK, Trendelkamp-Schroer B, Paul F, et al. PyEMMA 2: a software package for estimation, validation, and analysis of Markov models. *J Chem Theory Comput.* 2015;11:5525–5542.
159. Grossfield A. WHAM: the weighted histogram analysis method, version 2.0.9. http://membrane.urmc.rochester.edu/wordpress/?page_id=126.
160. Case DA, Cheatham IIITE, Darden T, et al. The amber biomolecular simulation programs. *J Comput Chem.* 2005;26:1668–1688.
161. Salomon-Ferrer R, Case DA, Walker RC. An overview of the amber biomolecular simulation package. *WIREs Comput Mol Sci.* 2013;3:198–210.
162. Van Der Spoel D, Lindahl E, Hess B, Groenhof G, Mark AE, Berendsen HJC. GROMACS: fast, flexible, and free. *J Comput Chem.* 2005;26:1701–1718.
163. Abraham MJ, Murtola T, Schulz R, et al. GROMACS: high performance molecular simulations through multi-level parallelism from laptops to supercomputers. *SoftwareX.* 2015;1-2:19–25.
164. Phillips JC, Braun R, Wang W, et al. Scalable molecular dynamics with NAMD. *J Comput Chem.* 2005;26:1781–1802.
165. Eastman P, Friedrichs MS, Chodera JD, et al. OpenMM 4: a reusable, extensible, hardware independent library for high performance molecular simulation. *J Chem Theory Comput.* 2013;9:461–469.
166. Eastman P, Swails J, Chodera JD, et al. OpenMM 7: rapid development of high performance algorithms for molecular dynamics. *PLoS Comput Biol.* 2017;13:1–17.
167. Plimpton S. Fast parallel algorithms for short-range molecular dynamics. *J Comput Phys.* 1995;117:1–19.
168. Van der Kamp MW, Mulholland AJ. Combined quantum mechanics/molecular mechanics (QM/MM) methods in computational enzymology. *Biochemistry.* 2013; 52:2708–2728.
169. Spiwok V, Lipovová P, Králová B. Metadynamics in essential coordinates: free energy simulation of conformational changes. *J Phys Chem B.* 2007;111:3073–3076.
170. Palazzesi F, Valsson O, Parrinello M. Conformational entropy as collective variable for proteins. *J Phys Chem Lett.* 2017;8:4752–4756.
171. Pérez-Hernández G, Noé F. Hierarchical time-lagged independent component analysis: computing slow modes and reaction coordinates for large molecular systems. *J Chem Theory Comput.* 2016;12:6118–6129.
172. Sultan MM, Pande VS. tICA-Metadynamics: accelerating metadynamics by using kinetically selected collective variables. *J Chem Theory Comput.* 2017;13:2440–2447.
173. Bottaro S, Di Palma F, Bussi G. The role of nucleobase interactions in RNA structure and dynamics. *Nucleic Acids Res.* 2014;42:13306–13314.
174. Branduardi D, Gervasio FL, Parrinello M. From A to B in free energy space. *J Chem Phys.* 2007;126:054103.
175. Suh D, Jo S, Jiang W, Chipot C, Roux B. String method for protein-protein binding free-energy calculations. *J Chem Theory Comput.* 2019;15:5829–5844.
176. Singh N, Li W. Recent advances in coarse-grained models for biomolecules and their applications. *Int J Mol Sci.* 2019;20:3774.

177. Chen PY, Tuckerman ME. Molecular dynamics based enhanced sampling of collective variables with very large time steps. *J Chem Phys*. 2018;148:024106.
178. Hopkins CW, Le Grand S, Walker RC, Roitberg AE. Long-time-step molecular dynamics through hydrogen mass repartitioning. *J Chem Theory Comput*. 2015;11: 1864–1874.
179. Ribeiro JML, Bravo P, Wang Y, Tiwary P. Reweighted autoencoded variational Bayes for enhanced sampling (RAVE). *J Chem Phys*. 2018;149:072301.
180. Noé F, Olsson S, Köhler J, Wu H. Boltzmann generators: sampling equilibrium states of many-body systems with deep learning. *Science*. 2019;365:eaaw1147.
181. Bonati L, Zhang YY, Parrinello M. Neural networks-based variationally enhanced sampling. *Proc Natl Acad Sci USA*. 2019;116:17641–17647.
182. Sultan MM, Pande VS. Automated design of collective variables using supervised machine learning. *J Chem Phys*. 2018;149:094106.
183. Trapl D, Horvacanin I, Mareska V, Ozcelik F, Unal G, Spiwok V. Anncolvar: approximation of complex collective variables by artificial neural networks for analysis and biasing of molecular simulations. *Front Mol Biosci*. 2019;6:25.

CHAPTER FIVE

Long-time methods for molecular dynamics simulations: Markov State Models and Milestoning

Brajesh Narayan[a,b], Ye Yuan[a,b], Arman Fathizadeh[c], Ron Elber[c,d], Nicolae-Viorel Buchete[a,b,*]

[a]School of Physics, University College Dublin, Dublin, Ireland
[b]Institute for Discovery, University College Dublin, Dublin, Ireland
[c]Oden Institute for Computational Engineering and Sciences, University of Texas at Austin, Austin, TX, United States
[d]Department of Chemistry, University of Texas at Austin, Austin, TX, United States
*Corresponding author: e-mail address: buchete@ucd.ie

Contents

Abstract

Molecular dynamics (MD) studies of biomolecules require the ability to simulate complex biochemical systems with an increasingly larger number of particles and for longer time scales, a problem that cannot be overcome by computational hardware advances alone. A main problem springs from the intrinsically high-dimensional and complex nature of the underlying free energy landscape of most systems, and from the necessity to sample accurately such landscapes for identifying kinetic and thermodynamic states in the configurations space, and for accurate calculations of both free energy differences and of the corresponding transition rates between states. Here, we review and present applications of two increasingly popular methods that allow long-time MD simulations of biomolecular systems that can open a broad spectrum of new studies. A first approach, Markov State Models (MSMs), relies on identifying a set of configuration states in which the system resides sufficiently long to relax and loose the memory of previous transitions, and on using simulations for mapping the underlying complex energy landscape and for extracting accurate thermodynamic and kinetic information. The Markovian

Progress in Molecular Biology and Translational Science, Volume 170
ISSN 1877-1173
https://doi.org/10.1016/bs.pmbts.2020.01.002

independence of the underlying transition probabilities creates the opportunity to increase the sampling efficiency by using sets of appropriately initialized short simulations rather than typically long MD trajectories, which also enhances sampling. This allows MSM-based studies to unveil bio-molecular mechanisms and to estimate free energy barriers with high accuracy, in a manner that is both systematic and relatively automatic, which accounts for their increasing popularity. The second approach presented, Milestoning, targets accurate studies of the ensemble of pathways connecting specific end-states (e.g., reactants and products) in a similarly systematic, accurate and highly automatic manner. Applications presented range from studies of conformational dynamics and binding of amyloid-forming peptides, cell-penetrating peptides and the DFG-flip dynamics in Abl kinase. As highlighted by the increasing number of studies using both methods, we anticipate that they will open new avenues for the investigation of systematic sampling of reactions pathways and mechanisms occurring on longer time scales than currently accessible by purely computational hardware developments.

Abbreviations

Abl Abelson tyrosine kinase
Aβ amyloid beta peptide
CPP cell-penetrating peptide
DFG Asp-Phe-Gly motif
DOPC 1,2-dioleoyl-sn-glycero-3-phosphocholine
DTC direct transition counting
FF diphenylalanine peptide
FPT first passage time
HPC high-performace computing
KFFE Arg-Phe-Phe-Glu peptide
MD molecular dynamics
MFPT mean first passage time
MSM(s) Markov state model(s)
NNQQ Asn-Asn-Gln-Gln peptide
RelRMSD relative root mean square deviation
REMD replica-exchange molecular dynamics
RMSD root mean square deviation
ST simulated tempering
TBA transition-based assignment
WKW Trp-Lys-Trp peptide

1. Introduction

Although the availability of high-performance computing (HPC) hardware capacity is likely to continue its remarkable growth, the high complexity of accurate kinetic and thermodynamic studies based on molecular dynamics (MD) simulations of biomolecular systems requires novel modeling methods that can be accurate yet simple and relatively easy to implement. One approach, that takes advantage of the intrinsically parallel architecture HPC facilities or the highly distributed computing projects such as Folding@Home[1], relies of mapping the complex underlying dynamics of otherwise complex biomolecular systems on relatively simple networks with nodes corresponding to stable conformational states that are interconnected through Markovian transitions. Thus, both the design and the goals of MD studies are targeted toward identifying the regions of the corresponding configuration space that are characterized by a sufficient conformational stability such that subsequent inter-state transition are independent from each other (i.e., Markov states). This approach has strong roots in the statistical mechanics of biomolecules being related to the projection operator formalism.[2]

Once the Markovian transition probabilities between different states are inferred from either long or short trajectories (see Fig. 1), they can be used to

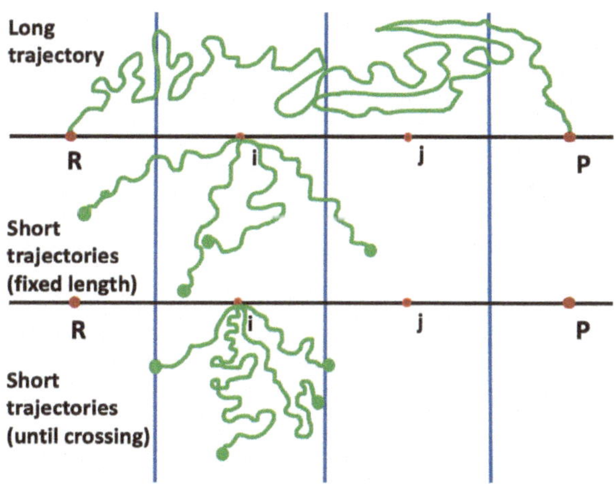

Fig. 1 Markov State Models (MSMs) aim to sample accurately the underlying free energy landscape by use of either long or short equilibrium trajectories. Consider the sampling of the conformational space region containing a typical transition pathway between two end points (e.g., reactants, R, and products, P). Using short, rather than long equilibrium trajectories brings major sampling advantages. In typical MSM simulations, sets of short trajectories are initialized from intermediate conformations between R and P regions that are candidates for Markovian states. During analysis, transition probabilities between candidate states are extracted at different lag times and the states can be either divided or combined (clustered) until the transitions between them can be shown to be truly Markovian and a proper MSM is built. Short trajectories can be simulated either for the same duration (fixed length) or only until crossing into nearby regions (macrostates).

build transition probability matrices corresponding to different lag times and, ultimately, to estimate time-independent rate matrices that contain a complete representation (in the limits of the underlying Markovian model) of the thermodynamics and kinetics of the corresponding biomolecular system.[3] Due to their popularity and usefulness,[4] several packages are available to assist both the implementation and the analysis of MSMs from molecular trajectories as well as several useful reviews.[4–10] Here, we highlight the basic theoretical considerations behind extracting Markovian transition probabilities and rate matrices from molecular simulations, including a brief discussion and comparison with the increasingly popular, related method of Milestoning[11] that was recently applied to quantitative studies of transition pathways in increasingly complex biomolecular systems.[12–14] We note that several recent studies have highlighted the usefulness of Milestoning both in developing new simulation and analysis methodological approaches and in actual applications.[12,15,16]

2. Markovian approach: Markov state models for MD simulations

Markov State Models (MSMs) are becoming increasingly popular as they have proven to be a useful approach to both generating and analyzing the results of a broad range of molecular dynamics (MD) simulations, from folding/unfolding of proteins and studies of conformational dynamics under applied forces,[1,2] to binding/unbinding of peptides.[17,18] MSM-based studies allow for the convenient combination of several MD trajectories in to a single model of the underlying network of kinetic transitions between Markovian states from which experimental observables can be estimated, often to a high degree of accuracy.[3–10]

Master equation and rate matrix: Assuming ideal and Markovian states, a master equation accurately describes the rate of change of the population of the states in terms of fluxes in and out of the states. In a real system the states are highly dependent on the reaction coordinate chosen and on the lag time used to observe the system, and as such the system, for very short lag times, cannot be seen as Markovian. However, beyond certain lag times, most complex systems become "memoryless" and therefore the Markovian assumption holds, and the master equation can accurately represent the kinetics of the system. Say the system has N states and the probability of a particular state m is given by p_m. The rate constant, k_{mn}, is the rate from state m to state n and thus the master equation of the system is given by:

$$\frac{dp_m(t)}{dt} = \sum_{n=0}^{N-1} [k_{mn}p_n(t) - k_{nm}p_m(t)]$$

This can be written in simpler terms by using matrix notation:

$$\frac{d\mathbf{p}(t)}{dt} = \mathbf{K}(t)\mathbf{p}(t) \tag{1}$$

where $\mathbf{p}$(t) is the probability matrix and $\mathbf{K}$(t) is the $N \times N$ rate matrix of the system.

At equilibrium the rate of change of the population is zero and therefore the equilibrium population, $\mathbf{p}^o$, can be defined as:

$$\mathbf{K}\mathbf{p}^o = 0 \tag{2}$$

which is normalized according to the relation $\sum_{n=0}^{N-1} p_n^o = 1$ and positive $p_n^o > 0$, $n = \{0, 1, \ldots, N-1\}$.

The components of the rate matrix $\mathbf{K}$ are given by:

$$k_{nm} = \begin{cases} -\sum_{i=0(i\neq n)}^{N-1} k_{in}, & n = m \\ k_{mn}p_n^o / p_m^o, & n > m \end{cases}$$

This allows detailed balance to be maintained, which is a requirement for a rate matrix when the system is at equilibrium

$$k_{nm}p_m^o = k_{mn}p_n^o$$

Interestingly, one can express the likelihood Λ of a system of N states being represented by the rate matrix $\mathbf{K}$ as a function of the transition probabilities between states, and thus of the transition rates in the system.[3] Propagators, or the corresponding Green's functions $G(n, \Delta t | m, 0)$ are the probability of being in the state n at time Δt after having been in state m at time 0. These propagators are weighted by the number of transitions from m to n observed in an MD trajectory, during the interval of Δt, and cumulating the observed numbers of transitions in a transition matrix $T_{nm}(\Delta t)$ such that

$$\Lambda = \prod_{n=0}^{N-1} \prod_{i=1}^{N_{\text{int}}} [G(n, \Delta t | m, 0)]^{T_{nm}(\Delta t)} \tag{3}$$

where N_{int} is the number of time intervals of equal time length Δt such that the total length of the trajectory is $t_{total}=N_{\text{int}}\Delta t$. Eq. (3) makes a quantitative connection between the observed number of transitions and the likelihood of these observations to occur in an MD trajectory, and thus allows the extraction of the rate matrix of the system by maximizing the likelihood (e.g., by using simulated annealing of the free parameters of the rate matrix until convergence upon the maximum likelihood).[3] In practice, rather than maximizing the likelihood value, it is often convenient to minimize the corresponding log-likelihood.[3]

Eigenspectrum properties of rate matrices: The eigenvectors of the rate matrix **K** satisfy the following eigenvalue equations

$$\mathbf{K}\varphi_i = \lambda_i\varphi_i, \text{ and } \chi_i\mathbf{K} = \lambda_i\chi_i$$

where φ_i and χ_i are the left and right eigenvectors **K**, respectively, and the eigenvalues λ_i can be sorted such that $\lambda_0=0>\lambda_1\geq\lambda_2\geq\ldots\geq\lambda_{N-1}$.

In practice, rather then working numerically with **K**, it is often more convenient computationally to use the symmetrized rate matrix, **H**, which is given by

$$\mathbf{H} = \mathbf{P}_0^{-1/2}\mathbf{K}\mathbf{P}_0^{1/2}$$

where $\mathbf{P}_0= diag\left[p_0^o,\ldots,p_{N-1}^o\right]$ is the equilibrium population matrix consisting of the equilibrium populations of each state and such that the trace is unity, i.e., $tr(\mathbf{P}_o)=1$.

The components of the symmetric matrix **H** are given by

$$h_{nm} = \begin{cases} k_{nn} = -\sum\limits_{i=0(i\neq n)}^{N-1} k_{in}, & n = m \\ \sqrt{k_{nm}k_{mn}}, & n \neq m \end{cases}$$

Accordingly, the eigenvalues, λ_i, of **H** are the same as the eigenvalues of **K** and they satisfy the equation

$$\mathbf{H}\psi_i = \lambda_i\psi_i$$

where the functions ψ_i are the orthonormal eigenvectors of **H**. We note that in many practical applications it is better to work with the symmetrized rate matrix **H** rather than with **K**, in order to avoid confusion, and for increased numerical accuracy, the eigenvectors of both matrices being connected analytically to each other and to the vector of equilibrium population of states as described below.

Autocorrelation functions: The autocorrelation function of a time-dependent observable $\mathbf{a}(t)$, which depends on the state $s(t)$ of the system at time t such that $s(t) \in \{0, 1, \ldots, N-1\}$, can be written using spectral decomposition

$$\langle \mathbf{a}(t)\mathbf{a}(0)\rangle = \sum_{i=0}^{N-1} \left[\sum_{i=0}^{N-1} a_n \psi_0(n)\psi_i(n) \right]^2 exp\,(\lambda_i t)$$

A useful identity can be derived if the case of $\mathbf{a}(t) = \psi_i(s(t))/\psi_0(s(t)) = \chi_i(s(t))$ is considered. This gives

$$\left\langle \frac{\psi_i(s(t))}{\psi_0(s(t))} \cdot \frac{\psi_i(s(0))}{\psi_0(s(0))} \right\rangle = \exp\,(\lambda_i t) \tag{4}$$

Eq. (4) is often used to validate the extracted master equation against actual simulation trajectories of Markovian dynamic systems.

Relation between symmetrized and non-symmetrized rate matrices: The right and left eigenvectors of the original rate matrix $\mathbf{K}$ can be recovered from the eigenvectors of $\mathbf{H}$ in the manner below:

$$\varphi_i^2(n) = p_n^o . \psi_i^2(n),$$
$$\chi_i^2(n) = \psi_i^2(n)/p_n^o$$

and also $\varphi_i^2(n) = (p_n^o)^2 . \chi_i^2(n)$

The first right eigenvector, φ_0, of $\mathbf{K}$, corresponding to the eigenvalue $\lambda_0 = 0$ is found to be the equilibrium population from Eq. (2). Therefore, $\varphi_0 = p_n^o$. From this it can be seen that $\chi_0(n) = 1$ for all $n = \{0, 1, \ldots, N-1\}$. Therefore Eq. (4) shows that

$$p_n^o = \psi_0^2(n), \quad \forall n \in \{0, 1, \ldots, N-1\}$$

Further relations can be extracted by using the orthonormality of the eigenvectors $\psi_i(n)$ of $\mathbf{H}$

$$\sum_{n=0}^{N-1} \psi_i(n)\psi_j(n) = \sum_{n=0}^{N-1} \chi_i(n)\chi_j(n)p_n^o = \sum_{n=0}^{N-1} \varphi_i(n)\varphi_j(n)/p_n^o$$
$$= \sum_{n=0}^{N-1} \varphi_i(n)\chi_j(n) = \delta_{ij}$$

where δ_{ij} is the Kronecker delta. Filling in for $j = 0$ results in:

$$\sum_{n=0}^{N-1} \psi_i(n)\sqrt{p_n^o} = \sum_{n=0}^{N-1} \chi_i(n)p_n^o = \sum_{n=0}^{N-1} \varphi_i(n) = \delta_{i0}$$

Deriving the likelihood function: We can write an analogous equation to Eq. (1) for the symmetrized rate matrix $\mathbf{H}$ if the substitution $\pi = \mathbf{P}_0^{-1/2}\mathbf{p}$ is used.

$$\frac{d\pi(t)}{dt} = \mathbf{H}(t)\pi(t) \tag{5}$$

The solution of Eq. (5) is exponential along with the eigenvectors and eigenvalues of $\mathbf{H}$ and has the form

$$\mathrm{p_n(t) = p_n(0)\sum_{i=0}^{N-1} \psi_i^2(n)\exp(\lambda_i t)}$$

The propagators of the system (i.e., the conditional probability of being in state n at time t given that the system was in state m at time $t_0 = 0$) are the Green's functions

$$G(n, t|m, 0) = \left[e^{\mathbf{K}t}\right]_{nm} = \frac{\psi_0(n)}{\psi_0(m)} \sum_{i=0}^{N-1} \psi_i(n)\psi_i(m)\exp(\lambda_i t)$$

The propagators of the system are used to form the likelihood function Λ of the Markovian trajectory such that

$$\Lambda = \prod_{i=1}^{N_{\mathrm{int}}} G(s(i\Delta t), \Delta t | s((i-1)\Delta t), 0) \tag{6}$$

where N_{int} is the number of time intervals of equal time length Δt such that the total length of the trajectory is $t_{total} = N_{\mathrm{int}}\Delta t$. The likelihood function can be factorized into a product of products given the Markovian nature of the system leading to the Eq. (3) described above.

In typical MSM studies, as illustrated in Fig. 1, a set of either long or short[19] trajectories are used to sample the underlying free energy landscape. The trajectories are first processed to assign Markovian states, using a variety of possible approaches such as, for example, the transition-based assignment (TBA, see Refs. 3, 5 and a recent computational implementation in Ref. 10). Subsequently, the transition probabilities, or propagator functions described above are used to estimate either transition matrices at certain finite lag time values or, the time-independent rate matrix $\mathbf{K}$ (e.g., by likelihood maximization using Eq. (3)).[3]

2.1 Relative RMSD for state assignment

Here, we illustrate a simple yet effective method, the use of the relative RMSD (RelRMSD or $RMSD^{rel}$) collective variable to assign efficiently the structures of a molecular system to correct conformational states in biomolecular simulations. Let $RMSD_i$ be the RMSD value calculated along a trajectory after aligning it with the set of atomic coordinates for the molecular conformation i. Similarly, $RMSD_j$ is the RMSD value calculated along a trajectory after aligning it with the set of atomic coordinates for the molecular conformation j (see Fig. 2A). Using the $RMSD_i$ and $RMSD_j$ values, we can define the time-dependent relative RMSD ($RMSD^{rel}$).

$$RMSD_{ij}^{rel}(t) = \frac{RMSD_i(t)}{RMSD_i(t) + RMSD_j(t)}$$

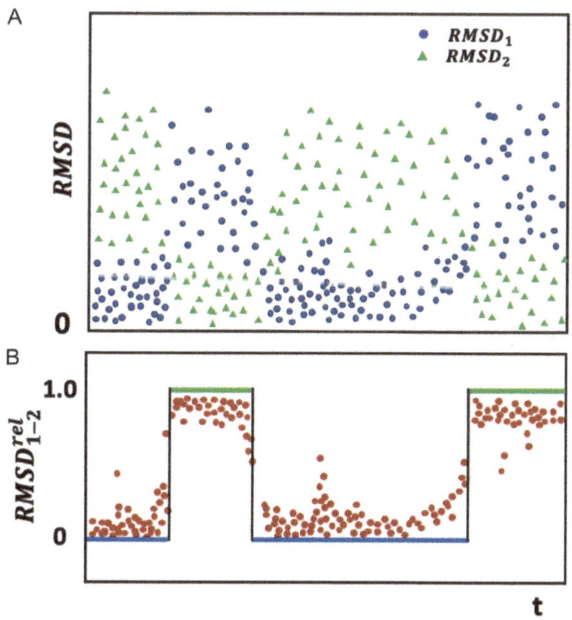

Fig. 2 Using the relative RMSD (RelRMSD) reaction coordinate to assign trajectories. (A) $RMSD_1$ (blue, circles) and $RMSD_2$ (green, triangles) are two illustrative values obtained by aligning a trajectory to structures corresponding to representative conformations for state 1 and state 2, respectively. As expected, these are most informative only for RMSD values close to zero. (B) By combining the two signals in $RelRMSD_{12}$ (red, circles) the overall ability to discriminate between the two states is improved, and the assignment of states can be done more accurately (for example, by using the transition-based assignment method) resulting in a discretized trajectory (horizontal lines).

The $RMSD^{rel}$ will always have values bound between 0 and 1. When the trajectory is close to conformation "*i*," the $RMSD^{rel}$ function takes positive values close to zero, and when the trajectory is close to the conformation "*j*," RelRMSD takes close to one but always less than one (see Fig. 2B). Combination of pairwise RelRMSD can be used in cases with more than two main conformations. We refer the readers to Ref. 20 for a more detailed description.

In Fig. 2 it is illustrated the use of the relative RMSD (RelRMSD) reaction coordinate to assign trajectories. Here, $RMSD_1$ (blue, circles) and $RMSD_2$ (green, triangles) are two illustrative values obtained by aligning a trajectory to structures corresponding to representative conformations for state 1 and state 2, respectively (Fig. 2A). As expected, these are only informative for RMSD values close to zero. In Fig. 2B it is shown how by combining the two signals in $RelRMSD_{12}$ (red, circles) the signal is improved, and the assignment of states can be done more accurately (for example, by using transition-based assignment) resulting in a discretized trajectory (lines).

Interestingly, the RelRMSD calculation can be used multi-dimensionally in the state assignment step for discriminating between multiple MSM states (in this case three conformational states), as illustrated in Figs. 3 and 4. Using the RelRMSD to assign states (e.g., in conjunction with the TBA method) is particularly useful in cases in which well-defined representative biomolecular

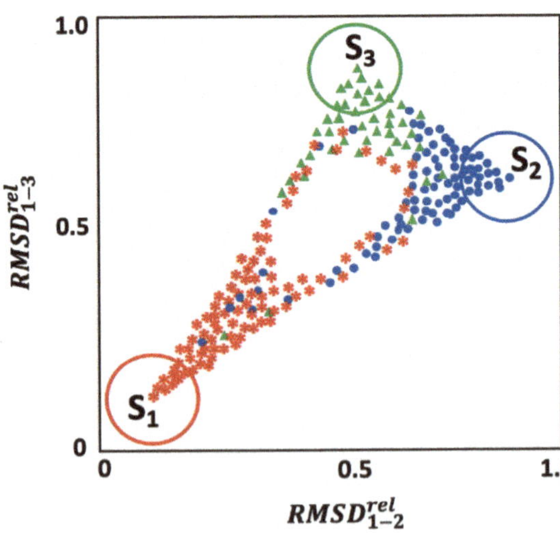

Fig. 3 The RelMSD calculation can be used multi-dimensionally in the state assignment step for discriminating between multiple MSM states (in this case three conformational states).

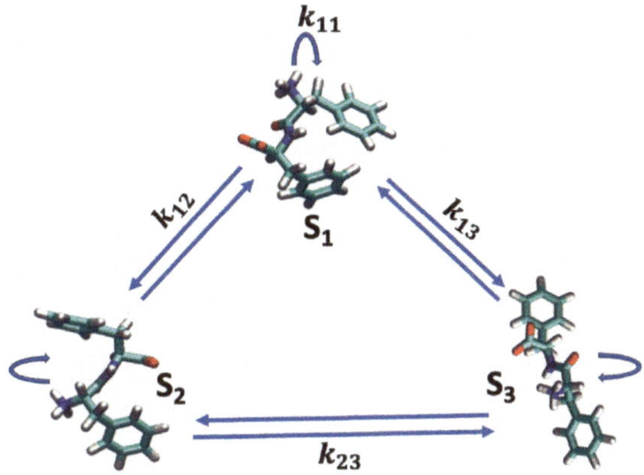

Fig. 4 Example of using RelRMSD for MSM analysis of the conformational dynamics of a small diphenylalanine (FF) peptide (replica exchange MD simulation with explicit water molecules).

conformations exist for the corresponding Markov states and can be used to make the thermodynamic and kinetic MSM analysis of the underlying conformational dynamics more automatic.

Fig. 4 shows an example of using RelRMSD for MSM analysis of the conformational dynamics of a diphenylalanine peptide (replica exchange MD simulation with explicit water molecules), which is shown to be essentially three-state in Ref. 20.

2.2 Applications of MSMs to enhance sampling in simulations of folding and binding of amyloid peptides

While current MD simulations are becoming capable of reaching millisecond-scale simulations, including by use of computers with hardware specialized in MD simulations,[21] and through distributed computing projects such as Folding@Home,[1] typical computational resources are still limited in their ability to sample properly standard MD simulations with explicit solvent molecules on even relatively small systems, such as amyloid peptides, due to the complexity of their underlying conformational landscape. This has resulted in the development of more efficient ways to extract the thermodynamic properties of the system that take better advantage of highly parallel architectures and the faster dynamics for certain thermodynamic parameters, such as temperature. Such enhanced sampling methods have become commonplace in modern simulations of proteins and biomolecules.

Replica-exchange MD (REMD)[22,23] and simulated tempering (ST)[24,25] have recently become popular modern methods of choice used to cross high energy barriers, and to map the free energy landscape of biomolecular systems, being currently available to use in many popular MD simulation packages. An important observation is that in temperature REMD simulations atomistic coordinates are typically saved at each temperature in data files for which we use the term "*T*-trajectories" (i.e., corresponding to one temperature *T*, and thus to all the different replicas evolving at this *T*). To fully characterize the REMD simulation, the history of replica exchange events is also saved separately in corresponding exchange data files. The information from both these two types of files can be used to generate what we denote as "*R*-trajectories" (i.e., corresponding to a replica *R* as it progresses at different temperatures after accepted exchange events).

Importantly, unlike the more typical *T*-trajectories, *R*-trajectories are continuous in time and can thus be used to assign states with our TBA method. It is only after this step, by using again the exchange history data, that states can be also assigned accurately along the more typical REMD *T*-trajectories, enabling thus the temperature-dependent investigation of the dynamics.[18]

The early aims of REMD simulations were focused mainly on enhanced conformational sampling in order to achieve faster the accurate estimation of the populations of conformational states of a molecular system. However, the initial studies were hampered by the fact that extracting quantitative information on kinetic transitions from REMD trajectories regarding the jumps between the various conformational states is made difficult by the replica exchanges themselves.[26–32] Proposed methods relied initially on either assumptions that transition rates tend to increase with temperature in an Arrhenius-like manner,[33,34] on algorithms using likelihood maximization and multi-dimensional optimization methods[18,26] or on complex statistical analysis of transition paths.[35] Recently, a new direct transition counting (DTC) method for calculating transition rates from REMD trajectories has been demonstrated[17] that was relatively simple to implement, while reaching similarly accurate temperature-dependent rates as compared to the alternative, more complex and indirect methods.[18,26] The DTC results were compared to those of the maximum likelihood propagator based (MLPB) method on the same data from all-atom REMD simulations of dimerization of computationally and experimentally-relevant amyloid-forming NNQQ tetrapetides,[36] in explicit water.[17,18]

Fig. 5 illustrates the network of conformational transitions extracted from REMD simulations with explicit water molecules of NNQQ peptide

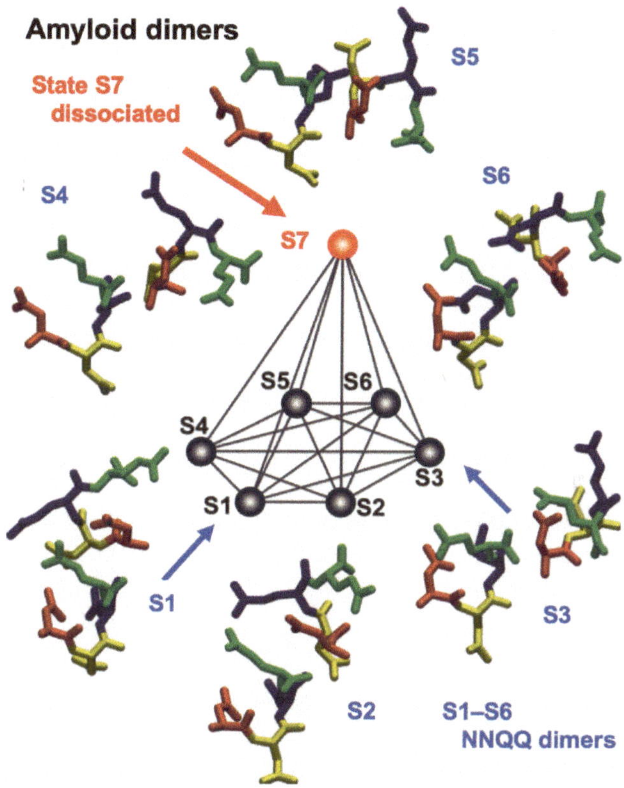

Fig. 5 Network of conformational transitions built using a Markov-based coarse master equation approach and REMD simulations with explicit water molecules, for quantifying the equilibrium dimer formation-dissociation of an amyloid forming NNQQ peptide. State S7 represents the ensemble of dissociated states of the dimer. In this case, the study revealed six main (states labeled S1–S6) association modes, with varying populations and transition rates depending on temperature.[18]

dimers,[36] using a Markov-based coarse master equation approach, with the goal of quantifying the equilibrium dimer formation-dissociation of an amyloid forming NNQQ peptide.[18] The conformational ensemble of dissociated state of the NNQQ dimer is represented by state 7. In this case, the REMD study revealed six main association modes (states labeled S1–S6), and allowed the direct calculation from REMD trajectories of their relative populations and transition rates, as a function temperature, for a broad temperature range between 310 and 369 K.[18] While this is one of the smallest two-state-like systems featuring peptide-peptide interactions that is, nevertheless, challenging to analyze systematically using REMD.[18] The rates extracted using the DTC method were validated both by comparison to the corresponding MLPB rates and by analyzing the decay rates of the state autocorrelation functions of the system.[3,5,37] Interestingly, the MSM approach in conjunction with the DTC method allowed the assess the corresponding REMD efficiency,[38] showing that one can obtain remarkably good agreement with the theoretically predicted errors in estimating the dimer and dissociated populations on the binding dynamics of NNQQ amyloid tetrapetides.[17,18]

We highlight here that the calculation of transition probabilities between Markovian conformational states, springing from the use of MSMs in conjunction with enhanced sampling methods such as REMD, has allowed the extension of REMD results beyond accurate calculation of thermodynamic properties alone (e.g., equilibrium populations and corresponding free energies) to accurate kinetic studies and the estimation of transition rates from short segments of REMD trajectories for MD simulations with explicit water molecules ranging from folding of short peptides[26] to unveiling association-dissociation pathways and binding constants.[18,31,39]

We note that both REMD and MD simulations of short amyloid peptides were also performed in other studies of amyloid conformational transitions.[20,27,40–42] However, REMD simulations are restricted to relatively small systems, as larger systems lead to broader potential energy distributions and thus require more replicas for the same temperature range. Also, conformational studies of larger amyloid peptides such as Alzheimer's Aβ peptides[43] or diabetes-related amylin[44,45] require a more challenging multiscale approach and new modeling methods.[28] Notably, in a recent study of Alzheimer's Aβ42,[46] new MD simulations were shown to lead to accurate calculations of both equilibrium state populations and transition probabilities in conjunction with a novel transition network analysis method.[46]

More recently, an automated method based on MSM analysis[47] has been developed and used for simulations of molecular self-assembly and aggregation of KFFE peptides, a subsequence of Alzheimer's amyloid-β peptide and one of the smallest peptides known to aggregate into amyloid fibrils in vitro, by constructing collective coordinates from molecular descriptors that are invariant to permutations of molecular indexing.[47]

3. Milestoning

For systems where the sampling problem can be stated in terms of estimating the thermodynamics and kinetics along transition pathways that connect two known metastable states (e.g., "reactant," R, and "product," P, see Figs. 1 and 2) one could target the sampling in order to identify and characterize the pathway(s) (typically with the maximum reaction flux) and the corresponding intermediate states that are relevant to the underlying reaction mechanism.

In Milestoning, an initial pathway between R and P conformations can be used to first select a set of intermediate conformations along any candidate reaction path that may be available (e.g., from a high-temperature simulation, steered MD). The intermediate representative conformations (Fig. 6, red points) are referred to as anchors. Milestones are further introduced as hypersurfaces in the free energy landscape that have an equal minimal distance from at least two anchors (blue lines in Figs. 6 and 7). Finally, once the energy landscape is divided along the reaction coordinate using milestones, one can run sets of short trajectories from each milestone (shown in Figs. 6 and 7). These short trajectories are terminated as soon as they reach a new milestone. The statistical information on local transitions between each pair of milestones (e.g., first passage times), can be combined to obtain the global statistical information on possible R–P transition pathways, free energy profiles and relative transition times.

Here we briefly illustrate the Milestoning method. To initiate Milestoning, we need a set of sample configurations, that we call anchors, from the coarse space between R and P (red dots in Figs. 6 and 7). These anchors are the centers of Voronoi cells in space of several course variables. A milestone between two anchors, *i* and *j*, is thus a set of points with equal distance from anchors *i* and *j*.

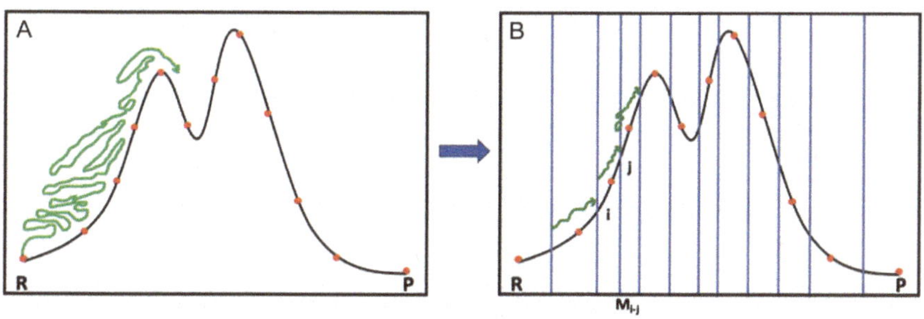

Fig. 6 Schematic 1d representation of two typical approaches to sampling free energy barriers in molecular simulations using (A) one or a few typically long trajectories, and (B) short trajectories initialized along the transition pathway between two end points (e.g., reactants, R, and products, P). In (B), central in the implementation of the Milestoning method, the aim is to achieve higher computational efficiency by initializing the short trajectories from multi-dimensional hypersurfaces (vertical lines, blue) that are located between anchors (red dots) along a low dimensional reaction coordinate (see also Fig. 2).

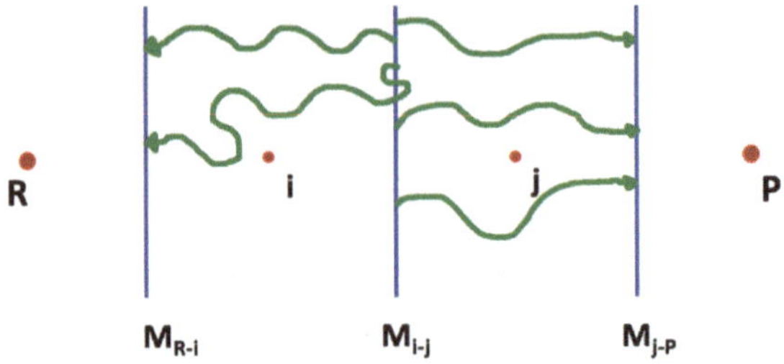

Fig. 7 In Milestoning, the sampling is initiated by first defining a set of "anchors" (red) conformations that span a (typically low dimensional) transition pathway between two end points (e.g., reactants, R, and products, P). The underlying kinetic and thermodynamic information is extracted by analyzing sets of short trajectories initiated in conformational space regions (blue, milestones, where M_{ij} is the milestone between anchors *i* and *j*, respectively). The short trajectories (green) are terminated as soon as they reach another milestone, different from the one at which they were initiated.

This distance is smaller than the separation from all other anchors. Milestone between anchors *i* and *j* is denoted by M_{ij} (shown in blue in Fig. 6) or M_α.

At this stage, one needs to sample configurations on each milestone (e.g., by using MD simulations constrained to corresponding milestone coordinates). Finally, we launch unbiased MD trajectories from the sampled configurations on each milestone. These trajectories are terminated as soon as they reach a milestone different from the trajectory's starting milestone. We record the starting and terminating milestone index and the time of termination. Using this information, we can estimate transition probability between milestones, also called kernel, **W**, and the corresponding lifetime of the milestone. Let n_α be the total number of trajectories initiated at the milestone M_α. Let $n_{\alpha\beta}$ be the number of trajectories which started at milestone M_α and terminated at milestone M_β. The transition probability, $W_{\alpha\beta}$, is estimated as $n_{\alpha\beta}/n_\alpha$. The lifetime of the milestone, M_α, is $t_\alpha = \frac{1}{n_\alpha}\sum_{l=1,\ldots,n_\alpha} t_l$, where l is the index of the trajectory and t_l is the time length of the trajectory l.

Using the transition probability or kernel matrix, **W**, and the milestone lifetime, **t**, we can compute important kinetic and thermodynamic observables such as the flux through a milestone, free energy, mean first passage time (MFPT), etc. The stationary flux of trajectories through a milestone is denoted by the eigenvector, **q**, of the matrix **W**, with an eigenvalue of one. It can be shown that the free energy of a milestone α is $F_\alpha = -k_B T \log[q_\alpha t_\alpha]$ and the mean first passage time (MFPT, $\langle\tau\rangle$) can be written as $\langle\tau\rangle = \mathbf{p}_0(\mathbf{I}-\mathbf{W})^{-1}\mathbf{t}$, where $\mathbf{p}_0$ is the vector of the initial distribution and **I** is the identity matrix.[11]

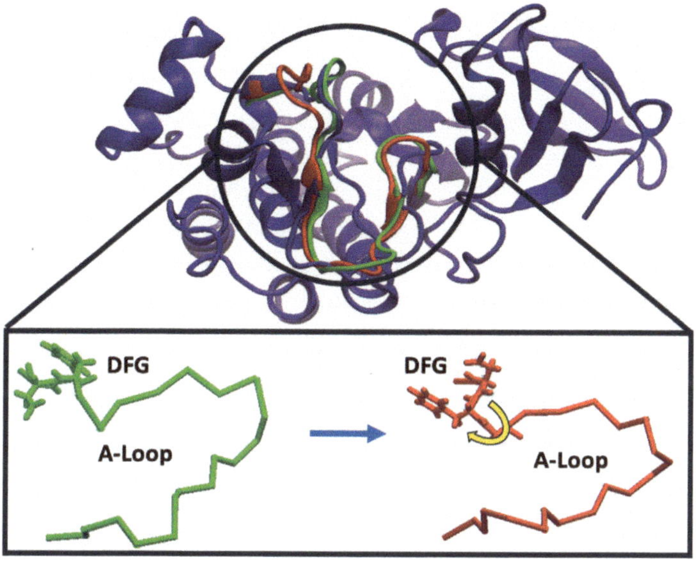

Fig. 8 Example of using a multi-dimensional Milestoning approach in a large and more complex molecular system: the characterization of the DFG-flip dynamics in Abl kinase.

Fig. 8 illustrates the use of a multi-dimensional Milestoning approach in a large and more complex molecular system: the characterization of the DFG-flip dynamics in Abl kinase.[15] Here, the Milestoning method to study the transition between active and inactive conformations of Abl kinase (shown in Fig. 8). Note that it would have been impractical to study a complex system like this with regular MD simulations using comparable HPC resources.[15]

Another example of an application of Milestoning is to peptide permeation through a biological membrane. Membranes separate life processes from aqueous solution and are doing so by presenting a hydrophobic barrier between the exterior and interior of cells. Molecules that are soluble in water are less likely to be found in hydrophobic environment and therefore are unlikely to cross the membrane barrier. Nevertheless, the barriers to and from cells and between cell compartments are not absolute. They allow for some passive permeation (permeation which is not assisted by channels or pumps). Nature uses a class of peptides, named Cell Penetrating Peptides (CPP), to efficiently translocate cargo between compartments. Cell penetration by peptides is an important process, with immediate application in drug development and subject to many recent MD and bioinformatics studies.[48] Interestingly, CPPs are frequently positively charged.[48,49] The charge makes them soluble in water, but is expected to reduce their permeability through membrane. A recent study[50] showed quantitatively how two positive charges interact strongly with the head groups of the membrane, create significant membrane distortion and narrowing, and thus enable the peptide permeation through the membrane. The peptide studied here is WKW. It has one positive charge at the lysine side chain and a second positive charge at the N terminal (see Fig. 9). The C terminal was amidated and therefore neutral. The overall time scale of the translocation of the peptide through the membranes is measured in tens of minutes and therefore an enhanced sampling method for the kinetics is necessary. The mechanism was discovered using Milestoning simulations and is sketched in the figure.

The free energy profile (Fig. 10) that follows the mechanism depicted in Fig. 9 identifies state D as a metastable state in which the opposite leaflets of the membrane are distorted and interact with the charges of the peptide. The DOPC membrane in which the simulations and experiments were

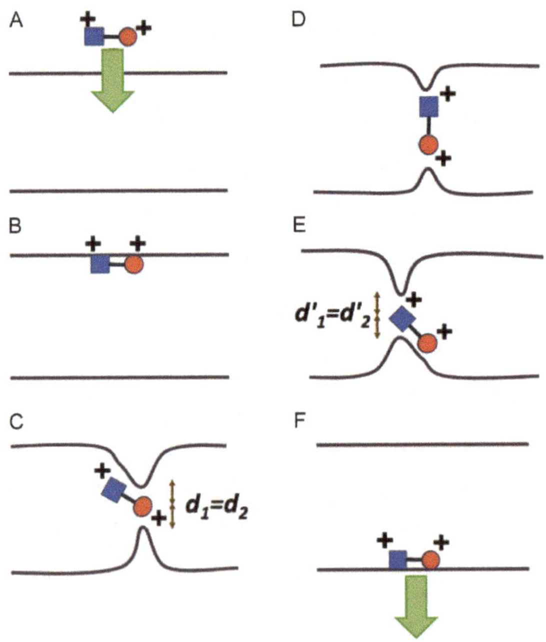

Fig. 9 Schematic representation of the permeation mechanism of the peptide WKW with two positive charges. The membrane permeation pathway includes two local minima at states B and F and two saddle points at states C and E. At state D, each charge is interacting with phosphate groups from one leaflet, making it a metastable state. The charge on the N-terminus is shown as a red circle and the charge on the lysine side chain is shown as a blue square.

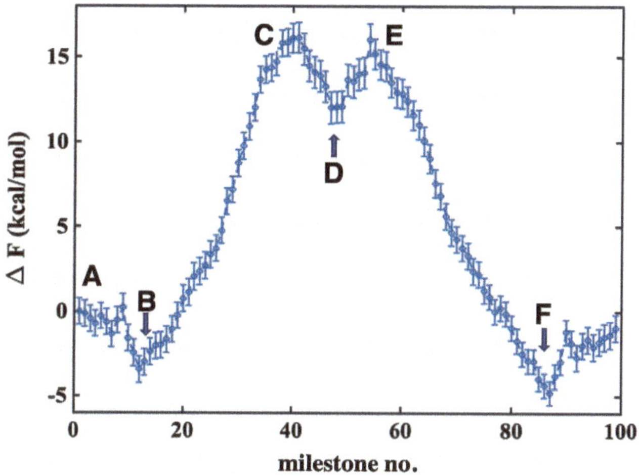

Fig. 10 Free energy profile for the permeation of the peptide WKW in a DOPC membrane. The approximate positions of the different states are labeled according to the mechanism depicted in Fig. 9. The slight asymmetry of the free energy on the right and the left sides of the D state may indicate different permeation of the lysine or the N terminus (i.e., whose charge permeates first). However, the differences are within the error bars.

conducted is neutral overall, but the head groups are highly polarized. The positive charges of the choline groups are embedded in the aqueous solution and are screened from direct interactions with the peptide. The negatively charged and the polarized groups such as the phosphates and carbonyls are more buried in the membrane compared to the chole and therefore assist the permeation of the positively charged peptide.

Comparison to experimental data reported[50] is encouraging, suggesting WKW to be one of the few peptides that are fully characterized by experiment and simulations. These interesting results highlight the broad range of applicability of Milestoning to systematic studies that can make direct connection with experimental measurements.[50]

4. Conclusions

In spite of significant and sustained advances in computational hardware allowing for larger size and longer scale computations, molecular dynamics studies of a large majority of biomolecular systems remain outside the reach of traditional simulation methods. A main problem springs from the intrinsically high-dimensional and complex nature of the underlying free energy landscape of most systems, and from the necessity to sample accurately such complex landscapes for identifying kinetic and thermodynamic states in the configurations space, and for accurate calculations of both free energy differences and of the corresponding transition rates between states. Here, we reviewed two modern methods that allow longer-time MD studies of biomolecular systems that can open a broad spectrum of applications.

A first approach, Markov State Models, relies on identifying a set of configuration states in which the system resides sufficiently long to relax and loose the memory of previous transitions, and on using simulations for mapping the underlying complex energy landscape on a network of Markovian transitions. The independence of the underlying transition probabilities creates the opportunity to increase the sampling efficiency by using appropriately initialized sets of short simulations rather than more typical long MD trajectories, which leads to both enhanced sampling and higher accuracy. This allows MSM studies to unveil bio-molecular mechanisms and to estimate free energy barriers with high accuracy, in a manner that is both systematic and relatively automatic, which accounts for their increasing popularity.

The second approach presented here, Milestoning, is focused on accurate studies of the ensemble of pathways connecting two specific end-states (e.g., reactants and products) in a similarly systematic and highly automatic and highly accurate manner. Conceptually, both methods are identical for transition paths between Markovian states,[12] however, Milestoning can be generalized and applied to studies of non-Markovian transitions as well. Applications presented here range from studies of conformational dynamics and binding of amyloid-forming peptides, cell-penetration of peptides, and a study on the atomistic characterization of the DFG-flip dynamics in Abl kinase.

As highlighted by the increasing number of studies using both methods, we anticipate that they will open new avenues for the investigation of systematic sampling of reactions pathways and mechanisms occurring on longer time scales than currently accessible by purely computational hardware and parallelization-related advances.

Acknowledgments

B.N. and N.V.B. acknowledge the financial support received from Thomas Preston Ph.D. Scholarship Fund and from the European Union's Horizon 2020 research and innovation programme "NanoinformaTIX" (topic NMBP-14-2018, grant No 814426). R.E. acknowledges support from the NIH GM59796 and a Welch grant F-1896.

References

1. Shirts M, Pande VS. Computing—screen savers of the world unite!. *Science.* 2000;290(5498):1903.
2. Zwanzig R. From classical dynamics to continuous time random walks. *J Stat Phys.* 1983;30(2):255.
3. Buchete NV, Hummer G. Coarse master equations for peptide folding dynamics. *J Phys Chem B.* 2008;112(19):6057.
4. Husic BE, Pande VS. Markov state models: from an art to a science. *J Am Chem Soc.* 2018;140(7):2386.
5. Buchner GS, Murphy RD, Buchete NV, Kubelka J. Dynamics of protein folding: probing the kinetic network of folding-unfolding transitions with experiment and theory. *BBA-Proteins Proteomics.* 2011;1814(8):1001.
6. Chodera JD, Noé F. Markov state models of biomolecular conformational dynamics. *Curr Opin Struct Biol.* 2014;25:135.
7. Noé F, Fischer S. Transition networks for modeling the kinetics of conformational change in macromolecules. *Curr Opin Struct Biol.* 2008;18(2):154.
8. Wang W, Cao S, Zhu L, Huang X. Constructing Markov state models to elucidate the functional conformational changes of complex biomolecules. *WIREs Comput Mol Sci.* 2018;8(1):e1343.
9. Bowman GR, Beauchamp KA, Boxer G, Pande VS. Progress and challenges in the automated construction of Markov state models for full protein systems. *J Chem Phys.* 2009;131(12):124101.
10. de Sancho D, Aguirre A. MasterMSM: a package for constructing master equation models of molecular dynamics. *J Chem Inf Model.* 2019;59(9):3625.
11. Bello-Rivas JM, Elber R. Exact milestoning. *J Chem Phys.* 2015;142(9):094102.
12. Berezhkovskii AM, Szabo A. Committors, first-passage times, fluxes, Markov states, milestones, and all that. *J Chem Phys.* 2019;150(5):054106.
13. Grazioli G, Andricioaei I. Advances in milestoning. I. Enhanced sampling via wind-assisted reweighted milestoning (WARM). *J Chem Phys.* 2018;149(8):084103.
14. Grazioli G, Andricioaei I. Advances in milestoning. II. Calculating time-correlation functions from milestoning using stochastic path integrals. *J Chem Phys.* 2018;149(8):084104.
15. Narayan B, Fathizadeh A, Templeton C, et al. The transition between active and inactive conformations of Abl kinase studied by rock climbing and Milestoning. *Biochim Biophys Acta Gen Subj.* 2019;1864:129508. https://doi.org/10.1016/j.bbagen.2019.129508.
16. Schütte C, Noé F, Lu J, Sarich M, Vanden-Eijnden E. Markov state models based on milestoning. *J Chem Phys.* 2011;134(20):204105.
17. Leahy CT, Kells A, Hummer G, Buchete N-V, Rosta E. Peptide dimerization-dissociation rates from replica exchange molecular dynamics. *J Chem Phys.* 2017;147(15):152725.
18. Leahy CT, Murphy RD, Hummer G, Rosta E, Buchete N-V. Coarse master equations for binding kinetics of amyloid peptide dimers. *J Phys Chem Lett.* 2016;7(14):2676.
19. Chodera JD, Swope WC, Pitera JW, Dill KA. Long-time protein folding dynamics from short-time molecular dynamics simulations. *Multiscale Model Simul.* 2006;5(4):1214.
20. Narayan B, Herbert C, Yuan Y, Rodriguez BJ, Brooks BR, Buchete N-V. Conformational analysis of replica exchange MD: temperature-dependent Markov networks for FF amyloid peptides. *J Chem Phys.* 2018;149(7):072323.
21. Lindorff-Larsen K, Piana S, Dror RO, Shaw DE. How fast-folding proteins fold. *Science.* 2011;334(6055):517.
22. Sugita Y, Okamoto Y. Replica-exchange molecular dynamics method for protein folding. *Chem Phys Lett.* 1999;314(1–2):141.

23. Garcia AE, Sanbonmatsu KY. Alpha-helical stabilization by side chain shielding of backbone hydrogen bonds. *Proc Natl Acad Sci U S A*. 2002;99(5):2782.
24. Lyubartsev AP, Martsinovski AA, Shevkunov SV, Vorontsovvelyaminov PN. New approach to Monte Carlo calculation of the free energy: method of expanded ensembles. *J Chem Phys*. 1992;96(3):1776.
25. Marinari E, Parisi G. Simulated tempering: a new Monte Carlo scheme. *Europhys Lett*. 1992;19(6):451.
26. Buchete NV, Hummer G. Peptide folding kinetics from replica exchange molecular dynamics. *Phys Rev E*. 2008;77(3):4.
27. Strodel B, Whittleston CS, Wales DJ. Thermodynamics and kinetics of aggregation for the GNNQQNY peptide. *J Am Chem Soc*. 2007;129(51):16005.
28. Buchete NV. Unlocking the atomic-level details of amyloid fibril growth through advanced biomolecular simulations. *Biophys J*. 2012;103(7):1411.
29. Frank AT, Andricioaei I. Reaction coordinate-free approach to recovering kinetics from potential-scaled simulations: application of Kramers' rate theory. *J Phys Chem B*. 2016;120(33):8600.
30. Carballo-Pacheco M, Strodel B. Comparison of force fields for Alzheimer's a β42: a case study for intrinsically disordered proteins. *Protein Sci*. 2017;26(2):174.
31. Stelzl LS, Hummer G. Kinetics from replica exchange molecular dynamics simulations. *J Chem Theory Comput*. 2017;13(8):3927.
32. De Sancho D, Best RB. What is the time scale for α-helix nucleation? *J Am Chem Soc*. 2011;133(17):6809.
33. van der Spoel D, Seibert MM. Protein folding kinetics and thermodynamics from atomistic simulations. *Phys Rev Lett*. 2006;96(23):238102.
34. Piana S, Lindorff-Larsen K, Shaw DE. Protein folding kinetics and thermodynamics from atomistic simulation. *Proc Natl Acad Sci U S A*. 2012;109(44):17845.
35. Zhu F. Calculating transition rates from durations of transition paths. *J Chem Phys*. 2017;146(12):124128.
36. Sawaya MR, Sambashivan S, Nelson R, et al. Atomic structures of amyloid cross-beta spines reveal varied steric zippers. *Nature*. 2007;447(7143):453.
37. Bicout DJ, Szabo A. Entropic barriers, transition states, funnels, and exponential protein folding kinetics: a simple model. *Protein Sci*. 2000;9(3):452.
38. Rosta E, Hummer G. Error and efficiency of replica exchange molecular dynamics simulations. *J Chem Phys*. 2009;131(16):12.
39. Andrec M, Felts AK, Gallicchio E, Levy RM. Protein folding pathways from replica exchange simulations and a kinetic network model. *Proc Natl Acad Sci U S A*. 2005; 102(19):6801.
40. Martini L, Kells A, Covino R, Hummer G, Buchete N-V, Rosta E. Variational identification of Markovian transition states. *Physical Review X*. 2017;7(3):031060.
41. Barz B, Strodel B. Thermodynamics and kinetics of amyloid aggregation from atomistic simulations. In: *Physical Biology of Proteins and Peptides: Theory, Experiment, and Simulation*. Springer; 2015:29–45. https://doi.org/10.1007/978-3-319-21687-4_2.
42. Barz B, Wales DJ, Strodel B. A kinetic approach to the sequence-aggregation relationship in disease-related protein assembly. *J Phys Chem B*. 2014;118(4):1003.
43. Tywoniuk B, Yuan Y, McCartan S, et al. Amyloid fibril design: limiting structural polymorphism in Alzheimer's Aβ protofilaments. *J Phys Chem B*. 2018;122(49):11535.
44. Tofoleanu F, Yuan Y, Pickard FC, Tywoniuk B, Brooks BR, Buchete N-V. Structural modulation of human amylin protofilaments by naturally occurring mutations. *J Phys Chem B*. 2018;122(21):5657.
45. Murphy RD, Conlon J, Mansoor T, Luca S, Vaiana SM, Buchete N-V. Conformational dynamics of human IAPP monomers. *Biophys Chem*. 2012;167:1.

46. Barz B, Liao Q, Strodel B. Pathways of amyloid-β aggregation depend on oligomer shape. *J Am Chem Soc.* 2018;140(1):319.
47. Sengupta U, Carballo-Pacheco M, Strodel B. Automated Markov state models for molecular dynamics simulations of aggregation and self-assembly. *J Chem Phys.* 2019; 150(11):115101.
48. Duffy F, Maheshwari N, Buchete N-V, Shields D. Goetz G, ed. *Cyclic Peptide Design.* New York, NY: Springer New York; 2019. https://doi.org/10.1007/978-1-4939-9504-2_5.
49. Guidotti G, Brambilla L, Rossi D. Cell-penetrating peptides: from basic research to clinics. *Trends Pharmacol Sci.* 2017;38(4):406.
50. Fathizadeh A, Kogan M, Anderson CM, Webb LJ, Elber R. Defect-assisted permeation through a phospholipid membrane: experimental and computational study of the peptide WKW. *J Phys Chem B.* 2019;123(31):6792.

CHAPTER SIX

Protein thermal stability

Stepan Timr[a,b], Dominique Madern[c], Fabio Sterpone*[,a,b]
[a]CNRS, Université de Paris, UPR 9080, Laboratoire de Biochimie Théorique, Paris, France
[b]Institut de Biologie Physico-Chimique-Fondation Edmond de Rothschild, PSL Research University, Paris, France
[c]Univ. Grenoble Alpes, CEA, CNRS, IBS, Grenoble, France
*Corresponding author: e-mail address: fabio.sterpone@ibpc.fr

Contents

Abstract

Proteins, in general, fold to a well-organized three-dimensional structure in order to function. The stability of this functional shape can be perturbed by external environmental conditions, such as temperature. Understanding the molecular factors underlying the resistance of proteins to the thermal stress has important consequences. First of all, it can aid the design of thermostable enzymes able to perform efficient catalysis in the high-temperature regime. Second, it is an essential brick of knowledge required to decipher the evolutionary pathways of life adaptation on Earth. Thanks to the development of atomistic simulations and ad hoc enhanced sampling techniques, it is now possible to investigate this problem in silico, and therefore provide support to experiments. After having described the methodological aspects, the chapter proposes an extended discussion on two problems. First, we focus on thermophilic proteins, a perfect model to address the issue of thermal stability and molecular evolution. Second, we discuss the issue of how protein thermal stability is affected by crowded in vivo-like conditions.

1. Introduction

Proteins are soft matter entities that, in general, acquire a well-defined three-dimensional structure in the environment where they perform their function (a water solution or a membrane space). This structure, called

Progress in Molecular Biology and Translational Science, Volume 170
ISSN 1877-1173
https://doi.org/10.1016/bs.pmbts.2019.12.007

native or folded, is marginally stable. In fact, the free energy difference between the folded state and the ensemble of highly disordered unfolded conformations is of only a few kcal/mol.[1,2] At ambient conditions, depending on this free energy difference, a thermodynamic equilibrium exists between the folded population and the unfolded one. The unfolded population can be favored by perturbing the environmental conditions, e.g., by changing the temperature, the pressure, or the pH. In this chapter, we focus on the effect of temperature.

The thermal stability of a protein can be described by the Gibbs–Helmholtz equation, which relates the free energy difference between the folded and unfolded states as a function of temperature, $\Delta G_{f/u}(T)$, to the temperature variation of the enthalpy ($\Delta H_{f/u}$) and entropy of unfolding ($\Delta S_{f/u}$).[1,3] The temperature variation of these quantities depends on the specific heat, C_p, which is the amount of energy that must be provided to a solution of unit mass to increase the temperature by one unit. It is experimentally verified that for a given protein and at a given temperature the specific heat of the unfolded solution is larger than that of a folded equivalent. This difference, called the specific heat of unfolding ΔC_p, is found to be only weakly temperature dependent and can be considered constant. If the characteristic thermodynamic parameters, $\Delta H_{f/u}$ and $\Delta S_{f/u}$, are evaluated at the melting temperature, where the populations of the folded and unfolded states are equal, the modified Gibbs–Helmholtz equation reads as follows:

$$\Delta G_{f/u} = \Delta H_m\left(1 - \frac{T}{T_m}\right) + \Delta C_p[(T - T_m) - Tln(T/T_m)] \quad (1)$$

The associated curve looks like an inverted parabola, characterized by a maximum, the temperature at which the folded state is the most stable, and two zeros, where the population of the unfolded state matches that of the folded one. The first zero refers to the high-temperature melting, while the second one, which is located at a low temperature, predicts cold unfolding.

The thermal stability of a protein can be experimentally characterized by using differential scanning calorimetry. Using this technique, it is possible to access the variation of the specific heat of a protein solution as a function of temperature. From the resulting variation, it is possible to individuate the unfolding temperature (T_m), and ideally to deduce the specific heat of unfolding (ΔC_p) and the enthalpy of unfolding (ΔH_m). Other approaches, more or less direct, can also be used to derive the thermodynamic parameter of the Gibbs–Helmholtz equation. According to data reported in literature, ΔH_m can vary between 50 and 100 kcal/mol (although smaller or higher values were found for specific cases), and the specific heat of unfolding can be in the order of a few tenths of kcal/(mol K) or as large as a few kcal/(mol K).[1] Depending on the organism they belong to, the range of the thermal stability of proteins is broad,[2,3] varying from cold adapted proteins to hyperthermophilic species.

The study of thermal stability has helped to single out the molecular factors that determine the energetics of the folded state over the unfolded configurations, namely the contribution of hydrophobic interactions[4–6] and desolvation of charged groups,[7] as well as of hydration.[8,9] This is a source of inspiration for protein design that is aimed to enhance the stability of a protein scaffold so as to exploit the catalytic activity in a broader range of temperatures[10] or to make a protein, e.g., an antibody,[11] more robust for therapeutical purposes.

In this chapter we first present computational approaches based on atomistic simulations that can be deployed to investigate protein thermal stability. We then consider extremophilic proteins that evolved to be stable and functional in extreme temperature conditions, up to 100°C. By comparing their behavior with that of the mesophilic homologues, it is possible to dissect the main molecular factors of thermal stability. Finally we address the issue of protein stability under in vivo-like conditions, where macromolecular crowding can affect protein stability because of the excluded volume effect and specific quinary interactions.

2. Thermal stability in silico

The possibility to predict thermal stability in silico and, eventually, to design mutations to enhance it, is of vital importance. Several approaches have been experimented. The most direct one is to define, and optimize, a functional that can estimate the folding free energy from the sequence and the structure of a protein.[12] This is not trivial since, besides ensuring an accurate description of the energetics, it is necessary to quantify the entropic contribution to stability. Other approaches involve Monte Carlo (MC) simulations[13] or the discrete molecular dynamics technique[14] and are coupled with a simplified, implicit-solvent representation of the protein.

A more time-consuming approach is to use molecular dynamics (MD) simulations based on atomistic force fields with an explicit representation of the solvent and to estimate the unfolding free energy via long simulations or ad hoc enhanced sampling techniques. However, when the problem is to quantify thermal stability, one needs to access the unfolding free energy at different temperatures and, ideally, reconstruct the stability curve $\Delta G_{f/u}(T)$. This can be done by applying the replica exchange MD (REMD) technique, which we will describe shortly.[15–18]

The general concept of the replica exchange methodology was proposed in different contexts, e.g., in MC simulations,[19] and then specialized for MD.[16] The technique is intended to help sample protein free energy landscapes that are characterized by many local minima separated by high enthalpic barriers. The idea is to create several copies of the molecular system under consideration and run them independently each at a different temperature. To enhance conformational sampling, exchanges of geometries are randomly attempted between two systems evolving at two different temperatures, e.g., T_i and T_j. An illustration of the process is presented in Fig. 1A. Because it is desired that the configurations collected at a given temperature represent a correct sample of the canonical ensemble, the probability of an exchange must respect the detailed balance condition. The resulting exchange probability will depend ultimately on the energy difference between the two systems, $U(\mathbf{x}_i) - U(\mathbf{x}_j)$, where U is the potential

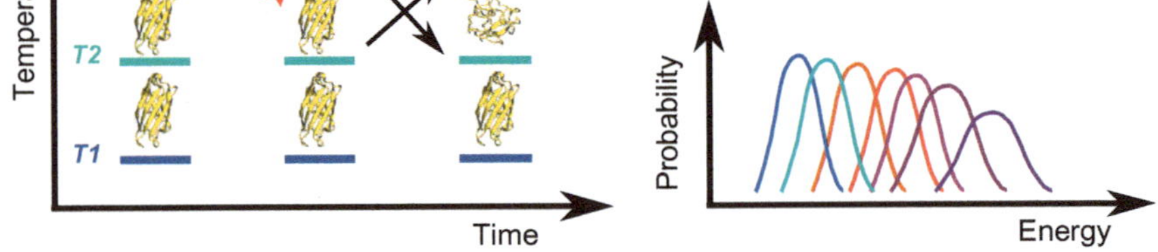

Fig. 1 Replica exchange MD. Panel (A): Schematic view of the REMD scheme. Protein conformations generated by the time evolution at a given temperature are exchanged during the simulation. This allows thermally excited conformations to populate configurational spaces at lower temperatures, thus enhancing the sampling. Panel (B): In the *top chart*, we illustrate a characteristic random walk in the temperature space for a selected protein conformation (*star*). The *bottom chart* shows the probability distribution of the potential energies for different reference temperatures. The overlap between the distributions ensures a good acceptance rate for the Monte Carlo moves in the temperature space.

energy controlling the dynamics of each system and $\mathbf{x}_{i(j)}$ indicate the coordinates of all particles of the system in replica $i(j)$. More precisely, the transition probability for the system in replica i to go evolving at the temperature of replica j and for the system in replica j to move to the temperature of replica i reads as $\min(1, \exp(\Delta))$, where $\Delta = (U(\mathbf{x}_i) - U(\mathbf{x}_j)) \cdot (1/k_b T_i - 1/k_b T_j)$. This choice, named the Metropolis criterion, ensures compliance with the detailed balance condition. In most implementations of the REMD method, only exchanges between neighboring temperatures are attempted (see Fig. 1A). At variance with the MC method, in REMD special care must be given to the treatment of the kinetic energy of the system upon exchange, and an ad hoc strategy has been devised for this.[16,17] The resulting dynamics of a system starting at a given replica at time zero is a random walk in the temperature space, and this erratic behavior helps populate several minima of the complex free energy landscape; see the schematic representation in Fig. 1B. The exploration of the high-temperature regime allows crossing barriers that otherwise would limit the sampling. As mentioned before, the transition probability depends on the energy difference between the two replicas, and in order to make the methodology efficient, it is necessary to maintain a good acceptance rate of the exchange moves. This requires, for instance, that the energy distributions collected for two neighboring replicas should overlap, see Fig. 1C. This condition limits the computational applicability of the REMD method to small molecular systems since by increasing the size of a molecular system, the energy separation between two replicas would scale proportionally to the number of degrees of freedom N while the fluctuations of the energy in each replica only as $\sqrt{N}$. Therefore, when increasing the system size and keeping fixed the range of temperatures to be covered in the simulations, the number of replicas for an efficient sampling must be increased as $\sqrt{N}$, see Ref. 17 for a detailed discussion.

The application of REMD simulations was crucial in that it allowed starting a systematic exploration of the thermodynamics of protein folding and stability at the atomistic level. REMD played an important role in assessing the capability of the current atomistic force fields to correctly find the folded native state of peptides and small proteins. However, a very clear picture emerged from the application of several models: the developed force fields have a tendency to over-stabilize the native conformation of a protein and to generate high melting temperatures, as well as to produce residual structures at high temperatures,[20–22] see examples in Fig. 2A–C. In general, the unfolding enthalpy ΔH_m computed at the melting was smaller than the experimental value. For example, in recent studies of Trp thermal folding/

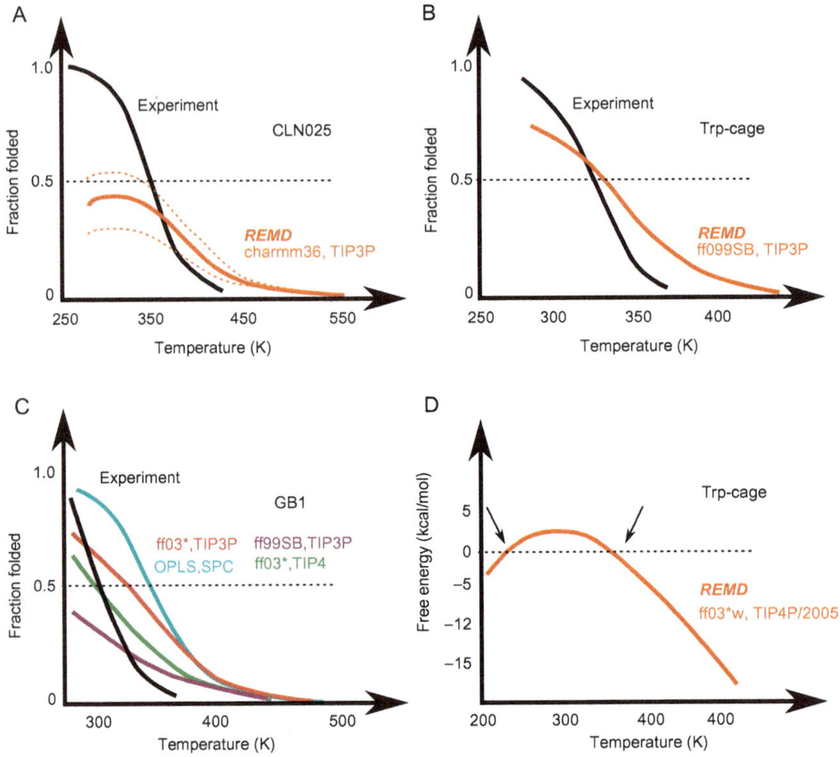

Fig. 2 Replica exchange MD. Panels (A) and (B): Temperature-dependent fraction of folded configurations sampled by the chignolin CLN025 peptide[22] and Trp-cage[21] in REMD simulations. The computational results are compared to experimental data. Panel (C): Fraction of folded structures sampled by the GB1 β-hairpin in REMD simulations run with different force field combinations and compared to experiments.[25] Panel (D): Stability curve obtained by REMD simulation of the Trp-cage, indicating the presence of a cold unfolding temperature.[23]

unfolding based on recent versions of the Amber force field,[21,23] even if the extracted melting temperature is very close to the experimental value $T_m \sim 320$ K, the enthalpy difference is smaller, namely $\Delta H_m \simeq 5 - 7$ kcal/mol (depending on the force field used) versus $\Delta H_m \simeq 12$ kcal/mol (experiments). A similar discrepancy was observed when ΔH_m was estimated by running very long folding/unfolding unbiased MD simulations close to the melting.[24] A possible source of this underestimate is an incorrect balance between protein–protein and protein–solvent interactions arising from a poor behavior of the water model at different temperatures.[25] It is important to mention that the application of REMD allowed exploring the cold unfolding process, too,[23,26] see Fig. 2D.

In order to face the issue of system size and be able to explore the thermal perturbation within the enhanced sampling REMD scheme also for larger systems, a variant of the method was proposed that helps reduce the number of replicas required to explore thermal transitions. The method, originally referred to as replica exchange with solute tempering[27] and as replica exchange with solute scaling (REST2) in its updated version,[28] relies on the corresponding state principle. Let us clarify: the dynamics of a system governed by the potential energy U samples the configurational space at a given high temperature T with weights given by the Boltzmann factor $e^{-U(\mathbf{x})/k_bT}$. With respect to a lower reference temperature T_{ref}, for instance, the effect of raising the temperature is formally identical to scaling down the potential energy, $\lambda \cdot U(\mathbf{x})$ with $\lambda = T_{ref}/T$, and performing the sampling at the lower T_{ref}. In the REST2 scheme, a set of independent simulations are run at the same reference temperature T_{ref} but using a different scaling λ value in each replica. The interesting manipulation proposed in the REST2 method is to scale down only some terms of the overall potential energy function so as to ease the sampling of the protein conformational space while reducing the number of replicas required for this purpose. Namely, for a protein in water solution, the total potential energy of the molecular system can be separated into three contributions, a first term representing the interactions between all the atoms of the protein (U_{pp}), a term representing the cross interactions between the protein's atoms and the solvent (U_{pw}), and a final term representing the interactions between solvent's particles (U_{ww}). The bath contribution (U_{ww}) is left unscaled and is therefore canceled out in the Metropolis criterion adapted to REST2.[28] Thus, it is only the U_{pp} and U_{pw} energies that determine the exchange probabilities. Since for a solvated protein the majority of particles come from the solvent, the REST2 scheme leads to a significant reduction in the number of degrees of freedom to be effectively heated and, in turn, to a substantial decrease in the number of replicas needed to span a given temperature range. The drawback of the approach is that because of the mixed scaling of the energy terms, the rigorous thermodynamics is recovered only at the replica associated with the unscaled energy potential, while the others become a reservoir of system configurations generated via an accelerated dynamics. As a consequence, there is no direct correspondence between the scaling factors $\{\lambda\}$ and the physical temperature scale. A mean field strategy has been proposed to solve the issue[22] and the method has been successfully applied to compare the thermal stability of relatively large proteins (e.g., lysozyme,[29] cold shock proteins,[30] SOD1[31]). The REST2 method has also been applied to explore the behavior of intrinsically disordered proteins[32] and ligand binding.[33] To be noted that other variants of REMD have been developed to increase efficiency.[34–36]

3. Thermophilic proteins

An excellent model to investigate the molecular and thermodynamic basis of protein thermal stability is represented by thermophilic enzymes. In nature organisms exist that thrive in very extreme environmental conditions. For example, some of them are adapted to live in very hot places like hot springs or deep sea hydrothermal vents.[37] The molecular machinery that sustains all the biological activities of these organisms has evolved to function optimally under such a thermal stress well above ambient conditions, up to 100°C.[37–39] It is far to be clear how this happened, and the study of hyperthermophilic organisms is key for understanding life evolution on Earth. It is convenient to start discussing the realization of this extreme thermal stability from the thermodynamic point of view.[3] As we discussed above, a measure of thermal stability is given by the melting temperature, the temperature at which, in a given reference sample, e.g., a protein/water solution, half of the proteins are folded and half unfolded. In a more complete thermodynamic description, other parameters help in describing the stability of a protein as a function of temperature changes. Namely, considering the Gibbs–Helmholtz equation that we have introduced before, the stability curve $\Delta G_{f/u}(T)$ is fully described by three parameters, the melting temperature (T_m), the specific heat of unfolding (ΔC_p), and the unfolding enthalpy at the melting (ΔH_m). By collecting experimental data, it was possible to individuate how thermophilic proteins achieve high thermal resistance, and why they possess higher melting temperatures when compared to their homologues optimally adapted to ambient conditions (referred to as mesophilic proteins). From looking at the typical shape of the stability curve, it follows that a higher T_m can be achieved by manipulating the curve.[37,40] Three elementary manipulations can be foreseen,[3] (i) an up-shift, (ii) a right-shift, and (iii) a broadening of the curve. In the first case, the protein gains stability by increasing the folding free energy difference by a given constant value. In the second case, the protein reaches the maximal stability at a higher temperature, and from this it derives an equivalent shift of the melting. The final manipulation corresponds to a decrease in the curvature of the curve in a such a way to reduce the influence of the temperature on the unfolding free energy; this corresponds to a decrease in the specific heat of unfolding.

According to thermodynamic analysis, in the majority of cases, hyperthermophilic proteins generally combine the first and the third mechanisms. The reader can find a detailed discussion in several reviews on the topic, e.g., Refs. 41 and 3. For the sake of clarity, and as an example, we have reconstructed the stability curves of two pairs of homologues, the RNaseH and the Csp proteins, see Fig. 3. The parameters inputted in the Gibbs–Helmholtz equation were taken from Ref. 41. In the first case, top panel, we can appreciate the combination of the up-shift and broadening mechanisms. In fact, the higher thermal stability of the thermophilic species from *Thermus thermophilus* (TtRNaseH) with respect to the mesophilic variant from *Escherichia coli* (EcRNaseH) is the consequence of a larger ΔH_m (131 vs 120 kcal/mol) and a smaller specific heat of unfolding ΔC_p (1.8 vs 2.7 kcal/(mol K)). In the second case, the cold shock protein from *Thermotoga maritima* achieves a higher melting temperature because of a considerably larger ΔH_m (62.6 vs 43.3 kcal/mol), which, however, is compensated by a higher specific heat of unfolding ΔC_p (1.1 vs 0.8 kcal/(mol K)), and the resulting curve is substantially up-shifted and right-shifted. Since the free energy of unfolding $\Delta G_{f/u}(T)$ is the result of a temperature-dependent delicate enthalpy–entropy compensation, it is necessary to understand which contribution drives the increased stability of hyperthermophiles. This can be disentangled experimentally, as done, for instance, for TaqDNA and other proteins by showing that the enhanced thermal stability of the thermophilic species arises because of a reduced entropic folding penalty. A strong support to this thermodynamic picture comes from an elegant computational model based on an ideal protein thermodynamics.[42,43] In this idealized model, the thermodynamic stability of a protein builds up from elementary contributions associated with the sequence, and it is ultimately linked parametrically to the sequence length. A reduced entropic folding penalty is generally related to the presence of a residual structure in the unfolding state,[44] but, in principle and we will discuss this later on, it could also result from an enhanced flexibility of the protein in its native state. The same microscopic vision—presence of residual secondary structure in the unfolded state or flexibility of the folded state—explains the observed decreased specific heat of unfolding ΔC_p.[45]

It is time now to attack a second problem, and to move from the pure thermodynamic description to the issue of protein function. This diversion, without entering into much detail, will help us to discuss what in literature is referred to as the stability–function trade-off, and to try to link thermodynamics to molecular features of proteins. In fact, hyperthermophilic proteins

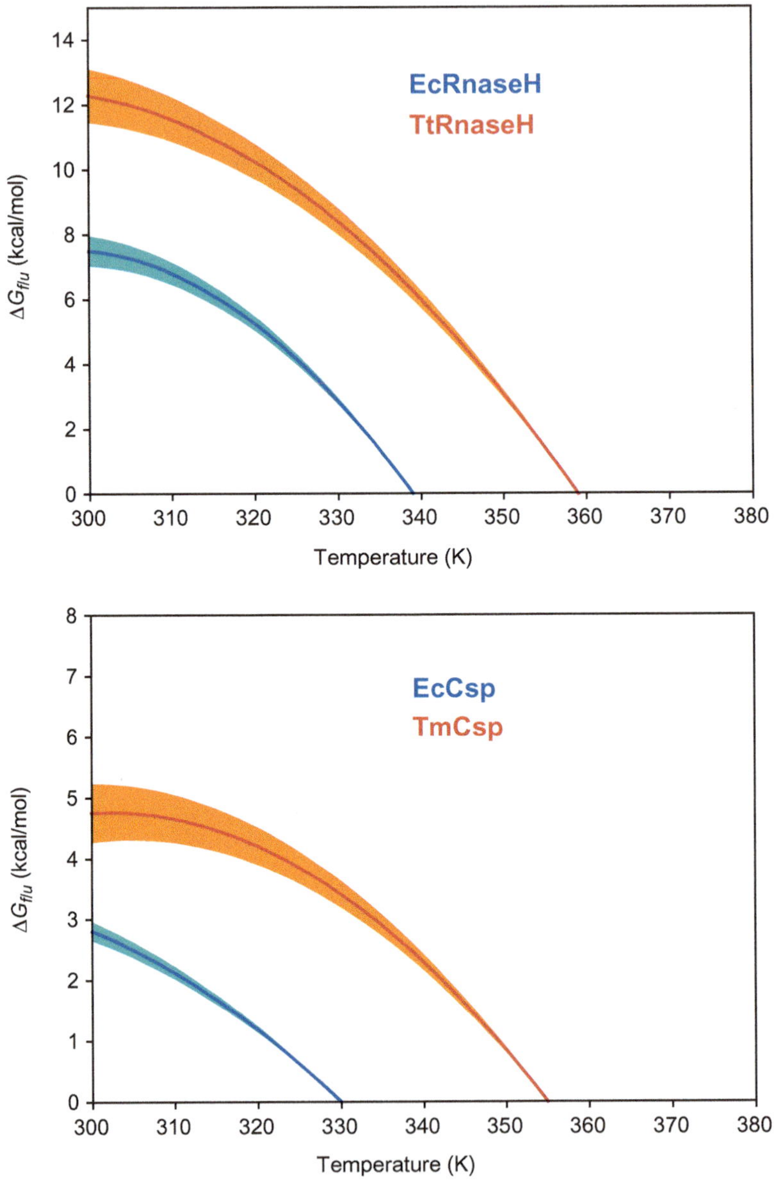

Fig. 3 Stability curves. *Top panel*: Stability curves for the RNaseH proteins from the mesophilic *E. coli* (*blue*) and the thermophilic *T. thermophilus* (*red*) organism. *Bottom panel*: Stability curves for the Csp proteins from the mesophilic *E. coli* (*blue*) and the thermophilic *T. maritima* (*red*) organism. The stability curves are computed using the Gibbs–Helmholtz equation and the parameters reported in Ref. 41. The incertitude (*shadow zone around the curves*) is computed considering the error associated with the estimate of the unfolding enthalpy at the melting (ΔH_m).

are adapted to work optimally at a much higher temperature than their mesophilic counterparts, and they show reduced activity at ambient conditions. Following the seminal work by Somero and others (see the historical perspective reported in Refs. 46 and 47), the main question one should answer is what causes this temperature shift in protein activity. For a thermophilic enzyme, a reduced activity at ambient temperature implies a larger free energy barrier for the reaction to occur, this being affected by the activation free energy of the chemical step ($k_{cat} \propto e^{-\Delta G^*}$) and by the Michaelis constant (K_M), which depends on the substrate binding affinity but also on the k_{cat}. Several studies compared the temperature effects on these parameters for homologous enzymes adapted to different temperatures, including psychrophiles, which are adapted to low temperatures, and they found for instance that the activation energy E_a extracted from the Arrhenius temperature dependence of the reaction rates correlates with the degree of *thermophilicity*. For enzymes adapted to low temperatures, a lower activation energy (E_a) implies that the reaction rate is less affected by the decrease in temperature, while the opposite occurs for thermophiles, which require a thermal boost to perform efficiently.[40,47] By relating the activation energy to the activation enthalpy of the reaction (ΔH^*), it was proposed that a small value of E_a means a decreased number of interactions to break in order to reach the transition state of the reaction, and this characterizes a looser state of the substrate/protein complex. Therefore, psychrophiles, which work at low temperatures, should have more flexible active sites than mesophilic variants, working at ambient conditions, and thermophilic enzymes should be even more tight and rigid. The role of flexibility was generalized to situations where the rate-limiting step of the catalysis was related to some large conformational shifts of the protein, e.g., the open/closed transition of a binding site loop. Since, from the perspective of the global conformation of a protein, a favorable enthalpic contribution is essential for the thermal stabilization of the folded state, a general concept was proposed according to which a thermophilic enzyme uses temperature to activate the flexibility necessary to the function. The acquired high-temperature flexibility, which is otherwise dumped by the enzyme's intrinsic enhanced rigidity, required to increase its thermal stability, would match that of mesophilies at ambient condition. In this view, the enzyme adaptation to different thermal environments would follow a corresponding state principle.[46–48]

The corresponding state principle was first corroborated by H/D exchange experiments able to probe the local exchange of hydrogen and deuterium in amino acids, which is linked to the amino acid exposure to the solvent.[49,50] The experiments measure the number of protein hydrogens that are exchanged as a function of time, and this relates to the capability of a protein matrix to protect the amino acids from solvent exposure and therefore from the exchange. This protection can be attributed to conformational rigidity.[49,50] As illustrated in the Fig. 4A, this fraction was compared for two proteins, the thermophilic GAPDH from *T. maritima* and the mesophilic homologue from *rabbit*, close to ambient conditions.[50] The thermophilic variant exhibits a larger protection factor, which can be

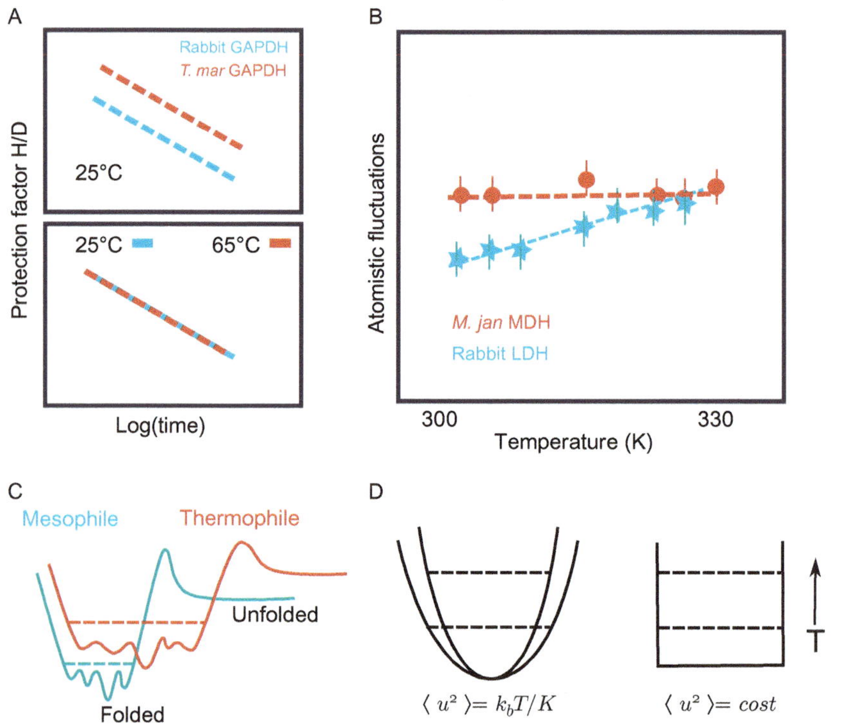

Fig. 4 Flexibility/rigidity. (A) Hydrogen/deuterium exchange experiments performed on the mesophilic (*rabbit*) and thermophilic (*T. maritima; T. mar*) GADPH homologues. The survival probability is recorded as a function of time for the two proteins at 25°C (*top chart*) and after placing the thermophilic species at its working condition 65°C. (B) Atomistic fluctuations recorded in neutron scattering experiments performed on a solution of the mesophilic LDH protein and its thermophilic homologue MDH from *M. jannaschii (M. jan)*. (C) Schematic representation of the conformational landscape of a mesophilic and a thermophilic homologues. The number of microstates in the folded configuration represents an indication of protein conformational flexibility. (D) Schematic representations of the potential energy controlling the atomic fluctuations.

interpreted as an enhanced rigidity of the protein matrix. However, when the thermophilic species is placed in its working conditions, the protection factor matches that of the mesophilic GAPDH at ambient *T*. This result illustrates the idea of the corresponding state principle according to H/D experiments and monitoring the global response of the protein fold. Some later H/D exchange experiments, however, demonstrated that hyperthermophilic proteins, such as rubredoxin from *Pyrococcus furiosus*, could be highly flexible.[48,51] It became clear that a general principle invoking protein rigidity/function trade-off needed to be tested at all pertinent length scales involved in protein motion and functionality. An extra challenge to the rigidity paradigm was posed by monitoring the thermal response of the atomistic fluctuations in a protein, as can be probed by neutron scattering (NS). It was observed that when focusing on local motion, a thermophilic protein is not necessarily more rigid than the mesophilic homologue, actually, it can be even more flexible.[52,53] In Fig. 4B we illustrate this finding. At ambient conditions, the atomistic fluctuations measured by NS, $\langle u^2 \rangle$, are large for the thermophilic malate dehydrogenase (MDH) *Methanococcus jannaschii* protein and smaller for the mesophilic rabbit lactate dehydrogenase (LDH). When increasing the temperature, the fluctuations of the mesophile increase linearly with *T* while those of the thermophile stay basically constant, showing a sort of resilience to the thermal perturbation. In line with this result, it was suggested that what characterizes better a thermophilic protein is the enhanced capability to mechanically resist the thermal increase, and not its absolute rigidity.[52] In terms of a simplified energy landscape, while the spatial fluctuations of a particle in a harmonic well would increase linearly with temperature, as it is measured by NS for the mesophilic rabbit LDH protein, if the confining potential is a square well, the fluctuations will be temperature independent, similarly to what was measured for the thermophilic MDH *M. jannaschii* To be noted that the interpretation of the mentioned NS experiments has been questioned by others.[54]

Let us now see what can be learned from computational studies when investigating thermophilic proteins. The first question to address is whether a computational model for proteins is apt to reproduce the different thermal stability of homologues. Let us start the discussion by considering atomistic simulations based on brute force MD. Because of the challenge, most of the investigations have focused on the kinetic aspect of thermal stability, and so it was verified that current force fields actually made it possible to probe the different thermal response of two homologues, eventually their unfolding, and therefore to access indirectly their thermal stabilities.[55–63]

However, in general, the thermal unfolding was caused by applying rather unphysical temperatures, e.g., higher than 400 K, and this could be a bias.[64] For instance, simulations at the microsecond timescale were needed in order to observe the early step of unfolding of a mesophilic G-domain at 360 K (*E. coli*, EcG) and to probe that at this temperature the hyperthermophilic variant was still well folded (*Sulfolobus solfataricus*, SsG).[63] To the best of our knowledge, the first work that reconstructed semiquantitatively the stability curves of two homologues was based on REST2 simulations and applied to two cold shock proteins.[30] It was verified that while the melting temperature of each of the proteins was higher than the experimental value, and force field dependent, the shift of the melting temperatures obtained in silico was relatively accurate with respect to experimental data.

In particular, MD simulations of homologues have been used to test the so-called rigidity paradigm. This was attempted by considering observables associated with different characteristic length- and timescales, e.g., the per residue atomistic fluctuations, which can be compared to what is probed by NS experiments,[60] or the compressibility, which can be accessed and related to the specific heat of unfolding.[65] By expanding the sampling, it is also possible to assess the conformational flexibility and count the number of microstates that the protein or a part of it can access. It was for instance shown by using brute force MD that while the mesophilic G-domain (EcG) is less stable than the hyperthermophilic variant (SsG), see above, the number of conformational states visited at ambient conditions by the hyperthermophile was larger, and therefore the SsG came out far more flexible.[63] This enhanced flexibility was caused by the motion of a specific structural region (an insertion of two small helices) that also acts as a protective element of the fold by screening the weak spot of the protein where the unfolding of the mesophilic species takes place. In the same spirit but using enhanced sampling simulations, it was shown that two Csp homologues, while having different thermal stabilities, sample actually the same number of conformational states at ambient conditions.[30] These examples indicate that thermophilic proteins can exhibit the same level of conformation flexibility as their mesophilic homologues, a finding that was put forward by several studies.[66] Recent long simulations at the microsecond timescales performed on a series of homologues support this view.[67]

While these studies focused on the issue of protein rigidity/flexibility by considering proteins in their apo state, one can monitor how the presence of the substrate in its reactant and product states affects enzyme flexibility. The key aspect is to understand whether the conformational changes during the catalytic turnover in a pair of mesophilic/thermopholic homologues are the same or dissimilar, and which functionally relevant modes are excited by temperature in both species. Again we present an illustrative example. By using enhanced sampling MD simulations, it was shown that when considering specifically the functional modes of G-domain proteins (EcG and SsG) in the holo state (with GTP or GDP bound), such as the opening of the binding site loop and the flexibility of the allosteric region, the corresponding state picture holds[68]: for these motions the proteins exhibit the same flexibility changes when simulated at their working temperature, see Fig. 5B. Moreover, it was shown that for these homologous G-domains the response to ligand binding is substantially different, and that the more sensitive region in the mesophile is also the weakest part of the protein where unfolding begins.[63,68] This supports the principle of a function/stability trade-off in these enzymes.

A very appealing approach to correlate mechanical and thermal stability is to probe the response of a thermophilic protein to a directional mechanical stress, as it is done, e.g., in atomic force microscopy pulling experiments. This is a hot topic and leading experimental teams are active on this issue[70,71] From the computational point of view, the force-induced unfolding can be compared to thermal unfolding by applying advanced molecular simulations. For example, we refer to a study performed on one variant from the *E. coli* bacterium and one from the thermophilic organism *T. maritima.* In this study, as mentioned above, the thermal stability was assessed by applying the REST2 method and using two different force fields. It was clearly shown that the thermophilic species, while being more thermostable in silico, does not exhibit any special mechanical resistance with respect to the less thermally stable mesophilic homologue, and that, and in the sampled force regime,[72] thermal and mechanical unfolding paths result quite different (see Fig. 5C).

The debate about protein rigidity has been complemented by other kinds of investigations more devoted to singling out the essential molecular features at the origin of the enhanced stability of hyperthermophiles. The interested reader can find several reports based on bioinformatic strategies applied to large sets of homologue sequences or structures and aimed to individuate special amino acid patterns or secondary structures features.[37] Among these, it is important to focus on the observed surplus of charged amino acids present in the thermophilic proteome.[73] This suggested that electrostatic interactions could be at the origin of thermal stabilization. It was demonstrated by using different types of calculations that

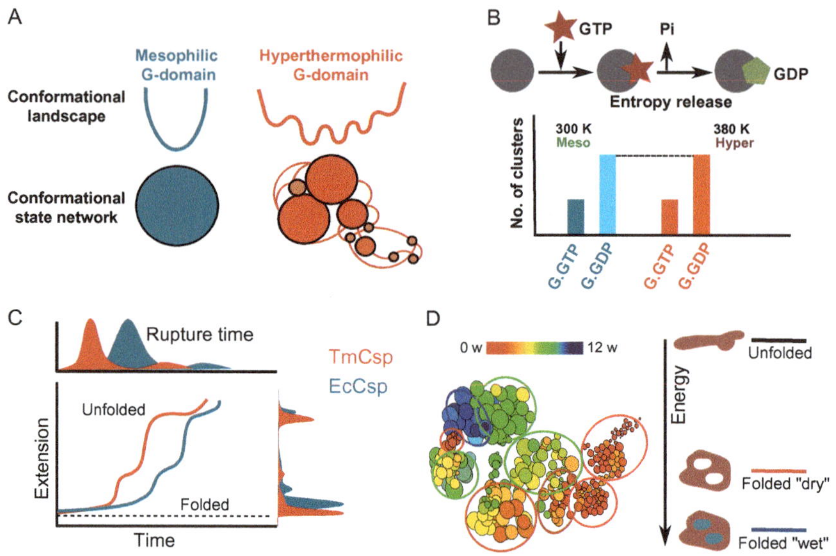

Fig. 5 Exploring protein flexibility. Panel (A): Conformational flexibility quantified by the number of microstates visited during MD simulation at the microsecond timescale at ambient temperature. The schematic plot illustrates the results published in Ref. 63 on homologue GTPase domains. A thermophilic protein (*red*) can exhibit larger conformation flexibility, thus higher conformational entropy, than the mesophilic homologue. Panel (B): The number of conformational states sampled by a key region of the GTPase domain are counted via enhanced sampling MD simulations for the holo protein associated with the reactant and product states during the GTP/GDP turnover.[68] The binding site loop of the thermophilic protein (*S. solfataricus), red and orange histograms*, acquires at its working temperature (380 K) the same flexibility of the binding site loop of the mesophilic homologue (*E. coli*) at ambient conditions. Panel (C): Unfolding dynamics of two Csp homologues under the action of a directional mechanical force applied at one extremity of the chain (C-terminus). The distribution of unfolding time for the mesophilic (*E. coli*, *blue*), and the thermophilic (*T. maritima*, *red*) are plotted as histograms at the top of the figure. The lateral side reports on the distribution of protein extensions during the unfolding trajectory collected from a large set of events. The plot illustrates schematically the results reported in Ref. 30. Panel (D): The conformational space of a protein can be linked to the number of internal water molecules penetrating the fold of the protein and providing a key stabilizing contribution (see Ref. 69).

electrostatic interactions actually form an essential contribution to the folding free energy.[66,74] This contribution is not simply related to the presence of local stabilizing interactions like ion pairs, but it is a consequence of the global distribution of charged groups in the protein matrix. This distribution was found to essentially control also the dielectric response of the protein, ensuring a favorable stabilization term for hyperthermophiles.

The optimization of electrostatic interactions at the protein surface was tested on real systems in order to generate more stable protein variants.[75,76] Moreover, the effect of charge decoration of the protein sequence was further investigated to assess the electrostatic contribution to stability by considering both the ensemble of folded and unfolded configurations.[67] The role of charges is also important for understanding the observed smaller heat capacity of unfolding observed in hyperthermophilic enzymes.[77] At a more local level, it is also worth mentioning that because of the temperature dependence of the water dielectric constant, ion pairs become more stable in the high-temperature regime, ensuring a contribution to the stability of the protein matrix.[78]

Let us consider specifically the problem of hydration.[62] Local hydration provides for instance an essential contribution to the specific heat of proteins.[45] In a recent study, by exploring many homologous pairs, it was systematically shown that hyperthermophiles are able to retain internal hydration at high temperature, gaining stability from it.[69,79] The result of this investigation suggested that the hydration of internal cavities is a potential molecular variable to be tuned in order to enhance thermal stability via site-directed mutagenesis, see the scheme in Fig. 5D. Moreover, it was proposed that protein conformational flexibility can be monitored by exploring water molecules penetrating the internal cavities of the protein. These water molecules have long residence times (>1 ns)[69,80] and give a free energy contribution to the stability of the protein. The protein conformational landscape could be characterized as a function of its internal hydration, as depicted in the left part of Fig. 5D.

We conclude the discussion by focusing on the interesting aspect of molecular evolution. For instance, protein evolutionary history can be recapitulated using ancestral sequence reconstruction (ASR) and protein resurrection as methodological approaches.[81] Several ancestral resurrected proteins were used as proxies to infer paleoenvironmental conditions on Earth. By studying the melting temperature of ancestral and modern bacterial elongation factors (Ef-tu), it was suggested that early organisms lived in thermophilic environments.[82] The study of enzymatic activity and conformational stability of ancestral 3-isopropylmalate dehydrogenase from bacteria revealed that its ancestral thermophilic phenotype decreased via roughness trajectories of thermal stability during adaption toward mesophily.[83] Another study based on the conformational stability of thioredoxin suggested that each last common ancestral enzyme of Archaeal and Bacterial domains was hyperthermophilic with melting temperature higher than 100°C.

The reconstructed thioredoxin of the last universal common ancestor of all forms of life was also hyperthermophilic.[84] The same conclusions were drawn using nucleoside diphosphate kinase, beta-lactamase, and imidazole glycerol phosphate synthase as models.[85] Recently, by studying ancestral adenylate kinase it was shown, however, that conformational stability and enzymatic activity can evolve independently during the thermoadaptation of enzymes.[86] In order to complement previous works based on ancestral and modern elongation factors, MD simulation were performed on these enzymes. The results showed that the conformational fluctuations of ancestral EF-tu are similar to those of its modern thermophilic counterpart.[87]

4. Protein stability in crowded environments

In this section we move our attention from the case of isolated proteins to the in vivo conditions. In living cells, proteins are immersed in a highly crowded and heterogeneous environment. Apart from proteins and peptides, the cellular interior is filled with many different molecular species, such as nucleic acids, lipids, and various metabolites. Depending on the cell type and subcellular localization, the fraction of volume occupied by all these molecules can reach up to 40%.[88] The densely packed and heterogeneous milieu inside cells, where proteins are constantly engaged in countless collisions and interactions with other biomolecules, differs drastically from dilute conditions under which protein stability assays are typically performed. Given this discrepancy between in vitro experiments and the actual in vivo conditions, one should ask whether measurements and simulations addressing protein stability in dilute conditions bear any relevance to the situation inside living cells. In other words, the question is whether and how crowding inside cells affects the thermal stability of proteins and, in turn, their function.

Initial theoretical considerations and experiments, mainly using synthetic polymers as crowders, suggested that the exposure to the crowded environment invariably leads to the stabilization of the native state. This observation was explained as a consequence of the excluded volume effect, according to which the reduction in the accessible volume entropically favors the more compact folded state over the more expanded unfolded conformations (see Fig. 6A).[89–91] In more mathematical terms, for a simplified hard-particle model, the contribution to the free energy difference between the folded and unfolded state arising from the excluded volume effect can be written as:

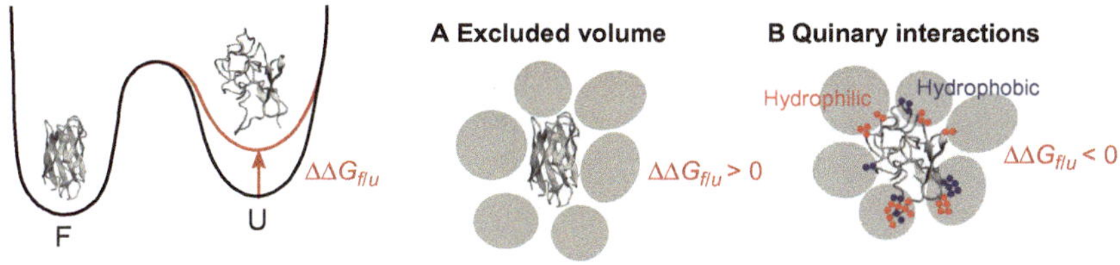

Fig. 6 Principal effects of macromolecular crowding on the conformational equilibrium between a folded native state (F) and an unfolded denaturated state (U). Panel (A): Excluded volume effect tends to stabilize the native state by penalizing the more extended denaturated state. Panel (B): Attractive quinary interactions with the crowders favor the expanded denaturated state, leading to native state destabilization.

$$\Delta\Delta G_{f/u}^{excl} = -RT \ln \frac{P_u}{P_f} \tag{2}$$

where P_u and P_f denote the probability that a randomly selected point in the crowded solution can accommodate a molecule in the unfolded and folded state, respectively.[90] However, more recent experiments, using proteins as crowders, underlined the importance of weak transient (quinary) interactions arising between the protein and the crowded environment. Quinary interactions modulate the free energy of protein unfolding by:

$$\Delta\Delta G_{f/u}^{int} = \Delta G_u^{int} - \Delta G_f^{int} \tag{3}$$

where ΔG_u^{int} and ΔG_f^{int} are the interaction free energies stemming from quinary interactions acting on the unfolded and folded state, respectively. Attractive quinary interactions ($\Delta G^{int} < 0$) can promote the unfolded state, which is expected to be more susceptible to favorable interactions with the crowders by exposing more residues on its surface than the folded state (Fig. 6B). This destabilizing effect of quinary interactions counteracts the excluded volume stabilization, and it can even lead to its full reversal.[92,93] As pointed out by Zhou,[94] the balance between the enthalpic component arising from quinary interactions and the excluded volume entropy component is temperature dependent, leading Zhou to postulate the existence of a crossover temperature for each crowder. Below this temperature, the crowders are stabilizing whereas they become destabilizing above it. As a consequence, the net stability effect of a crowded environment depends both on its chemical composition[95,96] and on temperature, and it can vary as a function of organelle localization as well as during cell differentiation and stress.[97,98]

The emerging picture is that, by tweaking the stability of proteins in the cytoplasm, crowding can have important physiological consequences. As already mentioned, many proteins are only marginally stable and destabilization by quinary interactions can promote their (partially) unfolded states with possible roles in cell function and in the spatial organization of the cell interior.[99] It has been suggested that proteins are evolutionarily adapted to the interaction with their host-cell environment.[95,100] Importantly, understanding the delicate interplay between protein conformation, stability, and quinary interactions with the cellular environment necessitates sensitive experimental techniques capable of in-cell measurements, as well as simulation approaches able to describe protein folding and unfolding in the crowded cellular environment.

Several experimental methods are currently available to inquire into the effects of crowding on protein stability in living cells and in model crowded solutions. While the thermal stability of whole proteomes can be measured using differential scanning calorimetry (DSC),[101] other methods offer more detailed insight into the stability of individual protein species in the crowded environment. First of all, in-cell NMR[102] allows determining folded and unfolded populations of an isotope-labeled protein delivered to the cytoplasm. Assuming a two-state model of the unfolding transition, this technique can be used to reconstruct the thermal stability curve $\Delta G_{f/u}(T)$ of a given protein in crowded conditions. Using in-cell NMR, Danielsson et al.[95] were able to determine the stability curves of SOD1 barrels in bacterial and mammalian cells, as well as in several different model crowded solutions. They observed that in the cells, the β-barrels were destabilized relative to the dilute solution. This destabilization manifested itself not only by a downward shift of the stability curve, but also by its narrowing, reflecting an increase in ΔC_p. As a consequence, the melting temperature T_m decreased, and at the same time, the cold unfolding temperature T_c moved upwards to the physiological regime (see Fig. 7A). An analogous, but even stronger destabilization effect was detected for the SOD1 barrels in a lysozyme protein solution. In contrast, synthetic crowders such as PEG and ficoll led to the stabilization of the β-barrel. Furthermore, the NMR-detected backbone amide ^{1}H exchange can detect both global and local unfolding by measuring the rates of H–D exchange of individual backbone amides.[103] This technique allows discriminating stability effects of crowding on a per-residue basis (see Fig. 7B), as was demonstrated, for example, for the protein chymotrypsin inhibitor 2 (CI2).[92] In agreement

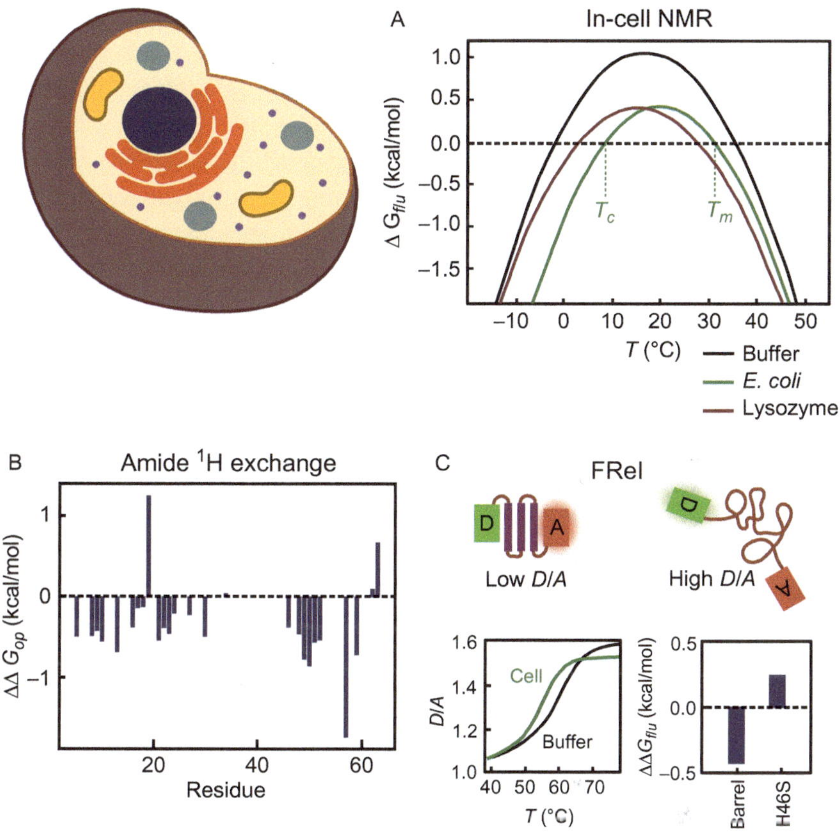

Fig. 7 Crowding effects from the experimental viewpoint. Panel (A): In-cell NMR experiments allow extracting the stability curve ($\Delta G_{f/u}(T)$) of an isotope-labeled protein in the cellular environment. SOD1 barrels are found to be destabilized by the cytoplasm as well as by lysozyme proteins relative to dilute conditions.[95] T_m and T_c denote the melting temperature and the cold denaturation temperature, respectively. Panel (B): NMR-detected backbone amide ^{1}H exchange provides a per-residue picture of the (de)stabilization by the crowded environment. This (de)stabilization is expressed by changes $\Delta\Delta G_{op}$ in the free energy associated with the exposure of the given residue to the solvent. For chymotrypsin inhibitor 2 (CI2) immersed in a 100 g/L lysozyme protein solution, destabilizing interactions predominate.[92] Panel (C): Fast relaxation imaging (FRel) is a fluorescence technique based on Förster resonance energy transfer (FRET). It discriminates between folded and unfolded states by detecting the ratio *D*/*A* of donor and acceptor fluorescence. This ratio allows constructing the thermal stability curve. Experiments performed using this technique demonstrated that a single mutation in the SOD1^{A4V} barrel can revert the destabilizing effect of the crowded environment.[31]

with the previous example, the authors reported the strongest destabilization by lysozyme, while artificial polymeric crowders were found stabilizing. Moreover, the study highlighted the important role of electrostatic interactions in the destabilization. Finally, the stability of proteins in cells and model crowded solutions has been probed by fluorescence techniques such as fast relaxation imaging (see Fig. 7C).[104] Recent experiments relying on this technique demonstrated that crowding effects can be highly sensitive to the amino acid sequence.[31] In particular, it was shown that the mutation of a single amino acid residue can be sufficient to revert the destabilizing effects of the crowded environment. Taken together, the findings obtained with the different experimental techniques converge to the same conclusion, namely that the stability effect of crowding can be strongly shaped by quinary interactions. Therefore, when investigating the crowding effects on protein stability, it is important to carefully consider the chemical composition of the crowded environment as well as the sequence of the protein in question.

Stability measurements of a single protein species immersed in a complex crowded environment constitute a tremendous experimental achievement. Still, interpreting the experimental data may necessitate a number of simplifying assumptions, such as describing the unfolding process using a two-state model. Moreover, residue-level details on interactions with the crowder environment can only be obtained indirectly from experiments. In these respects, molecular simulations—with their unprecedented spatial and temporal resolution—have a great potential to complement experimental observations. Simulations have the capacity to reveal a detailed picture of the interplay between protein conformation and quinary interactions, providing a rationale for fine stabilization or destabilization effects detected in experiments. Moreover, they should be able to elucidate the general factors governing protein stability in crowded conditions.

Pioneering molecular simulations[105,106] of large portions of the bacterial cytoplasm have demonstrated that, with the help of rapidly expanding proteomics data, it is now possible to build mechanistic models of the cellular interior with residue-level, or even atomistic resolution. The progress in the modeling of cell interior, as summarized in Refs. 107 and 108, moves us closer to the dream of simulating the entire cell. However, the present-day large-scale simulations still struggle to reach timescales relevant to the physiological processes occurring in the cytoplasm, including protein folding and conformational changes.

This brings us to the principal challenge of the computational evaluation of protein stability in a crowded environment. The task to sample protein conformational landscape is already a highly difficult one, as was discussed in a section above. On top of that, here one has to deal with the surrounding crowder molecules. As a consequence, the size of the system and the number of degrees of freedom to include in the simulation grow substantially, making such a simulation even more expensive in terms of computational resources. For this reason, most of the simulations performed so far resorted to a simplified representation of the crowder molecules, giving rise to multiscale schemes with varying resolutions of both the crowders and the protein of interest.[109]

In the first molecular simulations of protein folding under crowding, the crowders were modeled as repulsive spheres.[110–112] By employing this description in replica exchange simulations of the 34-residue WW domain, which was described by a simplified, Go-like model, Cheung et al.[110] confirmed that the excluded volume effect leads to the destabilization of the denaturated state and, therefore, to the stabilization of the native state. The authors found a monotonically increasing dependence of the melting temperature on crowder concentration (Fig. 8A). Moreover, they observed, in line with theoretical predictions, a decrease in the size of the denatured state with increasing crowder concentration, as less and less volume was available to the peptide chain. Interestingly, the folding rate was found to behave nonmonotonically, reaching a maximum at a certain crowder

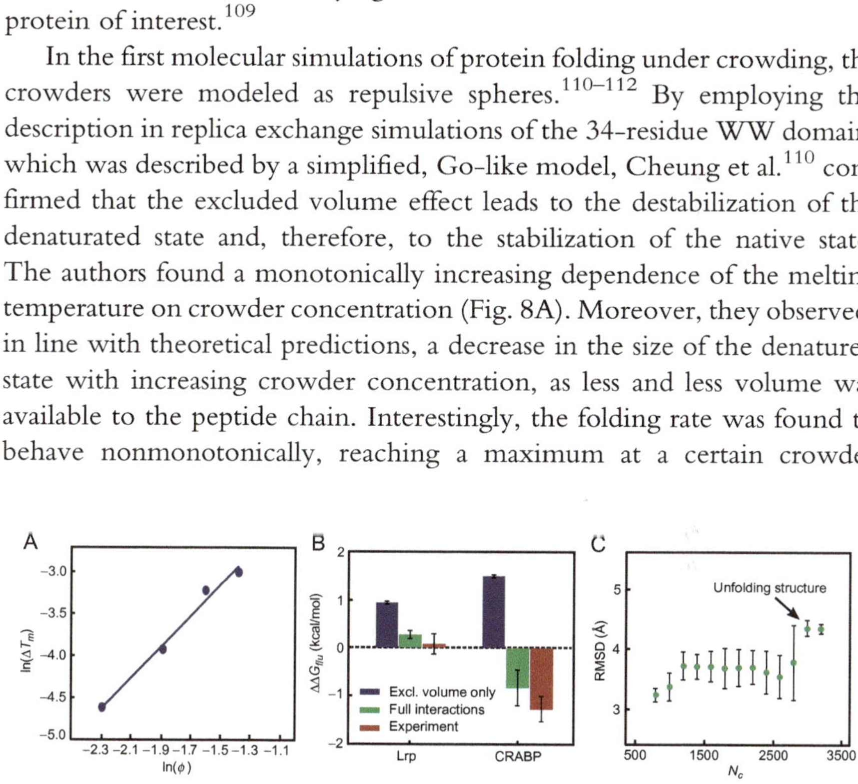

Fig. 8 Insights into the stability effects of macromolecular crowding from simulations. Panel (A): When crowders are modeled as hard spheres, the excluded volume effect leads to native state stabilization. Dependence of the melting temperature T_m of the WW domain as a function of the volume fraction ϕ occupied by the spherical crowders.[110] Panel (B): The insertion method reveals that the bacterial cytoplasm has a distinct effect on the stability of two proteins. The agreement with the experimental values of $\Delta\Delta G_{f/u}$ is only recovered if enthalpic interactions with the crowders are explicitly taken into account.[105] Panel (C): All-atom simulation of a bacterial cytoplasm shows that interactions with the crowded environment, quantified as the coordination number N_c of crowder Cα atoms, increase the root-mean-square deviation (RMSD) of PDHA proteins immersed in the cytoplasm.[106]

concentration, above which even the denaturated state favored compact, globular conformations. Beyond this concentration, the conformational fluctuations that may be required to overcome the free energy barrier associated with folding are restricted, causing a decrease in the folding rate. A subsequent report, focusing on the stability of a larger α/β protein flavodoxin found evidence of crowding-induced compaction both for the native and the denaturated states.[112] These results indicate that crowding has the power to modulate protein conformational ensembles.

Other simulations showed that the conformational landscape of a polypeptide in a crowded solution can be modulated not only by the excluded volume effect, but also by an overall decrease in the dielectric constant, caused by the high concentration of macromolecules.[113] These simulations, using an atomistic protein force field coupled with a generalized Born model of the dielectric response, revealed that low-dielectric conditions tend to stabilize extended α-helical conformations over more compact peptide conformations.

The computational studies mentioned so far focused on highly nonspecific effects dictated by general physical properties of the crowded environment such as the occupied volume and the dielectric constant. Going beyond this drastically simplified description requires more detailed models of the crowded environment, bringing realistic molecular shapes and quinary interactions into play. A step in this direction was marked by multiscale simulations of the Trp-cage peptide in a crowded protein solution formed by the B1 domain of protein G (GB1).[114] While Trp-cage was represented at an atomistic level, the crowder molecules were simulated using a coarse-grained model describing each residue with several beads and including both van der Waals and electrostatic nonbonded interactions, as well as explicit hydrogen bonding terms. The macromolecules were immersed in an implicit solvent, modeled using the generalized Born approach. The REMD simulations revealed significant differences between the conformational distribution sampled in the system with explicitly represented crowder molecules and that obtained from additional simulations reducing the description of the crowded environment to a mere decrease in the dielectric constant. Namely, in the presence of explicit crowders, Trp-cage visited partially unfolded conformations, which were stabilized by attractive interactions with the surface of the crowder molecules. These conformations did not appear for the purely dielectric model. Thus, quinary interactions with explicitly modeled protein crowders shifted the conformational distribution toward more unfolded conformations, suggesting a destabilizing effect on the native state.

Subsequent computational works explored the thermodynamics of Trp-cage unfolding in crowded protein solutions by performing replica exchange MC simulations and by employing an atomistic implicit-solvent force field to describe both Trp-cage and the crowders.[115,116] The results confirmed the previously inferred destabilizing effects of the protein crowders on Trp-cage at ambient temperatures. The magnitude of the native state destabilization was found to be greater for the BPTI crowder than for GB1. The authors rationalized the stronger effects of BPTI to be a consequence of specific interactions with a sticky surface patch on the BPTI surface. Interestingly, BPTI had an apparent stabilizing effect on Trp-cage at higher temperatures by promoting its native α-helix. The results, contrasting with the effects of spherical repulsive crowders, exemplified that a protein crowder can exert distinct and temperature-dependent effects on different parts of the amino acid chain of the test protein. For proteins that lack a simplistic two-state folding behavior, this means that the apparent stability effect of the crowded environment may depend on the coordinate chosen to measure the foldedness of the test protein.

Other simulation studies[117,118] with an explicit treatment of attractive intermolecular interactions systematically looked at the thermal stability of an α-helical peptide when varying the hydrophobicity of the crowders. The authors used a coarse-grain model[119] for both the peptide of interest and the crowders in combination with the replica exchange statistical temperature molecular dynamics (RESTMD) algorithm.[120] The simulations showed that the attractive hydrophobic interactions counteracted the excluded volume effect and led to a decrease in the melting temperature of the α-helix. On the other hand, a distinct effect was observed for crowders with very high hydrophobicity depending on their secondary structure: while β-rich crowders strongly destabilized the α-helical native state, inducing the formation of nonnative structures with increased β-content, strongly hydrophobic α-helical crowders reverted the trend and promoted the α-helical native state through favorable side chain–side chain interactions.

Simulating the folding equilibrium of larger proteins in the crowded environment remains extremely challenging. To avoid sampling the vast conformational space of a protein of interest within a simulation of a large crowded system, approaches relying on the postprocessing of existing crowded trajectories have been proposed.[105,121,122] The idea is to draw conformations from an ensemble generated in dilute conditions and to calculate crowding-induced change in the free energies of such conformations by attempting their insertion into the crowded system at a multitude of random positions. When applied to investigate the stability of two proteins, Lrp and CRABP, in the model of a bacterial cytoplasm,[105] the scheme yielded stability shifts due to the bacterial environment that were in good agreement with experimental data (see Fig. 8B). Importantly, this agreement was not obtained when neglecting electrostatic and hydrophobic interactions with the crowders. A limitation of the otherwise elegant postprocessing scheme is related to the requirement that the conformational ensemble is thoroughly presampled, including conformations which only appear with nonnegligible probability in crowded conditions. Otherwise, the scheme cannot accurately capture crowding-induced shifts in the conformations of the protein's folded and unfolded states. Moreover, the method neglects potential changes in the crowder conformation arising from the interaction with the protein.

Ideally, the simulations of protein stability under crowding would be performed using an atomistic force field with an explicit representation of the solvent. Such a detailed description would be best suited for capturing fine effects of the crowded environment, including altered solvent properties, on protein conformational populations. However, the use of atomistic explicit-solvent simulations to investigate crowding effects on protein stability remains limited by the high computational cost given by the very high number of degrees of freedom in atomistically modeled crowded systems and a comparably short time step used to integrate Newton's equations of motion in such simulations.

A few atomistic explicit-solvent simulation studies have attempted to extract information on the effects of a crowded environment on protein-stability. Feig and Sugita[123] performed direct MD simulations of crowded binary protein solutions containing chymotrypsin inhibitor 2 (CI2) in the presence of protein crowders, namely BSA and lysozyme, separately. Monitoring structural fluctuations of CI2 along the trajectories provided a hint of destabilization for lysozyme, in accordance with a previous experimental study.[92] This observation was further backed by using the MMGB/SA continuum approach combined with entropy estimation to calculate the shift ΔG_f in the free energy of the folded state due to crowding. In contrast, analogous calculations for CI2 in BSA suggested a stabilizing effect. The limitation of this computational work lies in its unique focus on the native state, thus omitting possible shifts in the free energy of the denaturated state. Traces of native state destabilization (see Fig. 8C) were also reported in large-scale all-atom simulations of a bacterial cytoplasm[106] (12–104 million atoms), but here again, the sampling of protein conformations was severely restricted by the limited simulation time (20–140 ns) currently achievable for such huge systems.

Encouraging results were recently obtained from all-atomistic REST2 simulations of a barrel of superoxide dismutase 1 ($SOD1_{bar}$) immersed in BSA.[31] By starting atomistic trajectories from representative geometries of local packing around $SOD1_{bar}$, as sampled in large-scale lattice Boltzmann molecular dynamics (LBMD) simulations, the authors were able to explore how the conformation of a loop region of the SOD1 barrel affects the number of contacts of $SOD1_{bar}$ with the crowders. In combination with a predicted effect of single-point mutations of an adjacent histidine residue on the loop's conformational population, the computational results helped to rationalize how the mutation of a single amino acid residue can cause a reversal of the stability effect of the crowded environment, from destabilizing to mildly stabilizing.

As computers become more and more powerful, all-atom studies of protein stability under crowding will ultimately become fully feasible. It should be noted, however, that the high level of detail captured in all-atom simulations does not automatically lead to a correct description of the phenomena of interest. In fact, atomistic force fields are known to suffer from diverse issues. For example, it was shown that current force fields tend to exhibit excessive protein clustering and aggregation[124] as well as unrealistically compact unfolded states,[125,126] largely resulting from an incorrect balance between protein–protein and protein–water interactions. Attempts to mitigate these issues are currently under way.[127–129]

5. Conclusion

In this contribution we have focused on the study of protein thermal stability from the computational point of view. The adopted perspective privileges the use of atomistic molecular simulations. Namely we have discussed the enhanced sampling techniques, REMD et similia, that effectively help to monitor the thermodynamics of protein unfolding. The study of extremophilic proteins adapted to function in a high-temperature regime was presented with the purpose of characterizing the molecular mechanisms underlying thermal stability. Moreover, in order to enlarge the perimeter of the discussion, we have tackled the problem of the impact of the crowded environment inside living cells on the stability and thermal resistance of proteins.

Despite the encouraging methodological advances, we still face crucial problems in monitoring and predicting thermal stability. One of these relates to the accuracy of molecular force fields, which, while very performant for targeting the folded structures, remain approximate in sampling disordered unfolded conformations. Secondly, with the focus moving to large systems designed to mimic the cellular environment, the effective sampling of temperature-induced transitions, and the estimation of thermodynamic parameters to be compared with experiment, is still problematic.

However, intriguing problems related to thermal stability call for innovative approaches. One of these open questions is the understanding of the molecular aspects of protein stability and functional optimization during the trajectory of evolution. A second one is the role of protein stability in the overall process of cell death or in pathological aggregation in cells.

Acknowledgments

The research leading to these results has received funding from the ERC (FP7/2007-2013) Grant Agreement No. 258748. We acknowledge the financial support by ANR (grant "AlloAnc" ANR-16-CE11-0011), the "Initiative d'Excellence" program from the French State (Grant "DYNAMO", ANR-11-LABX-0011-01, and "CACSICE", ANR-11-EQPX-0008). This project has received funding from the European Union's Horizon 2020 research and innovation program under the Marie Sklodowska-Curie grant agreement No. 840395.

References

1. Robertson AD, Murphy KP. Protein structure and the energetics of protein stability. *Chem Rev*. 1997;97(5):1251–1268.
2. Pastore A, Martin SR, Temussi PA. Generalized view of protein folding: in medio stat virtus. *J Am Chem Soc*. 2019;141:2194–2200.
3. Razvi A, Scholtz JM. Lessons in stability from thermophilic proteins. *Protein Sci*. 2006;15(7):1569–1578.
4. Kellis JT, Nyberg K, Sail D, et al. Contribution of hydrophobic interactions to protein stability. *Nature*. 1988;333:784–786.
5. Privalov PL, Gill SJ. Stability of protein structure and hydrophobic interaction. *Adv Protein Chem*. 1988;39:191–234.
6. Scheilman JA. Temperature, stability, and the hydrophobic interaction. *Biophys J*. 1997;73:2960–2964.
7. Yang AS, Honig B. Electrostatic effects on protein stability. *Curr Opin Struct Biol*. 1992;2:40–45.
8. Papoian GA, Ulander J, Eastwood MP, et al. Water in protein structure prediction. *Proc Natl Acad Sci USA*. 2004;101:3352–3357.
9. Adrover M, Martorell G, Martin SR, et al. The role of hydration in protein stability: comparison of the cold and heat unfolded states of Yfh1. *J Mol Biol*. 2012;417:413–424.
10. Rigoldi F, Donini S, Redaelli A, et al. Engineering of thermostable enzymes for industrial applications. *APL Bioeng*. 2018;2:011501.
11. Rouet R, Lowe D, Christ D. Stability engineering of the human antibody repertoire. *FEBS Lett*. 2014;588:269–277.
12. Pucci F, Kwasigroch JM, Rooman M. Scoop: an accurate and fast predictor of protein stability curves as a function of temperature. *Bioinformatics*. 2017;33:3415–3422.

13. Irbäck A, Mitternacht S, Mohanty S. An effective all-atom potential for proteins. *PMC Biophys*. 2009;2(1):2. https://doi.org/10.1186/1757-5036-2-2.
14. Ding F, Tsao D, Nie H, et al. Ab initio folding of proteins with all-atom discrete molecular dynamics. *Structure*. 2008;16(7):1010–1018. https://doi.org/10.1016/j.str.2008.03.013.
15. Mitsutake A, Sugita Y, Okamoto Y. Generalized-ensemble algorithms for molecular simulations of biopolymers. *Biopolymers*. 2001;60:96–123.
16. Sugita Y, Okamoto Y. Replica-exchange molecular dynamics method for protein folding. *Chem Phys Lett*. 1999;314(1–2):141–151.
17. Nymeyer H, Gnanakaran S, García AE. Atomic simulations of protein folding, using the replica exchange algorithm. *Methods Enzymol*. 2004;383:119–149.
18. Zhou R. *Replica Exchange Molecular Dynamics Method for Protein Folding Simulation*. 350. Humana Press; 2007.
19. Marinari E, Parisi G. Simulated tempering: a new Monte Carlo scheme. *Europhys Lett*. 1992;19:451–458.
20. Paschek D, Nymeyer H, Garcıa AE. Replica exchange simulation of reversible folding/unfolding of the Trp-cage miniprotein in explicit solvent: On the structure and possible role of internal water. *J Struct Biol*. 2007;157:524–533.
21. Day R, Paschek D, Garcia AE. Microsecond simulations of the folding/unfolding thermodynamics of the Trp-cage miniprotein. *Proteins Struct Funct Bioinform*. 2010;78(8):1889–1899.
22. Stirnemann G, Sterpone F. Recovering protein thermal stability using all-atom Hamiltonian replica-exchange simulations in explicit solvent. *J Chem Theory Comput*. 2015;11:5573–5577.
23. Kim SB, Palmer JC, Debenedetti PG. Computational investigation of cold denaturation in the Trp-cage miniprotein. *Proc Natl Acad Sci USA*. 2016;113:8991–8996.
24. Piana S, Klepeis JL, Shaw DE. Assessing the accuracy of physical models used in protein-folding simulations: quantitative evidence from long molecular dynamics simulations. *Curr Opin Struct Biol*. 2014;24:98–105.
25. Best RB, Mittal J. Free-energy landscape of the GB1 hairpin in all-atom explicit solvent simulations with different force fields: similarities and differences. *Proteins Struct Funct Genet*. 2011;79:1318–1328.
26. Yang C, Jang S, Pak Y. A fully atomistic computer simulation study of cold denaturation of a b-hairpin. *Nature Commun*. 2014;5:5773.
27. Liu P, Kim B, Friesner RA, et al. Replica exchange with solute tempering: a method for sampling biological systems in explicit water. *Proc Natl Acad Sci USA*. 2005;102:13749–13754.
28. Wang L, Friesner RA, Berne BJ. Replica exchange with solute scaling: a more efficient version of replica exchange with solute tempering (REST2). *J Phys Chem B*. 2011;115:9413–9438.
29. Katava M, Stirnemann G, Zanatta M, et al. Critical structural fluctuations of proteins at the thermal unfolding: challenging the Lindemann criterion. *Proc Natl Acad Sci USA*. 2017;114:9361–9366.
30. Stirnemann G, Sterpone F. Mechanics of protein adaptation to high temperatures. *J Phys Chem Lett*. 2017;8:5884–5890.
31. Gnutt D, Timr S, Ahlers J, et al. Stability effect of quinary interactions reversed by single point mutations. *J Am Chem Soc*. 2019;141:4660–4669.
32. Hicks A, Zhou H-X. Temperature-induced collapse of a disordered peptide observed by three sampling methods in molecular dynamics simulations. *J Chem Phys*. 2018;149:072313.
33. Re S, Oshima H, Kasahara K, et al. Encounter complexes and hidden poses of kinase-inhibitor binding on the free-energy landscape. *Proc Natl Acad Sci USA*. 2019;116:18404–18409.

34. Kouza M, Hansmann UHE. Velocity scaling for optimizing replica exchange molecular dynamics. *J Chem Phys*. 2011;134:044124.
35. Bernhardt NA, Xi W, Wang W, et al. Simulating protein fold switching by replica exchange with tunneling. *J Chem Theory Comput*. 2016;12:5656–5666.
36. Lu J, Vanden-Eijnden E. Infinite swapping replica exchange molecular dynamics leads to a simple simulation patch using mixture potentials. *J Chem Phys*. 2013; 138:084105.
37. Vieille C, Zeikus GJ. Hyperthermophilic enzymes: sources, uses, and molecular mechanisms for thermostability. *Microbiol Mol Biol Rev*. 2001;65(1):1–43.
38. Kumar S, Nussinov R. How do thermophilic proteins deal with heat? *Cell Mol Life Sci*. 2001;58:1216–1233.
39. Sterpone F, Melchionna S. Thermophilic proteins: insight and perspective from in silico experiments. *Chem Soc Rev*. 2012;41:1665–1676.
40. Feller G. Protein stability and enzyme activity at extreme biological temperatures. *J Phys Condens Matter*. 2010;22(32):323101.
41. Kumar S, Tsai C-J, Nussinov R. Thermodynamic differences among homologous thermophilic and mesophilic proteins. *Biochemistry*. 2001;40:14152–14165.
42. Gosh K, Dill K. Computing protein stabilities from their chain lengths. *Proc Natl Acad Sci USA*. 2009;106:10649–10654.
43. Sawle L, Ghosh K. How do thermophilic proteins and proteomes withstand high temperature? *Biophys J*. 2011;101:217–227.
44. Robic S, Guzman-Casado M, Sanchez-Ruiz JM, et al. Role of residual structure in the unfolded state of a thermophilic protein. *Proc Natl Acad Sci USA*. 2003;100(20): 11345–11349. https://doi.org/10.1073/pnas.1635051100.
45. Ninad VP, Sharp KA. Heat capacity in proteins. *Annu Rev Phys Chem*. 2005;56: 521–548.
46. Somero GN. Temperature adaptation of enzymes. *Ann Rev Ecol Syst*. 1978;9:1–29.
47. Somero G. Adaptation of enzymes to temperature: searching for basic strategies. *Comp Biochem Phys B*. 2004;139:321–333.
48. Jaenicke R. Do ultrastable proteins from hyperthermophiles have high or low conformational rigidity? *Proc Natl Acad Sci USA*. 2000;97(7):2962–2964.
49. Závodszky P, Kardos J, Svingor A, et al. Adjustment of conformational flexibility is a key event in the thermal adaptation of proteins. *Proc Natl Acad Sci USA*. 1998;95: 7406–7411.
50. Jaenicke R, Böhm G. The stability of proteins in extreme environments. *Curr Opin Struct Biol*. 1998;8(6):738–748.
51. Hernandez G, Jenney FE, Adams MWW, et al. Millisecond time scale conformational flexibility in a hyperthermophile protein at ambient temperature. *Proc Natl Acad Sci USA*. 2000;97(7):3166–3170.
52. Tehei M, Madern D, Franzetti B, et al. Neutron scattering reveals the dynamic basis of protein adaptation to extreme temperature. *J Biol Chem*. 2005;280(49):40974–40979. http://www.jbc.org/content/280/49/40974.full.pdf+html.
53. Meinhold L, Clement D, Tehei M, et al. Protein dynamics and stability: the distribution of atomic fluctuations in thermophilic and mesophilic dihydrofolate reductase derived using elastic incoherent neutron scattering. *Biophys J*. 2008;94(12): 4812–4818.
54. Marcos E, Mestres P, Crehuet R. Crowding induces differences in the diffusion of thermophilic and mesophilic proteins: a new look at neutron scattering results. *Biophys J*. 2011;101(11):2782–2789.
55. Merkley ED, Parson WW, Daggett V. Temperature dependence of the flexibility of thermophilic and mesophilic flavoenzymes of the nitroreductase fold. *Protein Eng Des Sel*. 2010;23(5):327–336. https://doi.org/10.1093/protein/gzp090.

56. Huang X, Zhou H-X. Similarity and difference in the unfolding of thermophilic and mesophilic cold shock proteins studied by molecular dynamics simulations. *Biophys J.* 2006;91:2451–2463.
57. Wintrode PL, Zhang D, Vaidehi N, et al. Protein dynamics in a family of laboratory evolved thermophilic enzymes. *J Mol Biol.* 2003;327(3):745–757.
58. Colombo G, Merz KM. Stability and activity of mesophilic subtilisin E and its thermophilic homolog: insights from molecular dynamics simulations. *J Am Chem Soc.* 1999; 121(29):6895–6903.
59. Lazaridis T, Lee I, Karplus M. Dynamics and unfolding pathways of a hyperthermophilic and a mesophilic rubredoxin. *Protein Sci.* 1997;6(12):2589–2605.
60. Marcos E, Jiménez A, Crehuet R. Dynamic fingerprints of protein thermostability revealed by long molecular dynamics. *J Chem Theory Comput.* 2012;8(3):1129–1142.
61. Kalimeri M, Girard E, Madern D, et al. Interface matters: the stiffness route to stability of a thermophilic tetrameric malate dehydrogenase. *PLoS ONE.* 2014;9:e113895.
62. Sterpone F, Bertonati C, Briganti G, et al. Key role of proximal water in regulating thermostable proteins. *J Phys Chem B.* 2009;113(1):131–137. https://doi.org/10.1021/jp805199c.
63. Kalimeri M, Rahaman O, Melchionna S, et al. How conformational flexibility stabilizes the hyperthermophilic elongation factor G-domain. *J Phys Chem B.* 2013;117(44): 13775–13785.
64. Wang T, Wade R. On the use of elevated temperature in simulations to study protein unfolding mechanisms. *J Chem Theory Comput.* 2007;3:1476–1483.
65. Dadarlat VM, Post CB. Adhesive-cohesive model for protein compressibility: an alternative perspective on stability. *Proc Natl Acad Sci USA.* 2003;100(25):14778–14783. https://doi.org/10.1073/pnas.2434157100.
66. Dominy BN, Minoux H, Brooks 3rd CL. An electrostatic basis for the stability of thermophilic proteins. *Proteins.* 2004;57(1):128–141.
67. Sawle L, Huihui J, Ghosh K. All-atom simulations reveal protein charge decoration in the folded and unfolded ensemble is key in thermophilic adaptation. *J Chem Theory Comput.* 2017;13:5065–6075.
68. Katava M, Kalimeri M, Stirnemann G, et al. Stability and function at high temperature What makes a thermophilic GTPase different from its mesophilic homologue. *J Phys Chem B.* 2016;120:2721–2730.
69. Rahaman O, Kalimeri M, Melchionna S, et al. Role of internal water on protein thermal stability: the case of homologous G domains. *J Phys Chem B.* 2015;119:8939–8949.
70. Tych KM, Batchelor M, Hoffmann T, et al. Tuning protein mechanics through an ionic cluster graft from an extremophilic protein. *Soft Matter.* 2016;12:2688–2699.
71. Tych KM, Batchelor M, Hoffmann T, et al. Differential effects of hydrophobic core packing residues for thermodynamic and mechanical stability of a hyperthermophilic protein. *Langmuir.* 2016;32:7392–7402.
72. Kouza M, Lan PD, Gabovich AM, et al. Switch from thermal to force-driven pathways of protein refolding. *J Chem Phys.* 2017;146:135101.
73. Berezovsky IN, Chen WW, Choi PJ, et al. Entropic stabilization of proteins and its proteomic consequences. *PLoS Comput Biol.* 2005;1(4):e47. https://doi.org/10.1371/journal.pcbi.0010047.
74. Xiao L, Honig B. Electrostatic contributions to the stability of hyperthermophilic proteins. *J Mol Biol.* 1999;289(5):1435–1444.
75. Gribenko AV, Patel MM, Liu J, et al. Rational stabilization of enzymes by computational redesign of surface charge-charge interactions. *Proc Natl Acad Sci USA.* 2009; 106(8):2601–2606. https://doi.org/10.1073/pnas.0808220106.
76. Gvritishvili AG, Gribenko AV, Makhatadze GI. Cooperativity of complex salt bridges. *Protein Sci.* 2008;17(7):1285–1290.

77. Day MW, Hsu BT, Joshua-Tor L, et al. X-ray crystal structures of the oxidized and reduced forms of the rubredoxin from the marine hyperthermophilic archaebacterium Pyrococcus furiosus. *Protein Sci.* 1992;1(11):1494–1507.
78. Elcock AH. The stability of salt bridges at high temperatures: implications for hyperthermophilic proteins. *J Mol Biol.* 1998;284(2):489–502.
79. Chakraborty D, Taly A, Sterpone F. Stay wet, stay stable? How internal water helps the stability of thermophilic proteins. *J Phys Chem B.* 2015;119:12760–12770.
80. Rahamn O, Melchionna S, Laage D, et al. Effect of protein composition on hydration dynamics. *Phys Chem Chem Phys.* 2013;15:3570–3576.
81. Merkl R, Sterner R. Reconstruction of ancestral enzymes. *Perspect Sci.* 2016;9:17–23.
82. Gaucher EA, Thomson JM, Burgan MF, et al. Inferring the palaeoenvironment of ancient bacteria on the basis of resurrected proteins. *Nature.* 2003;425:285–288.
83. Hobbs JK, Shepherd C, Saul DJ, et al. On the origin and evolution of thermophily: reconstruction of functional precambrian enzymes from ancestors of bacillus. *Mol Biol Evol.* 2011;29:825–835.
84. Perez-Jimenez R, Inglés-Prieto A, Zhao ZM, et al. Single-molecule paleoenzymology probes the chemistry of resurrected enzymes. *Nat Struct Mol Biol.* 2011;18:592–596.
85. Akanuma S, Nakajima Y, Yokobori S, et al. Experimental evidence for the thermophilicity of ancestral life. *Proc Natl Acad Sci USA.* 2013;110:11067–11072.
86. Nguyen V, Wilson C, Hoemberger M, et al. Evolutionary drivers of thermoadaptation in enzyme catalysis. *Science.* 2017;355:289–294.
87. Okafor CD, Pathak MC, Fagan CE, et al. Structural and dynamics comparison of thermostability in ancient, modern, and consensus elongation factor Tus. *Structure.* 2018;26:118–129.
88. Ellis RJ, Minton AP. Join the crowd. *Nature.* 2003;425:27–28.
89. Minton AP. Excluded volume as a determinant of macromolecular structure and reactivity. *Biopolymers.* 1981;20:2093–2120.
90. Zimmerman SB, Minton AP. Macromolecular crowding: biochemical, biophysical, and physiological consequences. *Ann Rev Biophys Biomol Struct.* 1993;22:27–65. https://doi.org/10.1146/annurev.bb.22.060193.000331.
91. Minton AP. Influence of excluded volume upon macromolecular structure and associations in 'crowded' media. *Curr Opin Biotechnol.* 1997;8(1):65–69.
92. Miklos AC, Sarkar M, Wang Y, et al. Protein crowding tunes protein stability. *J Am Chem Soc.* 2011;133(18):7116–7120. https://doi.org/10.1021/ja200067p.
93. Monteith WB, Cohen RD, Smith AE, et al. Quinary structure modulates protein stability in cells. *Proc Natl Acad Sci USA.* 2015;112:1739–1742. https://doi.org/10.1073/pnas.1417415112.
94. Zhou HX. Polymer crowders and protein crowders act similarly on protein folding stability. *FEBS Lett.* 2013;587(5):394–397. https://doi.org/10.1016/j.febslet.2013.01.030.
95. Danielsson J, Mu X, Lang L, et al. Thermodynamics of protein destabilization in live cells. *Proc Natl Acad Sci USA.* 2015;112(40):12402–12407. https://doi.org/10.1073/pnas.1511308112.
96. Cohen RD, Pielak GJ. Electrostatic contributions to protein quinary structure. *J Am Chem Soc.* 2016;138:13139–13142. https://doi.org/10.1021/jacs.6b07323.
97. Dhar A, Girdhar K, Singh D, et al. Protein stability and folding kinetics in the nucleus and endoplasmic reticulum of eucaryotic cells. *Biophys J.* 2011;101(2):421–430. https://doi.org/10.1016/j.bpj.2011.05.071.
98. Gnutt D, Sistemich L, Ebbinghaus S. Protein folding modulation in cells subject to differentiation and stress. *Front Mol Biosci.* 2019;6:1–5. https://doi.org/10.3389/fmolb.2019.00038.

99. Ribeiro S, Ebbinghaus S, Marcos JC. Protein folding and quinary interactions: creating cellular organisation through functional disorder. *FEBS Lett.* 2018;592:3040–3053. https://doi.org/10.1002/1873-3468.13211.
100. Mu X, Choi S, Lang L, et al. Physicochemical code for quinary protein interactions in Escherichia coli. *Proc Natl Acad Sci USA.* 2017;114(23):E4556–E4563. https://doi.org/10.1073/pnas.1621227114.
101. Lepock JR. Measurement of protein stability and protein denaturation in cells using differential scanning calorimetry. *Methods.* 2005;35:117–125. https://doi.org/10.1016/j.ymeth.2004.08.002.
102. Inomata K, Ohno A, Tochio H, et al. High-resolution multi-dimensional NMR spectroscopy of proteins in human cells. *Nature.* 2009;458:106–110.
103. Miklos AC, Li C, Pielak GJ. Using NMR-detected backbone amide 1H exchange to assess macromolecular crowding effects on globular-protein stability. *Methods Enzym.* 2009;466:1–18. https://doi.org/10.1016/S0076-6879(09)66001-8.
104. Ebbinghaus S, Dhar A, McDonald D, et al. Protein folding stability and dynamics imaged in a living cell. *Nat Methods.* 2010;7:319–323.
105. McGuffee SR, Elcock AH. Diffusion, crowding & protein stability in a dynamic molecular model of the bacterial cytoplasm. *PLoS Comput Biol.* 2010;6(3):e1000694. https://doi.org/10.1371/journal.pcbi.1000694.
106. Yu I, Mori T, Ando T, et al. Biomolecular interactions modulate macromolecular structure and dynamics in atomistic model of a bacterial cytoplasm. *eLife.* 2016;5:1–22. https://doi.org/10.7554/eLife.19274.
107. Feig M, Sugita Y. Reaching new levels of realism in modeling biological macromolecules in cellular environments. *J Mol Graph Model.* 2013;45:144–156. https://doi.org/10.1016/j.jmgm.2013.08.017.
108. Yu I, Mori T, Ando T, et al. Biomolecular interactions modulate macromolecular structure and dynamics in atomistic model of a bacterial cytoplasm. *eLife.* 2016;5:e19274.
109. Ostrowska N, Feig M, Trylska J. Modeling crowded environment in molecular simulations. *Front Mol Biosci.* 2019;6:1–6. https://doi.org/10.3389/fmolb.2019.00086.
110. Cheung MS, Klimov D, Thirumalai D. Molecular crowding enhances native state stability and refolding rates of globular proteins. *Proc Natl Acad Sci USA.* 2005;102(13):4753–4758.
111. Cheung MS, Thirumalai D. Effects of crowding and confinement on the structures of the transition state ensemble in proteins. *J Phys Chem B.* 2007;111:8250–8257. https://doi.org/10.1021/jp068201y.
112. Stagg L, Zhang S-Q, Cheung MS, et al. Molecular crowding enhances native structure and stability of a α/β protein flavodoxin. *Proc Natl Acad Sci USA.* 2007;104(48):18976–18981.
113. Tanizaki S, Clifford J, Connelly BD, et al. Conformational sampling of peptides in cellular environments. *Biophys J.* 2008;94(3):747–759. https://doi.org/10.1529/biophysj.107.116236.
114. Predeus AV, Gul S, Gopal SM, et al. Conformational sampling of peptides in the presence of protein crowders from AA/CG-multiscale simulations. *J Phys Chem B.* 2012;116(29):8610–8620. https://doi.org/10.1021/jp300129u.
115. Bille A, Linse B, Mohanty S, et al. Equilibrium simulation of Trp-cage in the presence of protein crowders. *J Chem Phys.* 2015;143(17):175102. https://doi.org/10.1063/1.4934997.
116. Bille A, Mohanty S, Irbäck A. Peptide folding in the presence of interacting protein crowders. *J Chem Phys.* 2016;144:175105. https://doi.org/10.1063/1.4948462.

117. Macdonald B, Mccarley S, Noeen S, et al. Protein-protein interactions affect alpha helix stability in crowded environments. *J Phys Chem B*. 2015;119:2956–2967. https://doi.org/10.1021/jp512630s.
118. Macdonald B, Mccarley S, Noeen S, et al. β-hairpin crowding agents affect α-helix stability in crowded environments. *J Phys Chem B*. 2016;120:650–659. https://doi.org/10.1021/acs.jpcb.5b10575.
119. Bereau T, Deserno M. Generic coarse-grained model for protein folding and aggregation. *J Chem Phys*. 2009;130:235106. https://doi.org/10.1063/1.3152842.
120. Kim J, Straub JE, Keyes T. Replica exchange statistical temperature molecular dynamics algorithm. *J Phys Chem B*. 2012;116:8646–8653. https://doi.org/10.1021/jp300366j.
121. Qin S, Zhou H-X. Atomistic modeling of macromolecular crowding predicts modest increases in protein folding and binding stability. *Biophys J*. 2009;97(1):12–19. https://doi.org/10.1016/j.bpj.2009.03.066.
122. Qin S, Zhou H-X. FFT-based method for modeling protein folding and binding under crowding: benchmarking on ellipsoidal and all-atom crowders. *J Chem Theory Comput*. 2013;9:4633–4643. https://doi.org/10.1021/ct4005195.
123. Feig M, Sugita Y. Variable interactions between protein crowders and biomolecular solutes are important in understanding cellular crowding. *J Phys Chem B*. 2012;116(1): 599–605. https://doi.org/10.1021/jp209302e.Variable.
124. Petrov D, Zagrovic B. Are current atomistic force fields accurate enough to study proteins in crowded environments? *PLoS Comput Biol*. 2014;10(5):e1003638. https://doi.org/10.1371/journal.pcbi.1003638.
125. Piana S, Klepeis JL, Shaw DE. Assessing the accuracy of physical models used in protein-folding simulations: quantitative evidence from long molecular dynamics simulations. *Curr Opin Struct Biol*. 2014;24:98–105. https://doi.org/10.1016/j.sbi.2013.12.006.
126. Piana S, Donchev AG, Robustelli P, et al. Water dispersion interactions strongly influence simulated structural properties of disordered protein states. *J Phys Chem B*. 2015;119:5113–5123. https://doi.org/10.1021/jp508971m.
127. Best RB, Zheng W, Mittal J. Balanced protein-water interactions improve properties of disordered proteins and non-specific protein association. *J Chem Theory Comput*. 2014;10:5113–5124. https://doi.org/10.1021/ct500569b.
128. Robustelli P, Piana S, Shaw DE. Developing a molecular dynamics force field for both folded and disordered protein states. *Proc Natl Acad Sci USA*. 2018;115(21): E4758–E4766. https://doi.org/10.1073/pnas.1800690115.
129. Nawrocki G, Wang Ph, Yu I, et al. Slow-down in diffusion in crowded protein solutions correlates with transient cluster formation. *J Phys Chem B*. 2017;121(49):11072–11084. https://doi.org/10.1021/acs.jpcb.7b08785.

CHAPTER SEVEN

Computer simulations of protein–membrane systems

Jennifer Loschwitz[a,b], Olujide O. Olubiyi[b,c], Jochen S. Hub[d], Birgit Strodel[a,b], Chetan S. Poojari[d,*]

[a]Institute of Theoretical and Computational Chemistry, Heinrich Heine University Düsseldorf, Düsseldorf, Germany

[b]Institute of Biological Information Processing (IBI-7: Structural Biochemistry), Forschungszentrum Jülich, Jülich, Germany

[c]Department of Pharmaceutical Chemistry, Faculty of Pharmacy, Obafemi Awolowo University, Ile-Ife, Nigeria

[d]Theoretical Physics and Center for Biophysics, Saarland University, Saarbrücken, Germany

*Corresponding author: e-mail address: chetan.poojari@uni-saarland.de

Contents

Abstract

The interactions between proteins and membranes play critical roles in signal transduction, cell motility, and transport, and they are involved in many types of diseases. Molecular dynamics (MD) simulations have greatly contributed to our understanding

Progress in Molecular Biology and Translational Science, Volume 170
ISSN 1877-1173
https://doi.org/10.1016/bs.pmbts.2020.01.001

of protein–membrane interactions, promoted by a dramatic development of MD-related software, increasingly accurate force fields, and available computer power. In this chapter, we present available methods for studying protein–membrane systems with MD simulations, including an overview about the various all-atom and coarse-grained force fields for lipids, and useful software for membrane simulation setup and analysis. A large set of case studies is discussed.

Abbreviations

Computational methods

AA all-atom or atomistic
AMBER assisted model building and energy refinement
CHARMM Chemistry at Harvard Macromolecular Mechanics
COM center of mass
CLI command line interface
CG coarse-grained
ELBA FF electrostatic-based FF
FF force field
GROMOS Groningen molecular simulation package
GUI graphical user interface
HMMM highly mobile membrane mimetic
LJ Lennard-Jones
MD molecular dynamics
MM molecular mechanics
MM/PBSA molecular mechanics Poisson–Boltzmann surface area
OPLS optimized parameters for liquid simulations
PME particle-mesh Ewald
QM quantum mechanics
RESP restricted electrostatic potential
TMD targeted MD
UA united atom
vdW van der Waals

Experimental methods and databases

AFM atomic force microscopy
cryo-EM cryo-electron microscopy
EPR electron paramagnetic resonance
FCS fluorescence correlation spectroscopy
FRAP fluorescence recovery after photobleaching
FRET Förster resonance energy transfer
HDX-MS hydrogen–deuterium exchange MS
MS mass spectroscopy
NMR nuclear magnetic resonance
OPM orientations of proteins in membranes
PDB protein data bank
PPM position of proteins in membrane
SFG sum frequency generation

Lipids, membranes, and related structures

ALA	α-linolenic acid
ARA	arachidonic acid
CHOL	cholesterol
CL	cardiolipin
DB	double bond
DCLE	1,1-dichloroethane
DGDG	digalactosyldiacylglycerol
DHA	docosahexaenoic acid
DLPC	dilauroylphosphatidylcholine
DMPC, DMPG	dipalmitoylphosphatidyl-choline, -glycerol
DOPC, DOPS	dioleoylphosphatidyl-choline, -serine
DPhPC	diphytanoylphosphatidylcholine
DPPC, DPPG	dipalmitoylphosphatidyl-choline, -glycerol
DSPC, DSPE	distearoylphosphatidyl-choline, -ethanolamine
DVPC, DVPS	divalerylphosphatidyl-choline, -serine
ER	endoplasmic reticulum
ER-phagy	ER-autophagy process
GPI	glycosylphosphatidylinositol
GPL	glycerophospholipid
GPMV	giant plasma membrane vesicle
GUV	giant unilamellar vesicle
GluCer	glucosylceramide
HB	hydrogen bond
LA	linoleic acid
LD	lipid droplet
LPC	lysophosphatidylcholine
LPS	lipopolysaccharides
MGDG	monogalactosyldiacylglycerol
PC, PE, PG	phosphatidyl-choline, -ethanolamine, -glycerol
PI	phosphatidyl-inositol
PIP	phosphatidylinositol-(3,4 or 5)-monophosphate
PIP(4,5)$_2$	phosphatidylinositol-4,5-biphosphate
PIP(3,4,5)$_3$	phosphatidylinositol-3,4,5-triphosphate
POPA, POPC	palmitoyloleoylphosphatidyl-acid, -choline
POPG, POPS	palmitoyloleoylphosphatidyl-glycerol, -serine
PS	phosphatidyl-serine
PUFA	polyunsaturated fatty acid
ROS	reactive oxygen species
SAPC	stearoylarachidonoylphosphatidylcholine
SDPC	stearoyldocosahexaenoylphosphatidylcholine
SDS	Sodium dodecyl sulfate
SLAS	sodium lauroyl sarcosine
SM	sphingomyelin
SOPC	stearoyloleoylphosphocholine
***sn*-1**	acyl chain of GPLs at the position 1 of glycerol
***sn*-2**	acyl chain of GPLs at the position 2 of glycerol
TG	triglyceride

Proteins, cofactors, and other biomolecules

$A_{2A}R$	adenosine A receptor
Aβ	amyloid-β
ABP	actin-binding protein
AD	Alzheimer's disease
apo A-1	apolipoprotein A-1
APP	amyloid precursor protein
AQP1	aquaporin-1
aSyn	α-synuclein
ATP	adenosine triphosphate
β_2AR	β_2-adrenergic receptor
CcO	cytochrome c oxidase
COX1	prostaglandin H2 synthase
δ-OPR	delta opioid receptor
DNA	deoxyribonucleic acid
hDAT	human dopamine transporter
hIAPP	human islet amyloid polypeptide
EGFR	epidermal growth factor receptor
FGF	fibroblast growth factor
GEF	guanine nucleotide exchange factor
Glc	glucose
GluA2	AMPA-sensitive glutamate receptor 2
GLUT1	glucose transporter 1
GPCR	G protein-coupled receptors
GRP1	general receptor of phosphoinositides 1
GTP	guanosine triphosphate
HDL	high-density lipoprotein
Kir2.1	inwardly rectifying potassium
Kiv1.2	voltage-dependent Shaker K^+ channel 1.2
LC3	light chain 3
LIR	LC3-interacting region
MSP	membrane scaffolding protein
NAC	nonamyloid-β component
NADH	nicotinamide adenine dinucleotide
Neu5Ac or NANA	N-acetylneuraminic acid
OMPF	outer membrane porin F
Pgp	P-glycoprotein
PH	pleckstrin homology
PMPs	peripheral membrane proteins
RHD	reticulon-homology-domain
RVFV	Rift valley fever virus
T2D	type 2 diabetes
TM	transmembrane
TMH	transmembrane helix
TLR4	toll-like receptor 4

1. Introduction

The complexity of biological systems constitutes one of the greatest fascinations for the research world, a complexity most easily exemplified in biological membranes. The interest in membrane structures has progressively risen in the last 50 years for various reasons, including the realization that about ~50% of drug targets are membrane proteins.[1,2] Thus, developing new drug compounds against membrane proteins requires an understanding of the membrane-based macromolecular targets both in terms of composition and the specific and nonspecific interactions formed within the matrix constituted by biological membranes. The kingdom of life can be divided into prokaryotes (bacteria and archaea) and eukaryotes (animals, plants, and fungi), which exhibit different kind of cell structure and can be seen in Fig. 1A.

Membranes are characteristic of individual biological cells as well as their subcellular structures. For instance, they determine cell identity and in the eukaryotic cells membrane constitutions vary among organelles like the endoplasmic reticulum (ER), mitochondria, vesicles, or chloroplasts in plants. In contrast prokaryotes (bacteria or archaea) are unicellular and feature a plasma membrane and no organelles (Fig. 1A). By serving as a barrier between the cell and its environment as well as between subcellular compartments, membranes are crucial for cell survival and most cellular processes. For all cell types, the plasma membrane defines the cellular barrier, controlling the transport of molecules across the membrane structure and aids in signal transmission, e.g., in response to stress. In the eukaryotes, lipids and proteins synthesized in the ER and Golgi, respectively, are packaged inside and transported to their destinations by membrane-bound vesicles. Membrane-enclosed nucleus stores the cells genetic information as deoxyribonucleic acid (DNA), while the mitochondria produce energy in form of adenosine triphosphate (ATP) required for cellular processes and regulation. Vesicles transport nutrients, lipids, and proteins to different cell compartments or extracellularly secrete certain important substrates. Lysosomes, which are also membrane-enclosed, host different hydrolytic enzymes for the degradation of various classes of biomolecules (e.g., proteins, lipids, and ligands of receptors). An exception are the human erythrocytes (red blood cells) for the transport of oxygen and carbon dioxide via carrier protein hemoglobin throughout the body. In these cells, the nuclei and most of other organelles are missing to provide more space for hemoglobin.[2a,b]

Bacteria exhibit an additional cell wall, which consists of branched peptidoglycans and no membrane lipids are included. They can be divided in two subclasses: Gram-positive bacteria have a thick cell wall and are positive

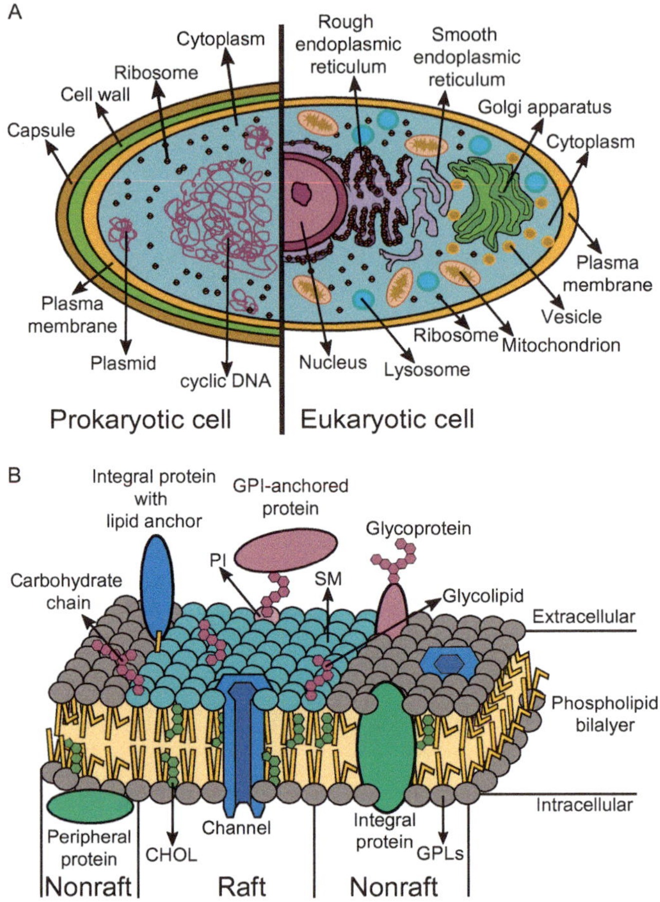

Fig. 1 Illustration of cell structures and membrane model. (A) Comparison between prokaryotic and eukaryotic cells. In prokaryotes, like bacteria and archaea, the cell exhibits no compartmentalization, thus, only the plasma membrane exists as barrier shielding the single circular DNA molecule as well as small plasmids along with all other metabolites in cytoplasm. Bacteria also have a cell wall of branched peptidoglycans in addition to the cell membrane. Most eukaryotes feature a membrane-bound nucleus and other organelles like ER, mitochondria, and the Golgi apparatus. In contrast to animals, plant cells have intracellular chloroplasts for photosynthesis, huge water vacuoles and a cell wall that contains plasmodesmata for a special cell–cell communication with the neighbor cells (not shown). (B) The fluid mosaic model describes a lipid bilayer which separates two different environments (intracellular/extracellular or luminal/cytosolic). The bilayer contains lipids that provide a structural scaffold for the membrane system. Integral and peripheral proteins that may contain posttranslational modifications (e.g., glycosylation by several sugars, leading to glycoproteins or -lipids) are found in membranes, serving as transporters and channels for substrates as well as receptors for transmembrane signal transfer. In the last years, the lipid-raft theory has become popular, in which the membrane is separated in nonraft (disordered region, L_d) with polyunsaturated glycerophospholipids (GPLs) and raft-like regions (ordered region, L_o) with high contents of cholesterol (CHOL) and sphingomyelin (SM) in the outer leaflet. Furthermore, proteins and lipids with glycosylation modification as well as glycosylphosphatidylinositol (GPI)-anchored proteins are enriched in L_o regions.

to the Gram stain test with crystal violet, which can be seen under the microscope. In contrast, Gram-negative bacteria exhibit a thin peptidoglycan cell wall that is sandwiched together with the periplasm between the inner cytoplasmic cell membrane and an additional outer membrane. The outer membrane protects the Gram-negative bacteria from many antibiotics and contains lipopolysaccharides (LPS) which can have toxic properties. Most archaea are extremophiles which can live in nonphysiological conditions, e.g., very high or low temperatures, at high salt concentrations or low pH values. Some of them exhibit a cell wall with pseudopeptidoglycans, which are structurally and physically similar to the bacterial peptidoglycans. In contrast to bacterial and eukaryotic cell membranes, their lipids are not ester-linked but more stable ether-linked, thus, which might be the reason of survival in extreme environments (see Section 1.1 for more information).

The total lipid composition of the cell membranes, called also as lipidomics, are required together with the corresponding proteins for fulfilling cellular functions and for providing the right environment for the chemical reactions unique to each cell. The plasma membrane in multicellular organisms exhibits a relative lipids to proteins composition of roughly 1:1 ratio, with the more precise ratio depending on the cell type. The membrane is crowded with about 15–30% of proteins[3] that are asymmetrically distributed laterally within the membrane structure and between the two leaflets. This contributes to the high complexity of biological membranes. The complex and dynamic structure of membrane systems was described by the fluid mosaic model (Fig. 1B) first introduced by Singer and Nicolson.[4] This model illustrates the organization of heterogeneous sets of components, including the lipid bilayer, integral and peripheral proteins (enzymes, channels and transporters), and modifications, like glycosylation of proteins and lipids. Over four decades, the model has successively been extended to a more complex model, capturing important phenomena such as membrane curvature, lateral and vertical heterogeneity of lipids and proteins distribution, differential occupancy of transmembrane, peripheral and extramembrane proteins as well as the cytoskeleton structure of the cell.[5–7]

In Sections 1.1–1.4, a short overview of the complexity of biological membranes is provided, including a brief introduction of the different types of lipids, proteins, and their modifications found in membranes. In Section 2, we examine the contributions of all-atom and coarse-grained force fields for modeling protein–membrane assemblies using molecular dynamics (MD) simulations. In Section 3, we present the general setting up and analysis of protein–membrane MD simulations. And finally in Section 4, we present selected case studies employing atomistic and/or coarse-grained force field-based MD simulations in studying different aspects of protein–membrane systems.

1.1 Lipid diversity: The scaffold of biological membranes

Biological membranes bounding cells and organelles typically comprise a lipid bilayer, forming a scaffold and usually consisting of more than a hundred lipid species. The dynamic barrier formed by this structural scaffold plays an important role for communication between different cells, between the intracellular and extracellular environments of a cell, as well as between an organelle's lumen and the cytosol. The lipid membranes additionally regulate the transport of substances, like nutrients into the cell and potentially toxic components out of the cells. Lipids are also involved in signaling processes, serving as energy resources, posttranslational modifications, and recruitment of proteins.

The diverse roles played by the membrane largely derive from the structural features of the lipids. Biological lipids share a common amphipathic character, containing a long hydrocarbon or acyl tail, that is distinctively hydrophobic, and a hydrophilic head group, which may possess a net charge. This amphipathic nature allows for a thermodynamically efficient spontaneous formation of a bilayer driven by the hydrophobic effect within a predominantly aqueous environment.[8] The inability of the nonpolar hydrocarbon chains to form hydrogen bonds with water contributes to the orientation of the hydrophobic tail away from water molecules, a process that is associated with an entropic term. The entropy S term defines the disorder of a system and is connected to the free energy change ΔG, the absolute temperature T and the enthalpy change ΔH according to Eq. (1).

$$\Delta G = \Delta H - T\Delta S \tag{1}$$

The lack of hydrogen bonds between the hydrophobic lipid tails and water would impose a cage-like ordering of water molecules around the hydrophobic parts of lipids, which would cause a decrease in ΔS of water (i.e., ΔS would become negative) and thus lead to a positive and thermodynamically unfavorable ΔG. In order to minimize contact with water and thus the entropic decrease, the hydrocarbon chains self-assemble and orient away from the aqueous medium. This effect is known as the hydrophobic effect and constitutes the fundamental mechanism governing the formation of lipid membrane structures. In this orientation the polar head groups interact with water molecules and other polar head groups that together form the contact area with the aqueous environment. This picture of membrane lipid assembly represents a highly simplified illustration of the membrane structure, with structural properties of different lipid types. An overview of the different types of membrane lipids found in biological membranes is presented in Fig. 2.

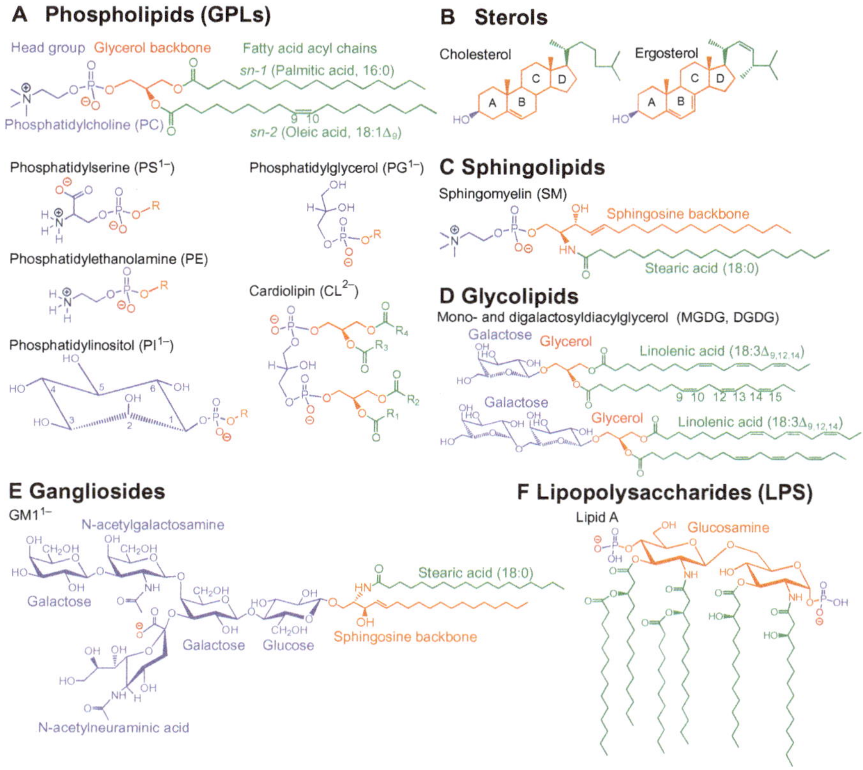

Fig. 2 Structure of membrane lipids. (A) GPLs or phospholipids represent the major components of membranes. The chemical structure of a palmitoyloleoylphosphatidylcholine (POPC– 16:0,18:1) molecule is shown, containing the polar head group indicated in *blue*, the glycerol backbone in *orange*, and two hydrophobic acyl chains in *green*. The positively charged atoms are shown in *dark blue* and the negatively charged atoms in *red*. Additionally, the different kinds of head groups, typically found in cell membranes, are displayed and in their corresponding abbreviation and net charge given in *parentheses*. (B) Sterols consist of a planar, four-ring system shown in *orange*, a hydrophobic acyl chain in *green*, and a polar hydroxyl group in *blue*. The common sterols cholesterol and ergosterol are shown. (C) Similar to GPLs, sphingolipids exhibit a polar head group (choline, serine, or ethanolamine), which is bound to the sphingosine backbone shown in *orange*, and a hydrocarbon chain with a typical length of 16–26 carbon atoms fused to sphingosine via N-acylation. (D) Glycolipids exhibit a sugar as polar head group instead of a phosphatidyl moiety in GPLs. The sugar head can contain one or two saccharides, e.g., mono- and digalactosyldiacylglycerol (MGDG, DGDG). (E) Another and complex class of sphingolipids comprises the gangliosides, containing sphingosine as backbone bound to an acyl chain and sugar acids as polar head group. One or more of the sugar groups are N-acetylneuraminic acid (NANA or Neu5Ac). To show the complexity of gangliosides, GM1, where G refers to ganglioside and M is for monosialic, is depicted. (F) Lipid A is an example for lipopolysaccharides (LPS).

The glycerophospholipids (GPLs, popularly referred to as phospholipids) represent the major component (up to ~75%) of most biological membranes. Like the name suggests, the head group of GPLs contains a phosphate and a glycerol group. The phosphate group is esterified at the *sn*-3 position of glycerol, whereas the fatty acid (or acyl) chains are esterified at *sn*-1 and *sn*-2 positions of the glycerol thus forming the hydrophobic region of the membrane (Fig. 2A). Usually, GPLs bear a total of one, two, or four fatty acid chains, with each chain containing between 12 and 24 carbons and sometimes featuring up to six double bonds. For instance, palmitic acid, an hexadecanoic acid that exhibits 16 carbons and no double bonds (DB), codified as 16:0 to represent the $N_{\text{carbons}}:N_{\text{DB}}$ ratio, whereas oleic acid has 18 carbon atoms and one double bond at the carbon atom 9 (i.e., octadecenoic acid) and is therefore labeled as $18:1\Delta_9$ or $18:1_{\omega 9}$. Unsaturated fatty acids in *cis* conformation introduce kinks in the chain and cause voids in the hydrophobic part of the membrane (packing defects) with the result that the bilayer is more fluid and flexible. In Table 1 the common fatty acids found in biological membranes are listed.[9]

The class of polyunsaturated fatty acids (PUFAs), which exhibit more than one double bond, include the essential omega-3 (ω3) and omega-6 (ω6) fatty acids. Here, the number following the omega characterizes the position of the first double bond with respect to the terminal methyl group in the chemical structure. For example, in the (ω3)-PUFAs the first unsaturation bond appears at the third carbon atom from the end of the acyl chain. For humans, linoleic ($18:2_{\omega 6}$, LA) and α-linolenic acid ($18:3_{\omega 3}$, ALA) belong to the essential PUFAs, which have to be obtained via food like seeds, nuts, and oils and from which longer PUFAs like arachidonic ($20:4_{\omega 6}$, ARA) and docosahexaenoic acid ($22:6_{\omega 3}$, DHA) can be synthesized. However, the biosynthesis of DHA from ALA is rather endogenous and so limited in humans to a maximum of 4%,[10] thus, DHA is mainly derived from food like fish oil. Many of the PUFAs are located in biological membranes. For instance, DHA is contained in the retina and in brain membranes, where it has a modulation function on membrane proteins, the transport of choline, glycine, and taurine, as well as on the response of rhodopsin[11] (see Section 4.2.1 for a case study, in which DHA is involved in Alzheimer's disease).

Table 1 Some native saturated and unsaturated fatty acids found in biological membranes.

Symbol	Condensed formula	Common name (systematic name)	Melting point (°C)
12:0	$CH_3(CH_2)_{10}COOH$	Lauric (*n*-Dodecanoic)	44.2
14:0	$CH_3(CH_2)_{12}COOH$	Myristic (*n*-Tetradecanoic)	53.9
16:0	$CH_3(CH_2)_{14}COOH$	Palmitic (*n*-Hexadecanoic)	63.1
18:0	$CH_3(CH_2)_{16}COOH$	Stearic (*n*-Octadecanoic)	69.6
20:0	$CH_3(CH_2)_{18}COOH$	Arachidic (*n*-Eicosanoic)	76.5
24:0	$CH_3(CH_2)_{22}COOH$	Arachidic (*n*-Tetracosanoic)	86.0
16:1	$CH_3(CH_2)_5CH{=}CH(CH_2)_7COOH$	Palmitoleic ($16{:}1_{\omega 9}$)	−0.5
18:1	$CH_3(CH_2)_7CH{=}CH(CH_2)_7COOH$	Oleic ($18{:}1_{\omega 9}$)	13.4
18:2	$CH_3(CH_2)_4CH{=}CHCHCH_2CH{=}CH(CH_2)_7COOH$	Linoleic ($18{:}2_{\omega 6}$, LA)	−5.0
18:3	$CH_3CH_2CH{=}CHCH_2CH{=}CHCH_2CH{=}CH(CH_2)_7COOH$	Linoleneic ($18{:}3_{\omega 3}$, ALA)	−11.0
20:4	$CH_3(CH_2)_4(CH{=}CHCHCH_2)_3CH{=}CH(CH_2)_3COOH$	Arachidonic ($20{:}4_{\omega 6}$, ARA)	−49.5
22:6	$CH_3(CH_2CH{=}CH)_6(CH_2)_2COOH$	Docosahexaenoic ($22{:}6_{\omega 3}$, DHA)	−44.0

The melting point for each fatty acid is given to assess the phase state at room temperature. Normally, the saturated fatty acids are solid and unsaturated species exhibit the liquid state at room temperature.

The complexity of the GPLs derives not only from the diversity of their head groups but also from the structural varieties of the acyl chains in terms of length, stereochemistry, degree of saturation and hydroxylation of the lipid tails, and the presence of other lipid types. As a consequence, different combinations of the head group structures and acyl chains can lead to different polymorphic shapes of lipids (Fig. 3). In the crystalline liquid state

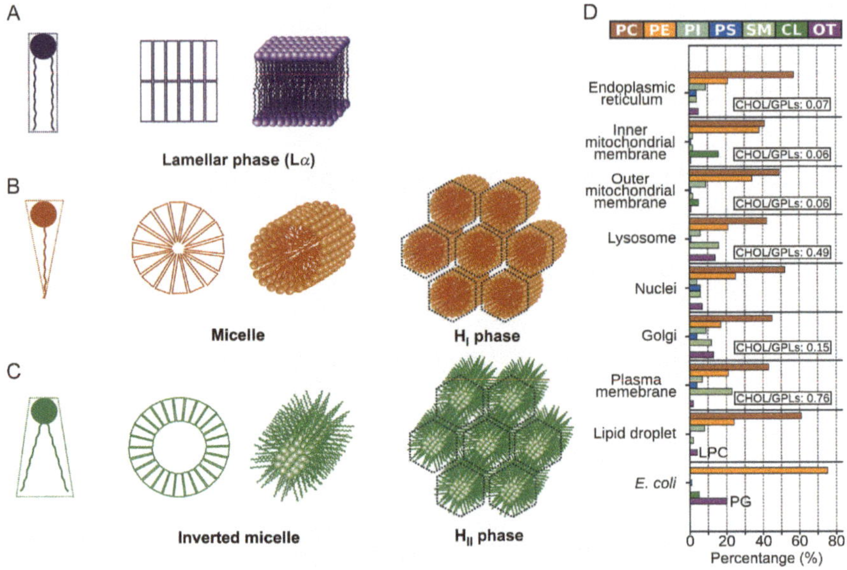

Fig. 3 The polymorphism of lipid assemblies in aqueous solution and the lipid composition of certain organelles. (A) The lamellar phase describes the typical lipid bilayer which is found in biological membranes. It can exist in two major states: First, the crystalline liquid phase (L_α) with the unsaturated and saturated acyl chains in *gauche* conformation providing free spaces within the hydrophobic part to accommodate protein conformational changes and a certain degree of passive permeability. The second extreme phase is the gel phase (L_β, not shown), involving all saturated acyl chains in *trans* conformation and the barest minimum levels of flexibility. This phase is generally not favored in biological membranes. The lipids PC, SM, PS, CL, PA, PG, and PI can form lipid bilayers as they have a rectangular shape. (B) Micellar and H_I shapes are induced by lipids with one acyl chain (lysophospholipids), in which the bulky polar groups are oriented toward the surrounding water, while the acyl chains associate with one another at the core. Lysophospholipids and detergents exhibit an inverted cone shape. (C) Lipids with polar head groups much smaller in comparison to the acyl chains (e.g., PE) or in low pH conditions (e.g., PA and PS under pH 3) or in the presence of divalent cations like Ca^{2+} (CL, PA) favor the inverted micellar or H_{II} structure.[16] Here, the polar groups form an own aqueous phase inside the inverted micellar structure and the acyl chains face the water phase. (D) The bar plot represents the relative abundance of each class of GPLs/SM in organelles based on composition data from different sources,[17] for lipid droplets from murine hepatocytes[18] and *Escherichia coli*.[19] The bar for other lipids (OT) includes PG, PA, and lysophospholipids (e.g., lysophosphatidylcholine, LPC), if it not otherwise mentioned. All data are given in mol%. Additionally, the molar ratio of cholesterol (CHOL) to GPLs is shown for certain organelles.[17] *Panels A–C: Reprinted with the permission from Lladó V, López DJ, Ibarguren M, et al. Regulation of the cancer cell membrane lipid composition by NaCHOleate: effects on cell signaling and therapeutical relevance in glioma. Biochim Biophys Acta Biomembr 2014;1838(6):1619–1627, https://doi.org/10.1016/j.bbamem.2014.01.027. Copyright 2014 Biochimica et Biophysica Acta (BBA)—Biomembranes.*

(lamellar phase or L_α, Fig. 3A), biological membranes adopt the shape of a lipid bilayer, which enhances mobility of the membrane components and passive permeability for water and other small molecules. In this phase, the acyl chains can exist in the all-*trans* or one of the two *gauche* conformations. The latter causes a kink in the acyl chain and thus generates space in the hydrophobic bilayer region, increasing its passive permeability. In contrast, the nonnative ordered gel phase of a lipid bilayer (L_β) forms when most of the acyl chains adopt the very rigid all-*trans* state. Generally, GPLs contain an unsaturated fatty acid with one or two double bonds at the *sn*-2 and a saturated or monounsaturated fatty acid at the *sn*-1 position of glycerol.[12,13] GPLs with only one acyl chain at the *sn*-1 position are called lysophospholipids and act as signaling molecules for the activation of membrane proteins. They are also involved in the generation of micellar structures, in which the polar head groups orient to the water phase and the acyl chains associate in the core (Fig. 3B).

The classification of GPLs is based on the head groups and the fatty acid content. The simplest GPL is phosphatidic acid (PA), which is negatively charged at physiological pH and exists as an intermediate during the biosynthesis of other lipids. It is generally not highly present in biological membranes. Mammalian cell membranes are mostly populated by GPLs with phosphatidylcholine (PC) head groups, in which the phosphate is connected with the alcohol choline. Interestingly, PCs are almost absent in prokaryotic cells. Under physiological conditions, PC is zwitterionic ($^+N(CH_3)_3$, PO_4^-) with a zero net charge. Zwitterionic phosphatidylethanolamines (PE) are found in all cell membranes, but in higher quantities in prokaryotic cells. In contrast, they are almost restricted to the inner leaflet of the plasma membrane (cytosolic side) in eukaryotes. The polar group of PE ($^+NH_3$, PO_4^-) displays stronger intramolecular interactions than PC and is more nonpolar in the aqueous phase, with the result that an hexagonal H_{II} arrangement (Fig. 3C) is more stable in an inverted orientation than in a micelle. In the inverted orientation, the polar PE groups are oriented toward the core region, whereas the acyl chains are extended outwards into the surrounding water. PE is the major GPL class in bacteria cell membranes together with phosphatidylserine (PS), phosphatidylglycerol (PG), and cardiolipin (CL or diphosphatidylglycerol). While the head group of PS is negatively charged ($^+NH_3$, PO_4^-, COO^-) and is mostly located in the inner leaflet of eukaryotic and prokaryotic cells, PG (also negatively charged) and CL (four fatty acid chains, twice negatively charged) have only been found in mitochondrial membranes of eukaryotes and in most of the bacterial plasma membranes.

The localization of CL to mitochondria suggests the evolutionary fusion of prokaryotes, as endosymbionts, with eukaryotic cells in the past. Phosphatidylinositols (PI) represent another class of GPLs that have been found to be involved with lipid and cell signaling, as well as regulation of membrane proteins and membrane trafficking.[14,15] PIs show a high diversity in the head group structure, especially since the hydroxyl groups of inositol can be phosphorylated at positions 3, 4, and 5 to generate any of three PI monophosphates (PI(3)P, PI(4)P, PI(5)P), three bisphosphates (PI(3,4)P_2, PI(3,5)P_2, PI(4,5)P_2), or one triphosphate (PI(3,4,5)P_3). PIs are mostly absent in prokaryotic cells.

The next class of membrane lipids are sterols, which play an essential role in eukaryotic cell, however, they are not found in bacteria and archaea. Cholesterol, also abbreviated as CHOL, is the major component in mammalian cells, which are not able to grow without cholesterol.[9] The chemical structure of cholesterol (Fig. 2B) consists of four fused rings with a double bond in ring B, a hydrocarbon chain at ring D, and a polar hydroxyl group at ring A, leading to a planar conformation with an amphipathic character. The hydroxyl group interacts with the polar head group of GPLs and sphingolipids and is orientated to the water side, whereas the planar ring system and the hydrocarbon chain are in contact with the acyl chains. Interestingly, although ergosterol from yeast shares a very similar structure with cholesterol, they cannot replace each other.[9] The fundamental role of cholesterol is the maintaining and controlling the membrane fluidity by increasing the membrane packing and thickness. Thus, cholesterol decreases the packing defects in the hydrophobic part and encourages the all-*trans* conformation in GPLs, which leads to a decrease of the membrane permeability for ions and small molecules.[9] Furthermore, cholesterol modulates membrane proteins in special conformations by decreasing the free volume in the membrane and mostly inhibits these proteins or can directly bind membrane proteins. Moreover, cholesterol can cause lateral membrane phase separation between cholesterol enriched and poor regions in order to regulate membrane proteins. The concept of lipid rafts will discussed later in this chapter (see Section 1.3).

Similar to GPLs, the sphingolipids in eukaryotic cells represent a further major component of biological membranes. Instead of a glycerol backbone, sphingosine serves as scaffold, in which the acyl chain is N-linked and the head group is O-linked. Sphingomyelin (SM, Fig. 2C), consisting of phosphatidylcholine and stearic acid, is the major sphingolipid in biological membranes (e.g., in human blood cells and membranous myelin sheaths of nerve cell axons) and is normally located in the outer leaflet membrane of cells. Like PC, SM is able to form bilayers in aqueous environment and interacts with cholesterol to build lipid rafts[20,21] (see Section 1.3). Instead of a phosphatidyl group at the position 1 of glycerol, one neutral sugar (typically, galactose, glucose, and lactose) can be covalently bound via glycosidic linkage to form cerebrosides. Additionally, the sugar group can be connected with a sulfate to build the sulfatides. They are found in all kind of species, especially located in brain membranes of mammalians.[22] Fig. 3D shows the distribution of GPLs and SM over the organelles.

Another class of sphingolipids is given by the complex gangliosides, which are based on the sphingosine backbone and a complex sugar acid head group. They play important roles in cell recognition and regulation of membrane proteins and are highly populated in nerve cells.[23] An example is the negatively charged GM1 (Fig. 2E), containing an N-acetylneuraminic acid (Neu5Ac or NANA), N-acetylgalactosamine, two galactose and one glucose (Glc) moiety. It is located in the outer leaflet of nerve cells and is involved in the development of Alzheimer's disease[24,25] (see also Section 4.2.1).

In the plant kingdom, glycerolglycolipids like mono- and digalactosyldiacylglycerol (MGDG, DGDG) have a major role in regulation of the photosynthesis in chloroplasts (Fig. 2D), whereas Lipid A (Fig. 2F) represents a basic component of the outer membrane in Gram-negative bacteria. The chemical structure of Lipid A reveals a glucosamine backbone, containing six acyl groups with chain length between 12 and 14 carbon atoms, as well as two phosphate groups. Lipid A represents the basic part of the lipopolysaccharides (LPS) which contains a core region and a long polysaccharide moiety termed O-antigen. As Lipid A is the main component of the outer membrane, Gram-negative bacteria have a greater intrinsic resistance to antibiotics. In addition, Lipid A acts as endotoxin upon destruction of bacteria cells which provokes immune response in host cells.

The final lipid class, which we want to briefly introduce, are the tetraether bolalipids of the archaea, which feature an ether-linkage between glycerol instead of the ester-linkage typical for eukaryotes and prokaryotes. Moreover, this kind of lipids occurs as dimers by the covalent binding of the end of acyl chains of two lipids to form one molecule with two polar head groups and one or two connected tails. Due to the ether-linkage and the connecting tails, they form highly stable membranes that can sustain incredibly hostile environments, such as extremely high temperatures and pressures, low or high pH values.[26]

With their different head groups, the charge behavior of membrane bilayers has an important impact on the regulation of membrane proteins, an effect which depends on the concentration of Na^+, Ca^{2+}, and Mg^{2+} for anionic GPLs (e.g., CL and PA). The existing potential difference between the two leaflets of the membrane is also very important for survival at the cellular level. Changes in membrane potential are employed in signal transduction by nerve cells and in regulating both cardiac automaticity and contraction by cardiomyocytes. Moreover, the phospholipids, glycolipids, CHOL, and PUFA's are all susceptible to oxidation due to excessive levels of reactive oxygen species (ROS). The oxidized lipids change membrane properties and indirectly affect protein–lipid interactions and other cellular processes, thus leading to cardiovascular, neurodegenerative, and pulmonary diseases. For more information on lipid composition of membranes, the reader is referred to the books[9,16] or the reviews.[12,27,28]

1.2 Membrane proteins: The complexity of biological membranes

We have already examined the complexity of membranes deriving from the diversity of lipids, especially from the different combinations of head groups and acyl chains. Now we want to present the complexity of biological membranes deriving from different classes of membrane proteins, which are known to perform diverse and essential functions. Membrane proteins are involved in protein–lipid interactions for cell–cell recognition, molecular transport across membrane barriers, energy synthesis, maintenance of concentration gradients, signal transduction, as well as formation of vesicles (see Fig. 4A for an overview and specific examples). Membrane proteins can be broadly divided into two classes:

(1) Peripheral proteins directly interacting with the membrane or with integral membrane proteins.

(2) Integral proteins embedded within the membrane or associating with a lipid anchor.

Peripheral proteins are usually not in full contact with the hydrophobic part of the membrane and can be washed off by changing the ionic concentration or pH of the system. In contrast, integral proteins directly interact with the hydrophobic core of membranes with a part of their structure. Examples include lipid anchorage via posttranslational lipidation of amino acids, and interactions involving hydrophobic sequences in transmembrane (TM) α-helices or β-barrel structures (Fig. 4B and D). Membrane proteins with

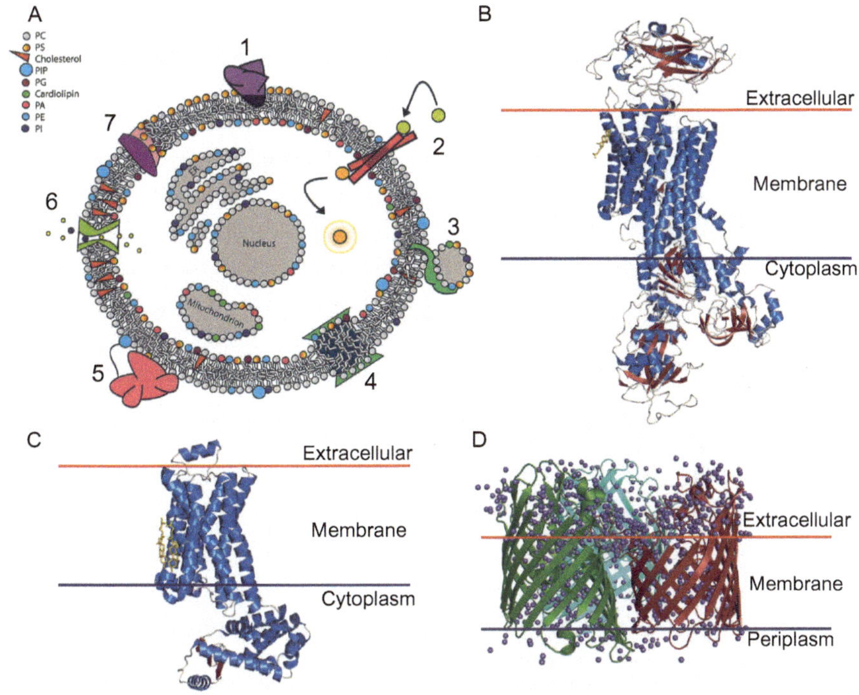

Fig. 4 Membrane protein functions and examples of membrane proteins. (A) Membrane proteins interact with lipids in biological membranes in different ways: (1) A peripheral membrane protein binds with a lipid-specific hydrophobic anchor (PS here). (2) An integral protein, contributing to TM signaling, binds ligands from the outside. (3) A membrane protein binds to and promotes the fusion of a vesicle with the cell membrane. (4) An integral protein induces local curvature via hydrophobic mismatch. (5) A peripheral protein binds a lipid, while its globular domain interacts with the membrane surface. (6) Protein channels regulate ion transport across the membrane via interaction with cholesterol. (7) Phospholipid scramblase transports lipids across the membrane. (B) The structure of the transmembrane $Na^{+}K^{+}$-ATPase (PDB code: 4HYT[36]), which actively (via ATP hydrolysis) pumps three Na^{+} ions extracellularly and two K^{+} into the cytoplasm against its concentration gradient. The pump maintains membrane resting potential, regulates cellular transport, and modulates cell volume. Cholesterol is displayed as *yellow sticks*. (C) The crystallographic structure of the human β2-adrenergic receptor (PDB code: 3D4S[37]), a G-protein-coupled receptor (GPCR) expressed in pulmonary and cardiac myocyte tissue featuring cholesterol (shown as *yellow sticks*) binding sites. (D) The porin OMPF (outer membrane protein F, PDB code: 3POX[38]) shows a β-barrel conformation, which is highly expressed in the outer membrane of *E. coli*. It serves to facilitate the selective entry of hydrophilic molecules (e.g., water molecules, shown as *blue spheres*) into the periplasm. The 3D structures were generated with PyMol version 1.8. *Panel A: Adapted with the permission from Muller MP, Jiang T, Sun C, et al. Characterization of lipid-protein interactions and lipid-mediated modulation of membrane protein function through molecular simulation. Chem Rev. 2019;119(9):6086–6161, https://doi.org/10.1021/acs.chemrev.8b00608. Copyright 2019 American Chemical Society.*

TM α-helices (TMH) are generally found in the plasma membrane and the inner membrane of bacteria cells, while β-barrels (e.g., outer membrane porin F from *Escherichia coli*, OMPF) are located in outer membranes of Gram-negative bacteria, cell walls of gram-positive bacteria, as well as the outer membrane leaflets of mitochondria and chloroplasts. Membrane lipids can influence membrane proteins via direct binding that allosterically affects the conformational preferences. Furthermore, membrane proteins can exhibit posttranslational modifications in form of sugars and GPLs, as seen for instance in glycosylphosphatidylinositol (GPI)-anchored proteins, which have functional roles in membrane protein transport, cell adhesion, cell wall synthesis, and cell surface protection. We will discuss different protein–lipid interactions in the case studies in Section 4.

1.3 Lipid rafts and hydrophobic mismatch: The regulation and organization of biological membranes

In the 1970s, membranes were shown to exhibit lateral heterogeneity using an experimental protocol that extracted membrane proteins and lipids in detergents. Within the past few years, the theory of lipid rafts developed by Simons and Ikonen has become established[29] and involves the accumulation of less flexible lipids for the building of ordered nanodomains, called rafts (L_o, <200 nm, especially via cholesterol, PC and SM), which are surrounded by disordered and more fluid lipid areas (L_d, polyunsaturated acyl chains). This activates membrane proteins in terms of conformational changes or oligomerization. Cholesterol has been observed to play a key role as at concentrations below 10 mol% of cholesterol no domain separation was reported between the phases. This was demonstrated for ternary membrane systems using in vitro methods and in both synthetic giant unilamellar vesicles (GUVs) and giant plasma membrane vesicles (GPMVs)[30,31] extracted from cells, as well as via MD simulations.[32,33] Until now, no experimental technique is available to visualize the process of raft formation in living cells since conventional optical microscopy is limited to a resolution of ~250 nm. Nevertheless, there are evidences for their existence originating from the presence of lipid raft markers which are located in lipid rafts as well as from other biophysical methods (see Refs. 21, 34).

Functional raft-like domains are involved in cellular processes like immune signaling. For instance, the SRC family kinases are located in raft-like domains, where they phosphorylate immune receptor complexes, whereas the phosphatases reside in nonraft domains. Another example is the host–pathogen interactions featured in bacterial or viral infections, involving the binding of specific bacterial toxins and viruses to special membrane proteins in raft-like domains.[21] With the help of microsecond-scale MD simulations, a three-component raft-like system of cholesterol and PC with saturated and unsaturated acyl chains has been successfully built. However, it remains to elucidate the mechanism of assembly of membrane components in different phase domains since this has not been unambiguously shown in living cells with the attendant level of structural complexity.

Another driving force to organize membrane domains derives from the hydrophobic interactions between TM proteins and the different kinds of lipids. In the case of hydrophobic mismatch, which is the difference between the thicknesses of the hydrophobic regions of a TM protein and of the biological membrane, it spans leading to unfavorable exposure of hydrophobic surfaces to water, membrane sorting and orientation of membrane proteins are affected.[21,39] TM proteins and surrounding lipids can adapt to the hydrophobic mismatch by different means. When the hydrophobic part of a TM protein is longer than the hydrophobic bilayer thickness, the protein might aggregate in the membrane to minimize the exposed hydrophobic area, it could tilt to reduce its effective hydrophobic thickness, or adopt a different conformation. Lipids in turn can modulate the membrane thickness by stretching their acyl chains or even assemble into another membrane state. When the hydrophobic part of a TM protein is too small to match the hydrophobic bilayer thickness, again this might result in protein aggregation or changes to its orientation or conformation. Lipids, on the other hand, could decrease the effective bilayer thickness by becoming more disordered or disrupting the bilayer organization to form an inverted nonlamellar structure.

1.4 Role of MD simulations in investigating protein–membrane systems

The biological membranes, as discussed in Sections 1.1–1.3, in composition, organization as well as protein–lipid interactions are highly complex systems. Thus, this complexity constitutes a unique kind of challenge for structural study. While experimental methods can be used to detect protein–lipid interactions, unraveling structural mechanisms at the atomistic level often remains problematic. Many membrane proteins are involved in disease pathologies (e.g., cystic fibrosis, Parkinsonism, Alzheimer's disease, atherosclerosis, and cancer) and to properly understand the molecular events associated with such disease states, their structural dynamics needs to be resolved. For drug development purposes, there is also continual need to investigate protein–lipid interactions.[40] For atomistic insight into protein–lipid interactions, experimental techniques like X-ray crystallography, nuclear magnetic resonance (NMR) spectroscopy, and cryo-electron microscopy (cryo-EM) are often used. X-ray structures in the 1–3 Å resolution range represent an often dynamic system with single snapshots and the crystallization process sometimes involves harsh treatments capable of generating nonnative protein conformations. Furthermore, flexible regions (e.g., loops) cannot be detected by X-ray crystallography. On the other hand, NMR can describe protein dynamics but it is limited to small protein systems of ~50 kDa. Cryo-EM has in recent years become an important tool for investigating protein–lipid association in membranes systems with a resolution of 2–4 Å.[41] Every resolved protein structure is available in the RCBS Protein Data Bank (PDB)[42,43] (https://www.rcsb.org), while the orientation of membrane proteins in biological membranes is provided by the Orientations of Proteins in Membranes (OPM) database[44] (https://opm.phar.umich.edu/).

MD simulations provide a computational complement to these experimental techniques. The history of lipid bilayers studied by MD simulations with no explicit water molecules began in the 1980s with investigations of few lipid molecules.[45–47] After about a decade, MD simulations were successfully employed in the study of increasingly complex and more realistic lipid bilayers, including PCs and PSs with explicit solvents,[48–52] and including embedded proteins.[53–56] These initial studies, however, were limited to conformational sampling on the picosecond timescale. Following the developments in hardware and software, MD simulation have increasingly begun to capture the structural dynamics of larger macromolecular systems, like ribosomes,[57] protein–membrane systems (e.g., transporters and channels),[58,59] virus membranes,[60] and lipid rafts.[32,33] By overcoming the resolution limits of the experimental methods and exploring the fine molecular details of macromolecular events, MD simulations can now serve as *molecular microscope*[35,61] to investigate lipid–lipid and protein–lipid interactions of more realistic cell membranes on submillisecond timescale (Fig. 5A). Different MD simulation techniques apart from traditional all-atom (AA) models have been developed for this purpose. These include coarse-grained (CG) force fields,[62,63] the highly mobile membrane mimetic (HMMM) model,[64] and multiscale models (Fig. 5B). To decrease the calculation time during the MD simulation, CG models represent multiple atom groups as fewer pseudoparticles (so-called beads), while the HMMM model employs a membrane core of organic solvent combined with truncated lipids for the head group region for increasing the lateral diffusion of membrane lipids. However, these approaches imply a loss in resolution such that in CG model small protein conformation changes cannot be captured, while the HMMM model is restricted to modeling peripheral protein–membranes systems.[64]

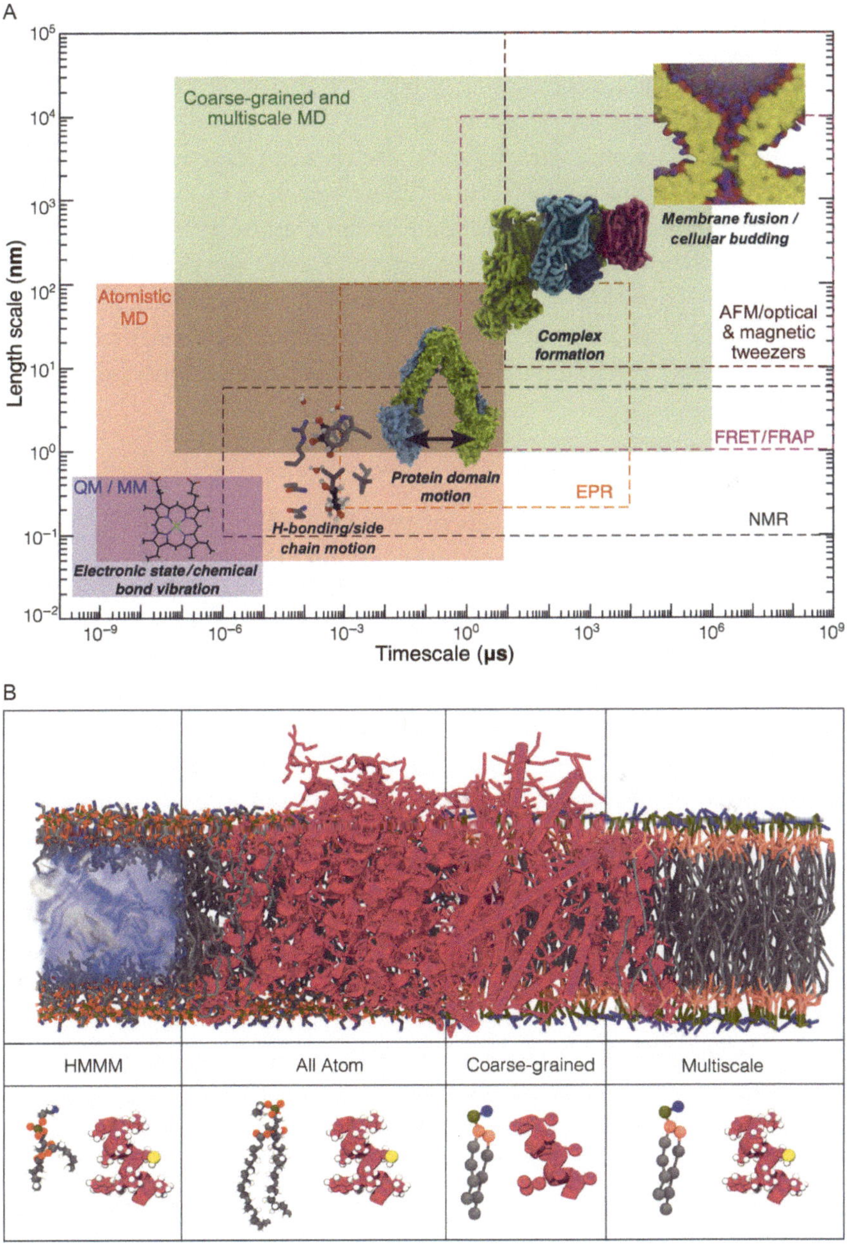

Fig. 5 See legend on next page.

Multiscale methods comprise the use of different levels of representation in performing MD simulations, e.g., starting with CG simulations to extract different large-scale conformations, which are then converted to atomistic structure to study at a better resolution.[65]

In the next Section 2, the different all-atom and coarse-grained models for lipids are discussed, while Section 3 describes how to set up and analyze MD simulations of protein–membrane systems. In Section 4, we describe different case studies of selected AA- and CG-MD simulations to grant a perspective into how MD simulations are employed in studying interactions between lipids and different membrane proteins. The presented systems include associated and integral membrane proteins in interaction with lipids.

2. Lipid force fields

In this section, three groups of lipid FFs will be reviewed. The all-atom (AA) FFs represent the most detailed description, where all atoms of a molecule are described in an explicit manner. The united-atom (UA) FFs merge the nonpolar hydrogens connected to heavy atoms to one interaction site. For instance, the methyl (CH_3) and methylene (CH_2) groups will

Fig. 5 Length- and timescale of experimental and MD simulation methods. (A) Application of methods to analyze the dynamics of chemical and biological processes. The effective length and timescales of QM/MM (quantum mechanics/molecular mechanics), AA, and CG/multiscale simulations are displayed in the *shaded boxes*. The experimental techniques nuclear magnetic resonance (NMR), electron paramagnetic resonance (EPR), Förster resonance energy transfer (FRET), fluorescence recovery after photobleaching (FRAP), atomic force microscopy (AFM), and magnetic and optical tweezers are indicated by the *dashed boxes*. (B) The resolution of different MD simulation techniques in protein–lipids interaction studies is shown. In the *upper panel*, the chloride ion channel (PDB code: 1OTS[66]) is represented in different resolutions, while in the *lower panels* a PS lipid and a short α-helix are displayed with different resolution. All-atom resolution implies explicit treatment of all interactions, whereas in CG simulation certain atoms have been merged into fewer beads. Multiscale simulations involve combining of multiple levels of resolutions either at the same time or in sequence. The highly mobile membrane mimetic (HMMM) model uses truncated lipids and a membrane core of organic solvent for increasing the lateral diffusion of membrane lipids. *Panels A and B: Reprinted with the permission from Muller MP, Jiang T, Sun C, et al. Characterization of lipid-protein interactions and lipid-mediated modulation of membrane protein function through molecular simulation. Chem Rev. 2019;119(9):6086–6161, https://doi.org/10.1021/acs.chemrev.8b00608. Copyright 2019 American Chemical Society.*

each be displayed by a single interaction site, resulting in a decrease of interaction sites from 4 to 1 and 3 to 1, respectively. This leads to lower costs of computing time, while still reaching a similar accuracy as with the AA FFs. The next level of reduction is the summarizing of 3–4 heavy atoms into a single interaction site (known as bead), which is typically called a coarse-grained (CG) model. Important FFs that were developed for lipids are listed in Table 2 and are shortly discussed in this section.

Table 2 All-atom (AA), united-atom (UA), and coarse-grained (CG) force fields for lipids.

Developer	Lipid FF	Resolution	Years published
	CHARMM22 (C22)[67,68]	AA	1996/1997
	CHARMM27 (C27)[69]	AA	2000
CHARMM	CHARMM27 (C27r)[70,71]	AA	2005
	CHARMM36 (C36)[72,73]	AA	2010
	CHARMM36 united atom FF (C36-UA)[74]	UA	2014
	GAFF[75,76]	AA	2007/2008
AMBER	Lipid11[77]	AA	2011
	Lipid14[78]	AA	2014
Jambeck & Lyubartsev	Stockholm (S)lipids[79,80]	AA	2012/2013
OPLS	OPLS-UA[81,82]	UA	1999/2000
	OPLS-AA[83]	AA	2014
	GROMOS (Berger)[84]	AA	1997
	GROMOS (Chiu)[85]	UA	1999
GROMOS	GROMOS (45A3)[86]	UA	2003
	GROMOS (53A6)[87]	UA	2004
	GROMOS (53A6-CPK)[88–90]	UA	2009/2011
	GROMOS54A7/54B7[91]	UA	2011
MARTINI	MARTINI[62,63]	CG	2004,2007
Shinoda, DeVane, & Klein	SDK[92,93]	CG	2007

2.1 Atomistic force fields for lipids

2.1.1 General description of classical force fields

MD simulations capture the different types of interactions defined by a potential energy function $U(r^N)$, called the force field. An MD simulation is based on Newton's second law of motion (Eq. 2) for a system of interacting particles, with the interactions being defined by $U(r^N)$, according to which velocities and positions are calculated for every time step. The parameters needed for the definition of $U(r^N)$ are derived from either quantum mechanical (QM) calculations or spectroscopic data. Eqs. (3) and (4) show the basic form of an AA-FF, which is divided into bonded and nonbonded interaction contributions and approximate a QM energy surface as a function of the atomic coordinates:

$$F = -\frac{\partial U(r^N)}{\partial r_{i,n}} = m_i \cdot \dot{v}_{i,n} \quad (2)$$

$$\text{with } i = 1, 2, 3, \ldots, N \text{ atoms}; \quad n = x, y, z; \; m = \text{mass}$$

$$U(r^N)_{\text{bonded}} = \underbrace{\frac{k_{ij}}{2}(r_{ij} - r_{ij,0})^2}_{\text{bonds}} + \underbrace{\frac{k_{ijk}}{2}(\theta_{ijk} - \theta_{ijk,0})^2}_{\text{angles}} + \underbrace{\frac{V_n}{2}(1 + \cos(n\Phi_{ijkl} - \Phi_{ijkl,0}))}_{\text{torsion}} + \underbrace{\frac{k_{ijkl}}{2}(\Phi_{ijkl} - \Phi_{ijkl,0})^2}_{\text{improper torsion}} \quad (3)$$

$$U(r^N)_{\text{nonbonded}} = \underbrace{4\varepsilon_{ij}\left[\left(\frac{\sigma_{ij}}{r_{ij}}\right)^{12} - \left(\frac{\sigma_{ij}}{r_{ij}}\right)^{6}\right]}_{\text{Lennard-Jones}} + \underbrace{\frac{1}{4\pi\varepsilon_0\varepsilon_r}\frac{q_i q_j}{r_{ij}}}_{\text{Coulomb}} \quad (4)$$

For computational efficiency, classical FFs employ some approximations in the treatment of bonded and nonbonded interactions. The bond lengths, angles, and improper torsion angles are often described by harmonic potentials, and for this reason, and unlike to QM descriptions, bond cleavage is not possible in classical MD simulations. With harmonic potentials even small deviations from the equilibrium values ($r_{ij,0}$, $\theta_{ijk,0}$, $\Phi_{ijkl,0}$) can cause considerable increases in the energy of the system, which also depend on the force constants (k_{ij}, k_{ijk}, k_{ijkl}). To model the rotation around bonds in accordance with thermodynamic data, a cosine function of the dihedral angle is typically used, where V_n is the force constant belonging to the torsion, $\Phi_{ijkl,0}$ is the angle where the potential passes through its minimum value, and n is the multiplicity, which indicates the number of minima as the bond is rotated through 360 degrees.

To describe van der Waals (vdW) and electrostatic interactions, the Lennard-Jones (LJ) 12/6 potential and Coulomb's law are employed, respectively. The attractive term of the LJ potential (r^{-6}) represents the component of the interactions based on London dispersion forces originating from induced dipole–dipole interactions. The repulsive term (r^{-12}) derives from the overlap of electron orbitals and describes the Pauli repulsion at short ranges, which can be efficiently calculated by squaring the attractive term. For each combination of atoms i and j, the well depth ε_{ij} and collision diameter σ_{ij} are required for the calculation of the LJ potential, in addition to the interatomic distance r_{ij}. The last important contribution to a FF is the electrostatic interaction between atoms, which is calculated using a Coulomb potential that depends on the distance r_{ij} and the partial charges of the atoms q_i and q_j as well as contains the permittivity of vacuum ε_0. With Eqs. (3) and (4), electrons are not explicitly considered. The atomic partial charges are derived from the electron distribution in a molecule as determined by QM calculations. In this approximation, the continuous function of the electron density becomes distributed point charges. Moreover, the partial charges, which are dependent on molecular conformations, are usually also fixed in most of the FFs. The interested reader is referred to chapter "Pairwise-additive and polarizable atomistic force fields for molecular dynamics simulations of proteins" by Lemkul of this book, which provides a more detailed discussion of classical FFs, including polarizable FFs that go beyond the fixed-charge approximation. In most MD simulations employing periodic boundary conditions, the particle-mesh Ewald (PME) method[94,95] is employed for treating the sum over electrostatic interactions, which are long-range interactions and thus require special care in periodic systems. In this method, the electrostatics interactions are divided into two parts: (1) a short-range contribution and (2) a long-range contribution. While the short-range contribution is calculated in real space in combination with a cutoff value (0.8–1.2 nm), the long-range contribution is computed using a Fourier transformation. This approach guarantees rapid convergence of the electrostatic energy calculation as compared with that of a direct summation.

The FF expression in Eqs. (3) and (4) was proven to be robust from thousands of applications in the past years, principally because it is sufficiently accurate in most cases for describing biomolecular interactions in a computationally efficient manner. The parameters needed as input for a FF are either derived from QM calculations or by empirical fitting to reproduce experimental observables, which are common strategies used for the parameterization of FFs. Moreover, atom types are defined to represent the different atoms and their chemical environment in the FF. As the different FFs usually followed diverse kinds of parameterization strategies, they usually employ slightly different atom types along with distinct values for the force constants as well as equilibrium distances and angles.

2.1.2 Experimental observables for the validation of lipid force fields

To study lipids and membrane proteins, various experimental techniques like X-ray, NMR, and different fluorescence spectroscopy methods were constantly developed and improved, with which characteristic properties of membranes can be determined. For instance, X-ray spectroscopy can be applied for the determination of membrane thicknesses, areas per lipid, electron density profiles, and lateral diffusion constants. To prove the phase state of a bilayer, structural and dynamic properties of the acyl chains are measured by NMR spectroscopy (e.g., area per lipid and order parameters). Moreover, various fluorescence spectroscopic techniques are available for determining the lateral diffusion of bilayer lipids, where the components of interest are marked with a fluorescent molecule for the measurements. Thus, besides expensive QM calculations experimental observables can be used to validate and improve lipid FFs. Here, we briefly discuss a set of membrane properties which are commonly used for benchmarking lipid FFs. Some of these observables are presented in Fig. 6. For more details on membrane properties used for the validation of FFs, we recommend the reviews by Poger et al.,[96] Leonard et al.,[97] and Moradi et al.[98]

2.1.2.1 Area per lipid

The first quality check of a lipid FF can be easily done by the calculation of A_L (Fig. 6A). It is related to the degree of fluidity of the membrane in a certain phase and, at the same time, a good indicator for whether equilibrium of an MD simulation has been reached. This common membrane parameter can be experimentally determined using X-ray or neutron scattering factors, volumetric methods, as well as NMR spectroscopy. From MD simulations, it can be calculated with Eq. (5):

$$A_L[nm^2] = \frac{B_x B_y}{n} \tag{5}$$

Here, the lateral dimensions B_x and B_y of the simulation box (unit cell) are along the x- and y-axes, respectively, while n is the number of lipids per leaflet. However, this equation can only be used for homogeneous membranes

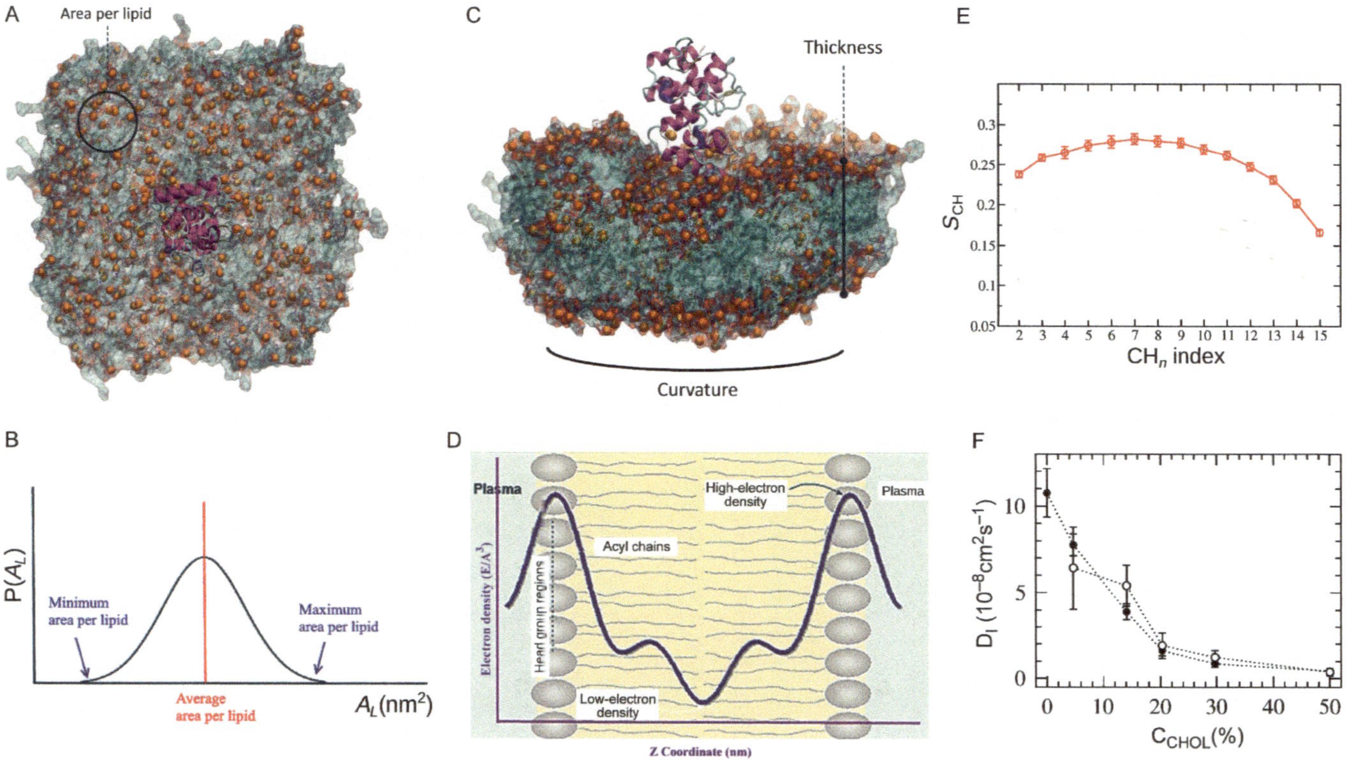

Fig. 6 See legend on next page.

with one lipid species. In the case of mixed bilayers, the membrane area must be partitioned into segments, for example by using two-dimensional Voronoi tessellation[100,101] or grid-based methods based on atomic vdW radii,[99,102] where the size difference between the different lipids and sterols are taken into account. A_L is tightly related to other membrane properties, like acyl chain ordering, compressibility, and molecular packaging. The time-average of A_L can be used to judge the equilibrium state of a bilayer during an MD simulation. An equilibrated bilayer shows only small fluctuations around an average value of A_L. However, this can only be assessed if the MD simulation is conducted in the *NpT* ensemble where the simulation box dimensions are allowed to fluctuate. In this ensemble, the number of molecules *N*, the pressure *p*, and the temperature *T* are constant, which is a more realistic ensemble than the *NVT* ensemble (where the volume *V* is kept constant) for comparison to experiments.[98] A higher value of A_L corresponds to a more disordered state, whereas low A_L values reflect highly packed lipids in the membrane. The probability distribution $p(A_L)$ for the area per lipid molecule can also be calculated (see Fig. 6B for an example distribution), allowing to determine the minimum, maximum and average values of A_L, which can be important for processes such as lateral diffusion.[98]

Fig. 6 Membrane properties commonly used for the validation of lipid force fields. (A) The average area per lipid (A_L) is computed by dividing the area of the *xy*-plane of the simulation box by the number of lipids per leaflet. (B) A representative probability distribution for the area per lipid. (C) The membrane thickness can determined by the mean distance between the head groups, marked by the positions of the phosphorus atoms (*orange*) of the outer and inner leaflet of the bilayer (D_{HH}). (D) An example for the electron density along the bilayer normal, from which the membrane thickness D_{HH} can be calculated. (E) A typical S_{CH} order parameter profile for a saturated acyl chain as obtained from an MD simulation. (F) Changes of lateral diffusion coefficient D_l of DPPC (•) and cholesterol (○) molecules depending on cholesterol (CHOL) concentration. *Panels A and C: Adapted with permission from Goossens K, De Winter H. Molecular dynamics simulations of membrane proteins: an overview. J Chem Inf Model. 2018;58 (11):2193–2202, 10.1021/acs.jcim.8b00639. Copyright 2018* American Chemical Society; *Panels B, D, and E: Adapted with permission from Moradi S, Nowroozi A, Shahlaei M. Shedding light on the structural properties of lipid bilayers using molecular dynamics simulation: a review study. RSC Adv. 2019;9(8):4644–4658, 10.1039/C8RA08441F. Copyright 2019* Royal Society of Chemistry (Creative Commons Attribution 3.0 Unported Licence)*; Panel F: Adapted with permission from Falck E, PatraM, Karttunen M, Hyvönen MT, Vattulainen I. Lessons of slicing membranes: interplay of packing, free area, and lateral diffusion in phospholipid/cholesterol bilayers. Biophys J. 2004;87(2):1076–1091, https://doi.org/10.1529/biophysj.104.041368. Copyright 2004* Elsevier.

The comparison of A_L obtained from MD simulations to experimental values can be difficult because this quantity cannot directly be determined by experimental techniques. Nonetheless, several methods exist to estimate A_L from experiments, for instance by using volumetric data and the electron density profiles of homogeneous bilayers,[103] or the lipid volume V_L and the Luzzati thickness D_B,[104] or the average NMR order parameter of carbon–deuterium bonds of the acyl chains.[105] Depending on which method and bilayer model were used, the resulting A_L values can significantly vary, which in turn generates some uncertainty with regard to the true value of A_L for a lipid bilayer in a given phase. Moreover, the determination of A_L is affected by thermal fluctuations in the lateral and transverse directions of lipid bilayers. In the last decades, it occurred that different A_L values for the same kind of bilayer were obtained with the same experimental technique or calculated by the same researcher group. Nevertheless, A_L determined from simulation studies using the same FF usually result in very similar values as opposed to those derived from experiments, which can be explained by the fact that the FFs were parameterized to produce a certain A_L value (see Table 1 in Ref. 96). It should further be considered that undulations or fluctuations in the lipid bilayer structure, which may result from insufficient sampling or from simulations artifacts (e.g., due to finite system size or periodic boundary conditions) may also cause uncertainties in A_L. As a consequence, A_L alone is not sufficient to evaluate the quality of a lipid FF or to assess the degree of fluidity of a lipid bilayer; it should be considered together with the electron density profile across the bilayer and the bilayer thickness for judging the phase state of a membrane.

2.1.2.2 Membrane thickness and electron density profile

While A_L facilitates information about the lateral structure of a lipid bilayer, the membrane thickness provides information on the orthogonal direction (Fig. 6C). Generally, the membrane thickness is in the range of 3–5 nm. There are three slightly different quantities commonly used to represent the thickness of a membrane: (1) the head-to-head thickness (D_{HH}) derived from the distance between the averaged z-positions of the phosphorus atoms of opposing leaflets, (2) the bilayer thickness obtained from the water density probability at half-maximum (D_B), and (3) the hydrophobic thickness ($2D_C$) calculated from the half-maximum of the density of the atoms in the hydrophobic acyl chains. The quantity D_{HH} can be determined by X-ray scattering as the distance between the two maxima of the electron density profile of the lipid bilayer (see Fig. 6D). With this approach, the accuracy of D_{HH} is limited by the Fourier truncation error. For measuring D_B, neutron scattering experiments can be applied by making use of the high contrast between the protiated lipid and deuterated water accurately defining the bilayer thickness, which is correlated to the penetration of the lipid bilayer by water molecules. Both D_{HH} and D_B can be easily determined from MD simulations. D_{HH} is usually computed as the average distance between the phosphate planes of the two leaflets. D_B is calculated using Eq. (6) (with the bilayer normal in z-direction and d being the simulation box length in that direction) as the distance between the points along the membrane normal, where the water density probability $p_w(z)$ is one half:

$$D_B[nm] = d - \int_{-d/2}^{d/2} p_w(z)dz \tag{6}$$

The third measure for the membrane thickness, $2D_C$, refers to the thickness of the hydrophobic core of a bilayer[96] and is determined based on the positions of the CH_2 groups in each leaflet of a lipid bilayer. These positions can be derived from X-ray or neutron scattering data, or from NMR making use of the carbon-deuterium bond order parameters in the acyl chain. However, $2D_C$ is highly dependent on the underlying approximations (e.g., for the volume of a methylene group or the estimate for the average tilt of the acyl chains with respect to the bilayer normal), thus the outcomes can significantly differ and are only comparable to the values determined from MD simulations as long as the same assumptions are employed.[96] Together with A_L, the membrane thickness allows one to assess the state the bilayer in question. In fact, a decrease in A_L almost always relates to an increase in the lipid bilayer thickness, which is further accompanied by a stronger ordering of the acyl chains in the membrane.

An electron density corresponds to the probability of the electrons being present at a specific location. With the help of X-ray crystallography, the electron density of a membrane can be experimentally estimated. In MD simulations, where electrons are not explicitly considered, the time-averaged positions of electrons can be inferred with good accuracy from the atomic positions. Fig. 6D shows a typical electron density profile across a membrane, which is characterized by following observations: The profile exhibits a global minimum in the middle of the bilayer (near the terminal CH_3 groups of the acyl chains), two maxima corresponding to the positions of the head groups (phosphate groups) in each leaflet, and a local minimum reflecting the water layer outside the membrane. The thickness D_{HH} is defined as the distance between the two maxima of the electron density profile, which changes upon the insertion or interaction of a molecule with a membrane. In the case of mixed membranes involving cholesterol, the peaks of the phosphate groups are slightly shifted depending on the cholesterol concentration, whereas the rest of the profile exhibits no significant changes.[98] The electron density profile is symmetric if the two leaflets have the same composition and when the bilayer is in its equilibrium state. The main aspects of an electron density profile are required to be reproduced by a FF, like the overall profile shape and the position of the peaks corresponding to specific groups of atoms (e.g., head groups, glycerols, carbonyls, methylenes in the hydrocarbon chains). In addition, these profiles can be also calculated from the mass density across the bilayer, which shows how the mass is distributed along the membrane z-axis and exhibits the same shape as the electron density distribution. In the case of a typical hydrated membrane system, the mass density is high at the beginning of the lipid chain, similar to the mass density of several soft polymers (0.9–1.3 g cm^{-3}). At carbon position 9, the density is reduced to the one of liquid hexadecane (0.753 g cm^{-3}) and it is considerably lower in the middle of the membrane (0.60 g cm^{-3}). This shows that the acyl chain region of the bilayer in the liquid crystalline phase is far from being homogeneous.

2.1.2.3 Acyl chain order parameters

Lipid bilayers can be characterized by their degree of orientational lipid order. Experimentally, the measurement of the carbon–deuterium (or hydrogen) bond order parameters S_{CD} (S_{CH}) can be accomplished by NMR spectroscopy using ^{2}H (deuterium, D) selectively labeled lipids. More recently also ^{1}H–^{13}C labeled lipids have been employed. The order parameters S_{CD} depend on a combination of local, molecular, and collective motions as well as static disorder, and the decomposition the overall value into individual contributions is difficult. ^{2}H NMR spectroscopy has been widely applied to determine the S_{CD} values of the acyl chains of GPLs, which have become an important target property to validate lipid FFs. The order parameter S_{CD} of a C–D bond of an acyl chain measured by NMR can be directly compared with values for S_{CH} calculated from a simulation using the following equation:

$$S_{CH} = \frac{1}{2}\langle 3\cos^2(\theta_z) - 1\rangle \quad (7)$$

Here, θ_z is the (time dependent) angle between the C–H bond vector and the bilayer normal (experimentally the magnetic field or z-axis of the simulation box). The angular brackets denote a time and ensemble average. The order parameter adopts values between -0.5 and 1, where a value of -0.5 means that the two vectors are perpendicular to each other, while a value of 1 implies that the two vectors are parallel. The order parameters do not only depend on the orientation, but also (dis)order of the system. For instance, $S_{CH} = 0$ can either arise from C–H bonds isotropically disordered with respect to the bilayer normal, or all corresponding bond vectors are perfectly oriented at the magic angle relative to the bilayer normal (that is 54.7 degrees, corresponding to $\cos^2(\theta_z) = 1/3$).

The general focus is on the values of S_{CD} as a function of the position of the carbon atom in an acyl chain, but S_{CD} (or S_{CH} in simulations) can be also determined for other lipid moieties, like the glycerol or the head group. The S_{CH} order parameter profiles that are obtained for saturated acyl chains look qualitatively similar to the one shown in Fig. 6E. It is characterized by some degree of disorder in the alkyl groups close to the water interface, which is followed by an increase in order with a order maximum at about the middle of the acyl chains. Toward the center of the membrane, corresponding to increasing C atom number positions along the acyl chain, the disorder increases (i.e., decreasing S_{CH}). This results from the greater degree of conformational flexibility of the acyl chains in the core region of a membrane. Furthermore, the length of an acyl chain influences the S_{CD} values, with longer chains reducing the order parameter. Similarly, temperature affects S_{CH}, with lower temperatures implying an increase in S_{CH} due to the closer packing of the lipids.

Differences between the experimental and calculated order parameters can be a result of incomplete sampling during the MD simulation in question. Moreover, membrane undulations and fluctuations in both experiments and simulations can contribute to the value of S_{CD} (S_{CH}). In simulations, large enough lipid bilayers are required to limit the artifacts due to a finite system size and periodic boundary conditions. If a UA-FF was used during an MD simulation, the C–H bond vector needs to be reconstructed. This is done based on the positions of three successive CH_2 groups and assuming tetrahedral geometry for the central C atom. It should be noted that this calculation only holds when the hydrogen atoms of a methylene group are equivalent and, more importantly, it is not valid around unsaturated bonds due to their different geometry.

2.1.2.4 Membrane area compressibility

Compressibility, which describes the relative volume change of a given substance in response to stress, is extensively studied for lipid bilayers. Membrane area compressibility, in particular, indicates the resistance of a lipid bilayer to isotropic area dilation and is defined by an area compressibility modulus K_A:

$$K_A[Nm^{-1}] = A\left(\frac{\partial\gamma}{\partial A}\right)_T = \frac{Ak_B T}{\sigma_A^2} = \frac{A_L k_B T}{N_l \sigma_A^2} \tag{8}$$

Here, γ represents surface tension, A is the average total area, k_B is the Boltzmann constant, T is the temperature, A_L is the area per lipid (see above), N_l is the number of lipid molecules, and σ_A^2 is the mean square fluctuation of A_L. As the area compressibility modulus is inversely proportional to the area fluctuations, K_A is a reliable parameter to estimate the rigidity of a lipid bilayer.

2.1.2.5 Lateral diffusion coefficient

In the last several years, lateral diffusion processes in biological membranes have been studied with experimental and computational methods, but the underlying molecular mechanism is still unknown. The lateral diffusion of lipids and proteins depends on the neighboring molecules. It is a fast and spontaneous movement relative to translational motions between the two leaflets, where special membrane proteins, called flippases and floppases, are required for this lipid transport. The experimental techniques for determining lateral diffusion constants range from fluorescence techniques (e.g., FRAP, FRET and fluorescence correlation spectroscopy, FCS) to EPR and NMR spectroscopy (e.g., ^{1}H, ^{2}H, ^{19}F, and ^{31}P). As for other membrane observables, the resulting values for the lateral diffusion constants can vary even of one uses the same method, as it is sensitive to experimental conditions, like the temperature, hydration content, pH, ionic strength and experimental setup.[96] Furthermore, EPR and fluorescence methods have to use labeling molecules (e.g., spin labels or fluorophores) for tracking the lipid diffusion. Here, the measured lateral diffusion coefficient derives from the labeling molecules instead from the lipid itself, which is not negligible in some cases depending on the type of labeling, its concentration, and whether the measurement was done in a bilayer or monolayer.[96] Neutron scattering can be used for fast, short-range diffusion (motions for 0.1–10 nm in <1 ns), while the other techniques sample slower, medium-to-long-range diffusion processes (motions of >0.1 μm on the ns-to-ms timescale).[106]

For the movement of lipids and other membrane molecules during MD simulations, the Einstein relation based on the average lateral mean squared displacement (MSD) $\langle[\Delta r(t)]^2\rangle$ can be used for computing the lateral diffusion constant by applying the following equation:

$$D_l[cm^2 s^{-1}] = \frac{1}{2n_f} \lim_{t\to\infty} \frac{\langle[\Delta r(t)]^2\rangle}{t} \tag{9}$$

Here, the n_f denotes as number of translational degrees of freedom ($n_f = 2$ for an in-plane motion or in the xy-plane) and $r(t)$ is the center of mass (COM) position of the molecule under study at time t. In general, the MSD is calculated as average over all lipids. As for experimental methods, the resulting lateral diffusion constants can vary as they depend on the applied FF, system size, the cutoff scheme for the truncation of long-range electrostatics, the time step, and the frequency used to update the nonbonded pair list. Thus, the comparison between the results from several studies is difficult; Table 2 in Ref. 96 lists some lateral diffusion constants obtained from simulations. Flack et al.[99] studied the influence of cholesterol on the lateral diffusion of lipids using 100 ns MD simulations of a DPPC bilayer with varying amounts of cholesterol, which demonstrated that the lateral diffusion coefficients of both DPPC and cholesterol decreased with increasing concentrations of cholesterol (see Fig. 6F). In general, the lateral diffusion can be evaluated only in the liquid crystal state.[98]

2.1.3 Comparison of the atomistic lipid force fields

In chapter "Pairwise-additive and polarizable atomistic force fields for molecular dynamics simulations of proteins" by Lemkul of this book, the historical development of the different protein FFs was described in detail. Thus, we concentrate here to give an overview about the FFs that are available for studying protein–membrane systems, including to list their advantages and shortcomings.

2.1.3.1 CHARMM

The CHARMM (Chemistry at Harvard Macromolecular Mechanics) all-atom force field originates from the lab of Martin Karplus at Harvard University and has an active community who developed a huge set of FFs with a focus on biomolecules, such as lipids,[72,73] carbohydrates,[107–110] nucleic acids,[111,112] and proteins.[113,114] Additionally, a general force field, known as the CHARMM General FF (CGenFF)[115] allows simulations with other biomolecules and drugs that conform to the general parameterization scheme applied in CHARMM. In addition to the common FF terms (see Eqs. 3 and 4), the intramolecular interactions include the so-called Urey–Bradley term for covalent angles, which is a cross term connecting the motions of bond angles and lengths:

$$U(r,\theta) = \underbrace{k_{ub,i(i+2)}\left(r_{i(i+2)} - r_{i(i+2),0}\right)^2}_{\text{Urey-Bradley}} \tag{10}$$

The Urey–Bradley term with force constant $k_{ub,i(i+2)}$ applies to atoms i and $i + 2$ connected by two bonds. However, this term is only used for older CHARMM parameters but not for newer ones. The parameterization strategy of CHARMM is based on fitting to reproduce experimental and ab initio QM data at the HF/6-31(g) level of theory of small model compounds (e.g., alkanes for the acyl chain) that are transferable to larger molecules (e.g., to lipids) with some further optimization. Furthermore, it is recommended to use the modified TIP3P water model—even though better water models like TIP4P are available—when employing the CHARMM pairwise additive FF,[97] since the parameters for interaction of biomolecules with water were optimized for this water model and the usage of another water model could lead to imbalances in the electrostatic and solvation energies. Another peculiarity of CHARMM is that this FF applies force-based cutoffs in which the force is smoothed to zero at a cutoff of 1–1.2 nm, affecting the overall potential. Therefore, this cutoff value must be used when using CHARMM. Moreover, the nonbonded interactions of atoms that are separated by four bonds, so-called 1–4 interactions are not scaled, which in turn implies that the dihedral potential parameters depend on the electrostatic and the LJ parameters. Thus, optimization of these coupled parameters is typically carried out in an iterative fashion, in which the dihedral parameters are fitted to the obtained LJ and electrostatics parameters per iteration step.

The first CHARMM parameters for lipids were published with the CHARMM22 parameter set (denoted as C22)[68] by the Pastor group. They were updated with the release of the CHARMM27 parameter set (C27),[69] which was the first widely available and thus used AA-FF for lipids. For a DPPC lipid bilayer C27 produced a good agreement with experimental deuterium NMR order parameters ($S_{\rm CD}$) for the acyl chains and the electron density profiles.[69] However, further investigations revealed non-negligible disagreements with experiment (e.g., for the A_L value)[116] as well as for dynamical properties obtained from NMR relaxation times, indicating that probably the C27 torsional barriers were too high.[117] For fully saturated lipids it was shown that the bilayers often exhibited a gel phase at conditions corresponding to liquid crystalline phases. To circumvent this problem, many MD simulations with C27 were performed in the *NVT* ensemble[118] with a fixed A_L value or under nonzero surface tension,[119] as under these conditions the *trans* conformation of the acyl chains is more preferred than the *gauche* conformations.[120] There was an update to the C27 parameter set[70,71] (called C27r), but the phase state problem of the lipid bilayers was still present and also the correct A_L could not be reproduced with MD simulations at zero surface tension (*NpT* ensemble).[70,121]

Five years later, the CHARMM36 parameter set (C36) was published[73] by Klauda et al. and verified with MD simulations of bilayers involving six saturated and monounsaturated lipids (DLPC–12:0,12:0; DMPC–14:0,14:0; DPPC–16:0,16:0; DOPC–18:1,18:1; POPC–16:0,18:1; POPE–16:0,18:1). Partial atom charges and torsion potentials were reparameterized by using ab initio computations, as well as revision of some LJ parameters for matching with semiempirical QM results of small bilayer assemblies, solvation free energies, and S_{CD} near the membrane interfacial region. Moreover, the description of the lipid head group structure and dynamics was improved in the same manner like as in the work by Sonne et al.[122] and compared to NMR data. Properties as average A_L at zero tension ($\gamma = 0$), structure factors, dipole electrostatic potential indicated good agreement with experiment and better than with the C27r parameter set.[73] The key improvement was that the C36 lipid FF accurately matched the X-ray and neutron form factors, bilayer and hydrocarbon thickness as well as S_{CD} of POPC and POPE bilayers. To sum up, the C36 FF allows to accurately represent bilayer properties of fully saturated and monounsaturated lipid molecules in the *NpT* ensemble.[73] In the last years, different kind of lipids were added to C36 using the same parameter strategy, such as a variety of acyl chains (PUFAs,[123] branched[124] chains and chains with cyclic moieties[125]) and head groups (PA, PG, PI, and PS),[72,126,127] sterols (cholesterol,[128] ergosterol[129] and plant sterols[130]), sphingolipids[131] and ceramides[132] as well as glycolipids[133] and LPS[126,134] making use of the CHARMM carbohydrate FF.

To accelerate the sampling during an MD simulation, the C36-UA FF was developed based on the C36-AA lipid parameters,[74] in which the acyl chains have the UA representation, while the head groups are described in an explicit manner. It performs well for lipid membranes and micelle formation of a surfactant. MD simulations of DMPC, DPPC, POPC, DOPC, and DMPC/CHOL bilayers produced good values for the A_L, X-ray and neutron form factors, and acyl chain order parameters S_{CD}.[74] However, the Klauda lab reported general issues (unpublished data) with too high values for the density of alkanes and too small values for the hydrophobic bilayer thickness; furthermore the head groups needed additional parameterization.[97]

2.1.3.2 AMBER

For a long time, no proper parameters for lipids were available for the AMBER FF (Assisted Model Building and Energy Refinement), although some attempts were done to simulate bilayers. However, the A_L values for DMPC and DOPC bilayers modeled with the Generalized Amber Force Field (GAFF)[135] with the extension to lipids[75] were underestimated[76] compared to the experimental values. A certain surface tension had to be applied to obtain the correct lipid area. In contrast to CHARMM, in AMBER the 1–4 LJ and Coulomb interactions are scaled by a factor of 0.833; moreover, the restricted electrostatic potential (RESP) method[136,137] is used for the calculation of the partial charges of whole molecules. With the goal to develop a lipid FF that is comparable to C36 FF, the Lipid11 FF[77] was developed, in which a careful reparameterization of LJ and charge values to improve modeling of lipid bilayers was undertaken. Bilayers of DOPC, POPE, and POPC were tested, but again a surface tension term had to be applied to obtain good agreement with experimental scattering profiles and NMR S_{CD} data. In an alternative approach, modifications to the GAFF LJ parameters for the simulation of acyl chains were made to generate accurate and stable simulations of pure lipid bilayers of DLPC, DMPC, DPPC, DOPC, POPC, and POPE, yielding the GAFFlipid FF.[138] For all six lipid types tested it achieved a higher level of agreement between simulation and experiment for numerous structural properties, like A_L, volume per lipid, bilayer thickness, order parameter, and head group hydration. Based on the GAFFlipid parameters, the Lipid11 FF was updated yielding the current standard AMBER lipid FF, named Lipid14,[78] which allows the simulation of a number of lipids without an additional surface tension term. As for C36, the Lipid14 FF has been tested by simulating bilayers of six different lipid types (DLPC, DMPC, DPPC, DOPC, POPC, and POPE) and no longer requires a surface tension to be applied for keeping the lipid area constant. Membrane properties, such as A_L, isothermal compressibility, bilayer thickness, and NMR S_{CD} parameters, favorably agree with experiment, reaching the same level of accuracy as C36. The only shortcoming is that S_{CD} for the head group and order splitting for carbon-2 on the *sn*-2 chain were not tested. Lipid14, which works well with the tensionless *NpT* ensemble, can be combined with the AMBER FFs for proteins, nucleic acids, carbohydrates,[139] and small molecules, and was extended to mixed membranes including cholesterol.[140] There is also an update to the Lipid17 FF for anionic head groups and polyunsaturated acyl chains[141]; however, it has not been applied yet, leaving it open how well it performs.

2.1.3.3 Slipids

Around the same time when the C27r FF was updated to the C36 FF and based on the suggestions in these papers,[120,122] a new AA-FF named Slipids (Stockholm lipids) for fully saturated phospholipids[142] and monounsaturated lipids[79] was published, which was extended a year later to include the lipid head groups PG, PE, PS, and SM as well as cholesterol.[80] As starting point the recent C36 parameter set was used and following optimizations performed: (1) The partial charges, LJ and torsion angle parameters of all lipid tails were updated by using high-level ab initio calculations, whereas the parameters for covalent bonds and angles along with some of the torsion angle and LJ parameters were used from C36. (2) The head group parameters were improved using QM calculations at a low level of theory and the water dielectric constant readjusted using the TIP3P water model. (3) Scaling factors for the LJ/Coulomb parameters for the 1–4 interactions were introduced, using a value of 0.5 between hydrogens of the lipid carbon chain and varying values for the rest, which is based on AMBER FFs with a scaling factor of 0.833. (4) The RESP method[136,137] was applied to determine the partial charges, as done for the AMBER parameters. For testing, the resulting Slipids FF was applied to DLPC, DMPC, DPPC, POPC, SOPC, POPE, and DOPE bilayers[79,142] as well as to mixed membranes with cholesterol.[80] A good agreement to the experimental values of many membrane observables, like A_L, NMR order parameters and structure factors was obtained. The most important aspect is that the parameters are able to reproduce structural properties of single and double component membranes without employing a surface tension, therefore, it allows to study complex systems containing many different lipids and proteins in a fully atomistic resolution in the *NpT* ensemble. Due to the parameterization close to the AMBER strategy, it is compatible with the AMBER FFs for proteins, which was tested for a WALP23 peptide (modeled by three versions of the AMBER protein FF),[79] and for drug-related compounds in combination with lipid bilayers.[143,144]

2.1.3.4 OPLS-AA

The development of the OPLS-AA FF (Optimized Parameters for Liquid Simulations All Atom) by Jorgensen et al. was mainly done for organic liquids to reproduce the experimental values for thermodynamic and partitioning properties.[145,146] Quite recently, lipids were parameterized[83] along with other biomolecules[147–149] for this OPLS-AA FF. The partial charges were obtained with the similar route as followed for AMBER FFs and the optimization of dihedral angles was accomplished as for the C36 lipid FF, using small molecules to represent the PC lipid head groups. Only a single 200 ns MD simulation of a DPPC bilayer at 323 K was performed for the parameterization that showed good agreement with experimental S_{CD} and X-ray form factors.[83] Furthermore, the inclusion of the dihedral optimization for the lipid head groups reproduced similar values as NMR spectroscopy for this region of the lipid. Since then, the OPLS-AA FF was extended to lipids with chain unsaturation,[150] cholesterol,[151] ether linkages,[152] PE,[152,152a] PI,[152b] PS, and SM[152a] with different saturation/unsaturation levels and acyl tail lengths as well as GM1.[152c] Apart from these, POPC oxidized lipids such as 1-palmitoyl-2-(9′-oxo-nonanoyl)-sn-glycero-3-phosphocholine (PoxnoPC) and 1-palmitoyl-2-azelaoyl-sn-glycero-3-phosphocholine (PazePC) were also paramterized for study with OPLS-AA FF.[152d] However, further developments are necessary to obtain the lipid diversity available in C36 and the AMBER FFs. There is also an OPLS-UA FF, which was initially developed for DMPC and resulted in good agreement with the experimental value of A_L at 303 K.[81] It was extended to cholesterol[82] and epicholesterol,[153] but lacks further lipids necessary to simulate more realistic biological membranes.

2.1.3.5 GROMOS

(Groningen molecular simulation package)[86] uses the UA approach to represent each nonpolar CH, CH_2, and CH_3 group of hydrocarbons as a single particle, thus reaching an about threefold acceleration in comparison to AA FFs. The main objective in the development of GROMOS FFs is the reproduction of thermodynamic properties of model compounds (small molecules representing biomolecular fragments), like enthalpies and free energies of solvation in polar and apolar environments. Therefore it became a popular choice for the simulation of membrane proteins. GROMOS FFs applied for bilayer simulations can be split into two groups: (1) using the original GROMOS nonbonded parameters (e.g., 45A3,[86] G53A,[87] and G54A[154] parameter sets) and (2) the Berger modification[84] (also called Berger lipid FF), in which the Ryckaert–Bellemans potential was implemented for describing torsion rotations of the hydrocarbon chains of lipids. Several studies with the 45A3 parameter set (e.g., DPPC)[86] and Berger lipids (e.g., DOPC and DMPC)[116,119,155,156] revealed that the GROMOS FF did not properly reproduce a number of experimental properties, like the electron density profile, structure factor, and A_L. Moreover, with GROMOS, MD simulations cannot produce the proper gel phase of bilayers at temperatures below the melting point.[157] Based on further ab initio computations and fitting to thermodynamical data for liquid alkanes, the 43A1-S3 parameter set[158] showed better results for A_L than the previous versions and was updated to the G53A6 parameter set,[87] which exhibited very good agreement with experimental membrane properties of DPPC bilayers (A_L and volume per lipid, electron density profiles, bilayer thickness and hydration, S_{CD} of acyl chains). Furthermore, it reproduces the structural and hydration values of common GPLs of varying length and degree of unsaturation of the acyl chains (e.g., DLPC, DMPC, DOPC, and POPC) in a liquid crystal phase.[159] In addition, the G53A6 parameter set can be applied for the MD simulation of PC bilayers in the liquid crystalline phase. The G53A6 FF was further improved with the Kukol parameters[88] and the parameters by Piggot[89] (known as the CKP parameter set) for saturated and unsaturated PC and PG lipids and later also for PG, PE, and CL.[90] Although it is known that the Berger parameters are not the best choice for reproduction of experimental lipid properties, they are still used in many recent papers in combination with older versions of the GROMOS FF or AMBER FFs. Finally, the last update is the G54A7 parameter set for a number of lipid types.[91] In addition to the G53A6 FF, the G54A7 parameters include bacterial lipids with branched fatty acids in their lipid chains,[160] with cyclopropane moieties,[161] as well as hopanoids and sterols.[162] However, the number of different lipids that has been tested for this FF is rather limited.

2.1.4 Limitations of atomistic lipid force fields

In spite of great hardware and software developments, which facilitated MD simulations of large protein–membrane systems on the microsecond timescale and with atomistic resolution, the sampling of protein conformations can become problematic due to high-energy barriers separating the different protein states. Enhanced sampling techniques, such as metadynamics or umbrella sampling may provide a means to overcome free energy barriers during MD sampling; they are discussed in detail in chapter "Enhanced sampling and free energy calculations for protein simulations" by Liao of this book. However, these methods are not often applied on lipid bilayers, except for surface tension replica exchange[163] adaptive biasing force[164] and metadynamics.[165] Another approach is given by reducing the degrees of freedom that need to be sampled, leading to CG-FFs, which allow to simulate larger systems for longer times compared to AA-MD simulations and are discussed in more detail in the next section. Before, we briefly introduce the highly mobile membrane mimetic (HMMM) model, which is not a CG model but nonetheless accelerates lipid dynamics by modeling the acyl tails in the membrane center as a fluid organic solvent.

2.1.4.1 The HMMM model

At physiological temperatures, the lateral diffusion of lipids in a membrane is typically on the order of 10^{-8} cm^2 s^{-1}, which can prevent the membrane association of peripheral proteins or spontaneous lipid mixing to occur in AA-MD simulations given their typical timescale of hundreds of nanoseconds. It would require at least orders of magnitude greater sampling to reach diffusive equilibration. A workaround to this problem was presented by Ohkubo et al. with the development of the HMMM model,[64] which uses an organic solvent layer to describe the hydrophobic core of the membrane while maintaining an atomic description of the lipid head groups and shortened GPL acyl tails. This leads to an acceleration of the lipid diffusion without sacrificing the atomistic level of detail for the head groups, which are essential for protein–membrane interactions. In the HMMM model, the GPLs are truncated in the acyl chains. The resulting space in the membrane core is then filled with an organic solvent, most often 1,1-dichloroethane (DCLE) to reproduce the characteristics of the hydrophobic core of lipid bilayer (Fig. 5B, left panel).

In the original HMMM model study,[64] the authors demonstrated the efficiency and robustness of their approach. They used the CHARMM36 General FF parameters[115] to perform the simulations. First, they demonstrated that the HMMM model is indeed able to form and maintain bilayers. To this end, an initial biphasic solvent box, containing water and DCLE was set up, and 54 divalerylphosphatidylserine (DVPS, 5:0,5:0) molecules, which were chosen to present the head groups and shortened acyl chains of PS lipids in this particular HMMM model, equally distributed in that box. A fully formed bilayer formed within 20 ns of MD and was stable in the remainder of that 40 ns simulation. Comparison to a DOPS (18:1,18:1) membrane simulated under the same conditions confirmed that the HMMM model was able to reproduce the atomic distribution of various chemical groups in the membrane, in particular in the head group region that is often essential for protein binding. Furthermore, the extent of hydration and the degree of counterion penetration was similar compared to the full-membrane system. As a second test, a set of five simulations were performed to calculate the lateral diffusion constant of the lipids in the HMMM model and its dependence on different configurations. The lateral diffusion constants of the DVPS molecules were found in the range of 10^{-5}–10^{-6} $cm^2\,s^{-1}$, presenting an increase of one to two orders of magnitude in comparison to the full-membrane system. This improvement enables more efficient sampling of head group configurations, e.g., during the binding and insertion of peripheral proteins. To test this, the authors employed the HMMM model for studying the membrane association of the coagulation Factor VII GLA domain, which is involved in the membrane binding of different peripheral proteins. The GLA domain is a common membrane-anchoring domain in coagulation proteins and predominantly binds to PS-rich regions of the cellular membrane, resulting in an increase of the catalytic activity of these proteins. At the time of the original HMMM model study, classical AA-MD simulations failed to reproduce the association of the GLA domain to membranes, while this event was sampled within 50 ns of MD simulations using a PS-HMMM model.[166]

Since its publication, the HMMM model has been applied to various membrane-associated proteins, including coagulation factor GLA domains,[167] cytochrome P450,[168] and TM domains.[169] Furthermore, the free energies from AA and CG simulations for the partitioning of amino acids into the interfacial membrane region could be reproduced by the HMMM model.[170] The application of the HMMM model to study membrane association of amyloid peptides[171,172] is discussed in Section 4.2.1. Despite its various successful applications, it should be noted that the HMMM model also exhibits several limits. First, it cannot be applied for accurately representing the energetics and processes at the membrane core because the behavior of DCLE significantly differs from that of acyl chains in the lipid bilayer core.[170,173] Second, simulations of TM proteins with the HMMM model are still difficult as the DCLE solvent destabilizes multihelix TM domains in the core region.[173] Finally, it still remains to be shown whether the HMMM model can also reproduce the characteristics of membranes involving other lipids, such as sphingolipids and sterols.

2.2 Coarse-grained force fields for lipids

To fully understand membrane systems, it is desirable for MD simulations to be able to study relevant timescales, which in practicality are beyond what most existing resources can currently handle at the atomistic level. For instance, subcellular processes involving lipid bilayers such as domain formation, spontaneous lipid aggregation into full micellar and vesicular structures, particle insertion, and other dynamical phenomena occurring at large time and length scales require CG models.[74,174,175]

In CG models, a certain number of atoms whose physical and chemical properties can be averaged to achieve the FF design philosophy is chosen and treated as a single particle (Fig. 7A). This reduces the degrees of freedom, and thus the computational effort for the integration of Newton's equation of motion (Eq. 2). With this, sampling larger timescales becomes increasingly possible, while bigger lipid systems (both homogeneous and mixed) that are closer to realistic biological membranes can be studied. CG-based MD simulations can more effectively sample conformational landscapes of molecular systems, in effect because they transform the typically rugged energy landscape into a much smoother surface; the level of *smoothness* depends on the level of coarse-graining (Fig. 7B). The coarser a model is, the smoother is the underlying energy landscape and the more efficiently can conformational and energy states be simulated. By using smooth potentials, most currently available CG models are able to employ integration steps much greater than the 2 fs time steps commonly employed in all-atom simulations. For instance, in MD simulations using the MARTINI FF time steps of 10–20 fs are typically used. However, this scale up in conformational sampling efficiency comes with a cost: atomistic resolution is lost and so is information about properties deriving directly from individual atomic constitution.

CG-FFs stand between AA-FFs and continuum models with respect to the level of resolution, speed of sampling the phase space, level of approximation, as well as properties that can be investigated. There is always a trade-off in resolution associated with the different scales of simulation, from QM to AA- and CG-MD, and to continuum models. The trade-off relates to different degrees of neglect in the explicit treatment of electronic (AA) and atomic properties (CG), and in each case it is deliberately accommodated in order to achieve more extensive sampling of time- and length-scales relevant to some biomolecular properties. And in the case of CG models of membrane systems, the founding philosophy of a particular model dictates both resolution and extent of sampling. Each CG model represents an attempt to address specific aspects of the membrane and it is important for the user to decide which CG model has been designed to provide answers to the kind of research questions that are being asked. Parameter sets available for most CG models are either the result of a top-down approach, a bottom-up approach, or a mix of both. In the top-down approach, the goal

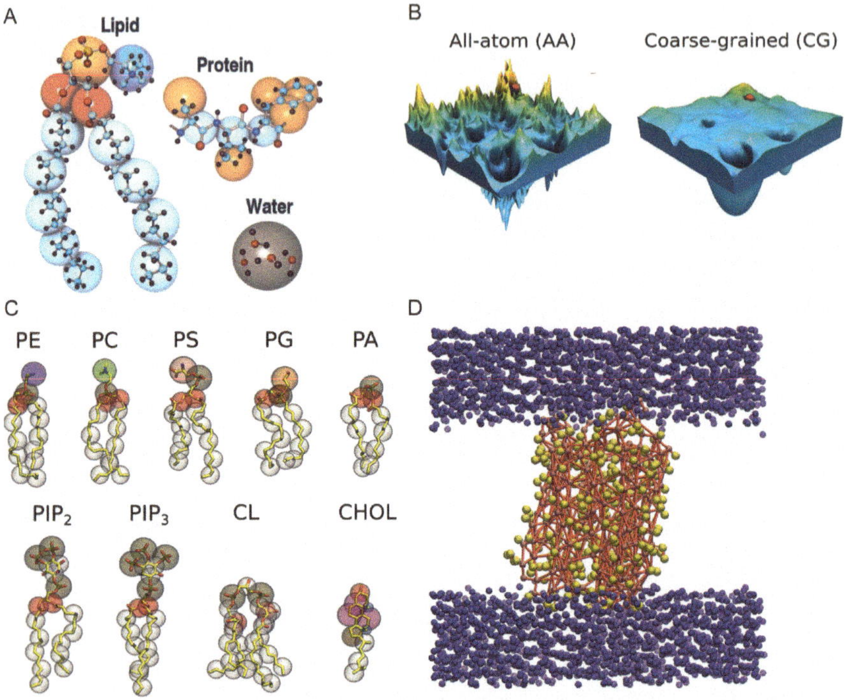

Fig. 7 MARTINI mapping and elastic network approach for proteins. (A) The AA-to-CG mapping underlying the MARTINI FF. *Balls* and *sticks* are used for the AA representation, while the CG beads are shown as large spheres. Each bead merges four heavy atoms with the associated hydrogen atoms. The mapping is exemplarily shown for a lipid (PSPC or POPC), a fragment of a protein chain, and a water molecule. (B) Comparison between AA and CG energy landscapes. Coarse-graining smoothens the energy landscape, facilitating a more efficient exploration of the energy landscape in an MD simulation. (C) The AA-to-CG mapping for common lipids in MARTINI, overlaid on the corresponding AA structures (hydrogen atoms are removed for clarity). The CG beads are displayed as transparent vdW spheres. (D) Schematic representation of the dummy bonds (*red*) in the elastic network approach used in MARTINI to keep the secondary structure of a protein (displayed with *yellow CG beads*) in its initial assignment. The water beads are shown as *blue spheres*, while the membrane bilayer is not shown for clarity. *Panels A and B: Reprinted with permission from Kmiecik S, Gront D, Kolinski M, Wieteska L, Dawid AE, Kolinski A. Coarse-grained protein models and their applications. Chem Rev. 2016;116(14):7898–7936, https://doi.org/10.1021/acs.chemrev.6b00163. Copyright 2016* American Chemical Society; *Panel C: Reprinted with permission from Muller MP, Jiang T, Sun C, et al. Characterization of lipid-protein interactions and lipid-mediated modulation of membrane protein function through molecular simulation. Chem Rev. 2019;119 (9):6086–6161, https://doi.org/10.1021/acs.chemrev.8b00608. Copyright 2019* American Chemical Society; *Panel D: Reprinted with permission from Goossens K, De Winter H. Molecular dynamics simulations of membrane proteins: an overview. J Chem Inf Model. 2018;58(11):2193–2202, https://doi.org/10.1021/acs.jcim.8b00639. Copyright 2018* American Chemical Society.

is to develop FFs that can satisfactorily reproduce specific sets of experimental data. In the bottom-up approach, the CG model employs building blocks capable of reproducing atomic-level properties obtained from AA-MD simulations or from experimental. It is also possible to mixed up both approaches. The developers of the MARTINI FF adopted the bottom-up approach. MARTINI represents the currently most widely employed CG model for studying membrane systems. Although originally designed in 2003 for lipids,[176] this FF has since gone through developments both vertically (improvement in parameters for studying the same class of system) and horizontally (development of parameter sets for studying new system classes). The MARTINI FF has since been extended to most major biomolecules like proteins,[177] carbohydrates,[23] and nucleic acids.[178,179] Therefore, our focus in this chapter is the MARTINI FF because of its wide extensibility to a range of other macromolecular systems.

2.2.1 The MARTINI CG model

2.2.1.1 The all-atom to coarse-grained mapping

One of the implications of coarse-graining is the replacement of the atomic representation with few pseudoparticles (called beads) and pseudo-bonds. The MARTINI model generally employs a mapping scheme that represents four atoms with one CG bead; there are some special situations that require the assignment of fewer atoms to a bead. The speed up achievable with the MARTINI model is immediately apparent if one considers the implication of the reduced degree of freedom. In AA-FFs, atom types are required to more accurately describe different chemical environments (see chapter "Pairwise-additive and polarizable atomistic force fields for molecular dynamics simulations of proteins" by Lemkul and also Section 2.1). For instance, in the GROMOS53A6 UA-FF, the 17 atoms commonly encountered in biomolecular systems have specially defined parameter sets. Carbon, oxygen, nitrogen, and hydrogen have 15, 9, 7, and 3 atom type definitions, respectively; in total, there are 53 atom types for all seventeen atoms. This directly leads to $53!/(2! \cdot (53 - 2)!) = 1378$ combinations of nonbonded atom type pairs. By mapping four heavy atoms into one CG current implementation (version 2.2) has four essential bead types:

(1) apolar bead C for modeling hydrophobic groups like butane;
(2) polar bead P for modeling groups that are readily soluble in water such as ethylene glycol;
(3) nonpolar bead N for modeling groups which are partly polar and partly apolar like propyl alcohol); and
(4) charged bead Q for modeling explicitly charged groups like ammonium.

A greater resolution is achieved by having subtypes for each of the four bead types: P_{1-5} and C_{1-5} represent increasing level of polar and apolar features, respectively, while Q_d/N_d, Q_a/N_a, Q_{da}/N_{da}, and Q_0/N_0 describe the hydrogen bonding (HB) ability of charged and nonpolar beads based on whether they are HB donor (d), HB acceptor (a), HB donor and acceptor (da), or unable to participate in HB formation (0). In total, 18 distinct types of CG beads are natively available in the MARTINI FF translating to $18!/(2! \cdot (18-2)!) = 153$ possible combinations of nonbonded bead pairs.

To correctly treat certain chemical groups, MARTINI employs special beads that are parametrically different from the 18 bead types. First, smaller beads are required to correctly model structural peculiarities, especially in ring systems where two or three atoms are mapped to one CG bead. Such beads are distinguishable in the topology by having an S prefix (e.g., SP for a small polar bead) with the LJ parameters scaled down. The normal MARTINI CG beads employ a vdW radius of $\sigma = 0.47$ nm (see Eq. 4), while a value scaled down by 91.5% is employed for the S beads, but only when computing interactions with another bead of the S subtype. Similarly, the ε value (corresponding to the LJ well depth and binding energy between two atoms/beads, see Eq. 4), associated with the main bead types is scaled by 0.75 when computing the interaction between two small beads. Second, specially big antifreeze particles, called BP4, with their vdW radius scaled up to $\sigma = 0.57$ nm and to $\varepsilon = 5.6$ kJ/mol for all interactions with P4 beads of the MARTINI CG water model were introduced. The large particle size of the BP4 beads as well as the strong interaction with water beads disrupt the otherwise neatly ordered lattice packing of the similarly sized water beads, which in earlier MARTINI version caused the water to freezing at temperatures higher than the freezing point of real water.[63]

The detailed presentation of the few bead types illustrates the basis of simulation speedup and improved sampling of the phase space achievable with a CG model like the MARTINI FF. Improved efficiency is obtained in many more ways, including simplified representation of bond types and the use of a uniform bead mass of 72 amu (equivalent to the mass of four water molecules). AA-FFs often use different force constants, equilibrium bond lengths, and angles for treating bonded potentials (Section 2.1). Instead, the MARTINI FF uses a 1.25×10^3 kJ mol^{-1} nm^{-2} harmonic force constant and a 0.47 nm equilibrium length for its pseudo-bonds. However, the FF retains the flexibility of tuning both force constant and equilibrium distance in cases, where more accurate description of molecular topology is desired. A particularly large increase in computational efficiency is achieved by the use of short-range potentials. The potential energy function employed by the MARTINI model closely mirrors the generally adopted expression in most AA-FFs (the full potential energy function is presented in Eqs. 3 and 4), i.e., the vdW interaction is modeled using the LJ 12/6 potential and electrostatic interactions are modeled with the Coulombic function. Beads i and j are separated by a distance r_{ij}, ε_{ij} is the LJ well depth and binding energy between beads i and j. The LJ interaction strength is discretized into 10 energy levels (0 to IX) parameterized to reproduce oil/water partition constants and varying from a most attractive $\varepsilon = 5.6$ kJ/mol interparticle binding energy to the least attractive $\varepsilon = 2.0$ kJ/mol as it may occur for a charged bead interacting with a strongly apolar bead.[63] The LJ parameter σ_{ij} is the vdW radius of the bead using 0.47 nm as the default. Special cases, involving aromatic rings require the use of CG beads with smaller bead radius, that is the S bead discussed above. Together, these result in a distance-dependent accounting of the LJ energy, which is continuously shifted in a way that it becomes zero at 1.1 nm cutoff. It should be noted that in MARTINI dipolar interactions, involving partial charges, are adequately captured by the LJ interaction matrix just explained, such that only the interactions involving explicitly charged beads of the Q type (with charges q_i and q_j) are modeled with a Coulombic expression. In order to correct the increase in hydration strength resulting from some of the recently introduced bead types, the dielectric constant in the latest version of the MARTINI FF was reduced from $\varepsilon_r = 20$ employed in the 2003 model[62] to $\varepsilon_r = 15$.[63] The Coulombic interactions are similarly screened to zero at 1.1 nm cutoff as done for the LJ potential.

2.2.1.2 CG mapping of lipids

Currently, the MARTINI FF provides parameter sets for different classes of lipids, which cumulated to over ~214 lipid models available at the MARTINI website.[180] The major classes of lipids include mainstream phospholipids, like PC, PE, PS, PG, PA, and PI, glycerols, lysophosphatidylcholine, SM, ceramides, glycosphingolipids, glycoglycerolipids, sterols, surfactants, and fatty acids (few examples are shown in Fig. 7C). Models for other smaller lipid classes like cardiolipin, bolalipid, and lipopolysaccharide are also provided. Using the CG mapping protocol discussed above and the lipid dipalmitoylphosphatidylcholine (DPPC– 16:0,16:0, see also Figs. 2A and 7A for the chemical structure of PC and palmitoyl moiety) as an example, we will describe the CG mapping scheme presented in Fig. 1A in Ref. 181.

As earlier discussed, the structure of most lipids is traditionally divided into a head group which is usually charged or polar, and one or more aliphatic tail groups. The differences among the thousands oflipids in existence essentially derive from the structural differences in the head groups and aliphatic tails. The head group of DPPC consists of a positively charged choline ($N^{+}(CH_3)_3$, dark blue bead in Fig. 7A lipid structure) which in the MARTINI FF is modeled with a Q_0 bead (charged without forming HB), and a negatively charged phosphate (PO_4, orange bead in Fig. 7A lipid structure) group modeled as Q_a (charged and acting as HB acceptor). With these assignments, the charged state and the hydrogen bonding property of the lipid head group are appropriately captured (the charge of each bead needs to be explicitly specified in the topology parameter file). The two uncharged C_{16}-aliphatic chains of palmitic acid (light blue beads in Fig. 7A lipid structure) are each represented with four C_1 beads, which are hydrophobic and with the least possible polarity. There are certain instances, however, where the chemistry of the compound demands some degree of uneven distribution in the CG mapping. In such a situation, the exact choice of mapping will ultimately depend on the user but importantly also on the property to be investigated. In a case where the the lipid aliphatic tails contain uneven number of carbon atoms, e.g., for a C_{15} saturated system, there is the additional need to maintain molecular symmetry in the mapping. For instance, using -C_1(3)-C_1(4)-C_1(4)-C_1(4) bead (number of atoms per bead in parentheses) mapping for the *sn*-1 tail and -C_1(4)-C_1(3)-C_1(4)-C_1(4) mapping for the *sn*-2 tail would introduce some level of dissymmetry in the two tails.[63] The charged PC head group and the hydrophobic lipid tail region are linked together by a glycerol backbone, which forms an ester bond with each aliphatic tail (Figs. 2A and 7A). This glyceryl ester bridge is represented with a nonpolar HB acceptor N_a bead (red beads in Fig. 7A lipid structure) for each of two the hydrophobic tails. Qualitatively, this representation suffices since the otherwise electronic (polar and HB) property has been coarse-grained into the N_a linker bead. This is especially appropriate since the standard bond length has been reduced to 0.37 nm for this particular bead in an attempt to capture its peculiarities.[63] However, if it is desirable to more explicitly reflect the polarity of the C_1 bead proximal to the polar linker bead, it is possible to tune MARTINI parameters to capture the specific property of interest.

Because of its versatility and acceptable accuracy at reproducing experimental and atomistic MD data, the MARTINI model has rapidly grown to become the most popular CG lipid FF that has been extended to many biomolecular and polymer systems. This was achieved in spite of its simplicity in the treatment of chemical systems and the unavoidable loss of atomic resolution.[62,63,176] By reducing the degrees of freedom in a physicochemically meaningful manner and, by adopting a bottom-up parameterization approach focused on reproducing experimental thermodynamic properties, the FF has been able to achieve a good degree of transferability to nonlipid biological systems, for which it was not originally designed. Good concordance with experimental data was obtained in the values of the partition free energy with various solvent systems as presented in Table 3 reproduced from reference.[63] In most cases, agreement between the MARTINI model (computed from the density distribution for the MARTINI model) was obtained within 1 $k_B T$ of experimental partition free energy values, and in some cases within 2 $k_B T$ (Boltzmann constant: $k_B = 1.381 \cdot 10^{-23}$ m^2 kg s^{-2} K^{-1}; Temperature: $T = 293$ K). A consequence of modeling nonbonded interaction with the LJ 12/6 potential is an overestimation of vaporization and hydration free energies, which was also observed for the MARTINI FF. This was shown to result in the underestimation of the water/vapor tension, especially in tested systems containing 1600 CG beads. Nevertheless, this is not of great concern as long as it is employed in studying condensed phases, where events are mostly governed by the partition free energies, which were correctly reproduced by the MARTINI FF. And of course, properties dependent on the degrees of freedom approximated out by the MARTINI FF cannot be directly investigated, expect by coupling to a multiscale modeling procedure. This is exemplified by the relative insensitivity at reproducing thermodynamic data for 1-propanol and 2-propranol[63] in Table 3.

The MARTINI FF described here refers to version 2.2. Since 2018, the beta version of MARTINI 3 is available, which contains topologies for over 120 lipids, includes new beads types, and revised parameters for some of the existing bead types. However, the MARTINI 2.2 lipid parameters are mostly unaffected by the transformation to version 3; rather the parameterization of the protein beads was revised as the protein–protein interactions are generally too attractive in MARTINI 2.2.[182]

2.2.2 Other lipid CG models

Some of the other popular CG-FFs available for simulation of membrane lipids include the SDK (named after its developers Shinoda–DeVane–Klein) FF,[93] later extended as SPICA FF,[183] the ELBA (stands for electrostatic-based) FF,[184] the SIRAH FF,[185] and the UNRES FF[186] (see also chapter "Scale-consistent approach to the derivation of coarse-grained force fields for simulating structure, dynamics, and thermodynamics of biopolymers" by Liwo of this book). The SDK CG lipid FF was originally developed for the study of surfactants and zwitterionic lipids.[93] Developed by Shinoda, DeVane, and Klein in 2010, it represents an extension of an earlier CG-FF by the same authors dedicated to the simulation of polyethylene

Table 3 Predicted versus experimental thermodynamics data (in bracket).

Bead type	Sample fragment	ΔG_{vap}	ΔG_{hyd}	$\Delta G_{H/W}$	$\Delta G_{C/W}$	$\Delta G_{E/W}$	$\Delta G_{O/W}$
C_1	Butane	−10 (−11b)	14 (9)	18 (18)	18 (NA)	14 (NA)	17 (16)
C_2	Propane	−10 (gas)	10 (8)	16 (NA)	15 (NA)	14 (NA)	16 (14)
C_3	1-Chloropropane	−10 (16)	5 (−1)	13 (12)	13	13	14 (12)
C_4	2-Butyne	−10 (−15)	5 (−1)	9 (NA)	13	13	9 (9)
C_5	Methyl ethyl sulfide	−10 (−17)	1 (−6)	5 (7)	10 (NA)	10 (NA)	6 (9)
N_0	Methoxyethane	−10 (−13)	−2 (−8)	−2 (1)	6 (NA)	6 (3)	5 (3)
N_a	2-Propanone	−13 (−17)	−9 (−16)	−7 (−6)	0 (1)	2 (−1)	3 (−1)
N_d	2-Propylamine	−13 (−17)	−9 (−18)	−7 (−6)	0 (1)	2 (−3)	3 (3)
N_{da}	1-Butanol	−16 (−25)	−9 (−20)	−7 (−5)	0 (2)	2 (4)	3 (4)
P_1	1-Propanol	−16 (−23)	−14 (−21)	−11 (−9)	−2 (2)	1 (0)	−1 (1)
P_1	2-Propanol	−16 (−22)	−14 (−20)	−11 (−9)	−2 (2)	1 (−1)	−1 (0)
P_2	Ethanol	−16 (−22)	−14 (−21)	−17 (−13)	−2 (−5)	1 (−3)	−2 (−2)
P_3	Acetic acid	−18 (−31)	−18 (−29)	−21 (−19)	−10 (−9)	−6 (−2)	−7 (−1)
P_4	Water	−18 (−27)	−18 (−27)	−23 (−25)	14 (NA)	−7 (−10)	−9 (−8)
P_5	Acetamide	sol (sol)	−25 (−40)	−28 (−27)	−18 (−20)	−13 (−15)	−10 (8)
$Q_0/Q_a/Q_d/Q_{da}$	Choline	–	−25 (NA)	−30 (<30)	−18 (NA)	13 (NA)	−18 (NA)

Data are shown for free energies of vaporization (ΔG_{vap}), hydration (ΔG_{hyd} (exp.)), and partitioning in hexadecane/water ($\Delta G_{H/W}$), chloroform/water ($\Delta G_{C/W}$), ether/water ($\Delta G_{E/W}$), and octanol/water ($\Delta G_{O/W}$) binary mixtures for different studied building blocks modeling different bead types. Extended data available in Ref. 63 from which the data were sourced.

glycol surfactants[92] and has been shown to reliably reproduce water/lipid interfacial tension. The ELBA FF was developed by Orsi and Essex and it employs two interdependent properties, explicitly modeled electrostatics and water molecules with single-point dipoles.[184] The SIRAH FF is another commonly employed CG-FF with parameters for a range of different molecules, including lipids, proteins and nucleic acids. The SIRAH FF is based on a top-down philosophy that aims to correctly reproduce lipid thickness and area per lipid among other membrane properties.[185]

2.3 Which lipid FF to choose for a simulation?

Choosing the right FF for the protein–membrane system to some extent depends on the availability of the right parameters for the desired lipids. The C36-AA, UA, GROMOS and MARTINI FF contain the largest variety of lipids as can be seen in Table 4.[97] Here, GROMOS currently has no FF parameters for chains with cyclic moieties, which are important in certain bacterial membranes, but in general GROMOS is able to simulate the same diversity of lipids as compared to C36-AA. Only parameters for ether lipids and lipid chains of bacteria are missing in the MARTINI FF, whereas Slipids and Lipid14 lack a variety of lipids. With the head group as AA, the C36-UA-FF would be similar to the C36-AA, however, it exhibits inaccuracies and can currently not be recommended for lipids other than PC and PE.[97] In general, the SDK FF could be an alternative to the MARTINI CG-FF, however, it has currently only a limited number of lipid and surfactant molecules implemented.

Table 4 Lipid parameters availability in the different FFs.

Lipids	C36-AA	Slipids	lipid14	GROMOS	OPLS-AA	C36-UA	MARTINI
PC/PE	✓	✓	✓	✓	✓	✓	✓
PG/PS/PA	✓	PG/PS	x	✓	PS	x	✓
PI/PIs	✓	x	x	✓	PI	x	✓
SM/Cer	✓	SM	x	✓	SM	x	✓
Sterols	✓	CHOL	CHOL	✓	CHOL	x	✓
Others[a]	✓	PUFAs	x	✓	ether, oxidized, GM1	x	✓

[a]Here, the less common lipids are summarized: CL, ether lipids, glycolipids, LPS, fatty acid, surfactants and PUFAs.

A tick means that all lipids of that group are available, while a cross indicates that none of the lipids in question are parameterized in the corresponding FF. Sterols include cholesterol, ergosterol and plant sterols if it is not marked otherwise.

3. MD simulation setup and analysis of protein–membrane systems

Biological membranes are complex and consist of different species of lipids, proteins and sugars. Recent developments in the field have enabled researchers to gain insights into the structure and function of biological membrane systems at atomistic resolution. In the following sections, we first present programs and tools that are available to build protein–membrane systems and prepare the input for corresponding MD simulations. Finally, we briefly discuss different tools for analyzing membrane properties from protein–membrane MD simulations.

3.1 CHARMM-GUI

To build initial protein–membrane systems for MD simulations in atomistic and CG resolution, the web-based and user-friendly CHARMM-GUI server,[126] developed by Im's group (http://www.charmm-gui.org/), has become widely popular over the last 10 years since the project started in 2006.[187,188] Currently, there are 434 different lipids available for the selection as a membrane component: GPLs, sterols, CLs, PUFAs, SM, detergents, ether-linked GPLs, fatty acids, LPS and glycolipids. There is the possibility to prepare both single- and multicomponent membranes, including the option of varying lipid concentration between the two leaflets. Moreover, the files that are generated by CHARMM-GUI are compatible with several MD engine programs,[164,188,189] including GROMACS,[190,191] AMBER,[192] NAMD,[193] GENESIS,[194] OpenMM,[195] CHARMM/OpenMM,[196] LAMMPS,[197] and Desmond.[198] Several *builder* and *maker* programs to create protein–membrane, protein–ligand, and protein–protein systems were developed within CHARMM-GUI. Here, we concentrate on setting up protein–membrane MD simulations via the *Membrane Builder* and *MARTINI Maker*. *Membrane Builder* offers different options to generate bilayers, monolayers, nanodisc, hexagonal phases and micelles with and without protein, as discussed in more detail below. At this stage, the force field parameters of CHARMM36 or CHARMM36m have to selected, which can be converted for use with several MD engines, including the change to another FF when running the MD simulations as explained in Section 3.2. To get further information about CHARMM-GUI, visit the http://www.charmm-gui.org/ website or consider this review.[188] All the lipid and protein files for the different CHARMM FFs can be downloaded from the MacKerell lab homepage (http://mackerell.umaryland.edu/charmm_ff.shtml#gromacs).

3.1.1 PDB loader and manipulator

Membrane Builder provides a PDB loader, which at the starting step imports a protein structure from either the OPM[44] database or the RCBS PDB.[42,43] In the next step, the user can decide to use the whole or only part(s) of the loaded protein, for instance, only a certain chain or certain residues of each available chain. Furthermore, the protein structure can be modified by the PDB manipulator to introduce cap groups, mutations, disulfide bridges, unnatural amino acids, or posttranslational modifications, including glycosylation, lipidation for a lipid anchor, phosphorylation, or attachment of a GPI-anchor. Additionally, the protonation state of each residue can be defined or missing residues can be added. Afterwards, the orientation of the protein in or on the membrane can be defined by the user if one does not want to use the preoriented protein coordinates provided by OPM. Several options, like protein flipping, translation along the z-axis, or rotation in the xy-plane are available. The Position of Proteins in Membrane (PPM) server may be of help here, as it calculates rotational and translational positions of transmembrane and peripheral proteins in membranes.[199] It is thus possible to place the protein in or on the membrane, or with a defined distance away from the membrane. At each step, the user can check the conformation of the protein and its position relative to the membrane. The next step is then to build the envisaged membrane type with the desired lipid composition.

3.1.2 Mono- and bilayer builder

Mono- and bilayer membranes can be built with CHARMM-GUI, where one can choose from 434 lipids that are currently available for creating realistic membranes. Next, the user can decide on the size of the membrane and the ratio of the chosen lipids in each leaflet. The desired protein can then be inserted into the membrane system, which will be done by a replacement or an insertion algorithm. In the last step, the user can choose the FF (CHARMM36m, CHARMM36, and most recently also AMBER lipid17), the MD software engine for which the input files shall be generated, and the MD parameters like temperature and ensemble.

3.1.3 Nanodisc builder

In many biophysical and biochemical studies, nanodiscs are used for studying membrane proteins in solution and are often better biological membrane mimics than micelles, liposomes, or bicelles. Nanodisc are stabilized by surrounding membrane scaffolding proteins (MSPs) to build a discoidal lipid bilayer. Most of the MSPs are derived from the apolipoprotein A-1 (apo A-1) with the long helical structure as it generates the discoidal shape of high-density lipoprotein (HDL), which is involved in cholesterol transport from the artery walls to the liver for degradation. In CHARMM-GUI, the *Nanodisc Builder* allows to create nanodisc systems involving proteins, which can also be inserted into the nanodisc. Depending on the size of the nanodisc, users can choose from twelve different MSPs, which are able to modulate the size of the nanodisc and includes a double-belt model for placing the MSP molecule around the nanodisc. For more details about MD simulation of nanodiscs the reader is referred to other references.[200,201]

3.1.4 Micelle and hex phase builder

In CHARMM-GUI, it is also possible to build micelles with the option to choose from more than 20 detergents and to insert proteins into the micelles.[202] The effects of detergents on protein structure are poorly understand. MD simulations of protein–micelle systems can help to elucidate the behavior of membrane proteins in certain detergents.[202] Similarly, the *Hex Phase Builder* can be used to assemble lipids in the hexagonal phase state, in which the polar head groups are located inside and form a water environment in the inverted micelle (Fig. 3C). This state can occur at high temperatures or low water concentration and has been applied in biophysical experiments and MD simulations to obtain information on lipids at high curvature.[203–205] Protein and/or interstitial alkane can be added to the system and all lipids are available for building this phase.

3.1.5 HMMM builder

CHARMM-GUI provides the possibility to build membrane systems with the HMMM model,[64,206] in which the lipid tails are replaced by small organic molecules, e.g., 1,1-dichloroethane (DCLE). This results in acceleration of lipid diffusion inside the membrane by 1–2 orders of magnitude. All available lipids are supported and it can also be chosen at which carbon position the acyl chains will be cut. There is also the possibility to convert the HMMM model back to the full-length lipid system.

3.1.6 MARTINI builder

Different CG models exist that allow to simulate large protein–membrane systems. One of them is MARTINI,[207] which is also available in CHARMM-GUI. *MARTINI Builder* enables the user to build solution, micelle, nanodisc, bilayer, and vesicle systems and also systems with randomly distributed detergent or lipid molecules for subsequent MD simulations with MARTINI. Furthermore, LPS are available and have been successfully applied with *MARTINI Builder* to model realistic bacterial outer membranes.[208] In CHARMM-GUI, the standard MARTINI version,[62,63,177] MARTINI with polarizable water,[209,210] Dry Martini,[211] and ElNeDyn[212] that applies an elastic network to the proteins are supported. However, the generated MD input files are only compatible with the latest GROMACS software.[190] The conversion from the CG MARTINI presentation to the AA level is available in CHARMM-GUI via the *backward.py* script.[213]

3.2 Setting up protein–membrane system with other programs

The CHARMM-GUI server facilitates setting up different kind of systems, but there are certain limitations. The choice of FF is mainly limited to the CHARMM FF. Chemical modifications of proteins or lipids are restricted to the options available in the server and it is not possible to insert several proteins or drug molecules into a membrane at the same time at varying depth. Recently, the server has been updated to provide AMBER FF files for simulations. To overcome the limitations with the CHARMM-GUI sever, we discuss here the set up of protein–membrane systems for other commonly used force fields.

3.2.1 Simulations with AMBER FF

Here, users can make use of CHARMM-GUI to build the initial protein–membrane system. For lipids, simply using Slipids FF files[79,80,142] is sufficient since it follows the same atom naming as CHARMM C36. For proteins, processing the protein PDB file (*pdb2gmx* tool in GROMACS) with the desired AMBER protein FF should work. CHARMM-GUI and PACKMOL-Memgen[214] can be used to generate PDB files compatible with AMBER Lipid17. Many structures, topology and preequilibrated mono- and multicomponent membrane patches for Slipids can be downloaded at: http://www.fos.su.se/~sasha/SLipids/Downloads.html.

3.2.2 Simulations with OPLS-AA FF

There is no freely available tool for setting up protein–membrane systems with OPLS-AA. Also, the atom naming does not follow that of CHARMM C36 lipids or AMBER Slipids, suggesting that converting a CHARMM-GUI output to OPLS-AA is not possible. Instead, the setup of a protein–membrane system can be done in two steps. First, tools such as

MemGen[215] and PACKMOL[216] can be used to build the pure lipid membrane from a PDB file of a single lipid. Second, the protein can be positioned above the membrane surface using VMD[217] or using the GROMACS tool *editconf*. For TM proteins, the protein may be embedded inside the membrane using VMD, *mdrun_hole*,[218] *InflateGRO*,[219] *InflateGRO2*,[220] or *g_membed*.[221] However, these tools remove lipids overlapping with the protein, which is problematic when preparing multicomponent systems with varying concentrations. Addressing the problem, Javanainen et al.[222] developed a universal protocol to insert proteins of various topology inside the membrane without having to remove the lipids. The method is independent of any FF and works by applying high lateral pressure. Thus, for simulations with OPLS-AA a combination of either *MemGen*[215] and PACKMOL[221] for building the initial membrane patch along with the method developed by Javanainen et al.[222] is probably the most efficient way for setting up a protein–membrane system. Topologies, structures, and parameter files for OPLS-AA can be obtained from Kulig et al.[150,151]

3.2.3 Simulations with GROMOS FF

The GROMOS united atom membrane patch can be prepared using *MemGen*,[215] PACKMOL,[221] *MemBuilder*[223] tools or by downloading preequilibrated bilayer patches. The *MemBuilder*[223] tool already includes structure and topology files for few selected lipids. With the method by Javanainen et al.[222], the protein can be placed into the membrane. If the protein has to be positioned above the membrane surface, either VMD[217] or the *editconf* tool of GROMACS can be used. Structures, topologies, and membrane patches can be downloaded from Peter Tieleman's webpage (http://wcm.ucalgary.ca/tieleman/downloads), Jochen Hub's webpage (https://biophys.uni-saarland.de/downloads.html), Mikko Karttunen's webpage (https://www.softsimu.net/downloads.shtml), Andreas Kukol webpage (https://sites.google.com/site/bioherts/home/lipid-topologies), Roland Feller's webpage (https://research.engineering.ucdavis.edu/multiscale/downloads/), and those uploaded on the GROMACS website (http://www.gromacs.org/Downloads/User_contributions/Molecule_topologies). Cholesterol-containing membrane structures are available at Jochen Hub's website at https://biophys.uni-saarland.de/downloads.html.

3.2.4 Simulations with MARTINI FF

The developers of MARTINI (http://cgmartini.nl/) also provide scripts to build protein–membrane systems, along with tutorials for the different applications to set up the desired systems and how to map new molecules from atomistic to CG resolution. Here, we briefly describe the possibilities of building protein–membrane systems with certain scripts. First, the Python script *martinize.py* automatically converts a PDB file of proteins, DNA or RNA from the AA to the CG model, where the user can decide to use standard MARTINI model for proteins, the elastic network model, or the ElNeDyn method. The elastic network ElNeDyn[212] maintains the overall conformation of a protein and limits the capability of changing the tertiary structure (Fig. 7D). Moreover, there is the possibility to change the protonation state of the histidines and to define disulfide bridges. Second, the python script *insane.py*[224] enables the user to setup homogeneous and heterogeneous membranes with the possibility to place and rotate the protein molecules above or on the membrane or move them inside.

3.2.5 Glycosylation

Proteins undergo several posttranslational modifications, such as covalent linkage of N- and O-bound glycans to proteins.[225–227] Similarly, many cell membranes contain glycolipids, which are decorated with sugar head groups, e.g., gangliosides. The software *doGlycans*[228] can be used to glycosylate proteins, or set up glycolipids and cellulose structures. The tool is user friendly and provides files compatible for use with OPLS-AA, AMBER, and CHARMM. CHARMM-GUI also enables users to add glycans to the proteins during system setup.[229]

3.3 MD simulation software packages

Numerous MD software packages have been developed by different research groups, whereby the most popular MD programs are summarized in Table 5. They differ in the availability of implemented algorithms, e.g., for the integration of Newton's equations of motions (see Eq. 2), the available thermostats and barostats, in their computational performance, in their support of various FFs, and in their availability of analysis tools. In GROMACS and NAMD, one can choose from a wide range of standard and custom FFs (e.g., all kinds of CHARMM, OPLS-AA, and AMBER FFs), whereas the MD engines CHARMM and AMBER only support their own FFs. Furthermore, some MD software packages are not freely available, such as AMBER for performing MD simulations but the corresponding analyzing package AmberTools[230] is freely available for academic purpose. Finally, the user has to consider which analysis tools are provided by the MD programs. For example, GROMACS has the big advantage that it comes

Table 5 Common MD software packages.

Software	Interface	License	Website
AMBER[192]	CLI	Proprietary	https://ambermd.org/
CHARMM[189]	CLI and GUI	Proprietary	https://www.charmm.org
Desmond[198]	GUI	Academic	https://www.deshawresearch.com/resources_desmond.html
GROMACS[190]	CLI	Open-source	http://www.gromacs.org/
NAMD[193]	GUI	Academic	https://www.ks.uiuc.edu/Research/namd/
LAMMPS[197]	CLI	Open-source	https://lammps.sandia.gov/

CLI means that the program is based on the command line, while the programs with GUI provides a graphical interface for the users.

with huge and diverse set of analysis tools, covering both analyses of protein and membrane properties, which can be easily modified and redistributed if a given tool should not exactly exhibit the required functionality. NAMD is implemented in VMD,[217] which provides powerful visualization capabilities, a graphical user interface (GUI), dynamic atom selections, and allows the reading of multiple file formats, including those obtained by all common MD packages.

3.4 Analysis tools for studying protein–membrane systems

3.4.1 Visualization and plotting tools

The most important output file obtained from an MD simulation is the trajectory file (GROMACS: *.xtc*/*.trr*, CHARMM/NAMD: *.dcd*, AMBER: *.netcdf*) which contains the coordinates of the atoms as a function of time. Trajectory files can be converted between formats using the *MDTraj* tool.[231] Programs such as VMD,[217] PyMOL,[232] and Chimera[233] visualize trajectories and provide options for measuring important biomolecular properties. In addition, numerical output files from various analyses can be plotted, for example, using Xmgrace, MATLAB,[234] and GnuPlot.[235]

3.4.2 Analysis tools

The MD simulations packages are already preloaded with several analysis tools, which can be used to analyze the structure and dynamics of the system. Here, we list several analysis software packages, which are widely used to characterize protein–membrane systems. The LOOS tool[236] supports dynamic atom selection and is compatible with many MD simulation software packages. The *g_lomepro* tool[237] measures local changes of membrane properties due to protein interaction and it is compatible with GROMACS files. *GridMAT-MD*[238] analyzes protein–membrane properties on a 2D grid and reads GROMACS format files. The *Membrainy* tool[239] easily determines several membrane properties, such as the fraction of the gel phase or the selection of annular lipids. *Membrainy* is compatible with GROMACS files and reads CHARMM36, Berger/GROMOS87, and MARTINI v2.0 FF files. The MEMBPLUGIN tool[240] works in combination with VMD. This tool measures membrane properties and accepts input files from all FFs and MD packages. The APL@Voro tool[101] calculates membrane properties based on Voronoi tessellation and is compatible with GROMACS format file. FATSLiM[241] computes properties of both planar and curved membranes and is compatible with GROMACS files. Several tools for analyzing membrane trajectories in GROMACS format are available at Luca Monticelli's website[242] (https://perso.ibcp.fr/luca.monticelli/tools/index.html) to analyze membrane trajectories. In addition, *MDAnalysis*[243] and *MDTraj*[231] libraries also include features for analyzing protein–membrane systems.

4. Case studies for protein–membrane systems

In this section, we will give an overview about different case studies with AA-FFs and three studies for CG-FFs, which are summarized in Table 6.

4.1 Atomistic simulations of integral membrane proteins

4.1.1 G-protein-coupled receptors

G-protein-coupled receptors (GPCRs) form the largest family of cell surface membrane proteins and all share a multipass seven-transmembrane helix (TMH) topology.[244] The activation of GPCRs occurs by extracellular binding of a ligand, which triggers a conformational change in the receptor structure, leading to dissociation of a receptor-bound heterotrimeric G protein subunit (Gα and G$\beta\gamma$). The dissociated subunits interact with intracellular substrates to resume the signal transduction cascade. After dissociation of the G protein, the receptor is available to bind to another heterotrimeric G protein, thereby continuing the signal transduction.[244] Due to their role in several physiological processes, these receptors are major targets for ~35% of the currently approved drugs refer to those available in the market.[245] Thus, understanding how GPCRs function at atomistic level is essential for therapeutic intervention.

Table 6 Summary of the different case studies.

Proteins	FF	Lipids	Water model	MD package	Reference
Transmembrane proteins and receptors					
β-2AR	OPLS-AA	DOPC, CHOL, CHS, Oxysterol	TIP3P	GROMACS 4.6	[253]
$A_{2A}R$	C36, C36c, C27	DPPC, DSPC, DOPC, SDPC, SM CHOL	TIP3P	ACEMD,[a] GROMACS 5	[266]
Opsin	C36	POPC, POPG	TIP3P	NAMD 2.1, ACEMD[a]	[263]
XylE	C36	DOPC, DOPE, DOPG, CL	TIP3P	NAMD 2.12	[277]
hDAT	C27, C36	POPC, POPE, POPS, CHOL, PIP_2	TIP3P	NAMD 2.9, ACEMD[a]	[280]
Kir2.1	C27	POPC	TIP3P	NAMD2	[58]
TLR4	OPLS-AA	POPC, POPE, POPS, SM, GluCer, LPS	TIP3P	GROMACS 4.6.7	[226]
CcO	C36	POPC, POPE, CL	TIP3P	GROMACS 5	[294]
10 different membrane proteins	MARTINI	POPC, POPE, SM, LPC, GM, CER, DAG, PI, PIP, PA, CHOL	MARTINI water	GROMACS 4.6.3	[431]
Amyloids					
$A\beta_{42}$	GROMOS96, G53A6, Kukol lipids	POPC, DPPC, POPG	SPC	GROMACS 4	[319, 320]
$A\beta_{40}$	GROMOS96, G53A6, Tieleman lipids	POPC, POPE, POPS SM, CHOL, GM1	SPC	GROMACS 4	[321]
$A\beta_{42}$	OPLS-AA	POPC, SM, CHOL, GM1	TIP3P	GROMACS 4.6	[322]
$A\beta_{1-28}$, $A\beta_{26-40}$	GAFF, Slipids	DMPC, SDPC, DSPE, DDPE, DPPC, DOPC	TIP3P	GROMACS 4.6.7	[323]
$A\beta_{29-42}$	AMBER99SB*-ILDN, Slipids	POPC, $SAPC_{\omega 6}$, $SDPC_{\omega 3}$	TIP3P	GROMACS 5.1.2	[324]

hIAPP	GROMOS87, Berger lipids	POPC, POPG	SPC	GROMACS 3.3.3	[330–332]
hIAPP	Amber03w, Slipids	DOPC, DOPS	TIP4P/2005	GROMACS 4.6.7	[333]
hIAPP	GROMOS87, Tieleman lipids	POPG	SPC	GROMACS 3.3.3	[334]
hIAPP	C36, C22*	DOPC, DOPS, DVPS, DVPC	cTIP3P[b]	GROMACS 5.0.2	[172]
hIAPP	C27	DOPC	TIP3P	NAMD	[335, 336]
hIAPP	GROMOS96, G53A6	DPPG	SPC	GROMACS 4.0	[338]
aSyn	C22, C36	POPC, POPA	TIP3P	CHARMM	[349]
aSyn	C36	DOPC SAPC, DPhPC	TIP3P	GROMACS 4	[351]
aSyn	C27, C36	DVPC, DVPS	TIP3P	NAMD 2.8	[171]
aSyn	AMBER99SB-ILDN Slipids	POPC, POPG, DMPC, DMPG	TIP3P	GROMACS 4.6	[352]
Peripheral membrane proteins					
FGF2	C36	POPC, CHOL, PI(4,5)P_2	TIP3P	GROMACS 5.1	[365]
GRP1-PHD	C36	POPC, POPS, PIP_3	TIP3P	NAMD2	[379]
ABPs	C36	POPC, POPE, POPS, PI(4,5)P_2	cTIP3P[b]	GROMACS 5.0.4	[380]
RVFV	AMBER99SB*-ILDN Slipids	DOPC, CHOL	TIP3P	GROMACS 4.6	[408]
Integral membrane proteins					
FAM134B-RHD	C36m, MARTINI	POPC	TIP3P, MARTINI water	GROMACS 4.6,5	[417]
Lipid droplets					
Seipin	SDK	POPC, DOPC, CHOL DPPC, DOPE, DAG	SDK water	LAMMPS	[419]

[a]ACEMD is a high performance MD code specifically designed for NVIDIA GPUs.
[b]cTIP3P refers to the CHARMM-modified TIP3P water model.

Recent technical advances have enabled structural biologists to solve structures of several GPCRs, which in turn has provided a wealth of information on ligand binding pockets, on key residues stabilizing the ligand–receptor interactions, and on functionally important conformational changes of GPRCs, such as the rotation and movement of helices during activation. However, GPCRs are embedded in lipid environment and our understanding on how phospholipids and sterols, which form the major components of cell membrane, modulate the receptor function is still limited. Therefore, insights into the lipid species interacting with the GPCRs is relevant to further understand the GPCRs activation mechanisms. Among the membrane components, the role of cholesterol in GPCR function has been intensively investigated.[246] In vivo studies have indicated cholesterol to be an important modulator of GPCR activity and several functions are attributed to cholesterol in regulating GPCR functions, such as ordering of lipid membrane, stability and oligomerization of receptor and in binding of ligands.[247–250] In addition, cholesterol is also required for crystallization of GPCRs.[37,251] Furthermore, GPCR structures have been resolved with cholesterol and its analogues bound to the receptor, indicating direct interactions with the receptor.

Rhodopsin is one of the best-studied GPCRs. Its interaction with lipids containing PE head group, polyunsaturated acyl chain (e.g, docosahexaenoic acid, DHA, 22:6), and with cholesterol is known to trigger visual processes.[252] To complement experimental findings and to gain further insights into the interactions between GPCRs and lipids, MD simulations of various GPCRs embedded in lipid membrane have been carried out, including β_1 and β_2-adrenergic receptors,[253,254] serotonin 1A and 2A receptors,[255,256] cannabinoid 2 receptor,[257] rhodopsin,[258] A_{2A} adenosine receptor,[259,260] δ opioid receptor,[261] sphingosine 1-phosphate receptor,[262] opsin receptor,[263] and dopamine D2 receptor.[264] Here, we will discuss key findings obtained from atomistic MD simulations for few selected GPCRs.

4.1.1.1 β2-adrenergic receptor (β_2AR)

The human β_2AR is primarily expressed in pulmonary, skeletal, and cardiac muscles. It is a major drug target for treatment of asthma, respiratory diseases and heart failure. Recent structural and biochemical studies suggest that the receptor exists in multiple conformations and that binding of cholesterol affects the conformation and signaling. Cholesterol could induce its effect either by altering the membrane properties or by direct specific interactions with the receptor. Although structures solved with cholesterol bound to

β_2AR indicate the availability of specific cholesterol binding sites on the receptor, the mechanism explaining the modulatory role of cholesterol is still unknown. To clarify the effect of cholesterol on β_2AR, Manna et al.,[253] performed extensive atomistic MD simulations (>100 μs) of β_2AR embedded in 1,2-dioleoyl-*sn*-glycero-3-phosphocholine (DOPC)/CHOL membrane, with cholesterol concentration ranging from 0 to 40 mol%. In addition, simulations were also carried out using the cholesterol analog cholesteryl hemisuccinate, since it is commonly used in experiments to mimic cholesterol. This study reported two binding sites on the intracellular side (IC1 and IC2) and one binding site on the extracellular side (EC1) for cholesterol (Fig. 8A and B). Binding at IC1/EC1 sites are supported by with the crystal structures,[37,251,265] however, the relevance of the additional binding at site IC2 has to be experimentally validated. Cholesteryl hemisuccinate also binds to β_2AR at the same sites as cholesterol, but the binding is weaker in comparison to cholesterol. The study also shows that in the absence of cholesterol, β_2AR adopts multiple conformations, whereas with cholesterol concentration >10 mol% the structural flexibility of β_2AR is reduced to one single conformation. These studies conclude that the modulatory effect of cholesterol functions via specific interactions with the receptor and not via modulating the mechanical membrane properties.

4.1.1.2 Adenosine A2A receptor (A_{2A}R)

The receptor is expressed across several tissues. The highest expression levels are found in brain and intermediate levels in heart, where it regulates blood flow and oxygen consumption. A_{2A}R is a target for drugs in the treatment of Alzheimer's and Parkinson's disease. Similar to β_2AR, cholesterol binding sites have been reported from both experimental (e.g., PDB code: 3EML[267])[265] and MD studies.[259,260] Also, a stabilizing effect of cholesterol on the receptor has been shown.[260] However, the effect of cholesterol on ligand binding has not been explored. To clarify this, Guixà-González et al.[266] combined both experiments and MD simulations to understand how cholesterol depletion influences ligand binding and dynamics of A_{2A}R. The MD simulations revealed that on average 12 cholesterol molecules are bound to the receptor. The cholesterol molecules were found preferentially bound to TMH2-3, 3-4-5 and 7-1 (Fig. 8C). Of the three regions, cholesterol binding to TMH2-3 is supported by with the experimental structure solved in complex with the ligand.[265] The bound cholesterol molecules are required to maintain the stability of the receptor as observed in experiments. Apart from confirming the experimentally known

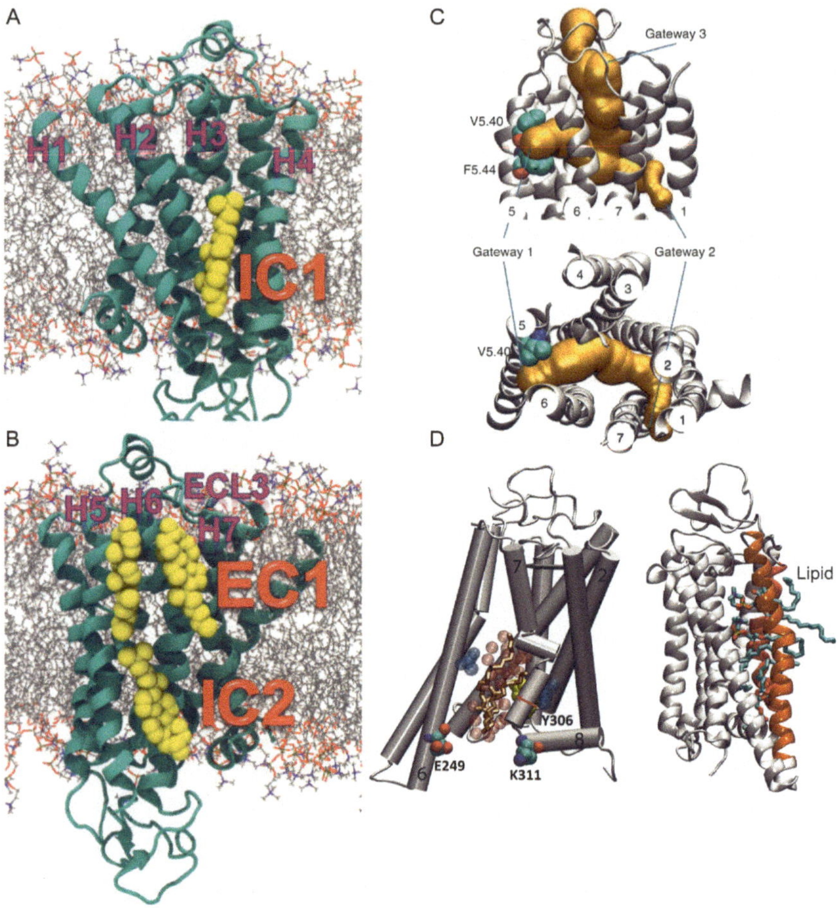

Fig. 8 GPCRs in action. (A) and (B) Cholesterol interactions with β_2AR. β_2AR embedded in a lipid membrane (*line representation*), highlighting two intracellular cholesterol binding sites IC1 (H1–H4) and IC2 (H5–H6) as well as one extracellular binding site EC1 (H5–H6 and H6–ECL3–H7). Cholesterol is rendered as *yellow spheres*. (C) Cholesterol gateways (*yellow surface*) in A_{2A}R. *Top* and *bottom* figures represent views from intracellular and extracellular sides, respectively. TMHs are marked from 1 to 7. (D) Lipid translocation triggered by opening on the intracellular side of the Opsin receptor. In the *left panel*, final snapshot showing disruption of E249–K311 interactions and conformational switch of Y306 inducing water pore (*transparent pink spheres*) and lipid penetration (rendered as *licorice*). In the *right panel*, snapshot showing translocation of lipids as a credit-card mechanism. TMHs involved in conformational rearrangement are labeled (H2/6/7/8). *Panels A and B: Reprinted with permission from Manna M, Niemelä M, Tynkkynen J, et al. Mechanism of allosteric regulation of* β_2*-adrenergic receptor by cholesterol. eLife. 2016;5:e18432, 10.7554/eLife.18432, eLife Sciences Publications. Copyright © 2016 (https://creativecommons.org/licenses/by/4.0/); Panel C: Reprinted with permission from Guixà-González R, Albasanz JL, Rodriguez-Espigares I, et al. Membrane cholesterol access into a G-protein-coupled receptor. Nat Commun. 2017;8:14505, Springer Nature. Copyright © 2017 Springer Nature (https://creativecommons.org/licenses/by/4.0/); Panel D: Reprinted with permission from Morra G, Razavi AM, Pandey K, Weinstein H, Menon AK, Khelashvili G. Mechanisms of lipid scrambling by the G protein-coupled receptor opsin. Structure. 2018;26(2):356–367, Elsevier. Copyright © 2017 Elsevier Ltd (https://creativecommons.org/licenses/by/4.0/).*

cholesterol binding sites, few additional sites were observed on the surface of the receptor, which could affect ligand binding. The studies also revealed a mechanism in which the cholesterol enters the receptor interior, thus competing with the orthosteric ligand binding site. The path taken by cholesterol to enter the $A_{2A}R$ interior is identical to the retinal pathway in the opsin receptor. These studies indicate that ligand binding to $A_{2A}R$ receptor is regulated by both orthosteric- and allosteric binding of cholesterol to the receptor.

4.1.1.3 Opsin receptor

These receptors are found in the photoreceptor cells of the retina and are involved in vision. Opsins and several other class A GPCRs facilitate trans-bilayer translocation of phospholipids and scramble >10,000 lipids per second when embedded in vesicles.[268–270] When reconstituted in lipid vesicles, opsins are known to accelerate the lipid transfer to a rate of $>10^4$ (i.e., 10,000 lipids per second per protein), which is faster than opsin unassisted lipid transfer occurring at a rate of $\sim 10^{-5}\ s^{-1}$.[271] But opsin-mediated lipid transfer is still a rare event taking ~100 μs per lipid per protein. The lipid flipping or the scramblase activity of GPCRs is not only necessary to maintain the photoreceptor cell homeostasis but also finds importance in various other physiological processes, such as in cell growth, blood clotting and apoptotic cell clearance.[271–274] Recently, the molecular events facilitating the receptor-mediated lipid scrambling were characterized by preinserting opsin into membranes composed of POPC/POPG (ratio of 9:1) and performing extensive AA-MD simulations (>50 μs).[263] The study revealed that TMH6 and TMH7 mediate translocation of a lipid from the intracellular to the extracellular side through three steps (Fig. 8D). First, the intracellular disruption of TMH6-E249–TMH7-K311 interaction and conformational switch of residue TMH7-Y306. Second, the changes observed in the first step lead to the increase of intracellular distance between TMH6 and TMH7 by ~10 Å which allows influx of water molecules. The created hydrophilic environment favors the lipid to tilt and move toward the membrane center. Third, the distance between TMH6 and TMH7 in the membrane center is increased by ~5 Å, thereby creating a continuous water channel enabling the lipid to completely flip. The lipid translocation event observed here occurs at a timescale of ~33 μ, which is consistent with experimental lipid transfer, taking place at ~100 μs or smaller.[269,275]

4.1.2 Membrane transporters and channels

Cell membranes permit the passage of various molecules such as water, ions, amino acids, sugars, and nucleotides, and most often make use of specialized proteins embedded within membranes for permeation. The membrane transport proteins exhibit both multipass TM topology (e.g., glucose transporters, GPCRs, monoamine transporter) or β-barrel topology (e.g., porin family, amyloid channels), as can been seen in Fig. 4. Most transport proteins are selective, i.e., they are permeated only by a single or a few chemically similar solutes. Membrane transporters and channels are typically grouped into two major classes of membrane proteins: (1) transporters strongly bind to the permeating solute and undergo a structural change to pass the solute across the membrane, leading to a slow permeation rate; (2) in contrast, channel proteins bind to molecules weakly, do not require a conformational change, hence allowing high permeation rates.

Over the last few years, several X-ray structures of transporters and channels have been solved and the rapid growth in the computing power has enabled MD simulations of complex biomolecules at timescales relevant to observe physiological phenomena. Although obtaining the time scale of entire transport process is still a major challenge, MD simulations have been successfully employed to describe the transition steps in various transport cycles. Here, we discuss results of three well documented transporters, lactose permease (LacY), xylose transporter (XylE), and human dopamine transporter (hDAT), and of the channel protein Kir2.1, whose functions are regulated by lipids.[276]

4.1.2.1 XylE/LacY transporters

MD simulations along with hydrogen–deuterium exchange mass spectrometry (HDX-MS) have proven to be robust methods to characterize dynamics of transporters in different membrane systems, such as liposomes, micelles, bicelles, and nanodiscs. Recently, Martens et al.,[277] combined these methods to understand how direct interactions with lipids modulate the conformational dynamics of secondary transporters, such as LacY and XylE. The two transporters (XylE, PDB code: 4JA4[278] and LacY, PDB code: 2CFQ[279]) in their inward-open conformation were embedded into bilayers, composed of DOPE/DOPG/CL (70/20/10 mol%) and DOPC/DOPG/CL (70/20/10 mol%) lipids, and AA-MD simulations were carried out to identify how lipids regulate the conformational switch at the molecular level. The simulations reveal that PE lipids localize between the TMH2 and TMH11, where they interacts with R84 and with the switch residue E153 (Fig. 9A). The interactions stabilize the transporter in the inward-open

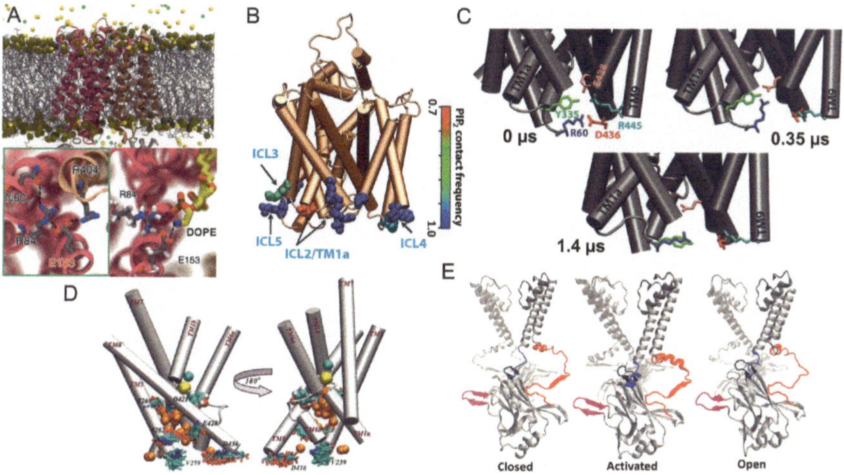

Fig. 9 Transporters and channel proteins. (A) DOPE–XylE interactions disrupt the charge-relay networks. In the *top panel*, snapshot showing XylE (PDB code: 4JA4[278]) in lipid membrane. Lipid acyl chains are shown as *lines*, phosphorus atoms rendered as *ochre spheres*, sodium and chloride ions as *yellow* and *green spheres*, respectively, and protein N- and C-domains colored as *pink* and *tan cartoon*, respectively. In the *bottom left panel*, close-up view of conserved network 1 connecting N- and C-domains stabilizes outward open state of XylE. In the *bottom right panel*, DOPE head group interacting with residues R84 and E153 disrupts network 1 connection and prevents association of TMH2 and 11. (B–D) hDAT undergoes PIP_2 mediated conformational shift to inward-open state. (B) Residues within 3 Å of PIP_2 lipid head group are rendered as *spheres*. TMHs, where PIP_2 aggregation is largest are labeled and shown by *black arrows*. (C) Snapshots showing initial formation of charge network (*top panel*) being disrupted by PIP_2 (*middle* and *bottom panel*) during the course of the MD simulation. (D) Snapshot depicting spontaneous inward release of sodium ion (*yellow sphere*) from its Na2 site and the release pathway is shown as *orange spheres*. The conformation of residues during the release process is displayed as *licorice*. The dopamine ligand and chloride ions are rendered as *VDW* and *cyan sphere*, respectively. The protein TMHs are shown as *white cartoon* and labeled, accordingly. (E) Regulation of Kir2.1 channel activity by PIP_2. PIP_2 mediated change in C-linker structure (*blue*) during conformational change from closed—activated—open states. Also, N-terminal and G-loop regions are displayed as *red* and *black*, respectively. *Panel A: reprinted with permission from Martens C, Shekhar M, Borysik AJ, et al. Direct protein-lipid interactions shape the conformational landscape of secondary transporters. Nat Commun. 2018;9(1):4151. Springer Nature. Copyright © 2018 Springer Nature. https://creativecommons.org/licenses/by/4.0/); Panel B–D: Reprinted with permission from Khelashvili G, Stanley N, Sahai MA, et al. Spontaneous inward opening of the dopamine transporter is triggered by PIP2-regulated dynamics of the N-Terminus. ACS Chem Neurosci. 2015;6(11):1825–1837, https://doi.org/10.1021/acschemneuro.5b00179, American Chemical Society. Copyright © 2015 American Chemical Society (https://pubs.acs.org/doi/10.1021/acschemneuro.5b00179). Further permissions related to the material excerpted should be directed to the ACS; Panel E: Reprinted with permission from Li J, Lü S, Liu Y, et al. Identification of the conformational transition pathway in PIP2 opening Kir channels. Sci Rep. 2015;1858(7, Pt. B):1610–1618, https://doi.org/10.1038/srep11289, Nature Publishing Group. Copyright © 2015, Macmillan Publishers Limited. (https://creativecommons.org/licenses/by/4.0/).*

state, preventing the conformational switch. In contrast, simulations in PC membrane shifts the conformation by closing the intracellular side. This conformation change is initiated by movement of TMH2/3 toward TMH10/11, which reduces the distance to ~9 Å in the PC bilayer, whereas in the PE bilayer it is at ~13 Å. Similar direct interactions between E139 and PE lipids were observed in LacY simulations. This study again highlights the role of lipids in regulating conformational changes.

4.1.2.2 Human dopamine transporter (hDAT)

Belonging to the neurotransmitter transport family, hDAT are responsible for clearance of dopamine neurotransmitter from the synaptic cleft into the cytosol, thereby terminating the dopamine signal. Increased activity or mutations in hDAT are associated with several disorders, such as depression, psychiatric and neurological disorders, and they also effect cognitive functions. To gain insights into the activation mechanism of hDAT, Khelashvili et al.,[280] performed extensive AA-MD simulations (>14 μs) on a full-length hDAT embedded in a lipid membrane composed of phosphatidylinositol 4,5-biphosphate (PIP_2) lipids. The study reveals the molecular details leading to the conformational shift from outward- to inward-open state (Fig. 9B and C). This shift also causes the release of a sodium ion from its Na2 site and the destabilization of dopamine from its S1 binding site (Fig. 9D). Closer examination of the membrane environment shows that PIP_2 mediates the association of the N-terminal segment and the intracellular loop 4 (ICL4) by forming electrostatic interactions with N-terminal residues K3, K5, K27, K35, R51 and with ICL4 residues H442, R443, H444. The N-terminal/ICL4 association stabilizes the inward-open state. To further validate the importance of PIP_2, additional simulations were carried out by replacing PIP_2 with POPE lipids. In absence of PIP_2 the stabilized N-terminal/ICL4 complex dissociates rapidly. The PIP_2-induced conformation shift and the transition mechanism reported for hDAT is similar to the activation mechanism of structurally similar Leucine transporter (LeuT).[281]

4.1.2.3 Inwardly rectifying potassium (Kir2.1) channel

Kir channels are found in several cell types, including neurons, kidney, cardiac, endothelial, blood cells, and they are involved in maintaining the membrane resting potential, cardiac excitability, and insulin release.[282,283] These channels allow more readily influx of potassium ions into the cells than outward-directed flux. The channel function is regulated by a conformational switch between the closed and open states. Moreover, several factors are involved in the process, including proteins and lipids. Among lipids, the negatively charged PIP_2 is required to activate all Kir channels. Apart from PIP_2, cholesterol is another important molecule involved in Kir regulation. Here, we discuss how PIP_2 and cholesterol regulates Kir2.1 activity. Li et al.[58] employed atomistic targeted MD (TMD) method to observe Kir2.1 channel opening from a closed state. By comparing X-ray structures of Kir channels, the authors recognized the conformational change of the C-linker (C-terminus region of the inner helix) loop to a helical conformation, and the structures further revealed PIP_2 binding sites on the C-linker. Guided by the hypothesis that PIP_2 could induce the conformational change of the C-linker, they carried out TMD on the C-linker residues (K185 to T192) with a goal to change the secondary structure from a loop to a helix. The change to the more rigid helical structure, which is stabilized by interactions with PIP_2, pulls the cytoplasmic domain (CD) ~5 Å toward the membrane, which enables contact with the TM domain. The first stage of TMD results in pushing the channel toward an active state. To achieve the fully active state, TMD was again applied to guide the C-linker–PIP_2 complex toward a partial helical conformation. The partially folded C-linker kinks, leading to a ~8 degrees rotation of the N- and C-termini, which shifts the CD downwards, thus, reaching a full open channel. The study reports all the keys necessary for the conformational shift, hence providing a detailed picture into PIP_2-induced opening of Kir channel.

Experimental studies have shown that cholesterol directly regulates the Kir2.1 channel activity by forming a network of specific interactions with the protein. However, the cholesterol binding sites were yet to be located. Avia et al.,[284] combined experimental as well as several computational methods such as docking, atomistic MD simulations and molecular mechanics Poisson–Boltzmann surface area (MM/PBSA) calculations to shed light on the cholesterol binding sites and their role in regulation of channel activity. Initial docking studies showed that cholesterol mainly binds at the TM domain interfaces at three typical binding poses. Two additional poses were located at the cytosolic domain. Subsequent MD simulations of Kir2.1 with bound cholesterol embedded in a lipid membrane revealed a rearrangement of cholesterol into two distinct binding sites. Site 1 corresponds to the center of the TM domain, whereas site 2 is near the cytosolic domain. To gain insights into the binding energies at these two sites, MM/PBSA approach was used, which indicated cholesterol to bind slightly stronger at the site 1.

The study concludes with the identification of two binding sites for cholesterol on Kir2.1 channel with large cholesterol flexibility within the two sites. Furthermore, mutation studies and functional analysis reveal that the identified binding sites may interfere with motion of the inner helix, which is required for keeping the channel in a closed state.

4.1.3 Glycoproteins

Glycosylation is one of the significant posttranslational modifications, during which glycans (oligosaccharides) are covalently added to an amino acid. Glycosylation occurs in all species. More than 50% of eukaryotic proteins undergo glycosylation,[285,286] with implications on various function and on protein activity.[287–291] Furthermore, all cells are covered by glycans necessary for cellular function.[286] Glycosylations are grouped into five classes[292]:

(1) *N-linked glycosylation*: Glycans are covalently linked to amide nitrogen of asparagine in the sequence N–X–S/T. Apart from proline or aspartic residues, X in the sequence can be any amino acid.

(2) *O-linked glycosylation*: Glycans are linked to hydroxyl group of serine or threonine residues, without a consensus sequence for O-glycan addition.

(3) *Phospho-serine glycosylation*: Glycans involves a linkage of sugars GlcNAc, xylose, mannose, and fructose to proteins by a phosphodiester bond.

(4) *C-mannosylation*: As an unusual type of glycosylation, sugar (mannose) is linked to a carbon atom rather than to amide or hydroxyl groups. Here, mannose is added to tryptophan in the sequence W–X–X–W where X can be any amino acid.

(5) *Glypiation*: Glycosylphosphatidylinositol (GPI) is formed, anchored to a lipid by glycan chain.[292]

Here, we focus on the membrane receptor toll-like receptor 4 (TLR4) which is glycosylated with a core N-linked glycan ($GlcNAc_2Man_3$). Mobarak et al.[226] carried out microsecond timescale simulations of LPS bound to glycosylated TLR4 complex, embedded in asymmetric membrane composed of CHOL/SM/GluCer/PC lipids (45/25/10/20 mol%) in the outer leaflet and CHOL/PS/PE/PC lipids (35/20/20/25 mol%) in the inner leaflet. To study the effect of Glucosylceramide (GluCer) on the receptor, additional simulations were performed without GluCer in the outer leaflet and the composition was adjusted to CHOL/SM/PC (45/35/20 mol%) in the outer leaflet and the inner leaflet lipids were unmodified. The simulations showed that LPS induced tilting of the TLR4 extracellular domain, however, tilting was reduced and TLR4 was stabilized in membrane containing GluCer. Quantification of tilt angles showed that the LPS/TLR4 complex tilted by ∼33 and 57 degrees in the presence and absence of GluCer, respectively (Fig. 10A). However, in the absence of LPS, TLR4 only tilted by 15 and 19 degrees in the presence and absence GluCer, respectively. It was unlikely that direct interactions between GluCer and TLR4 would reduce tilting since the GluCer head group is small and uncharged, whereas TLR4 is a negatively charged protein. The authors speculated that the reduced tilting could be due to altering of the membrane properties by GluCer. A subsequent analysis revealed that

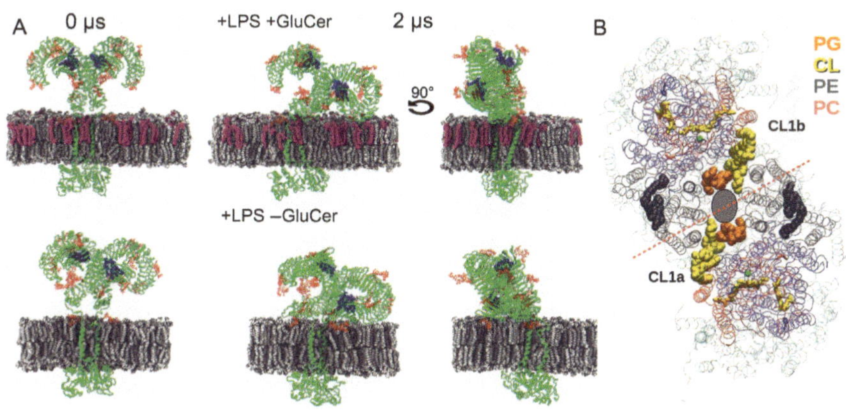

Fig. 10 Glycoprotein and respiratory complex IV. (A) Change in TLR4 dimer orientation. Snapshots show the TLR4 orientation at time 0 μs (*left panels*) and at time 2 μs (*middle* and *right panels*). Presence of GluCer (*purple*) in the membrane causes a change in the orientation of TLR4 extracellular domain (*green*). In the TLR4 dimer structure, LPS and glycans are *colored blue* and *red*, respectively. Rest of the membrane lipids are colored white. (B) X-ray structure of complex IV, Cytochrome c oxidase (PDB ID: 2DYR[293]) bound to PG, CL, PE, PC lipids (VDW representation, see legend for *color code*). Snapshot showing view of dimer structure from inter membrane space and the subunits I, II, and III are displayed as *blue*, *red*, and *gray ribbons*, respectively. The two monomers are separated by *red dotted lines* and the *gray filled circle* indicates empty region between the two monomers. Other structural components such as heme (*yellow licorice*), copper ion (*green spheres*), and some charged/polar residues (*licorice*) are shown as well. *Panel A: Reprinted with permission from Mobarak E, Håversen L, Manna M, et al. Glucosylceramide modifies the LPS-induced inflammatory response in macrophages and the orientation of the LPS/TLR4 complex in silico. Sci Rep. 2018;8(1):13600, Springer Nature. Copyright © 2018 Springer Nature (https://creativecommons.org/licenses/by/4.0/); Panel B: Reprinted with permission from Malkamäki A, Sharma V. Atomistic insights into cardiolipin binding sites of cytochrome c oxidase. Biochim Biophys Acta. 2019; 1860(3):224–232, 10.1016/j.bbabio.2018.11.004. Elsevier. Copyright © 2019, Elsevier (https://creativecommons.org/licenses/by/4.0/).*

the electrostatic potential of the membrane increased by 30% in presence of GluCer. Thus, the authors concluded that repulsive interactions between the LPS/TLR4 complex and the membrane reduced the tilting of LPS/TLR4 in the presence of GluCer. In addition, the authors observed that the effect of tilting on the extracellular side translated to the TM domains, where the conformation of the TM helix was altered in absence of GluCer, further emphasizing the stabilizing effect of GluCer on the LPS/TLR4 complex.

4.1.4 Mitochondrial membrane protein

Mitochondria are referred to as powerhouse of the cell due to synthesis of the energy carrier ATP.[295] Apart from ATP synthesis, mitochondria are required for the production of other molecules, such as NADH (nicotinamide adenine dinucleotide), GTP (guanosine triphosphate), amino acids, and lipids.[296,297] Mitochondria carry out many fundamental process, for instance in cellular respiration, cell and calcium signaling, as well as during stress signaling.[298–300] Thus, dysfunction of mitochondrial activity leads to several human diseases.[301] Mitochondria consist of a porous outer membrane and a tightly regulated inner membrane. At the inner membrane, ATP synthesis and oxidative phosphorylation take place in presence of membrane protein complexes. The complexes are (1) complex I (NADH:ubiquinone oxidoreductase), (2) complex II (succinate dehydrogenase), (3) complex III (cytochrome c reductase), and (4) complex IV (cytochrome c oxidase).[301] The structures of these complexes have been solved experimentally. In combination with computational techniques, the molecular details of electron transport are well-explored. However, how lipids regulate the activity of these complexes at the atomistic level much less understood. CL is one of the key components of the inner mitochondrial membrane, constituting to about 20% of the total lipid composition. CL has a highly charged small head group with four hydrocarbon tails (Fig. 2A).[302,303] All respiratory complex structures reveal CL binding sites.[304] Experiments have shown that the removal of CL reduces the activity of cytochrome c oxidase (CcO) by 50%,[305] whereas the activity is only mildly affected by the removal of PC or PE lipids. Though the effect of lipids on the CcO activity is well established, little is understood on the lipids' modulatory mechanism. Here, we discuss results from a recent study, which provides atomistic insights into the role of the mitochondrial membrane lipid CL in the functioning of the terminal enzyme complex IV, which is essential for the energy metabolism.

Malkamäki et al.[294] performed extensive AA-MD simulations (18 μs in total) of the bovine cytochrome c oxidase (CcO) dimer to gain insights into the dynamics of CL. The monomeric and dimeric structures were embedded into a lipid membrane composed of POPC/POPE/CL (ratio of 5:3:2). The X-ray structure (PDB code: 2DYR[293]) revealed that each CcO monomer has two CL binding sites, where one of the sites (site 2) is located at the interface of the two CcO monomers, suggesting a possible role in stabilizing the dimer structure (Fig. 10B). During MD simulations, CL occupied the crystallographically known site 1 throughout the simulations; however, at the buried site 2, CL was only weakly bound and dissociated within 1 μs. The dissociation of CL on site 2 occurred first by a movement of the CL head group away from the site, and then by tails, suggesting stronger interactions between the CL tails and protein hydrophobic interface. In contrast, CL at site 1, where it binds more strongly, is stabilized by strong ionic interactions. Additional simulations in the absence of CL at buried site 2, but with crystallographic bound PG/PE, revealed no effect on the stability of the dimer structure; however, removal of CL from site 1 not only destabilized the protein structure but also perturbed the strongly bound PG and PE lipids from their respective binding sites. This destabilization would potentially lead to loss of enzyme function. Thus, these simulation results provide an explanation at the atomistic level for the experimentally observed activity loss of CcO upon removal of CL.

4.2 Atomistic simulations of membrane–associated proteins

4.2.1 *Amyloids*

4.2.1.1 Amyloid β (Aβ) peptide

Alzheimer's disease (AD) characterized by loss of neuronal cells is primarily linked to neurofibrillary tangles and neuritic plaques.[306,307] Neurofibrillary tangles are associated with abnormal accumulation of tau proteins, whereas neuritic plaques (amyloid plaques) are associated with the aggregation of Aβ peptide.[308] So far, no consensus hypothesis of the molecular cause of AD has emerged, however, due to the presence of Aβ peptide in amyloid plaques, Aβ is widely accepted as a key molecule in AD. Aβ exists in two isoforms, Aβ_{40} and Aβ_{42}, with the latter being more prone to aggregation and more toxic.[309] Furthermore, several studies have shown that Aβ oligomers and not the fibrils are the causative agent of AD.[310–312] Since, Aβ peptide is a cleaved product of the transmembrane amyloid precursor protein (APP), it is also widely accepted that Aβ interactions with the membrane could play a role in the formation of toxic oligomers.[313–318] In spite of experimental and computational studies exploring and proposing several possible mechanisms toward Aβ induced cell toxicity, it is still difficult to draw any conclusion on the neurotoxicity mechanism. Here, we discuss findings from computational studies of Aβ-membrane interactions with a focus on lipids.

Poojari et al.[319,320] using atomistic MD simulations (>6 μs) have extensively investigated the behavior of monomeric $A\beta_{42}$ in helical and β-sheet conformations and tetrameric $A\beta_{42}$ in β-sheet conformations in POPC, DPPC, and POPG bilayers. In addition, it was investigated how various $A\beta_{42}$ mutants (K16M, E22G, D23G, and K28M) affect membrane stability. These simulations showed that the stability of the peptide inside the membrane is influenced by the degree of lipid acyl chain unsaturation and by the head group charge. Peptides were most stable in POPC bilayers, whereas pronounced peptide instability was observed in DPPC and POPG bilayers. Simulations also revealed that the β-sheet tetrameric structure increases water permeation into membranes (Fig. 11A), indicating that, indeed, $A\beta_{42}$ oligomers and not monomers are the toxic species. Mutation studies revealed that the "Arctic-type" D23G mutant exhibits an increased ability to disrupt membranes as compared to other studied mutants. However, neither wild type nor mutant monomers leave the membrane interior in the simulated time.

In order to understand how Aβ gets released from the membrane, Lemkul et al.[321] carried out atomistic MD simulations of $A\beta_{40}$ partially embedded (residues 29–40 inside the membrane) in POPC, POPS, POPC/POPE, POPC/SM/CHOL/GM1, and POPC/SM/CHOL membranes. The partially inserted $A\beta_{40}$ peptide anchors itself to the membrane through hydrogen bonding and electrostatic interactions with the lipid head groups. The interactions stabilize the peptide at the initial inserted depth in POPC, POPS, POPC/POPE, and POPC/SM/CHOL membranes.

In membranes containing GM1 lipid, $A\beta_{40}$ C-terminal residues (29–40) lose their α-helix structure and become more disordered (Fig. 11B). The C-terminal residues approach the polar interface were they form stable interactions with the GM1 sugars, glucose (Glc) and N-Acetylneuraminic acid (Neu5Ac). The interactions would eventually promote release of the peptide from the membrane interior, thereby exposing aggregation-prone hydrophobic regions of the peptide. The results provide mechanistic insights into the GM1-mediated release of membrane inserted $A\beta_{40}$.

Membrane mediated aggregation and disruption of cell membrane is one of the hypotheses linked to $A\beta_{42}$ induced toxicity. Neuronal cell membranes are complex with several lipid species and vary in head group charge, size and saturation/unsaturation level of acyl tails. To resolve how different lipids

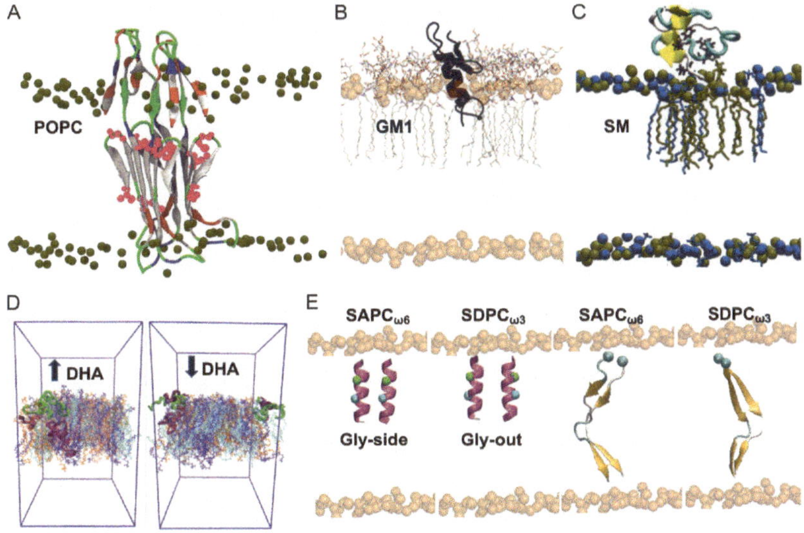

Fig. 11 Amyloid beta (Aβ) peptide. (A) Aβ_{42} β-sheet tetramer increasing water translocation by perturbing the POPC head group arrangement. *Pink spheres* refer to peptide residues involved in HB formation with water molecules. POPC phosphorous atoms are shown as *tan spheres* and lipid tails not shown for clarity. (B) GM1 lipids inducing the release of Aβ_{40} peptide from the membrane interior. GM1 lipids are rendered as *lines*, phosphorous atoms as *transparent orange spheres* and rest of the lipids are not shown for clarity. (C) SM mediated β-sheet formation in the Aβ_{42} peptide termini. SM lipids below the peptide are shown as *blue licorice*. For SM lipids further away, only the phosphorus atoms as *blue sphere* is displayed. POPC lipids are shown in tan. Rest of the lipids are not shown for clarity. (D) High DHA (healthy brain) content in the membrane favors deeper absorption of both Aβ_{1-28} (*green*) and Aβ_{26-40} (*purple*) peptides, whereas as low DHA content (AD brain) prevents deep absorption and promotes aggregation of the peptides into toxic oligomers on the membrane surface. (E) SAPC$_{\omega6}$ and SDPC$_{\omega3}$ membranes stabilize the helical dimer structure of Aβ_{29-42} with Gly-side and Gly-out orientations, respectively. However, with β-sheet dimer structures, SDPC$_{\omega3}$ reduces the β-sheet content. The change in orientation and secondary structure observed here is not sufficient to explain the SAPC$_{\omega6}$ triggered increase in Aβ production. *Panel A: Reprinted with permission from Poojari C, Kukol A, Strodel B. How the amyloid-β peptide and membranes affect each other: an extensive simulation study. Biochim Biophys Acta Biomembr. 2013;1828(2):327–339, https://doi.org/10.1016/j.bbamem.2012.09.001, Elsevier. Copyright © 2012 Elsevier B.V. All rights reserved, Elsevier. (https://creativecommons.org/licenses/by/4.0/); Panel B: Reprinted with permission from Lemkul JA, Bevan DR. Lipid composition influences the release of Alzheimers amyloid β-peptide from membranes. Protein Sci. 2011;20(9):1530–1545, https://doi.org/10.1002/pro.678. Copyright 2011* The Protein Society *(https://creativecommons.org/licenses/by/4.0/); Panel C: Reprinted with permission from Owen MC, Kulig W, Poojari C, Rog T, Strodel B. Physiologically-relevant levels of sphingomyelin, but not GM1, induces a β-sheet-rich structure in the amyloid-β(1-42) monomer. Biochim Biophys Acta Biomembr. 2018;1860(9):1709–1720, https://doi.org/10.1016/j.bbamem.2018.03.026, Elsevier. Copyright © 2018 Elsevier B.V. (https://creativecommons.org/licenses/by/4.0/); Panel D: Reprinted with permission from Ntarakas N, Ermilova I, Lyubartsev AP. Effect of lipid saturation on amyloid-beta peptide partitioning and aggregation in neuronal membranes: molecular dynamics simulations. Eur Biophys J. 2019;48(8):813–824, https://doi.org/10.1007/s00249-019-01407-x, Springer Nature. Copyright © 2019 The Author(s). (https://creativecommons.org/licenses/by/4.0/); Panel E: Reprinted with permission from Lu Y, Shi XF, Nguyen PH, Sterpone F, Salsbury FR, Derreumaux P. Amyloid-β(29-42) dimeric conformations in membranes rich in omega-3 and omega-6 polyunsaturated fatty acids. J Phys Chem B. 2019;123(12):2687–2696, https://doi.org/10.1021/acs.jpcb.9b00431, American Chemical Society. Copyright (2019) American Chemical Society.*

affect $A\beta_{42}$ structure, Owen et al.[322] carried out AA-MD simulations of $A\beta_{42}$ on membranes composed of POPC/CHOL (70/30 mol%), POPC/CHOL/SM (30/40/30 mol%) and POPC/CHOL/GM1 (70/25/5 mol%). The simulations provide molecular insights into the role of cholesterol (CHOL), sphingomyelin (SM) and ganglioside (GM1) lipids in conversion of $A\beta_{42}$ structure to toxic oligomers. The presence of SM, renders membrane highly rigid, which in turn reduces interactions between highly disordered N-terminus residues of $A\beta_{42}$ and the membrane. The only interaction observed is between residue R5 and SM, which prevents $A\beta_{42}$ from bouncing back to the aqueous solution. The lack of interactions promotes β-sheet formation in the N-terminus, thus, the final topology is a conformation with central loop connecting N- and C-terminal β-sheets (Fig. 11C). This final structure resembles experimentally known $A\beta$ fibril structure. In comparison, GM1 has small effect on the membrane property but does not induce secondary structure change. The membranes with or without GM1 had no effect on the structure of the $A\beta_{42}$. Thus, the results provide new insight into the role of SM ceramide in altering membrane property and in formation of toxic $A\beta_{42}$ oligomers.

Apart from the lipids discussed above, neuronal membranes also contain large amount of polyunsaturated fatty acids (PUFAs) such as docosahexaenoic acid (DHA, 22:6). AD has been associated with reduction of DHA lipids from 23% in normal brain to 12% in AD brain.[325–327] In vitro studies have shown that DHA inhibits $A\beta_{42}$ fibril formation.[328] To better understand the inhibitory role of DHA, Ntarakas et al.,[323] carried out MD simulations of $A\beta$ peptide fragments $A\beta_{1-28}$ and $A\beta_{26\text{-}40}$ in single component bilayers (DMPC– 14:0,14:0 or SDPC– 18:0,22:6) and in mixed bilayers (DSPE– 18:0,18:0, DDPE– 22:6,22:6, DPPC– 16:0,16:0, DOPC– 18:1,18:1) with varying composition. The mixed bilayer is divided into two types based on lipid composition. The mixed bilayer with high fraction of DHA is referred to as "normal bilayer" and the other bilayer with low fraction of DHA and high fraction of saturated lipids is referred to as "AD bilayer" (Fig. 11D). The simulation results from single component bilayer show that in DMPC bilayer, both the $A\beta$ fragments prefer to stay on the membrane surface. In contrast, in presence of 18:0–22:6 PC, $A\beta_{1-28}$ partially enters the membrane and the $A\beta_{26-40}$ fragment even fully inserts into the membrane. The behavior of the $A\beta$ fragments in DMPC bilayer is consistent with the experimental data.[329] Similar trends were observed for $A\beta$ interaction when studied in mixed bilayer. The $A\beta_{1-28}$ fragment fully enters the normal bilayer with high fraction of DHA and stays on the membrane surface of an AD bilayer. However, the $A\beta_{26\text{–}40}$ fragment penetrates both the normal and AD bilayers. The authors next looked into the aggregation propensity of these peptides, and they observed that both $A\beta$ fragments clustered on the membrane surface in DMPC and AD bilayer, but they were more dispersed in normal bilayer. Thus, the study suggests the presence of DHA to favor internalization of $A\beta$ peptide, which in turn prevents aggregation and formation of toxic oligomers on the membrane surface.

In another related study using PUFAs, Lu et al.[324] looked into the role of omega-3 ($\omega3$) and omega-6 ($\omega6$) PUFAs in stabilizing $A\beta_{29\text{–}42}$ dimer structures by performing extensive AA-MD simulations ($\sim$7 μs). The $A\beta_{29\text{–}42}$ dimers in both helical and β-sheet conformations were fully preinserted into a bilayer of POPC, 1-stearoyl-2-arachidonoyl-*sn*-glycero-3-phosphocholine ($SAPC_{\omega6}$ 18:0,20:4) and 1-stearoyl-2-docosahexaenoyl-*sn*-glycero-3-phosphocholine ($SDPC_{\omega3}$ 18:0,22:6). For the β-sheet conformation, an additional simulation was carried with the peptide fully inserted into a mixed bilayer composed of POPC/SAPC/SDPC lipids. The simulations show both $SDPC_{\omega3}$ and $SAPC_{\omega6}$ to stabilize the helical dimer structure (80% α-helix) with a parallel orientation and reorient the G33–X–X–X–G37 motif with $SDPC_{\omega3}$ favoring Gly-out and $SAPC_{\omega6}$ favoring Gly-side orientation (Fig. 11E). However, $SAPC_{\omega6}$ disorders the β-sheet dimer by reducing its β-sheet content from 59% to 44% and inducing α-helix conformation by 7% (Fig. 11E). $SDPC_{\omega3}$ and POPC bilayers behave similarly with no large effect on the β-sheet dimer. However, in a mixed membrane, the β-sheet dimer is stabilized with high β-sheet content ($\sim$63%), thus $SDPC_{\omega3}$ and POPC lipids suppress the individual effect of $SAPC_{\omega6}$ on the peptide. The study demonstrates how variation in PUFAs influence oligomerization of $A\beta$ inside membranes.

4.2.1.2 The human islet amyloid polypeptide (hIAPP)

The hIAPP or amylin is a small 37 residue peptide secreted by the pancreatic β-cells. hIAPP deposits are seen in 90% of the patients with type 2 diabetes (T2D) and are known to induce cytotoxicity and β-cell death. hIAPP monomers are not toxic to cells, however, they aggregate into toxic oligomers which disrupt the cell membrane. Several mechanisms have been proposed to explain hIAPP-induced cell toxicity, namely, (1) by the formation of channel structures, (2) by extraction of lipids, and (3) by membrane thinning. However, the exact membrane disruption mechanism is still not clear. Here, we present results from AA-MD simulations of hIAPP-membrane interactions and focus mainly on how different lipids affect hIAPP binding, orientation, aggregation, and insertion into membranes which provide insight into the membrane disruption mechanism.

Jia et al. and Qian et al.[330–332] investigated the initial binding of hIAPP monomers to zwitterionic POPC and anionic POPG bilayers using AA-MD simulations. As a starting structure, they used hIAPP peptide in α-helical conformation and placed the peptide 1.4 nm above the POPC or POPG membrane surface. Simulations show that hIAPP adopts four different orientations in POPC membrane. N-terminal residues localize at the hydrophilic head group region of the lipid, whereas the C-terminal residues penetrate deep into the membrane (Fig. 12A). In case of simulations with POPG membrane, the peptide adopts a single, surface-bound orientation. The lack of membrane penetration exposes the aggregation-prone C-terminal residues 20–37 to peptide–peptide interactions. Analysis of membrane properties revealed that the hIAPP monomer has a negligible effect on the membrane, indicating that the experimentally observed membrane disruption is caused by oligomeric structures and not by monomers.

Similarly, Dignon et al.,[333] carried out AA-MD simulations of hIAPP monomers with DOPC, DOPS and DOPC/DOPS membranes using the parallel tempering method. This study also allows a comparison with simulations in POPC or POPG membranes described above. The simulations generated a large ensemble of structures differing between the three membranes. The α-helical content is reduced and changed in DOPS and DOPC membranes, respectively, whereas in mixed membrane the α-helical content is increased (Fig. 12B). Also, the insertion of the residues 15–28 was same in DOPC membranes but deeper in DOPC/DOPS membranes. Here, the shallow insertion in the DOPS membrane was observed possibly due to a loss of secondary structure. No details were provided regarding the membrane structure, probably due to negligible effect of the peptide on the membrane property.

The above studies present case studies of effects of oligomeric structures on membrane properties. The formation of hIAPP dimers is the first step toward the formation of large oligomers. Zhang et al.[334] carried out AA-MD simulations of hIAPP dimers preinserted into POPG bilayer. During simulation, the dimers inserted deep into the membrane with residues K1 anchoring to the lipid head group (Fig. 12C). The interaction perturbs the phosphorous atoms around the dimer in both, the upper and the lower leaflets, which results in a reduction of membrane thickness in close proximity to the dimer. Analysis of the lipid order parameters revealed that

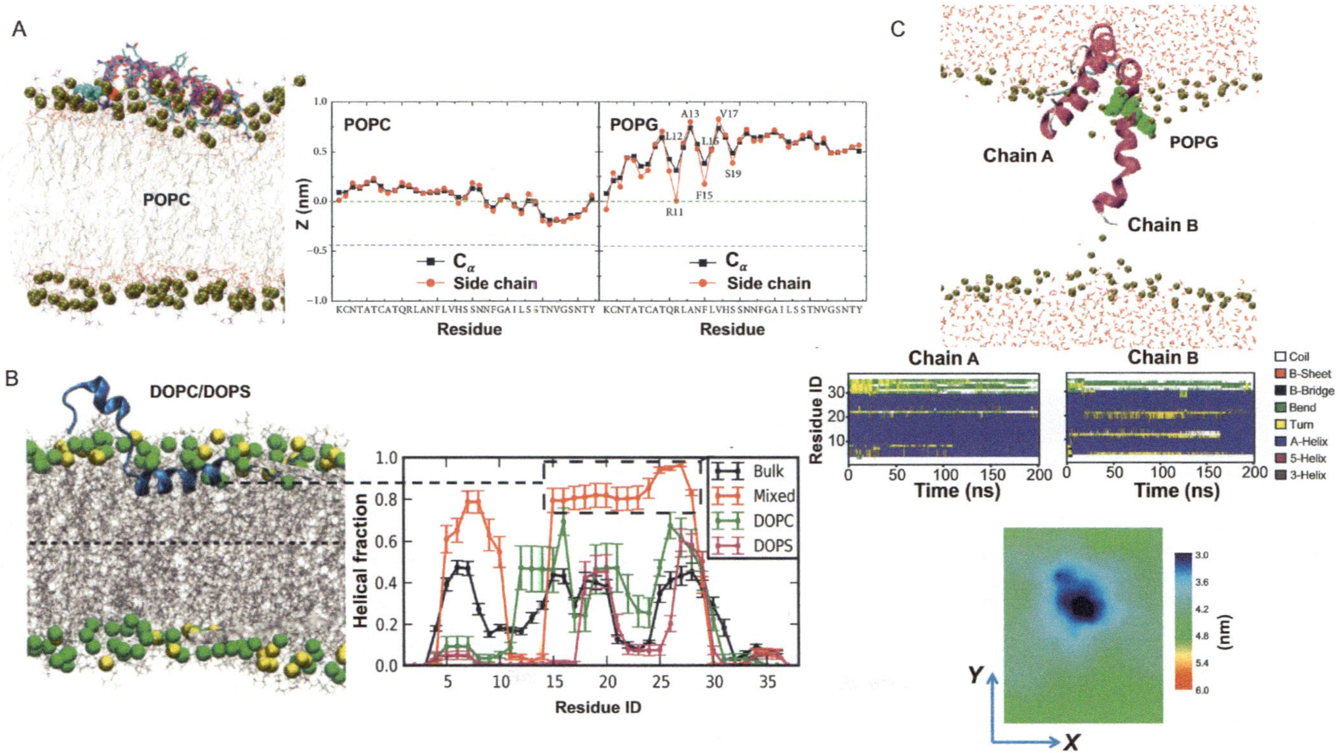

Fig. 12 See legend on next page.

the hIAPP dimer also decreases the ordering of carbon atoms close to the head group. During an additional simulation, the order parameters were also decreased at carbon atoms 5–12, close to the membrane center. These results provide evidence for a membrane disruption ability of hIAPP, and they also suggest dimers as the minimal aggregate required to alter membrane properties.

To understand how hIAPP monomers aggregate into toxic oligomers, Christensen et al.[172] carried out AA-MD simulations starting from a monomer already bound to the membrane surface. The monomeric system was multiplied to have four peptides on a membrane composed of divalerylphosphatidylserine (DVPS, 30%) and zwitterionic divalerylphosphatidylcholine (DVPC, 70%) lipids. To enhance the diffusion of lipids and peptides, the highly mobile membrane mimetic (HMMM) model was utilized. The simulations suggested that self-assembly takes place in the following order (Fig. 13A): (1) the monomers diffuse on the membrane surface with N-terminal residues (K1 and R11) anchoring the peptide to the membrane; (2) the aromatic residues (F15 and F23) in the central region of the peptide initiate peptide–peptide interactions. For the self-assembly to start, the interactions between peptide and membrane observed in step 1 must break, leading

Fig. 12 The human islet amyloid polypeptide (hIAPP). (A) Snapshot showing side view of hIAPP monomer interacting with POPC bilayer. The plot shows insertion depths for hIAPP Cα and side chain atoms in POPC and POPG bilayers. (B) Mixed bilayer containing DOPC/DOPS lipids allow deeper insertion of the peptide and stabilizes the helical structure from residues 15–28 of hIAPP. The plot shows reduction in helical content in pure DOPC and pure DOPS bilayers. (C) Final snapshot of hIAPP dimer in POPG bilayer. *Top*: The dimers (chain A and chain B) penetrate deeper into the bilayer with helical content preserved. Residues from 18 to 20 are rendered as *green VDW spheres* and lipid tails are not shown for clarity. *Center*: Plot showing secondary structure content for chain A and B peptides. α-helical content is preserved in both the peptides. *Bottom*: Plot showing reduced bilayer thickness in vicinity of the dimer. *Panel A: Reprinted with permission from Qian Z, Jia Y, Wei G. Binding orientations and lipid interactions of human amylin at zwitterionic and anionic lipid bilayers. J Diabetes Res. 2016;2016:1–13, Hindawi Publishing Corporation. Copyright © 2016 Zhenyu Qian et al. (https://creativecommons.org/licenses/by/3.0/); Panel B: Reprinted with permission from Dignon GL, Zerze GH, Mittal J. Interplay between membrane composition and structural stability of membrane-bound hIAPP. J Phys Chem B. 2017;121(37):8661–8668, https://doi.org/10.1021/acs.jpcb.7b05689, American Chemical Society. Copyright (2017) American Chemical Society; Panel C: Reprinted with permission from Zhang Y, Luo Y, Deng Y, Mu Y, Wei G. Lipid interaction and membrane perturbation of human islet amyloid polypeptide monomer and dimer by molecular dynamics simulations. PLoS ONE. 2012;7(5):1–10, https://doi.org/10.1371/journal.pone.0038191, Public Library of Science. Copyright © 2012 Zhang et al. (https://creativecommons.org/licenses/by/4.0/).*

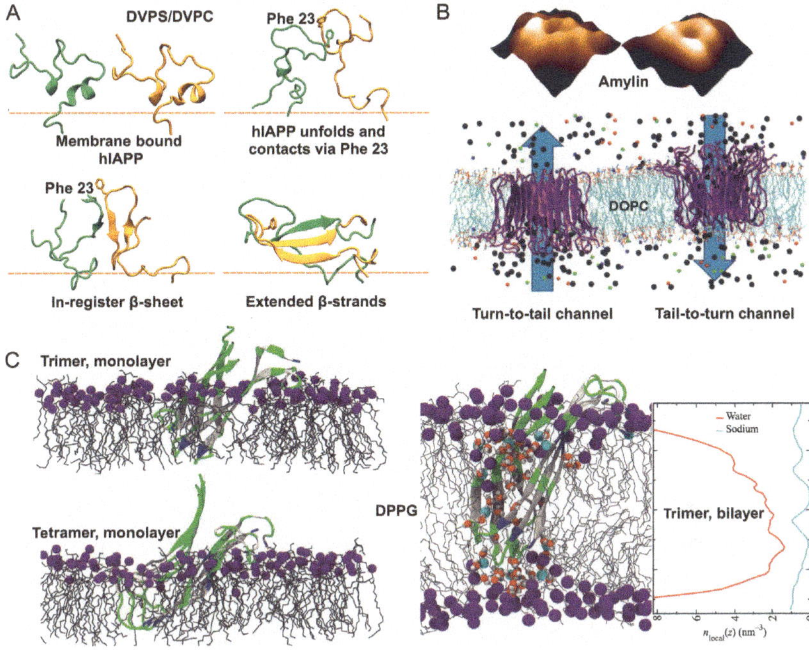

Fig. 13 Conventional MD and HMMM simulations of hIAPP. (A) Snapshots 1–4 represent various structural changes the peptides undergo in the formation of stable β-sheet dimers on the DVPS/DVPC membrane surface. The *orange dashed lines* indicate phosphorus atoms of the lipids. (B) Oligomerization of hIAPP into toxic ion channel structures. *Top*: Atomic force microscopy (AFM) image of amylin showing channel-like structures. *Bottom*: Double-annular-like ion channels (resembling AFM images) with β-strand-turn-β-strand topology embedded into DOPC bilayer. MD simulations show unregulated conductance of ions across the ion channels. (C) hIAPP protofibril structures in DPPG membrane. *Left*: Most populated central cluster structure of hIAPP trimer and tetramer in DPPG monolayer. *Right*: hIAPP trimer in fully membrane inserted state adopts a β-sandwich structure which allows influx of water and sodium ions (*cyan spheres*). The plot indicates the density of water and sodium ions inside the membrane. *Panel A: Reprinted with permission from Christensen M, Skeby KK, Schiøtt B. Identification of key interactions in the initial self-assembly of amylin in a membrane environment. Biochemistry. 2017;56(36):4884–4894, https://doi.org/10.1021/acs.biochem.7b00344, American Chemical Society. Copyright (2017) American Chemical Society; Panel B (Top): Reprinted with permission from Quist A, Doudevski I, Lin H, et al. Amyloid ion channels: a common structural link for protein-misfolding disease. Proc Natl Acad Sci. 2005;102 (30):10427–10432, https://doi.org/10.1073/pnas.0502066102, United States National Academy of Sciences. Copyright (2005) National Academy of Sciences, USA; Panel B (Bottom): Reproduced from Zhao J, Hu R, Sciacca MFM, et al. Non-selective ion channel activity of polymorphic human islet amyloid polypeptide (amylin) double channels. Phys Chem Chem Phys. 2014;16:2368–2377, https://doi.org/10.1039/C3CP53345J, with permission from the PCCP Owner Societies 2014. Royal Society of Chemistry (United Kingdom). Permission conveyed through Copyright Clearance Center, Inc.; Panel C: Reprinted with permission from Poojari C, Xiao D, Batista VS, Strodel B. Membrane permeation induced by aggregates of human islet amyloid polypeptides. Biophys J. 2013;105(10):2323–2332, https://doi.org/10.1016/j.bpj.2013.09.045, Elsevier. Copyright © 2013 Biophysical Society. Published by Elsevier Inc. All rights reserved. (https://creativecommons.org/licenses/by/4.0/).*

to a partial unfolding of the secondary structure; (3) the initial aromatic interactions allow the peptide segment 22-NFGAIL-27 to form stable interactions; (4) self-assembly induces *β*-sheet formation in segments 14-NFLVH-18 and 25-AILSST-30, which are involved in intra- and inter-peptide interactions. In these simulations, the H18 residue was modeled as electroneutral during the self-assembly process, since a charged H18 residue would have increased the interaction with the PS lipid head group, thereby stabilizing the secondary structure and repelling the peptides, which would inhibit the self-assembly process. This study highlights the role of lipids in aggregation and provides atomistic insight into the formation of oligomer structures.

The membrane-dependent self-assembly of hIAPP to oligomers is hypothesized to allow the formation of ion channel/pores in membranes thereby inducing cell toxicity. It is computationally challenging to simulate the spontaneous hIAPP oligomerization and ion channel formation in a membrane. However, to understand the mechanism of how oligomers transfer ion across the membrane, Zhao et al.[335,336] used the *β*-strand-turn-*β*-strand (U-turn) motif as starting topology and modeled several single and double annular-like ion channels resembling the channel morphology observed with atomic force microscopy (AFM)[337] (Fig. 13B). The channels were fully inserted into DOPC bilayer for studying the ion conductivity. MD simulations revealed a loss of the *β*-sheet secondary structure, which destabilized the channels into unstable loosely packed subunits. The unstable channel structures allowed unregulated flow of ions across the bilayer which would lead to ion imbalance and toxicity. These studies provided a first view of channel structures and their role in toxicity.

The unstructured toxic oligomers are dynamic and can transform to more ordered, fibril-like *β*-sheet rich structures called protofibrils, which are also toxic to cells. The structure of a mature fibril in its *β*-strand-turn-*β*-strand topology has been solved using solid-state NMR by Tycko's[339] and Langen's[340] groups. The structure provides the basis to construct and study protofibrils in a membrane environment. To study how protofibrils orient in a membrane, Xiao et al.[341] combined surface-selective sum frequency generation (SFG) spectroscopy and quantum chemistry techniques. These authors noticed that the protofibril is partially inserted into dipalmitoylphosphatidylglycerol (DPPG) monolayer with an tilt angle of 48 degrees. To check for the stability of protofibrils in lipid membranes, Poojari et al.,[338] carried out atomistic simulations of protofibrils (trimer and tetramer structures) partially inserted into DPPG monolayers and bilayers at an tilt angle of 48 degrees (Fig. 13C). The simulations show protofibrils to adopt a tilt angle of 40 degrees in DPPG monolayers, which is close to 48 degrees measured in experiments. However, in bilayers the protofibrils penetrate deep into the membrane with an angle of 60 degrees. The protofibril inside the membrane is stabilized by close interactions with the lipid head groups. Furthermore, trimer structures in both monolayer and bilayer were structurally less stable when compared to tetramer due to increased flexibility of the outer β-strands. The flexible trimer structure distorts into a β-sandwich structure which further perturbs the membrane integrity, thereby allowing the permeation of water and sodium ions. This study revealed for the first time key details of protofibril structures, such as the stability and the orientation of protofibril in lipid membranes, the formation of β-sandwich structure by protofibrils, membrane disruption, and conductance of water and ions.

4.2.1.3 α-synuclein (aSyn)

aSyn is a presynaptic neuronal protein linked to Parkinson's disease (PD).[342] The 140-residue protein is divided into three domains based on their function: (1) residues from M1–K60 are rich in lysine residues and are required for binding to membrane; (2) residues E61–V95 form the central "nonamyloid-β component" (NAC) segment of the protein and also contains the hydrophobic motif necessary for protein aggregation, and (3) residues K96–Ala140 involved in interactions with proteins, metal ions, small molecules, etc.[343–345] So far, two monomeric structures of aSyn (PDB code: 1XQ8,[346] 2KKW[347]) have been solved in sodium dodecyl sulfate (SDS) and sodium lauroyl sarcosine (SLAS) micelles. Recently, using cryoEM, Li et al.,[348] solved the structure of aSyn fibril in rod and twister polymorphs (PDB code: 6CU8[348]) at 3.7 Å resolution. Unlike Aβ and hIAPP, computational studies of aSyn–membrane interactions are limited. Nevertheless, we aim to discuss the key results from AA simulations of aSyn in membrane environment.

Experiments and AA simulations were initiated to probe the membrane-penetrating ability of the N-terminal residues.[349] Here, a short-length $aSyn_{1-15}$ was used for simulations in a POPA/POPC membrane, and the studies showed that residues M1, W4, L8 penetrate deep into the membrane, whereas residues D2, K6, S9, K10, and E13 remain solvent-exposed (Fig. 14A). These studies confirm that the N-terminal residues bind to the membrane. Further, aSyn–membrane interactions were investigated in

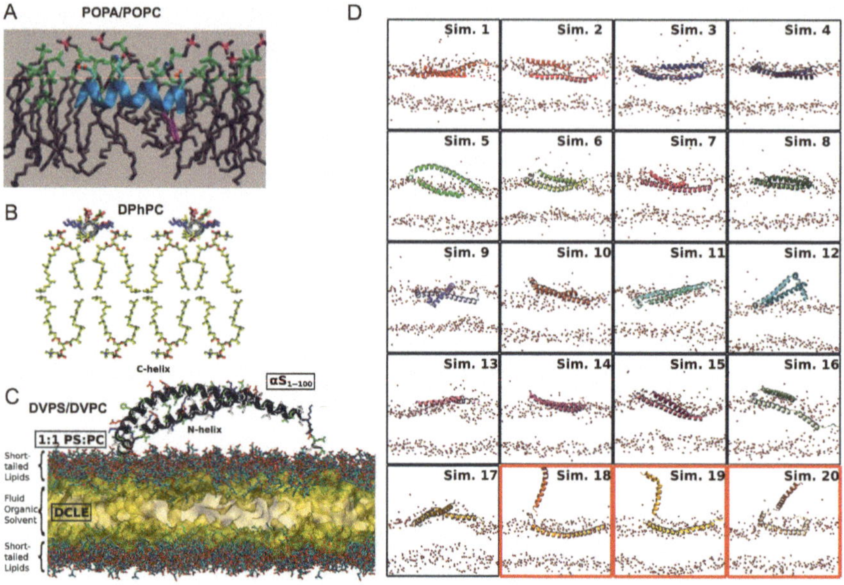

Fig. 14 *Conventional MD and HMMM simulations of* α-Synuclein (aSyn). (A) Peptide to penetrate POPA:POPC membrane and position below the phosphorus atoms of the lipids. The peptide orients such that hydrophilic residues face toward the head group-water region. Lipids tails are rendered as *gray licorice* and protein is displayed as *cartoon*. (B) In DPhPC membrane, peptide is surface bound and makes a shallow defect in the membrane. DPhPC lipids are rendered as *yellow licorice* and protein as *cartoon*. (C) and (D) Using HMMM to study aSyn dynamics on DVPS/DVPC membrane surface. (C) Initial snapshot showing aSyn above the HMMM surface and rendered as *black cartoon*. (D) Final snapshots from 20 simulations showing aSyn to adopt various conformations from extended helical conformation to broken-helix conformation. The structures in *red box* indicate the peptide interacting across the periodic boundary. Lipid phosphorus atoms are shown as *pink spheres* and lipid tails removed for clarity. *Panel A: Reprinted with permission from Pfefferkorn CM, Heinrich F, Sodt AJ, et al. Depth of α-synuclein in a bilayer determined by fluorescence, neutron reflectometry, and computation. Biophys J. 2012;102(3):613–621, https://doi.org/10.1016/j.bpj.2011.12.051, Elsevier. Copyright © 2012 Biophysical Society. Published by Elsevier Inc. All rights reserved (https://creativecommons.org/licenses/by/4.0/); Panel B: Reproduced from reference Garten M, Prévost C, Cadart C, et al. Methyl-branched lipids promote the membrane adsorption of α-synuclein by enhancing shallow lipid-packing defects. Phys Chem Chem Phys. 2015;17:15589–15597, https://doi.org/10.1039/C5CP00244C, with permission from the PCCP Owner Societies 2015. Royal Society of Chemistry (United Kingdom). Permission conveyed through Copyright Clearance Center, Inc.; Panels C and D: Reprinted with permission from Vermaas JV, Tajkhorshid E. Conformational heterogeneity of α-synuclein in membrane. Biochim Biophys Acta Biomembr. 2014;1838(12):3107–3117, https://doi.org/10.1016/j.bbamem.2014.08.012, Elsevier. Copyright © 2014 Elsevier B.V. All rights reserved (https://creativecommons.org/licenses/by/4.0/).*

different lipid bilayers such as DOPS, SAPC, DOPC and DPPC. Studies with DOPS[350] bilayer shows aSyn to localize below the lipid head group region and the protein orients such that the hydrophilic side of the protein faces the lipid head group/water. The peptide had no effect on the DOPS membrane structure. Garten et al.,[351] carried out atomistic simulations to understand how aSyn packs and alters mechanical properties of DOPC, DPhPC and SAPC bilayers. The simulations revealed that aSyn was absorbed into all the bilayers, and shallow packing defects were observed in both SAPC and DPhPC bilayers. In addition, major packing defects were observed in a DPhPC bilayer.

It is also important to understand the effect of lipids on aSyn structure and dynamics, since any change in aSyn conformation will affect protein function. Vermaas et al.[171] performed atomistic simulations to investigate the dynamics of aSyn on DVPS/DVPC membrane. Simulations revealed aSyn to adopt two conformations, one with a broken-helix conformation similar to the NMR structure solved in micelle and second a semiextended helix conformation. A lipid density analysis revealed that a semi-extended helix conformation increases the density of the PS head groups along its edge. Such lipid-mediated transition of protein conformation might be physiologically important for aSyn.

aSyn-membrane interactions are well documented by a study carried out from Viennet et al.[352], who combined several experimental and computational technique to characterize various properties of the membrane and of aSyn (Fig. 15). The atomistic simulations were carried out to investigate how $aSyn_{1-61}$ orients on membranes composed of POPC/POPG (50/50 mol%, fluid phase) and DMPG/DMPC (50/50 mol%, gel phase). The simulations confirm that lysine residues in the N-terminus interact strongly with the POPG head groups rather than with the DMPG head groups. Analysis of lipid order parameter reveal that aSyn has a small effect on the gel phase membrane (DMPC/DMPG), whereas the lipid order parameter is increased in fluid phase membranes (POPC/POPG). The increase in order parameter results in an increased bilayer thickness and a reduced area per lipid.

4.2.2 Peripheral membrane proteins

4.2.2.1 Fibroblast growth factor (FGFs)

FGFs are multifunctional mitogens involved in cell growth, tissue regeneration, wound repair, embryonic development, etc.[353] There are 22 structurally similar FGF members in the human family, and involved in signaling.[353]

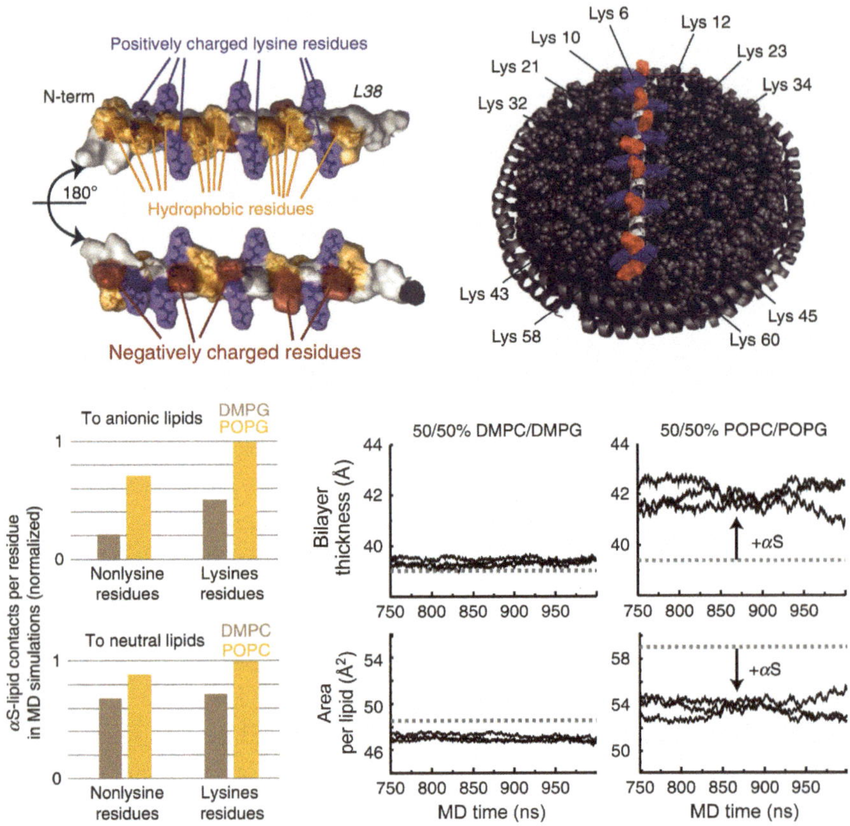

Fig. 15 Peptide–lipid interactions of aSyn in HMMM model. aSyn-DMPC/DMPG and POPC/POPG interactions are depicted. *Top left*: Surface representation of aSyn showing negative, positive, and hydrophobic residues as *licorice*. *Top right*: aSyn on nanodisc surface showing all the lysine residues (*blue*) and negatively charged residues (*red*). *Bottom left*: aSyn–PG/PC contacts within <4 Å are shown. Though both the lipid membranes contain same head group, aSyn makes more contacts with the fluid phase POPC/POPG membrane rather than gel phase DMPC/DMPG membrane. *Bottom right*: In gel phase membrane, aSyn has no or little effect on the membrane properties. But in fluid phase, aSyn increases the bilayer thickness and reduces the area per lipid. *Dashed lines* in the plots indicate membrane only system without aSyn. *Reprinted with permission from Viennet T, Wördehoff MM, Uluca B, et al. Structural insights from lipid-bilayer nanodiscs link α-synuclein membrane-binding modes to amyloid fibril formation. Commun Biol. 2018;1(1):44, Springer Nature. Copyright © 2018 Springer Nature (https://creativecommons.org/licenses/by/4.0/).*

Among them, the basic fibroblast growth factor 2 (FGF2), apart from carrying out its normal functions, also mediates tumor-induced angiogenesis and inhibits tumor cell apoptosis.[354–356] Further, the lack of a signal peptide makes FGF2 unique among its family members, and it is secreted into the extracellular space by an unconventional route, independent of ER/Golgi pathway.[357–359] In vitro studies suggest that the unconventional FGF2 translocation takes place through initial interactions with the inner leaflet PI(4,5)P_2 lipids.[358–361] The FGF2–PI(4,5)P_2 interactions trigger formation of large oligomeric structures on the membrane surface which induce pore formation. The membrane translocated FGF2 monomers are caught by the outer leaflet heparan sulfate proteoglycans.[357,362,363] Moreover, biochemical studies have pointed out that the residues K127, R128, and K133 are important for initial PI(4,5)P_2 binding, residues C77/C95 for oligomerization, and residue K133 for interacting with heparan sulfates.[357,364] Further, there is no structural data describing dimer and trimer interfaces that are a minimal requirement for oligomerization process.

Recently, Steringer et al.[365] tested the key predictions from biochemical studies using AA-MD simulations. To observe the spontaneous interaction of FGF2 monomers with PI(4,5)P_2, the protein was placed 1.5 nm above the mixed membrane composed of POPC/CHOL (70/30 mol%) in the outer leaflet and POPC/CHOL/PIP_2 (65/29.5/5.5 mol%) in the inner leaflet. The simulations revealed FGF2 to spontaneously bind to PI(4,5)P_2 through the known binding pocket residues K127, R128, and K133, thus, confirming with the biochemical studies (Fig. 16A). In addition, simulations showed that FGF2 adopts two different orientations on the membrane surface, called (1) high-affinity orientation evident by strong interaction energies and (2) low-affinity orientation evident by weak interactions. In the high-affinity orientation, FGF2 orients with C95 being accessible to form a disulfide bridge with the second monomer. Whereas in low-affinity, both C95 and C77 are available to form a disulfide bridge with the second monomer, but the interaction with PI(4,5)P_2 is weakened with only R128 making contact with PI(4,5)P_2. The weak interaction with the PI(4,5)P_2 might be a transient state and formation of stable dimers or higher ordered structures with such orientation might not be possible. To test the dependence of PI(4,5)P_2 on FGF2 orientation, the authors also simulated in presence of a single PI(4,5)P_2 in the membrane. The simulations showed a specificity of FGF2 for binding PI(4,5)P_2 with its known binding pocket residues, however, the protein orientation was never stable but instead continuously fluctuating between high- and low-affinity orientations. These results indicate that more than a single PI(4,5)P_2 is required to stabilize the protein on the membrane surface and to promote oligomerization. Having established the binding and orientation for a single FGF2, the next step was to identify the dimeric interface required for oligomer formation. Since trimer

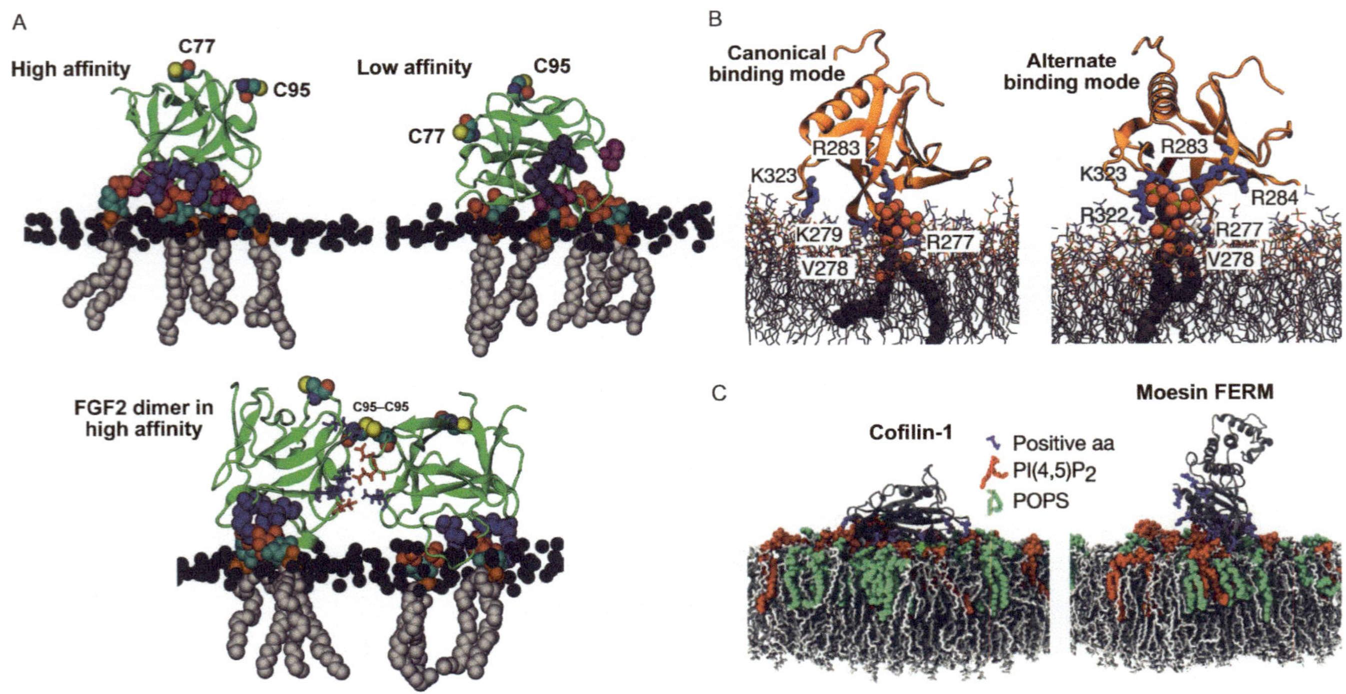

Fig. 16 See legend on opposite page.

structures were the minimal oligomer size observed in experiments, simulations were initiated with three monomers placed 0.5 nm above the membrane surface and arranged such that two monomers faced each other at 0.7 nm and a third monomer faced the other monomers at 1.5 nm distance. During simulations, all three monomers formed stable interactions with $PI(4,5)P_2$ and adopt the high-affinity orientation. The two nearby monomers dimerized with C95–C95 interactions at the interface, and the structure was further stabilized by four salt bridges formed by two pairs of E86–K118 and E99–K85. The third monomer approached the dimer with the C77–C77 residues at the interface and further stabilized by salt bridges R47–D49, D45–R41, and R80–E53. The dimer interfaces involving residues C95–C95 (interface 1) and C77–C77 (interface 2) observed in MD simulations are consistent with the in vitro studies. Thus, this study

Fig. 16 Peripheral membrane proteins in the case studies. (A) FGF2 orientation on the membrane surface. FGF2 binds to $PI(4,5)P_2$ in high-affinity and low-affinity orientations. High-affinity binding includes all the experimentally known binding pocket residues K127, R128, K133 (rendered as *blue spheres*), whereas in low-affinity, FGF2 loses contact with $PI(4,5)P_2$. Additional binding residues are rendered as *purple spheres*. High-affinity binding favors FGF2 dimerization via C95–C95 disulfide bridge, which is unavailable with low-affinity orientation. $PI(4,5)P_2$ tails are rendered as *white spheres*, $PI(4,5)P_2$ bisphosphates as *red spheres*, inositol ring as *cyan spheres*, phosphate linking PI(4,5) P_2 tails and the inositol ring as *orange spheres*, POPC phosphorous atom as *black sphere*, C95/C77 are shown as *yellow spheres* and protein as *green cartoon*. Rest of the lipids, water, and ions are not shown for clarity. (B) GRP1-PH domain binds to $PI(3,4,5)P_3$ in both canonical (C site) and alternate (A site) binding modes. In C site mode, PH domain $\beta 1/\beta 2$ loops bind to $PI(3,4,5)P_3$, whereas in A site mode, binding takes place with $\beta 5/\beta 6$ loops. $PI(3,4,5)P_3$ is rendered as *spheres*, rest of the lipids as *lines*, protein as *orange cartoon* and key residues are shown as *blue licorice*. (C) Conformation of Cofilin-1 and moesin FERM domain on membrane surface. Positively charged amino acids interacting with the lipids are shown as *blue licorice*, $PI(4,5)P_2$ and POPS lipids are displayed as *red* and *cyan spheres*, respectively. *Panel A: Reprinted with permission from Steringer JP, Lange S, Čujová S, Sinning I, Tews I, and Nickel W. Key steps in unconventional secretion of fibroblast growth factor 2 reconstituted with purified components. eLife. 2017;6:e28985, https://doi.org/10.7554/eLife.28985, eLife Sciences Publications. Copyright 2017 Steringer et al. (https://creativecommons.org/licenses/by/4.0/); Panel B: Reprinted with permission from Pant S, Tajkhorshid E. Microscopic characterization of GRP1 PH domain interaction with anionic membranes. J Comput Chem. 2019;9999:1–11, https://doi.org/10.1002/jcc.26109, John Wiley and Sons. Copyright © 2019 Wiley Periodicals, Inc. (https://creativecommons.org/licenses/by/4.0/); Panel C: Reprinted with permission from Senju Y, Kalimeri M, Koskela EV, et al. Mechanistic principles underlying regulation of the actin cytoskeleton by phosphoinositides. Proc Natl Acad Sci. 2017;114(43):E8977–E8986, https://doi.org/10.1073/pnas.1705032114, United States National Academy of Sciences.*

establishes a minimal molecular basis required for FGF2 to initially bind $PI(4,5)P_2$ and to aggregate to large oligomers, which would eventually create membrane pores and escape into extracellular space. The molecular mechanism described here for FGF2 could be related to other $PI(4,5)P_2$-dependent unconventionally secreting molecules, such as HIV-Tat[366–368] and Interleukin 1β.[369]

4.2.2.2 Pleckstrin homology (PH) domain

Cell signaling is predominantly initiated by the recruitment of peripheral membrane proteins (PMPs) to the cytosolic side of the membrane where they make specific interactions with negatively charged signaling lipids such as PIPs.[370,371] PIPs exist in various isoforms depending on the head group phosphorylation, $PI(4,5)P_2$, $PI(3,5)P_2$, $PI(3,4)P_2$, and $PI(3,4,5)P_3$.[372,373] The pleckstrin homology (PH) domains are a small important class of PMPs that recognize and bind to only specific PIPs on the cytosolic side of the membrane.[374–376] The PDB currently contains ~177 solved structures of PH domains, some of them solved with a PIP head group (inositol phosphates). All PH domains fold into a β-barrel topology with open ends capped by α-helix on one side and PIP binding motif on the opposite side. The domains comprise about 120 amino acids. The positively charged motif K–X–K/R–X–R in the loop connecting the β1 and β2 strands forms the binding site for the negatively charged PIP head group. However, not all PH domains have the canonical binding motif (K–X–K/R–X–R) for PIPs; for example, the (noncanonical) binding motif of β-spectrin is situated at the opposite face of the β1 and β2 strands,[377] and the ASAP1 PH domain was reported to bind PIP at both canonical and noncanonical binding sites.[378] Here, we present recent results for the general receptor for phosphoinositides isoform 1 (GRP1) PH domain, which has a canonical PIP binding site.

GRP1 is one of the well characterized $PI(3,4,5)P_3$ recognizing protein belonging to guanine nucleotide exchange factor (GEFs) family.[381] GRP1 is involved in insulin signaling,[382,383] receptor endocytosis[384] and actin dynamics.[385,386] Pant et al.[379] carried out several AA-MD simulations to investigate the spontaneous binding of the GRP1 PH domain to $PI(3,4,5)P_3$. The simulations were initiated by placing the PH domain 1.5 nm above a mixed membrane composed of PC, PS, and $PI(3,4,5)P_3$ lipids in varying concentrations. In a neutral pure PC membrane, the PH domain bounced back to the aqueous solution after initial contacts, demonstrating only weak interactions with the PC head groups. However, upon the addition of 20%

negatively charged PS lipid to the membrane, the PH domain was found to strongly interact with the membrane throughout the simulations. The interactions were mainly mediated through the loop connecting $\beta 1$ and $\beta 2$ strands, where the canonical binding motif is located. The side chains of the key residues R277, V278, and K279 stabilizing the interactions were localized to the phosphate head groups, and no deeper insertion was observed. On addition of highly negatively charged PI(3,4,5)P_3 lipids to the membrane, PI(3,4,5)P_3 recognized their canonical binding motif and PH domain orient at an angle of 36.9 $\pm$ 9.1 degrees relative to the membrane normal. The key residues at the canonical binding site forming hydrogen bonds with the phosphate head groups are K273, R277, K282, R284, and Y298. Simulations also revealed that PI(3,4,5)P_3 may recognize the loop connecting $\beta 5$ and $\beta 6$ strands, a new binding site previously not reported for the GRP1 PH domain. At this new site, the PH domain orients an angle of 29 $\pm$ 11.9 degrees with respect to the membrane normal (Fig. 16B). The key residues at the alternate new site forming hydrogen bonds with the phosphate head groups are R283, R322, and K323. Inspection of the local lipid density revealed that the orientation is influenced by the local concentration of PS lipids, which is higher when PI(3,4,5)P_3 binds to the PH domain at its canonical binding site. Furthermore, in presence of PI(3,4,5)P_3 V278 was observed to penetrate deep into the membrane. Additional free energy calculations suggested that both binding modes are favorable with higher affinity for binding at canonical binding site. The results from this study provide evidence for lipid-dependent binding and activity of PH domains.

4.2.2.3 Actin-binding proteins (ABPs)

Actin, the most abundant and dynamic protein in eukaryotic cells, participates in several cellular functions including the polymerization of filaments, preservation of membrane integrity, contraction of muscle cells, etc.[387] Polymerization of actin filaments leads to cell movement and morphogenesis.[388] The stability, dynamics, length of filaments are controlled by actin-binding proteins (ABPs), which are in turn regulated by PIPs.[389,390] Thus, membrane binding of ABPs, such as cofilin, vinculin, formins, and N-WASP are critical for their activity.[391–395] Among the PIPs, PI(4,5)P_2 interacts with ABPs and serves several functions, including the prevention of actin filament disassembly and capping of filament ends,[396–398] the activation of actin filament assembly. Hence, PI(4,5)P_2 serves as a link between the plasma membrane and actin filaments.[399–401] The presence of PI(4,5)P_2

is central in filament assembly, and any loss in membrane PI(4,5)P_2 triggers disassembly of actin filaments.[402–405] Despite the importance of ABPs interaction with PI(4,5)P_2, molecular insights into ABPs association and dissociation with lipid membranes has been largely unexplored. To gain molecular insights into dynamics of ABPs on interaction with PI(4,5)P_2, we discuss results for cofilin-1 and for the FERM domain of moesin.

Senju et al.[380] performed unbiased AA-MD simulations, initiated by placing the protein 0.5 nm above the membrane surface. During the simulations, both proteins interact with the PI(4,5)P_2-enriched membrane via their large positively charged surface (Fig. 16C). Cofilin-1 equilibrates within 50–60 ns with no change in the number of hydrogen bonds formed with the lipids, whereas moesin, due to its large size and slower diffusion, equilibrates within 150 ns. The large and highly negatively charged PI(4,5)P_2 head groups readily bind to experimentally identified binding sites on Cofilin-1 and moesin. Cofilin-1 and moesin is stabilized on the membrane surface by several lysine and arginine residues. In addition to PI(4,5)P_2 binding to the known binding site for moesin, binding to an additional site was identified, formed by a patch of nine cationic residues: K3, K35, R40, K72, R246, K254, K258, K262, and K26. On average, Cofilin-1 and moesin bind to 3.0 ± 0.1 and 6.0 ± 0.3 PI(4,5)P_2 head groups, respectively. The PI(4,5)P_2 binding specificity was not hindered by a large concentration of PS lipids in the membrane. None of the residues identified were able to penetrate deep into the membrane but they were instead localized near phosphate atoms. To estimate the binding free energy difference between these two proteins, umbrella sampling simulations were performed. The calculations estimated that cofilin-1 binds with 12 ± 2 kcal/mol and moesin wtih 33 ± 2 kcal/mol, further highlighting the weak interactions of cofilin-1 with the membrane. Thus, the study reveals that despite the interactions to be mainly electrostatic, the dynamics and affinities of ABPs are different which relates to their differential function.

4.2.3 Viral fusion proteins

The genetic material (DNA/RNA) of enveloped viruses is confined within a lipid bilayer, termed virus envelope. The release of the viral genome into host cells requires fusion of the virus envelope with the host cell membrane. However, membrane fusion is opposed by a high kinetic barrier. First, a repulsive dehydration barrier is developed as the distance between the two membranes is within ~30 Å. Second, during the fusion process, membrane strongly deform during fusion, which requires considerable free energy.[406] These kinetic barriers are overcome by the action of membrane fusion glycoproteins that are located at the surface of viruses, where they undergo series of conformational changes during fusion. Membrane fusion begins with the proteins in a high-energy, metastable, compacted state. Upon change in pH or upon sensing receptors on the host membrane, the fusion proteins open up and interact with the host membranes through fusion loops or fusion peptides. The protein releases its stored energy by collapsing, thereby forcing the two membranes into close proximity, hence favoring the formation a hemifused state (Fig. 17A). Finally, a stable low-energy fusion pore is formed, allowing the release of viral genome.

Based on structural differences, fusion proteins can be grouped into three classes, namely, class I, class II, and class III. Class I proteins are characterized by a central trimeric α-helical coiled-coil structure and contain hydrophobic fusion peptides on the N-terminal side (found in retrovirus, filovirus, coronavirus). A typical example is haemagglutinin from influenza A. Class II proteins exhibit β-sheets-rich domains that also contain the fusion loops (found in alphavirus, flavivirus, rubivirus, phlebovirus). Similar to class I proteins, class III proteins contain a central trimeric α-helical coiled-coil structure, while the fusion loops extend β-sheet domains, which is similar to class II (vesicular stomatitis virus, herpes simplex virus, Epstein–Barr virus, baculovirus). Here, we discuss recent work on a class II fusion protein from Rift Valley fever virus (RVFV), which revealed a previously unknown lipid recognition site.

RVFV is transmitted by mosquitos affecting both livestock and humans. The virus enters the host cell through membrane interactions mediated by glycoprotein Gc. However, little was known about the lipid specificity of protein–membrane interactions. Thus, Guardado-Calvo et al. solved a new X-ray structure of glycoprotein Gc that adopts a typical class II conformation.[408] Remarkably, electron density from a lipid mimic was detected in the crystal, at a site that is expected to be in contact with the host membrane during fusion. The structure was next subjected to MD simulations in DOPC/CHOL membranes to investigate specific lipid binding sites at the fusion loops. In agreement with biochemical studies, MD simulations showed that increase of cholesterol concentration resulted in deep insertion of the fusion loops and, hence, stronger binding to the membrane. The cholesterol-dependent stability was attributed to the ability of cholesterol to occupy space between the DOPC lipids, thereby preventing close

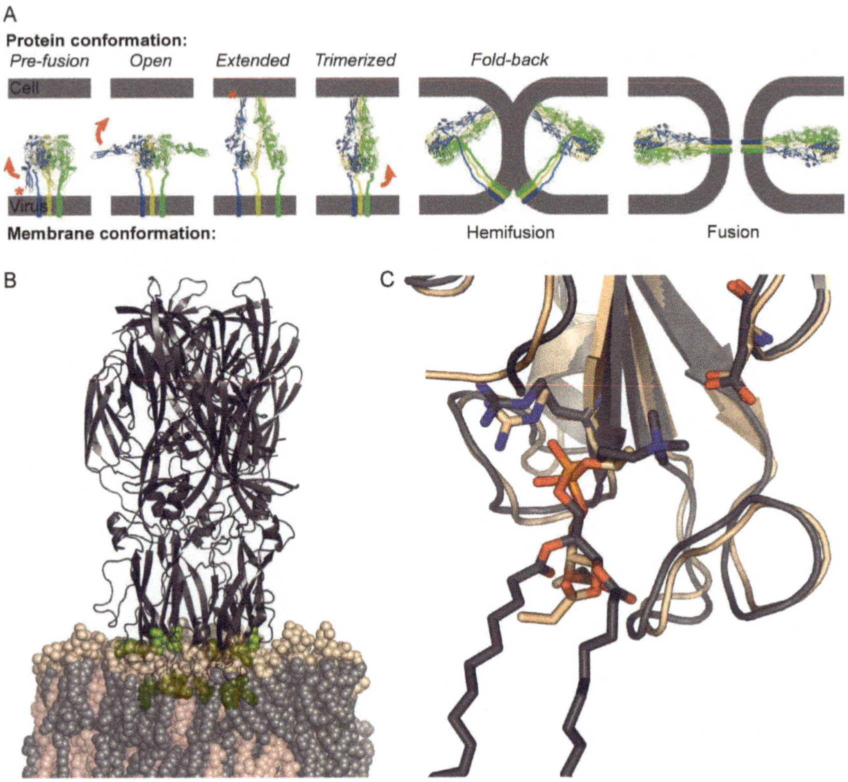

Fig. 17 Viral fusion proteins. (A) Sequential change in protein conformation during membrane fusion.[407] (B) Simulation system of the class II fusion protein Gc of Rift Valley fever virus in contact with a DOPC/CHOL membrane. (C) Closeup view of a DOPC lipid bound to the PC lipid recognition site of Gc. *Gray*: MD simulation. *Brown*: crystal structure with a short-tailed lipid mimic.[408] *Panel A: Reprinted with permission from Kim IS, Jenni S, Stanifer ML, et al. Mechanism of membrane fusion induced by vesicular stomatitis virus G protein. Proc Natl Acad Sci. 2017;114(1):E28–E36, https://doi.org/10.1073/pnas.1618883114, United States National Academy of Sciences.*

packing of DOPC head groups. The loosely packed DOPC head groups allowed the fusion loops to penetrate the membrane, thus forming stable interactions. Further, during the simulations DOPC was seen to bind to the same contact site as also observed in the crystal (Fig. 17B and C). Comparison of the RVFV structure with other class II family members (flavivirus E and alphavirus E1) revealed that the DOPC binding site is largely conserved, suggesting that the site may be widely used by viruses with class II proteins to sense the lipid composition of their host cells.

4.3 Coarse-grained simulations of membrane proteins

4.3.1 *Curvature induction and membrane remodeling by FAM134B reticulon homology domain assist selective ER-phagy*

The endoplasmic reticulum (ER) is an important cellular structure found in eukaryotic cells and composed of phospholipid bilayer delimited tubules and sheets. The ER is associated with different cellular functions depending on tissue location in the body.[409] It is major site of proteins synthesis and transport, protein folding, lipid and steroid synthesis, carbohydrate metabolism, and storage. Newly synthesized polypeptides are folded inside the ER lumen by chaperones, where also postsynthesis glycosylation takes place. Loss of ER function can result in the accumulation of wrongly folded proteins and other stress factors, resulting a major cellular response to deal with this is through an autophagy receptor-mediated degradative ER-autophagy process (ER-phagy).[410] The specific mechanism involved in ER-autophagy has been the subject of different research investigations and it is believed to involve membrane-modulated protein interactions, including the reticulon-homology-domain (RHD) of the ER. The RHD contains the FAM134 protein family (FAM134A, FAM134B, FAM134C) and two autophagy-related proteins, the microtubule associated light chain 3 (LC3) and GATE16/GABARAP.[411]

FAM134B interacts with LC3 using an LC3-interacting region (the LIR motif, sequence DDFELL) that is highly conserved across the FAM134 protein family. For instance, downregulation of the FAM134B[412] or substitution of the LIR motif with alanine residues prevented interaction with LC3 and ER autophagosomes. A number of diseases have been attributed to the loss of FAM134B function, including neuropathy which is associated with the truncation of the protein sequence. The function of the FAM134 family in the ER-phagy is believed to have a relationship with the membrane-shaping RHD structural motif that is conserved throughout the family and shared with other membrane shaping proteins.[413–416] The underlying mechanism, however, remains unsolved as a result of lack of atomic-level insight into its structural and dynamical processes.

FAM134 presents an exciting possibility to interrogate the initial sequence of events, deriving from protein–lipid interactions, that affect membrane structural integrity and drive ER-phagy. To understand these separate processes, Bhaskara et al. employed a multiscale simulation approach that generated structural models for CG-MD simulations from multiple μs-long explicit solvent AA-MD sampling.[417] The model protein consists of two TM helical hairpins that are connected to another two helical segments composed of amphipathic amino acids. The FAM134B-RHD model structures generated from the AA-MD were embedded in MARTINI lipids using both POPC and mixed lipid bilayers (to model the ER membrane) as well as bicelle model systems. Intracellular aqueous and ionic conditions of the physiological membrane were mimicked with MARTINI water, Na^+ and Cl^- models. In total, 95 × 1 μs independent FAM134B-RHD-lipid simulations were performed. As reference simulations, 97 × 1 μs with empty bicelles (no protein present) and 96 × 1 μs membrane systems with inserted $KALP_{15}$ protein were also performed. KALP proteins contain an alternating A/L sequence, flanked by lysine residues and serve as TM model. They exhibit a TM helical structure with different length, depending on the number of repeating A/L residues. The large-scale CG simulations with the MARTINI FF allowed the detection of very interesting structural phenomena which identified the detailed mechanism by which FAM134B (and by extrapolation other RHD proteins) induces membrane curvature in ER-phagy. First, the overall mechanism was found to proceed via an initial FAM134B RHD domain interaction with the membrane, followed by embedding events facilitated by the amphipathic nature of the RHD. The membrane insertion of the RHD motif perturbed the local symmetry of the membrane and created uneven and extensive asymmetric patches around the amphipathic protein segment. This was observed to directly stretch the lipid layer accompanied by increased area per lipid (curvature induction). This structural remodeling was observed in 97% of all independent FAM134B-RHD membrane simulations. And the membrane curvature was eventually observed to result in complete shape transition from an initial bicellar structure to vesicle formation (Fig. 18). In contrast, $KALP_{15}$ only marginally perturbed the membrane as seen in slight reduction in bilayer thickness in 5 out of total 96 independent CG-MD simulations. The dramatic and complete vesiculation observed in the CG simulations involving FAM134B-RHD shed some light on the role of FAM134B-RHD in provoking structural changes in ER membranes. The vesicles formation facilitates autophagosome clearance and firmly establishes the involvement of specific protein–membrane interactions in ER-phagy. The fact that membrane curvature was not observed with $KALP_{15}$ further highlights the high specificity of the process for the amphipathic motif of RHD proteins, which in the FAM134 protein family is structurally coupled to LIR motifs that serve as recognition site for autophagosomes.

In summary, the FAM134 family expressed in ER membranes actively induces formation of smaller vesicles from the ER membrane through their

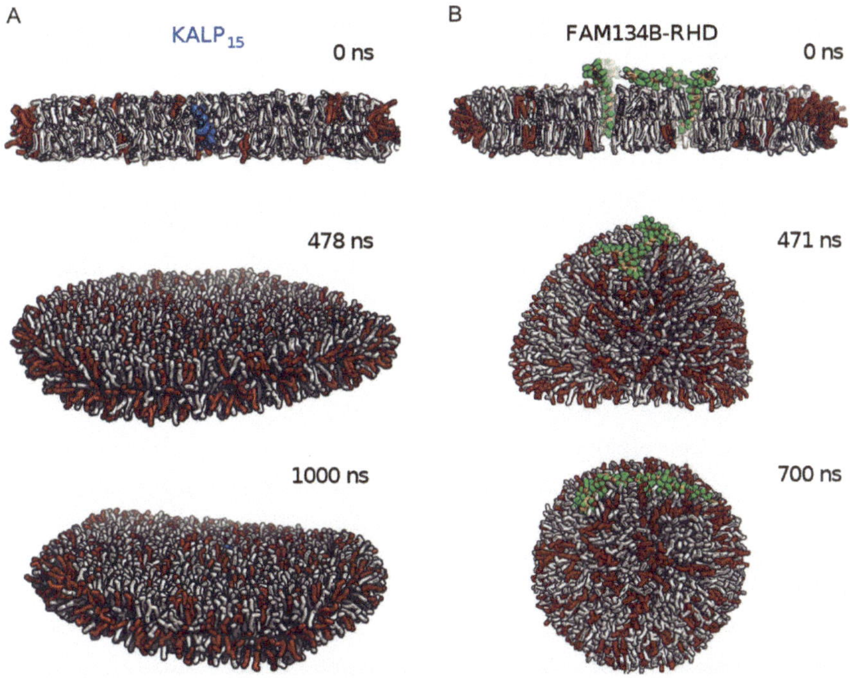

Fig. 18 Membrane remodeling induced by the FAM134B-RHD protein. Snapshots taken from different time points of CG-MD simulations of membrane systems with (A) $KALP_{15}$, which was unable to induce membrane vesiculation within 1 μs, and (B) FAM134B-RHD, which induced vesicle formation within 700 ns. *Panels A and B: Reprinted with the permission from Bhaskara RM, Grumati P, Garcia-Pardo J, et al. Curvature induction and membrane remodeling by FAM134B reticulon homology domain assist selective ER-phagy. Nat Commun. 2019;10(1):2370. Copyright 2019* Springer Nature Limited.

RHD motifs; the protein then employs its LIR motif for interacting with autophagosomes to facilitate ER clearance. This study demonstrates how MD simulations can serve as *molecular microscope* and help to elucidate mechanism of protein–membrane system, in this case the bilayer-to-vesicle transition via induced membrane curvature by the highly conserved amphipathic RHD.

4.3.2 Lipid droplet biogenesis is a liquid phase separation spatially regulated by seipin and membrane curvature

Lipid droplets (LD) are organelles and highly dynamic structurally and in terms of their composition. They are vesicular assemblies generated in the ER and contain neutral lipids, such as triglycerides (TGs) and sterol esters at their core. The neutral lipids are surrounded by a single phospholipid layer with protein molecules (integral and peripheral) decorating the phospholipid monolayer.[418] LDs are associated with functional roles, which include their role as energy storage compartments for excess neutral lipids from where they can be later mobilized for energy generation via enzymatic lipolysis. Moreover, LDs also provide a buffer area where otherwise toxic endogenous fatty acids are stored and prevented from triggering cell death. The biogenesis of LDs has recently been understood to proceed within the two ER membrane leaflets with an initial lens formation resulting from lipid demixing. For this, integral ER membrane proteins are believed to play a critical role. In addition, a critical concentration of neutral lipids within the ER membrane bilayer is crucial for lipid lens formation to get started, while differences in surface tension between the ER lumenal (the ER interior) and cytoplasmic (the ER exterior) lipid layers as well as the ER membrane protein seipin are important to ensure that nascent LDs bud off the ER membrane from the cytoplasmic lipid monolayer. However, the detailed mechanism of LD formation, including the specific contribution of membrane and protein interactions, remains poorly understood.

In the case study presented here, Zoni et al. employed two CG models and several μs-long MD simulations to provide insight into LD biogenesis.[419] In their simulation setups featuring only lipids, i.e., in the absence of proteins they employed the SDK model.[92,420–422] The SDK model is a CG-FF originally developed by Shinoda, DeVane, and Klein (SDK), which has been demonstrated to correctly model the interfacial interaction of lipids and water.[92,93,423] This model was also shown to reliably reproduce the energy barrier for diacylglycerol-water flip-flop and interfacial tensions. For the system setups with the membrane protein seipin, the MARTINI parameter sets were used. Seipin is mostly expressed in the ER and in its functional form, it is assembled into a homo-oligomeric barrel-shaped integral membrane protein with a central pore. It is mostly found concentrated at membrane junctions and in contact with LDs.

Multiple system setups were designed to investigate different structural involvements in LD biogenesis. First, the process of lens formation was examined by inserting variable amounts of triglyceride molecules (between 2% and 10%) randomly between the two leaflets of a bilayer membrane and simulated the systems for 3 × 1.5 μs or until spontaneous triglyceride lens formation occurred. The effect of different lipid types on lens formation dynamics was additionally examined using various binary combinations of DOPC (18:1) with either of DOPE (18:1), POPC (16:0,18:1), POPE (16:0,18:1), DPPC (16:0), CHOL, or diacylglycerol added to it, while an

ER-like membrane system was modeled using a constant composition of DOPC, DOPE, CHOL, and diacylglycerol. The simulation investigating the role of seipin-membrane interaction was initiated from a model, in which seipin was fitted into the available cryo-EM model of human seipin arranged into a ring shape[424] (PDB code: 6DS5). This was inserted into a DOPC membrane containing 2% triglyceride molecules. Three replicas of this setup were simulated for 4 μs. In total, about 75 different simulation systems were investigated with most of them simulated in three replicas, and each replica lasting for 0.5–9.5 μs depending on the particular setup.

In the first CG case study, the role of the RHD motif-containing FAM1343B protein in ER-phagy was examined. It was shown that RHD interaction with the ER bilayer induces membrane curvature and complete membrane remodeling into vesicles for presentation for degradation by autophagosomes. In the present case study, the ER integral protein seipin was also shown to induce membrane curvature in cytoplasmic ER membrane layer as part of the structural events in LD formation. Furthermore, interaction of triglycerides with seipin resulted in the reduction of the diffusion of the neutral lipids within the membrane bilayer, which was found to encourage the accumulation of triglycerides in the vicinity of the membrane protein. The concentration of triglycerides was observed to double around seipin. Interestingly, experimental studies reported seipin to preferentially localize at ER membrane sites that have been reported to be associated with LD formation. In the absence of seipin, LD biogenesis was significantly impaired.[425,426] LD lens formation occurred in the absence of seipin, but at a much reduced rate. Moreover, lens formation was seen to strongly depend on the concentration of free triglycerides, showing the involvement of a liquid–liquid phase separation process that is principally driven by the equilibrium concentration of triglycerides. By increasing the surface tension of the membrane lipids, an effect that was accomplished via a gradual scaling of the box size in *xy*-dimension, lens formation was inhibited. In effect, a lower membrane tension was found to favor lens formation, and thus LD generation. This agrees well with the observation that LDs are formed by the cytoplasmic monolayer of the ER membrane, characterized by a surface tension much lower than that of the lumenal ER membrane monolayer. Other factors found to promote liquid–liquid phase separation include the presence of precursor molecules for neutral lipids in the membrane, especially diacylglycerol, cholesterol, as well as unsaturated lipids. ER-like membrane systems were observed to form TG lenses at solute concentrations much lower than membrane systems without cholesterol and diacylglycerol. This was believed to be an adaptive protection mechanism that allows the ER to spontaneous mop up neutral lipids to prevent toxic aggregation.

Taken together, the use of CG-MD simulations allowed the sampling of critical structural events in LD biogenesis, showing that liquid–liquid phase separation initiated by demixing of neutral lipids within the ER bilayer is accentuated by interaction with the ER membrane protein seipin and the presence of neutral lipid precursor molecules, like cholesterol and diacylglycerol in the membrane. Differential ER-membrane surface tension caused by a higher osmotic pressure within the ER lumen ensures that following the formation of critical intra-ER membrane neutral lipid lenses, LDs are budded off the cytoplasmic monolayer, where a lower surface tension operates. For more detailed information about LD biogenesis, we recommend reading the review article.[427]

4.3.3 Lipid–protein interactions are unique fingerprints for membrane proteins

The emergence of CG-MD simulation in recent years has not just made access to longer timescales feasible, simulating more complex biophysical phenomena has increasingly become possible. In the first CG case study, protein–membrane interaction was investigated using a multiscale simulation approach that clearly demonstrated the influential roles played by a TM protein in membrane vesiculation. The use of a CG-FF made it possible to follow the entire structural process, starting with the interaction of lipid molecules with FAM134B-RHD, to curvature induction in the membrane architecture, and ultimately to vesicle formation. The fact that membrane remodeling was not observed with another TM protein, $KALP_{15}$, indicates that there exists a degree of specificity in the protein–membrane interaction. In other words, rather than merely providing a structurally dynamic matrix for membrane proteins, lipid membranes are capable of establishing specific binding interactions with proteins, both integral and peripheral. This realization suggests that biological membranes are active participants in biochemical processes rather than functioning as a passive bystander in cellular processes. With an average cell membrane containing many different membrane proteins of diverse structural families (constituting up to 30% of the membrane area), a highly complex interaction pattern is expected between the proteins and the membrane lipids.[27,428,429]

One class of membrane proteins are the peripheral proteins that interact with the polar head groups of the membrane lipid or with integral membrane proteins.[430] Peripheral proteins have no contact with the hydrophobic core of the membrane and are located both on the cytosolic surface of the cell membrane, where they are often involved in structural functions (e.g., actin) or in signal transduction (e.g., protein kinase C), and on the extracellular surface of the membrane (e.g., extracellular matrix proteins such as collagen). The integral membrane proteins directly interact with the hydrophobic membrane either via a TM protein domain traversing the membrane, or by covalent tethering through a membrane fatty acid. Interaction with the different protein classes allows the cell membrane to perform its varied biological functions and knowledge of the specific details of these interactions is crucial to understanding the basis of life at the subcellular level.

In the case study reviewed here, Corradi et al. developed a CG-based MD simulation strategy to probe how 10 structurally diverse plasma membrane proteins interact with a membrane[431]; these include glucose transporter 1 (GLUT1), aquaporin-1 (AQP1), AMPA-sensitive glutamate receptor 2 (GluA2), the dopamine transporter (DAT), the epidermal growth factor receptor (EGFR), the voltage-dependent Shaker potassium channel 1.2 (Kv1.2), the sodium/potassium pump (Na^+K^+-ATPase), the P-glycoprotein (Pgp), the delta opioid receptor (δ-OPR), and the prostaglandin H2 synthase (COX1). In this work, lipids of 63 different types were distributed asymmetrically between the two membrane leaflets to provide a realistical model of a plasma membrane.[33,224] In the extracellular membrane leaflet, the incorporated lipids feature a higher degree of lipid chain saturation and include gangliosides and PC, but also to some extent cholesterol, SM, ceramide, and diacylglycerol. The cytosolic membrane leaflet, on the other hand, included most of the charged lipids PA, PE, PS, PI, PIP, PIP_2, and PIP_3 and a higher content of unsaturated fatty acid chains. The inner leaflet further contained lysophosphatidylcholine (LPC) lipids. Using the MARTINI FFs, patches of the membrane model were set up, one patch for each of the 10 proteins. Each patch measured 42 nm in both x- and y-dimensions and contained about 6000 mixed lipids as described above. Four copies of each membrane protein were inserted in the membrane patch with each protein copy separated by a distance of about ~20 nm from the directly neighbored ones, all four of which were arranged in a square formation (Fig. 19). Simulating four copies of the same protein per membrane patch allows for improved statistics. Ten such assemblies were established for the ten protein–membrane assemblies, each of them was simulated for 30 μs. With the aid of weak position restraints, the protein

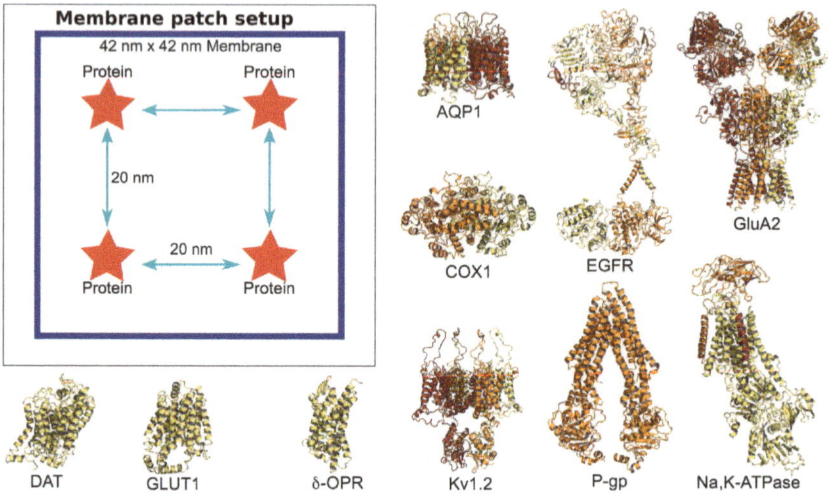

Fig. 19 Protein–membrane simulation setup for probing the lipid–protein interaction *fingerprints*. In the *inset panel*, the *red stars* mark the positions of the protein molecules held in place with a weak position restraining force of 1 kJ mol^{-1} nm^{-2}. Each protein molecule is separated from the other ones by a distance of at least 20 nm. The four protein molecules are embedded in the membrane system (*white background*), which was modeled as a mixture of 63 different lipid types. The assembly is replicated for each of the 10 protein molecules studied, which are shown one the right and below. These proteins are structurally diverse, ranging from single domain globular proteins like DAT, GLUT1, δ-OPR, AQP1, COX1, to more complex multidomain TM protein assemblies like Kv1.2, EGFR, P-gp, GluA2, and $Na^{+}K^{+}$-ATPase. *Reprinted with permission from Corradi V, Mendez-Villuendas E, Ingólfsson HI, et al. Lipid-protein interactions are unique fingerprints for membrane proteins. ACS Cent Sci. 2018;4(6):709–717, https://doi.org/10.1021/acscentsci.8b00143, American Chemical Society. Copyright 2015 American Chemical Society. Further permissions related to the material excerpted should be directed to the ACS.*

molecules were kept in place, preventing them from aggregating. In reality, protein aggregation within the membrane matrix remains a possibility. The use of protein position restraints allows for the lateral diffusion of lipid molecules to be studied; by decoupling this from protein aggregation, it becomes easy to observe how each lipid molecule interacts with the protein molecules.

Each system was then analyzed for any discernible pattern in protein–membrane interactions that is essentially unique to each protein (called *fingerprint*). For each protein, selective interactions with distinct lipid types were observed, resulting in a lipid enrichment pattern in the immediate vicinity of the protein molecules. With different protein macromolecules being present in the cell membrane, it immediately becomes apparent how such nonuniform interaction with the lipids might contribute toward influencing the global properties of the membrane. For this reason, it is also expected that membrane properties will vary with protein composition of the cell membranes, from a particular cell type to another, as well as between individual organisms based on the internal constitution. In other words, the *matrix experience* presented by the membrane continuum is not as uniform as it might first appear but in actual fact is dictated by specific structural phenomena, resulting from each protein and other membrane constituents. A radius of 0.7 nm was used in describing interaction shells around each protein molecule and the reported behavior represents average over four copies of each protein (Fig. 19).

From the complex array of interactions between membrane protein and lipids, the most glaring observations include a generalized depletion of SM, PCs, LPC (except for COX1), and ceramides (with the exception of COX1 and EGFR) in the immediate vicinity of all 10 protein molecules. On the other hand, diacylglycerol, PIP_2, PIP_3, gangliosides, and polyunsaturated lipids were specifically attracted to the local interaction shell around all 10 proteins. The lipid enrichment-depletion behavior was, however, not always symmetrical between the two membrane leaflets. This was most striking with respect to the gangliosides, PIP_2 and PIP_3, which are important lipids involved in cell signaling function. For instance, gangliosides play an active role in protein sorting, while PIP_2 and PIP_3 are crucial in modulating protein–protein interactions within the membrane matrix (e.g., in Kir2.1 channels, see Section 4.1.2). The enrichment of gangliosides around the proteins was most striking in the outer membrane leaflet, while the PIP_2 and PIP_3 lipids were selectively enriched in the membrane inner leaflet.

These results represent a necessary simplification of an extremely complex interaction behavior, especially considering that each protein represents a structural architecture that deviates from a uniform cylindrical geometry. Considering the no irregular shapes of the biological proteins and the varied composition of amino acids, each set of protein–membrane interaction constituted an interaction fingerprint, which strongly influences the lateral organization of lipids in the cell membrane. Since the model employed in this case study investigated each protein molecule in isolation, a more complex membrane with attendant further complication of membrane properties is likely to occur in biological membranes, where different protein molecules simultaneously colocalize in close proximity.

5. Conclusions and future directions

In this chapter, we provide molecular level insights into protein–membrane interactions in various cellular processes. For each case study, we have given a general overview of the system, followed by the most relevant findings from MD simulations. In Section 1, we provide a short introduction to prokaryotic and eukaryotic cells and then discussed in detail the structures of various lipid species which are basic elements of cell membranes. Next, we have introduced membrane proteins, which along with lipids are essential to carry out different cellular functions. We concluded the section discussing various experimental techniques used in the studies of protein–membrane interactions and the role of computational methods, such as MD simulations in further advancing our knowledge in the field. In Section 2, we have reviewed various FFs used to describe the lipids in MD simulations. The section includes a discussion on the availability of lipid species for each FF and explains the membrane properties commonly used to validate the lipid FFs against experimental data. For a more detailed background on FFs, readers are requested to refer to chapters "Pairwise-additive and polarizable atomistic force fields for molecular dynamics simulations of proteins" by Lemkul and "Scale-consistent approach to the derivation of coarse-grained force fields for simulating structure, dynamics, and thermodynamics of biopolymers" by Liwo of this book. In Section 3, we have provided some practical information on how to build protein–membrane systems, a list of MD simulation packages and tools available to analyze the output the data from MD simulations. The section begins with a detailed review on CHARMM-GUI functionalities to build various protein–membrane systems for use with the CHARMM FF, followed by methods to set up systems for use with AMBER, OPLS-AA, GROMOS, and MARTINI FFs. In addition, we have also covered the tools available to build glycoproteins and glycolipids for MD simulations. We have concluded this section listing the tools available to analyze the trajectory files created by protein–membrane MD simulations. Finally, in Section 4, we have reviewed MD studies of protein–membrane systems. The section is split into three subsections: first, we have presented case studies of integral membrane proteins elucidated by atomistic simulations; second, we have reported atomistic simulations of peripheral membrane proteins: third, we selected a few case studies that employed CG MD simulations to study biological process of large protein–membrane systems, which are computationally too demanding for an atomistic approach. The reviewed topics include protein-induced bicelle-to-vesicle transformation, spontaneous triglyceride lens formation in lipid droplet biogenesis, and identifying unique lipid fingerprints for membrane proteins of various topology.

The diverse cases reviewed in Section 4 provided key molecular insights into how specific lipids regulate protein activity, impressively demonstrating that MD simulations are a vital technique for studying protein–membrane systems. As seen in the presented cases, the recent advancements in the supercomputing facilities along with methodological developments regarding the MD codes has already enabled MD simulations at the microsecond timescale. With regular improvements of FFs to increase their accuracy along with developments of advanced algorithms to speed up MD simulations, the efficacy of the MD studies will further increase in the future. In addition, sustained improvement in experimental techniques, particularly in lipidomic studies, has benefited MD studies immensely in narrowing down from a large pool of lipid species to few specific lipids of interest.

In spite of MD simulations being able to complement experimental results, there are few limitations to be addressed. There are several FFs available for both proteins and lipids, but all have their own shortcomings in reproducing experimental data. Moreover, the simulation of protein–membrane systems requires protein and lipid FFs that are compatible with each other and reproduce the properties of both the protein and the lipid bilayer. The FF benchmarks to date are mostly limited to either proteins or to lipids, while their compatibility and ability to provide a balanced description of protein–lipid interactions has not yet been systematically checked. In terms of application challenges, it should be considered that biological membranes are complex with asymmetric distributions of different lipid species between the two leaflets. Thus, even in the absence of proteins, membranes can undergo curvature and separate into specialized lipid domains. However, most MD simulations are carried out with planar lipid membranes composed of a few selected lipids, rendering them incapable to mimicking the lipid complexity of biological membranes or the observed membrane curvature. Moreover, the traditional way of setting up infinite membrane simulations with periodic boundary conditions further hinders any stabilization of membrane curvature. While mimicking biological membrane composition would be a challenge, spontaneous membrane curvature or protein-induced curvature can be studied using discontinuous finite membranes, such as bicelles, an ideal system to study membrane curvature. Proteins undergo several posttranslational modifications before being transferred into the plasma membrane, and one of the modifications is the covalent attachment of sugar residues or glycans to the proteins. More than 50% of all the eukaryotic proteins are glycosylated, however, only few MD studies have considered protein glycosylation. With the availability of tools to identify glycan sites and build glycoproteins, future MD studies should increase the system complexity by considering glycosylation when required. Such studies would lead to several discussions in the glycobiology field in terms of structure, function and composition of glycans. Experimentally observed conformational changes in protein structure on ligand binding or observing spontaneous lipid-mediated protein structural change is still a challenge with conventional MD simulations. There are several enhanced sampling methods to overcome the limitations, but the systems have to be treated with several biases which would further influence the end results.

In spite of these limitations, MD simulations of protein-membrane systems have rapidly progressed over the last few years and complemented experimental studies in providing molecular and dynamic details at atomistic resolution for protein–lipid interactions.

Acknowledgments

B.S. and J.L. received funding for this project from the Deutsche Forschungsgemeinschaft (DFG, German Research Foundation, http://www.dfg.de/) through grant number 267205415 (CRC 1208, project A07). B.S. and O.O.O. acknowledge support through the Initiative and Networking Fund of the Helmholtz Association, grant number ERC-RA-0034. J.S.H. and C.P. acknowledge support through the Deutsche Forschungsgemeinschaft (SFB 1027/B7). We thank Kalina Atkovska for preparing parts of Fig. 17.

References

1. Bakheet TM, Doig AJ. Properties and identification of human protein drug targets. *Bioinformatics*. 2009;25(4):451–457. https://doi.org/10.1093/bioinformatics/btp002.
2. Yildirim MA, Goh KI, Cusick ME, Barabsi AL, Vidal M. Drug-target network. *Nat Biotechnol*. 2007;25:1119–1126. https://doi.org/10.1038/nbt1338. (a) Berg C, Engels I, Rothbart A, et al. Human mature red blood cells express caspase-3 and caspase-8, but are devoid of mitochondrial regulators of apoptosis. *Cell Death Differ*. 2001;8:1197–1206. https://doi.org/10.1038/sj.cdd.4400905. (b) Ji P, Murata-Hori M, Lodish HF. Formation of mammalian erythrocytes: chromatin condensation and enucleation. *Trends Cell Biol*. 2011;21(7):409–415. issn: 0962–8924. https://doi.org/10.1016/j.tcb.2011.04.003.
3. Lyman E, Hsieh CL, Eggeling C. From dynamics to membrane organization: experimental breakthroughs occasion a "modeling manifesto". *Biophys J*. 2018;115(4): 595–604. https://doi.org/10.1016/j.bpj.2018.07.012.
4. Singer SJ, Nicolson GL. The fluid mosaic model of the structure of cell membranes. *Science*. 1972;175(4023):720–731. https://doi.org/10.1126/science.175.4023.720.
5. Engelman D. Membranes are more mosaic than fluid. *Nature*. 2014;438:578–580. https://doi.org/10.1016/j.bbamem.2014.01.006.

6. Nicolson GL. The fluid-mosaic model of membrane structure: still relevant to understanding the structure, function and dynamics of biological membranes after more than 40 years. *Biochim Biophys Acta Biomembr.* 2014;1838(6):1451–1466. https://doi.org/10.1016/j.bbamem.2013.10.019.
7. Goñi FM. The basic structure and dynamics of cell membranes: an update of the Singer-Nicolson model. *Biochim Biophys Acta Biomembr.* 2014;1838(6):1467–1476. https://doi.org/10.1016/j.bbamem.2014.01.006.
8. Tanford C. The hydrophobic effect and the organization of living matter. *Science.* 1978;200(4345):1012–1018. https://doi.org/10.1126/science.653353.
9. Yeagle PL. Chapter 2—The lipids of biological membranes. In: Yeagle PL, ed. *The Membranes of Cells.* 3rd ed. Boston: Academic Press; 2016:27–56. https://doi.org/10.1016/B978-0-12-800047-2.00002-4.
10. Marventano S, Kolacz P, Castellano S, et al. A review of recent evidence in human studies of n-3 and n-6 PUFA intake on cardiovascular disease, cancer, and depressive disorders: does the ratio really matter? *Int J Food Sci Nutr.* 2015;66(6):611–622. https://doi.org/10.3109/09637486.2015.1077790.
11. Spector AA. Essentiality of fatty acids. *Lipids.* 1999;34(Pt. 1):S1–S3. https://doi.org/10.1007/BF02562220.
12. Harayama T, Riezman H. Understanding the diversity of membrane lipid composition. *Nat Rev Mol Cell Biol.* 2018;19(5):281–296. https://doi.org/10.1038/nrm.2017.138.
13. Yamashita A, Hayashi H, Nemoto-Sasaki Y, et al. Acyltransferases and transacylases that determine the fatty acid composition of glycerolipids and the metabolism of bioactive lipid mediators in mammalian cells and model organisms. *Prog Lipid Res.* 2014;53:18–81. https://doi.org/10.1016/j.plipres.2013.10.001.
14. Krauß M, Haucke V. Phosphoinositides: Regulators of membrane traffic and protein function. *FEBS Lett.* 2007;581(11):2105–2111. https://doi.org/10.1016/j.febslet.2007.01.089.
15. Viaud J, Mansour R, Antkowiak A, et al. Phosphoinositides: important lipids in the coordination of cell dynamics. *Biochimie.* 2016;125:250–258. https://doi.org/10.1016/j.biochi.2015.09.005.
16. Dowhan W, Bogdanov M, Mileykovskaya E. Chapter 1—Functional roles of lipids in membranes. In: Ridgway ND, McLeod RS, eds. *Biochemistry of Lipids, Lipoproteins and Membranes.* 6th ed. Boston: Elsevier; 2016:1–40. https://doi.org/10.1016/B978-0-444-63438-2.00001-8.
17. Vance JE. Phospholipid synthesis and transport in mammalian cells. *Traffic.* 2015;16(1):1–18. https://doi.org/10.1111/tra.12230.
18. Chitraju C, Trötzmüller M, Hartler J, et al. Lipidomic analysis of lipid droplets from murine hepatocytes reveals distinct signatures for nutritional stress. *J Lipid Res.* 2012;53(10):2141–2152. https://doi.org/10.1194/jlr.M028902.
19. Raetz CR, Dowhan W. Biosynthesis and function of phospholipids in Escherichia coli. *J Biol Chem.* 1990;265(3):1235–1238.
20. Ramstedt B, Slotte JP. Membrane properties of sphingomyelins. *FEBS Lett.* 2002;531(1):33–37. https://doi.org/10.1016/S0014-5793(02)03406-3.
21. Sezgin E, Levental I, Mayor S, Eggeling C. The mystery of membrane organization: composition, regulation and roles of lipid rafts. *Nat Rev Mol Cell Biol.* 2017; 18:361–374. https://doi.org/10.1038/nrm.2017.16.
22. Tan RX, Chen JH. The cerebrosides. *Nat Prod Rep.* 2003;20:509–534. https://doi.org/10.1039/B307243F.
23. López CA, Rzepiela AJ, de Vries AH, et al. Martini coarse-grained force field: extension to carbohydrates. *J Chem Theory Comput.* 2009;5(12):3195–3210. https://doi.org/10.1021/ct900313w.

24. Yanagisawa K. Role of gangliosides in Alzheimer's disease. *Biochim Biophys Acta Biomembr.* 2007;1768(8):1943–1951. https://doi.org/10.1016/j.bbamem.2007.01.018.
25. Matsuzaki K, Kato K, Yanagisawa K. Aβ polymerization through interaction with membrane gangliosides. *Biochim Biophys Acta.* 2010;1801(8):868–877. https://doi.org/10.1016/j.bbalip.2010.01.008.
26. Benvegnu T, Lemiègre L, Cammas-Marion S. Archaeal lipids: innovative materials for biotechnological applications. *Eur J Org Chem.* 2008;2008(28):4725–4744. https://doi.org/10.1002/ejoc.200800452.
27. van Meer G, Voelker DR, Feigenson GW. Membrane lipids: where they are and how they behave. *Nat Rev Mol Cell Biol.* 2008;9(2):112–124. https://doi.org/10.1038/nrm2330.
28. van Meer G, de Kroon AIPM. Lipid map of the mammalian cell. *J Cell Sci.* 2011;124(1):5–8. https://doi.org/10.1242/jcs.071233.
29. Simons K, Ikonen E. Functional rafts in cell membranes. *Nature.* 1977;387(43):569–572. https://doi.org/10.1038/42408.
30. Baumgart T, Hammond AT, Sengupta P, et al. Large-scale fluid/fluid phase separation of proteins and lipids in giant plasma membrane vesicles. *Proc Natl Acad Sci.* 2007;104(9):3165–3170. https://doi.org/10.1073/pnas.0611357104.
31. Levental KR, Levental I. Chapter Two–Giant plasma membrane vesicles: models for understanding membrane organization. In: Kenworthy AK, ed. *Lipid Domains. Current Topics in Membranes.* vol. 75. Academic Press; 2015:25–57. https://doi.org/10.1016/bs.ctm.2015.03.009.
32. Risselada HJ, Marrink SJ. The molecular face of lipid rafts in model membranes. *Proc Natl Acad Sci.* 2008;105(45):17367–17372. https://doi.org/10.1073/pnas.0807527105.
33. Ingólfsson HI, Melo MN, van Eerden FJ, et al. Lipid organization of the plasma membrane. *J Am Chem Soc.* 2014;136(41):14554–14559. https://doi.org/10.1021/ja507832e.
34. Lingwood D, Simons K. Lipid rafts as a membrane-organizing principle. *Science.* 2010;327(5961):46–50. https://doi.org/10.1126/science.1174621.
35. Muller MP, Jiang T, Sun C, et al. Characterization of lipid-protein interactions and lipid-mediated modulation of membrane protein function through molecular simulation. *Chem Rev.* 2019;119(9):6086–6161. https://doi.org/10.1021/acs.chemrev.8b00608.
36. Laursen M, Yatime L, Nissen P, Fedosova NU. Crystal structure of the high-affinity Na+, K+-ATPase-ouabain complex with Mg2+ bound in the cation binding site. *Proc Natl Acad Sci USA.* 2013;110(27):10958–10963. https://doi.org/10.1073/pnas.1222308110.
37. Hanson MA, Cherezov V, Griffith MT, et al. A specific cholesterol binding site is established by the 2.8 Å structure of the human β2-adrenergic receptor. *Structure.* 2008;16(6):897–905. https://doi.org/10.1016/j.str.2008.05.001.
38. Efremov RG, Sazanov LA. Structure of Escherichia coli OmpF porin from lipidic mesophase. *J Struct Biol.* 2012;178(3):311–318. https://doi.org/10.1016/j.jsb.2012.03.005.
39. Strandberg E, Esteban-Martín S, Ulrich AS, Salgado J. Hydrophobic mismatch of mobile transmembrane helices: merging theory and experiments. *Biochim Biophys Acta Biomembr.* 2012;1818(5):1242–1249. https://doi.org/10.1016/j.bbamem.2012.01.023.
40. Marinko JT, Huang H, Penn WD, Capra JA, Schlebach JP, Sanders CR. Folding and misfolding of human membrane proteins in health and disease: from single molecules to cellular proteostasis. *Chem Rev.* 2019;119(9):5537–5606. https://doi.org/10.1021/acs.chemrev.8b00532.
41. Murata K, Wolf M. Cryo-electron microscopy for structural analysis of dynamic biological macromolecules. *Biochim Biophys Acta Gen Subj.* 2018;1862(2):324–334. https://doi.org/10.1016/j.bbagen.2017.07.020.

42. Berman HM, Westbrook J, Feng Z, et al. The protein data bank. *Nucleic Acids Res.* 2000;28(1):235–242. https://doi.org/10.1093/nar/28.1.235.
43. Burley SK, Berman HM, Bhikadiya C, et al. RCSB protein data bank: biological macromolecular structures enabling research and education in fundamental biology, biomedicine, biotechnology and energy. *Nucleic Acids Res.* 2018;47(D1):D464–D474. https://doi.org/10.1093/nar/gky1004.
44. Lomize MA, Lomize AL, Pogozheva ID, Mosberg HI. OPM: orientations of proteins in membranes database. *Bioinformatics.* 2006;22(5):623–625. https://doi.org/10.1093/bioinformatics/btk023.
45. Kox A, Michels J, Wiegel F. Simulation of a lipid monolayer using molecular dynamics. *Nature.* 1980;287:317–319. https://doi.org/10.1038/287317a0.
46. van der Ploeg P, Berendsen HJC. Molecular dynamics simulation of a bilayer membrane. *J Chem Phys.* 1982;76(6):3271–3276. https://doi.org/10.1063/1.443321.
47. van der Ploeg P, Berendsen HJC. Molecular dynamics of a bilayer membrane. *Mol Phys.* 1983;49(1):233–248. https://doi.org/10.1080/00268978300101131.
48. Marrink SJ, Berendsen HJC. Simulation of water transport through a lipid membrane. *J Phys Chem.* 1994;98(15):4155–4168. https://doi.org/10.1021/j100066a040.
49. Tieleman DP, Berendsen HJC. Molecular dynamics simulations of a fully hydrated dipalmitoylphosphatidylcholine bilayer with different macroscopic boundary conditions and parameters. *J Chem Phys.* 1996;105(11):4871–4880. https://doi.org/10.1063/1.472323.
50. Heller H, Schaefer M, Schulten K. Molecular dynamics simulation of a bilayer of 200 lipids in the gel and in the liquid crystal phase. *J Phys Chem.* 1993;97(31):8343–8360. https://doi.org/10.1021/j100133a034.
51. Damodaran KV, Merz Jr KM, Gaber BP. Structure and dynamics of the dilauroylphosphatidylethanolamine lipid bilayer. *Biochemistry.* 1992;31(33):7656–7664. https://doi.org/10.1021/bi00148a029.
52. Damodaran KV, Merz Jr KM. Head group-water interactions in lipid bilayers: a comparison between DMPC- and DLPE-based lipid bilayers. *Langmuir.* 1993;9(5):1179–1183. https://doi.org/10.1021/la00029a005.
53. Edholm O, Johansson J. Lipid bilayer polypeptide interactions studied by molecular dynamics simulation. *Eur Biophys J.* 1987;14(4):203–209. https://doi.org/10.1007/BF00256353.
54. Zhou F, Schulten K. Molecular dynamics study of phospholipase A2 on a membrane surface. *Proteins.* 1996;25(1):12–27. https://doi.org/10.1002/(SICI)1097-0134(199605)25:13.0.CO;2-M.
55. Woolf TB, Roux B. Molecular dynamics simulation of the gramicidin channel in a phospholipid bilayer. *Proc Natl Acad Sci.* 1994;91(24):11631–11635. https://doi.org/10.1073/pnas.91.24.11631.
56. Zhong Q, Jiang Q, Moore PB, Newns DM, Klein ML. Molecular dynamics simulation of a synthetic ion channel. *Biophys J.* 1998;74(1):3–10. https://doi.org/10.1016/S0006-3495(98)77761-6.
57. Perilla JR, Goh BC, Cassidy CK, et al. Molecular dynamics simulations of large macromolecular complexes. *Curr Opin Struct Biol.* 2015;31:64–74. https://doi.org/10.1016/j.sbi.2015.03.007.
58. Li J, Lü S, Liu Y, et al. Identification of the conformational transition pathway in PIP_2 opening Kir channels. *Sci Rep.* 2015;1858(7, Pt. B):1610–1618. https://doi.org/10.1038/srep11289.
59. Kirsch SA, Böckmann RA. Membrane pore formation in atomistic and coarse-grained simulations. *Biochim Biophys Acta Biomembr.* 2016;1858(10):2266–2277. https://doi.org/10.1016/j.bbamem.2015.12.031.

60. Reddy T, Sansom MSP. Computational virology: from the inside out. *Biochim Biophys Acta Biomembr.* 2016;1858(7, Part B):1610–1618. https://doi.org/10.1016/j.bbamem.2016.02.007.
61. Marrink SJ, Corradi V, Souza PCT, et al. Computational modeling of realistic cell membranes. *Chem Rev.* 2019;119(9):6184–6226. https://doi.org/10.1021/acs.chemrev.8b00460.
62. Marrink SJ, de Vries AH, Mark AE. Coarse grained model for semiquantitative lipid simulations. *J Phys Chem B.* 2004;108(2):750–760. https://doi.org/10.1021/jp036508g.
63. Marrink SJ, Risselada HJ, Yefimov S, Tieleman DP, de Vries AH. The MARTINI force field: coarse grained model for biomolecular simulations. *J Phys Chem B.* 2007;111(27):7812–7824. https://doi.org/10.1021/jp071097f.
64. Ohkubo YZ, Pogorelov TV, Arcario MJ, Christensen GA, Tajkhorshid E. Accelerating membrane insertion of peripheral proteins with a novel membrane mimetic model. *Biophys J.* 2012;102(9):2130–2139. https://doi.org/10.1016/j.bpj.2012.03.015.
65. Kmiecik S, Gront D, Kolinski M, Wieteska L, Dawid AE, Kolinski A. Coarse-grained protein models and their applications. *Chem Rev.* 2016;116(14):7898–7936. https://doi.org/10.1021/acs.chemrev.6b00163.
66. Dutzler R, Campbell EB, MacKinnon R. Gating the selectivity filter in ClC chloride channels. *Science.* 2003;300(5616):108–112. https://doi.org/10.1126/science.1082708.
67. Schlenkrich M, Brickmann J, MacKerell Jr AD, Karplus M. An empirical potential energy function for phospholipids: criteria for parameter optimization and applications. In: Merz Jr. KM, Roux B, eds. *Biological Membranes: A Molecular Perspective From Computation and Experiment.* Boston, MA: Birkhäuser Boston; 1996:31–81. https://doi.org/10.1007/978-1-4684-8580-6_2.
68. Feller SE, Yin D, Pastor RW, MacKerell Jr AD. Molecular dynamics simulation of unsaturated lipid bilayers at low hydration: parameterization and comparison with diffraction studies. *Biophys J.* 1997;73(5):2269–2279. https://doi.org/10.1016/S0006-3495(97)78259-6.
69. Feller SE, MacKerell Jr AD. An improved empirical potential energy function for molecular simulations of phospholipids. *J Phys Chem B.* 2000;104(31):7510–7515. https://doi.org/10.1021/jp0007843.
70. Klauda JB, Brooks BR, MacKerell Jr AD, Venable RM, Pastor RW. An ab initio study on the torsional surface of alkanes and its effect on molecular simulations of alkanes and a DPPC bilayer. *J Phys Chem B.* 2005;109(11):5300–5311. https://doi.org/10.1021/jp0468096.
71. Klauda JB, Pastor RW, Brooks BR. Adjacent gauche stabilization in linear alkanes: implications for polymer models and conformational analysis. *J Phys Chem B.* 2005;109(33):15684–15686. https://doi.org/10.1021/jp0527608.
72. Zhuang X, Dávila-Contreras EM, Beaven AH, Beaven AH, Im W, Klauda JB. An extensive simulation study of lipid bilayer properties with different head groups, acyl chain lengths, and chain saturations. *Biochim Biophys Acta Biomembr.* 2016;1858(12):3093–3104. https://doi.org/10.1016/j.bbamem.2016.09.016.
73. Klauda JB, Venable RM, Freites JA, et al. Update of the CHARMM all-atom additive force field for lipids: validation on six lipid types. *J Phys Chem B.* 2010;114(23):7830–7843. https://doi.org/10.1021/jp101759q.
74. Lee S, Tran A, Allsopp M, Lim JB, Hnin J, Klauda JB. CHARMM36 united atom chain model for lipids and surfactants. *J Phys Chem B.* 2014;118(2):547–556. https://doi.org/10.1021/jp410344g.

75. Jójárt B, Martinek TA. Performance of the general amber force field in modeling aqueous POPC membrane bilayers. *J Comput Chem.* 2007;28(12):2051–2058. https://doi.org/10.1002/jcc.20748.
76. Rosso L, Gould IR. Structure and dynamics of phospholipid bilayers using recently developed general all-atom force fields. *J Comput Chem.* 2008;29(1):24–37. https://doi.org/10.1002/jcc.20675.
77. Skjevik ÅA, Madej BD, Walker RC, Teigen K. Lipid11: a modular framework for lipid simulations using amber. *J Phys Chem B.* 2012;116(36):11124–11136. https://doi.org/10.1021/jp3059992.
78. Dickson CJ, Madej BD, Skjevik ÅA, et al. Lipid14: the amber lipid force field. *J Chem Theory Comput.* 2014;10(2):865–879. https://doi.org/10.1021/ct4010307.
79. Jämbeck JPM, Lyubartsev AP. An extension and further validation of an all-atomistic force field for biological membranes. *J Chem Theory Comput.* 2012;8(8):2938–2948. https://doi.org/10.1021/ct300342n.
80. Jämbeck JPM, Lyubartsev AP. Another piece of the membrane puzzle: extending slipids further. *J Chem Theory Comput.* 2013;9(1):774–784. https://doi.org/10.1021/ct300777p.
81. Pasenkiewicz-Gierula M, Takaoka Y, Miyagawa H, Kitamura K, Kusumi A. Charge pairing of headgroups in phosphatidylcholine membranes: a molecular dynamics simulation study. *Biophys J.* 1999;76(3):1228–1240. https://doi.org/10.1016/S0006-3495(99)77286-3.
82. Pasenkiewicz-Gierula M, Róg T, Kitamura K, Kusumi A. Cholesterol effects on the phosphatidylcholine bilayer polar region: a molecular simulation study. *Biophys J.* 2000;78(3):1376–1389. https://doi.org/10.1016/S0006-3495(00)76691-4.
83. Maciejewski A, Pasenkiewicz-Gierula M, Cramariuc O, Vattulainen I, Rog T. Refined OPLS all-atom force field for saturated phosphatidylcholine bilayers at full hydration. *J Phys Chem B.* 2014;118(17):4571–4581. https://doi.org/10.1021/jp5016627.
84. Berger O, Edholm O, Jähnig F. Molecular dynamics simulations of a fluid bilayer of dipalmitoylphosphatidylcholine at full hydration, constant pressure, and constant temperature. *Biophys J.* 1997;72(5):2002–2013. https://doi.org/10.1016/S0006-3495(97)78845-3.
85. Chiu SW, Clark MM, Jakobsson E, Subramaniam S, Scott HL. Optimization of hydrocarbon chain interaction parameters: application to the simulation of fluid phase lipid bilayers. *J Phys Chem B.* 1999;103(30):6323–6327. https://doi.org/10.1021/jp983219x.
86. Chandrasekhar I, Kastenholz M, Lins RD, et al. A consistent potential energy parameter set for lipids: dipalmitoylphosphatidylcholine as a benchmark of the GROMOS96 45A3 force field. *Eur Biophys J.* 2003;32(1):67–77.
87. Oostenbrink C, Villa A, Mark AE, Van Gunsteren WF. A biomolecular force field based on the free enthalpy of hydration and solvation: the GROMOS force-field parameter sets 53A5 and 53A6. *J Comput Chem.* 2004;25(13):1656–1676. https://doi.org/10.1002/jcc.20090.
88. Kukol A. Lipid models for united-atom molecular dynamics simulations of proteins. *J Chem Theory Comput.* 2009;5(3):615–626. https://doi.org/10.1021/ct8003468.
89. Piggot TJ, Holdbrook DA, Khalid S. Electroporation of the E. coli and S. Aureus membranes: molecular dynamics simulations of complex bacterial membranes. *J Phys Chem B.* 2011;115(45):13381–13388. https://doi.org/10.1021/jp207013v.
90. Piggot TJ, Piñeiro A, Khalid S. Molecular dynamics simulations of phosphatidylcholine membranes: a comparative force field study. *J Chem Theory Comput.* 2012;8(11):4593–4609. https://doi.org/10.1021/ct3003157.

91. Schmid N, Eichenberger AP, Choutko A, et al. Definition and testing of the GROMOS force-field versions 54A7 and 54B7. *Eur Biophys J.* 2011;40(7):843. https://doi.org/10.1007/s00249-011-0700-9.
92. Shinoda W, DeVane R, Klein ML. Multi-property fitting and parameterization of a coarse grained model for aqueous surfactants. *Mol Simul.* 2007;33(1–2):27–36. https://doi.org/10.1080/08927020601054050.
93. Shinoda W, DeVane R, Klein ML. Zwitterionic lipid assemblies: molecular dynamics studies of monolayers, bilayers, and vesicles using a new coarse grain force field. *J Phys Chem B.* 2010;114(20):6836–6849. https://doi.org/10.1021/jp9107206.
94. T Darden DY, Pedersen L. Particle mesh Ewald: an N·Log(N) method for Ewald sums in large systems. *J Chem Phys.* 1993;98(12):10089–10092.
95. Essmann U, Perera L, Berkowitz ML. A smooth particle mesh Ewald method. *J Chem Phys.* 1995;103(19):8577–8593.
96. Poger D, Caron B, Mark AE. Validating lipid force fields against experimental data: progress, challenges and perspectives. *Biochim Biophys Acta Biomembr.* 2016;1858 (7, Pt. B):1556–1565. https://doi.org/10.1016/j.bbamem.2016.01.029.
97. Leonard AN, Wang E, Monje-Galvan V, Klauda JB. Developing and testing of lipid force fields with applications to modeling cellular membranes. *Chem Rev.* 2019;119(9):6227–6269. https://doi.org/10.1021/acs.chemrev.8b00384.
98. Moradi S, Nowroozi A, Shahlaei M. Shedding light on the structural properties of lipid bilayers using molecular dynamics simulation: a review study. *RSC Adv.* 2019;9(8):4644–4658. https://doi.org/10.1039/C8RA08441F.
99. Falck E, Patra M, Karttunen M, Hyvönen MT, Vattulainen I. Lessons of slicing membranes: interplay of packing, free area, and lateral diffusion in phospholipid/cholesterol bilayers. *Biophys J.* 2004;87(2):1076–1091. https://doi.org/10.1529/biophysj.104.041368.
100. Mori T, Ogushi F, Sugita Y. Analysis of lipid surface area in protein-membrane systems combining Voronoi tessellation and Monte Carlo integration methods. *J Comput Chem.* 2012;33(3):286–293. https://doi.org/10.1002/jcc.21973.
101. Lukat G, Krüger J, Sommer B. APL@Voro: a voronoi-based membrane analysis tool for GROMACS trajectories. *J Chem Inf Model.* 2013;53(11):2908–2925. https://doi.org/10.1021/ci400172g.
102. Poger D, Mark AE. The relative effect of sterols and hopanoids on lipid bilayers: when comparable is not identical. *J Phys Chem B.* 2013;117(50):16129–16140. https://doi.org/10.1021/jp409748d.
103. Petrache HI, Tristram-Nagle S, Nagle JF. Fluid phase structure of EPC and DMPC bilayers. *Chem Phys Lipids.* 1998;95(1):83–94. https://doi.org/10.1016/S0009-3084(98)00068-1.
104. Kučerka N, Gallová J, Uhríková D, et al. Areas of monounsaturated diacylphosphatidylcholines. *Biophys J.* 2009;97(7):1926–1932. https://doi.org/10.1016/j.bpj.2009.06.050.
105. Leftin A, Brown MF. An NMR database for simulations of membrane dynamics. *Biochim Biophys Acta Biomembr.* 2011;1808(3):818–839. https://doi.org/10.1016/j.bbamem.2010.11.027.
106. Vaz WL, Almeida PF. Microscopic versus macroscopic diffusion in one-component fluid phase lipid bilayer membranes. *Biophys J.* 1991;60(6):1553–1554. https://doi.org/10.1016/S0006-3495(91)82190-7.
107. Raman EP, Guvench O, MacKerell Jr AD. CHARMM additive all-atom force field for glycosidic linkages in carbohydrates involving furanoses. *J Phys Chem B.* 2010;114(40):12981–12994. https://doi.org/10.1021/jp105758h.
108. Guvench O, Hatcher E, Venable RM, Pastor RW, MacKerell Jr AD. CHARMM additive all-atom force field for glycosidic linkages between hexopyranoses. *J Chem Theory Comput.* 2009;5(9):2353–2370. https://doi.org/10.1021/ct900242e.

109. Guvench O, Mallajosyula SS, Raman EP, et al. Charmm additive all-atom force field for carbohydrate derivatives and its utility in polysaccharide and carbohydrate-protein modeling. *J Chem Theory Comput.* 2011;7(10):3162–3180. https://doi.org/10.1021/ct200328p.
110. Mallajosyula SS, Guvench O, Hatcher E, MacKerell Jr AD. CHARMM additive all-atom force field for phosphate and sulfate linked to carbohydrates. *J Chem Theory Comput.* 2012;8(2):759–776. https://doi.org/10.1021/ct200792v.
111. Hart K, Foloppe N, Baker CM, Denning EJ, Nilsson L, MacKerell Jr AD. Optimization of the CHARMM additive force field for DNA: improved treatment of the BI/BII conformational equilibrium. *J Chem Theory Comput.* 2012;8(1): 348–362. https://doi.org/10.1021/ct200723y.
112. Denning EJ, Priyakumar UD, Nilsson L, Mackerell Jr AD. Impact of 2′-hydroxyl sampling on the conformational properties of RNA: update of the CHARMM all-atom additive force field for RNA. *J Comput Chem.* 2011;32(9):1929–1943. https://doi.org/10.1002/jcc.21777.
113. Huang J, MacKerell Jr AD. CHARMM36 all-atom additive protein force field: validation based on comparison to NMR data. *J Comput Chem.* 2013;34(25):2135–2145. https://doi.org/10.1002/jcc.23354.
114. Huang J, Rauscher S, Nawrocki G, et al. Charmm36m: an improved force field for folded and intrinsically disordered proteins. *Nat Methods.* 2017;14(1):71.
115. Vanommeslaeghe K, Hatcher E, Acharya C, et al. CHARMM general force field: a force field for drug-like molecules compatible with the CHARMM all-atom additive biological force fields. *J Comput Chem.* 2010;31(4):671–690. https://doi.org/10.1002/jcc.21367.
116. Benz RW, Castro-Román F, Tobias DJ, White SH. Experimental validation of molecular dynamics simulations of lipid bilayers: a new approach. *Biophys J.* 2005;88(2):805–817. https://doi.org/10.1529/biophysj.104.046821.
117. Strader ML, Feller SE. A flexible all-atom model of dimethyl sulfoxide for molecular dynamics simulations. *J Phys Chem A.* 2002;106(6):1074–1080. https://doi.org/10.1021/jp013658n.
118. Jensen MØ, Mouritsen OG, Peters GH. Simulations of a membrane-anchored peptide: structure, dynamics, and influence on bilayer properties. *Biophys J.* 2004;86(6): 3556–3575. https://doi.org/10.1529/biophysj.103.029140.
119. Prakash P, Sankararamakrishnan R. Force field dependence of phospholipid headgroup and acyl chain properties: comparative molecular dynamics simulations of DMPC bilayers. *J Comput Chem.* 2010;31(2):266–277. https://doi.org/10.1002/jcc.21313.
120. Högberg CJ, Nikitin AM, Lyubartsev AP. Modification of the CHARMM force field for DMPC lipid bilayer. *J Comput Chem.* 2008;29(14):2359–2369. https://doi.org/10.1002/jcc.20974.
121. Klauda JB, Kučerka N, Brooks BR, Pastor RW, Nagle JF. Simulation-based methods for interpreting X-ray data from lipid bilayers. *Biophys J.* 2006;90(8):2796–2807. https://doi.org/10.1529/biophysj.105.075697.
122. Sonne J, Jensen MØ, Hansen FY, Hemmingsen L, Peters GH. Reparameterization of all-atom dipalmitoylphosphatidylcholine lipid parameters enables simulation of fluid bilayers at zero tension. *Biophys J.* 2007;92(12):4157–4167. https://doi.org/10.1529/biophysj.106.087130.
123. Klauda JB, Monje V, Kim T, Im W. Improving the CHARMM force field for polyunsaturated fatty acid chains. *J Phys Chem B.* 2012;116(31):9424–9431. https://doi.org/10.1021/jp304056p.
124. Lim JB, Klauda JB. Lipid chain branching at the iso- and anteiso-positions in complex chlamydia membranes: a molecular dynamics study. *Biochim Biophys Acta Biomembr.* 2011;1808(1):323–331. https://doi.org/10.1016/j.bbamem.2010.07.036.

125. Pandit KR, Klauda JB. Membrane models of E. coli containing cyclic moieties in the aliphatic lipid chain. *Biochim Biophys Acta Biomembr*. 2012;1818(5):1205–1210. https://doi.org/10.1016/j.bbamem.2012.01.009.
126. Wu EL, Fleming PJ, Yeom MS, et al. E. coli outer membrane and interactions with OmpLA. *Biophys J*. 2014;106(11):2493–2502. https://doi.org/10.1016/j.bpj.2014.04.024.
127. Kang H, Klauda JB. Molecular dynamics simulations of palmitoyloleoylphosphatidylglycerol bilayers. *Mol Simul*. 2015;41(10–12):948–954. https://doi.org/10.1080/08927022.2014.926548.
128. Lim JB, Rogaski B, Klauda JB. Update of the cholesterol force field parameters in CHARMM. *J Phys Chem B*. 2012;116(1):203–210. https://doi.org/10.1021/jp207925m.
129. Monje-Galvan V, Klauda JB. Modeling yeast organelle membranes and how lipid diversity influences bilayer properties. *Biochemistry*. 2015;54(45):6852–6861. https://doi.org/10.1021/acs.biochem.5b00718.
130. Zhuang X, Ou A, Klauda JB. Simulations of simple linoleic acid-containing lipid membranes and models for the soybean plasma membranes. *J Chem Phys*. 2017;146(21):215103. https://doi.org/10.1063/1.4983655.
131. Venable RM, Sodt AJ, Rogaski B, et al. CHARMM all-atom additive force field for sphingomyelin: elucidation of hydrogen bonding and of positive curvature. *Biophys J*. 2014;107(1):134–145. https://doi.org/10.1016/j.bpj.2014.05.034.
132. Wang E, Klauda JB. Molecular dynamics simulations of ceramide and ceramide-phosphatidylcholine bilayers. *J Phys Chem B*. 2017;121(43):10091–10104. https://doi.org/10.1021/acs.jpcb.7b08967.
133. Patel DS, Park S, Wu EL, et al. Influence of ganglioside GM1 concentration on lipid clustering and membrane properties and curvature. *Biophys J*. 2016;111(9):1987–1999. https://doi.org/10.1016/j.bpj.2016.09.021.
134. Kim S, Patel DS, Park S, et al. Bilayer properties of lipid A from various gram-negative bacteria. *Biophys J*. 2016;111(8):1750–1760. https://doi.org/10.1016/j.bpj.2016.09.001.
135. Wang J, Wolf RM, Caldwell JW, Kollman PA, Case DA. Development and testing of a general amber force field. *J Comput Chem*. 2004;25(9):1157–1174. https://doi.org/10.1002/jcc.20035.
136. Bayly CI, Cieplak P, Cornell WD, Kollman PA. A well-behaved electrostatic potential based method using charge restraints for deriving atomic charges: the RESP model. *J Phys Chem*. 1993;97(2):10269–10280.
137. Cornell WD, Cieplak P, Bayly CI, Kollman PA. Application of RESP charges to calculate conformational energies, hydrogen bond energies, and free energies of solvation. *J Phys Chem*. 1993;115(21):9620–9631.
138. Dickson CJ, Rosso L, Betz RM, Walker RC, Gould IR. GAFFlipid: a general amber force field for the accurate molecular dynamics simulation of phospholipid. *Soft Matter*. 2012;8:9617–9627. https://doi.org/10.1039/C2SM26007G.
139. Kirschner KN, Yongye AB, Tschampel SM, et al. GLYCAM06: a generalizable biomolecular force field. Carbohydrates. *J Comput Chem*. 2008;29(4):622–655. https://doi.org/10.1002/jcc.20820.
140. Madej BD, Gould IR, Walker RC. A parameterization of cholesterol for mixed lipid bilayer simulation within the amber Lipid14 force field. *J Phys Chem B*. 2015;119(38):12424–12435. https://doi.org/10.1021/acs.jpcb.5b04924.
141. Gould I, Skjevik A, Dickson C, et al. Lipid17: a comprehensive AMBER Force field for the simulation of zwitterionic and anionic lipids; 2018. [in prep.].
142. Jämbeck JPM, Lyubartsev AP. Derivation and systematic validation of a refined all-atom force field for phosphatidylcholine lipids. *J Phys Chem B*. 2012;116(10):3164–3179. https://doi.org/10.1021/jp212503e.

143. Jämbeck JPM, Lyubartsev AP. Exploring the free energy landscape of solutes embedded in lipid bilayers. *J Phys Chem Lett.* 2013;4(11):1781–1787. https://doi.org/10.1021/jz4007993.
144. Paloncýová M, Fabre G, DeVane RH, Trouillas P, Berka K, Otyepka M. Benchmarking of force fields for molecule-membrane interactions. *J Chem Theory Comput.* 2014;10(9):4143–4151. https://doi.org/10.1021/ct500419b.
145. Jorgensen WL, Maxwell DS, Tirado-Rives J. Development and testing of the OPLS all-atom force field on conformational energetics and properties of organic liquids. *J Am Chem Soc.* 1996;118(45):11225–11236. https://doi.org/10.1021/ja9621760.
146. Jorgensen WL, Tirado-Rives J. The OPLS [optimized potentials for liquid simulations] potential functions for proteins, energy minimizations for crystals of cyclic peptides and crambin. *J Am Chem Soc.* 1988;110(6):1657–1666. https://doi.org/10.1021/ja00214a001.
147. Damm W, Frontera A, Tirado-Rives J, Jorgensen WL. Opls all-atom force field for carbohydrates. *J Comput Chem.* 1997;18(16):1955–1970. https://doi.org/10.1002/(SICI)1096-987X(199712)18:163.0.CO;2-L.
148. Kaminski GA, Friesner RA, Tirado-Rives J, Jorgensen WL. Evaluation and reparametrization of the OPLS-AA force field for proteins via comparison with accurate quantum chemical calculations on peptides. *J Phys Chem B.* 2001;105(28):6474–6487. https://doi.org/10.1021/jp003919d.
149. Ulmschneider JP, Jorgensen WL. Monte Carlo backbone sampling for nucleic acids using concerted rotations including variable bond angles. *J Phys Chem B.* 2004;108(43):16883–16892. https://doi.org/10.1021/jp047796z.
150. Kulig W, Pasenkiewicz-Gierula M, Róg T. Cis and trans unsaturated phosphatidylcholine bilayers: a molecular dynamics simulation study. *Chem Phys Lipids.* 2016;195:12–20. https://doi.org/10.1016/j.chemphyslip.2015.07.002.
151. Kulig W, Pasenkiewicz-Gierula M, Róg T. Topologies, structures and parameter files for lipid simulations in GROMACS with the OPLS-aa force field: DPPC, POPC, DOPC, PEPC, and cholesterol. *Data Brief.* 2015;5:333–336. https://doi.org/10.1016/j.dib.2015.09.013.
152. Rog T, Koivuniemi A. The biophysical properties of ethanolamine plasmalogens revealed by atomistic molecular dynamics simulations. *Biochim Biophys Acta Biomembr.* 2016;1858(1):97–103. https://doi.org/10.1016/j.bbamem.2015.10.023. (a) Róg T, Orłowski A, Llorente A, et al. Data including GROMACS input files for atomistic molecular dynamics simulations of mixed, asymmetric bilayers including molecular topologies, equilibrated structures, and force field for lipids compatible with OPLS-AA parameters. *Data in Brief.* 2016;7:1171–1174. issn: 2352-3409. https://doi.org/10.1016/j.dib.2016.03.067. (b) Grabon A, Orłowski A, Tripathi A, et al. Dynamics and energetics of the mammalian phosphatidylinositol transfer protein phospholipid exchange cycle. *J Biol Chem.* 2017;292(35):14438–14455. https://doi.org/10.1074/jbc.M117.791467. (c) Manna M, Javanainen M, Martinez-Seara Monne H, et al. Long-chain GM1 gangliosides alter transmembrane domain registration through interdigitation. *Biochim Biophys Acta Biomembr.* 2017;1859(5):870–878. issn: 0005-2736. https://doi.org/10.1016/j.bbamem.2017.01.033. (d) 152d Owen MC, Kulig W, Rog T, et al. Cholesterol protects the oxidized lipid bilayer from water injury: an all-atom molecular dynamics study. *J Membr Biol.* 2018;251(3):521–534.
153. Róg T, Pasenkiewicz-Gierula M. Effects of epicholesterol on the phosphatidylcholine bilayer: a molecular simulation study. *Biophys J.* 2003;84(3):1818–1826. https://doi.org/10.1016/S0006-3495(03)74989-3.
154. Poger D, Van Gunsteren WF, Mark AE. A new force field for simulating phosphatidylcholine bilayers. *J Comput Chem.* 2010;31(6):1117–1125. https://doi.org/10.1002/jcc.21396.

155. Anézo C, de Vries AH, Höltje HD, Tieleman DP, Marrink SJ. Methodological issues in lipid bilayer simulations. *J Phys Chem B*. 2003;107(35):9424–9433. https://doi.org/10.1021/jp0348981.
156. Siu SWI, Vácha R, Jungwirth P, Böckmann RA. Biomolecular simulations of membranes: physical properties from different force fields. *J Chem Phys*. 2008;128(12):125103. https://doi.org/10.1063/1.2897760.
157. Tjörnhammar R, Edholm O. Reparameterized united atom model for molecular dynamics simulations of gel and fluid phosphatidylcholine bilayers. *J Chem Theory Comput*. 2014;10(12):5706–5715. https://doi.org/10.1021/ct500589z.
158. Chiu SW, Pandit SA, Scott HL, Jakobsson E. An improved united atom force field for simulation of mixed lipid bilayers. *J Phys Chem B*. 2009;113(9):2748–2763. https://doi.org/10.1021/jp807056c.
159. Poger D, Mark AE. On the validation of molecular dynamics simulations of saturated and cis-monounsaturated phosphatidylcholine lipid bilayers: a comparison with experiment. *J Chem Theory Comput*. 2010;6(1):325–336. https://doi.org/10.1021/ct900487a.
160. Poger D, Caron B, Mark AE. Effect of methyl-branched fatty acids on the structure of lipid bilayers. *J Phys Chem B*. 2014;118(48):13838–13848. https://doi.org/10.1021/jp503910r.
161. Poger D, Mark AE. A ring to rule them all: the effect of cyclopropane fatty acids on the fluidity of lipid bilayers. *J Phys Chem B*. 2015;119(17):5487–5495. https://doi.org/10.1021/acs.jpcb.5b00958.
162. Caron B, Mark AE, Poger D. Some like it hot: the effect of sterols and hopanoids on lipid ordering at high temperature. *J Phys Chem Lett*. 2014;5(22):3953–3957. https://doi.org/10.1021/jz5020778.
163. Mori T, Jung J, Sugita Y. Surface-tension replica-exchange molecular dynamics method for enhanced sampling of biological membrane systems. *J Chem Theory Comput*. 2013;9(12):5629–5640. https://doi.org/10.1021/ct400445k.
164. Lee CT, Comer J, Herndon C, et al. Simulation-based approaches for determining membrane permeability of small compounds. *J Chem Inf Model*. 2016;56(4):721–733. https://doi.org/10.1021/acs.jcim.6b00022.
165. Cao Z, Bian Y, Hu G, et al. Bias-exchange metadynamics simulation of membrane permeation of 20 amino acids. *Int J Mol Sci*. 2018;19(3). https://doi.org/10.3390/ijms19030885. https://www.mdpi.com/1422-0067/19/3/885.
166. Ohkubo YZ, Tajkhorshid E. Distinct structural and adhesive roles of Ca^{2+} in membrane binding of blood coagulation factors. *Structure*. 2008;16(1):72–81. https://doi.org/10.1016/j.str.2007.10.021.
167. Muller MP, Wang Y, Morrissey JH, Tajkhorshid E. Lipid specificity of the membrane binding domain of coagulation factor x. *J Thromb Haemost*. 2017;15(10):2005–2016. https://doi.org/10.1111/jth.13788.
168. Baylon JL, Lenov IL, Sligar SG, Tajkhorshid E. Characterizing the membrane-bound state of cytochrome P450 3A4: structure, depth of insertion, and orientation. *J Am Chem Soc*. 2013;135(23):8542–8551. https://doi.org/10.1021/ja4003525.
169. Blanchard AE, Arcario MJ, Schulten K, et al. A highly tilted membrane configuration for the prefusion state of synaptobrevin. *Biophys J*. 2014;107(9):2112–2121. https://doi.org/10.1016/j.bpj.2014.09.013.
170. Pogorelov TV, Vermaas JV, Arcario MJ, Tajkhorshid E. Partitioning of amino acids into a model membrane: capturing the interface. *J Phys Chem B*. 2014;118(6):1481–1492. https://doi.org/10.1021/jp4089113.
171. Vermaas JV, Tajkhorshid E. Conformational heterogeneity of α-synuclein in membrane. *Biochim Biophys Acta Biomembr*. 2014;1838(12):3107–3117. https://doi.org/10.1016/j.bbamem.2014.08.012.

172. Christensen M, Skeby KK, Schiøtt B. Identification of key interactions in the initial self-assembly of amylin in a membrane environment. *Biochemistry*. 2017;56(36): 4884–4894. https://doi.org/10.1021/acs.biochem.7b00344.
173. Vermaas JV, Pogorelov TV, Tajkhorshid E. Extension of the highly mobile membrane mimetic to transmembrane systems through customized in silico solvents. *J Phys Chem B*. 2017;121(15):3764–3776. https://doi.org/10.1021/acs.jpcb.6b11378.
174. Lyubartsev AP, Rabinovich AL. Force field development for lipid membrane simulations. *Biochim Biophys Acta Biomembr*. 2016;1858(10):2483–2497. https://doi.org/10.1016/j.bbamem.2015.12.033.
175. Skjevik ÅA, Madej BD, Dickson CJ, Teigen K, Walker RC, Gould IR. All-atom lipid bilayer self-assembly with the AMBER and CHARMM lipid force fields. *Chem Commun*. 2015;51:4402–4405. https://doi.org/10.1039/C4CC09584G.
176. Marrink SJ, Mark AE. Molecular dynamics simulation of the formation, structure, and dynamics of small phospholipid vesicles. *J Am Chem Soc*. 2003;125(49):15233–15242. https://doi.org/10.1021/ja0352092.
177. Monticelli L, Kandasamy SK, Periole X, Larson RG, Tieleman DP, Marrink SJ. The MARTINI coarse-grained force field: extension to proteins. *J Chem Theory Comput*. 2008;4(5):819–834. https://doi.org/10.1021/ct700324x.
178. Uusitalo JJ, Ingólfsson HI, Akhshi P, Tieleman DP, Marrink SJ. Martini coarse-grained force field: extension to DNA. *J Chem Theory Comput*. 2015;11(8):3932–3945. https://doi.org/10.1021/acs.jctc.5b00286.
179. Uusitalo JJ, Ingólfsson HI, Marrink SJ, Faustino I. Martini coarse-grained force field: extension to RNA. *Biophys J*. 2017;113(2):246–256. https://doi.org/10.1016/j.bpj.2017.05.043.
180. Marrink, S. J. Martini parameter for lipids. Accessed 05.11.10, http://cgmartini.nl/index.php/force-field-parameters/lipids.
181. Periole X. Interplay of G protein-coupled receptors with the membrane: insights from supra-atomic coarse grain molecular dynamics simulations. *Chem Rev*. 2017;117(1):156–185. https://doi.org/10.1021/acs.chemrev.6b00344.
182. Alessandri R, Souza PCT, Thallmair S, Melo MN, de Vries AH, Marrink SJ. Pitfalls of the martini model. *J Chem Theory Comput*. 2019;15(10):5448–5460. https://doi.org/10.1021/acs.jctc.9b00473.
183. Seo S, Shinoda W. SPICA force field for lipid membranes: domain formation induced by cholesterol. *J Chem Theory Comput*. 2018;15(1):762–774.
184. Orsi M, Essex JW. The ELBA force field for coarse-grain modeling of lipid membranes. *PLoS ONE*. 2011;6(12):e28637.
185. Barrera EE, Frigini EN, Porasso RD, Pantano S. Modeling DMPC lipid membranes with SIRAH force-field. *J Mol Model*. 2017;23(9):259. https://doi.org/10.1007/s00894-017-3426-5.
186. Liwo A, Baranowski M, Czaplewski C, et al. A unified coarse-grained model of biological macromolecules based on mean-field multipole-multipole interactions. *J Mol Model*. 2014;20(8):2306.
187. Jo S, Kim T, Iyer VG, Im W. CHARMM-GUI: a web-based graphical user interface for CHARMM. *J Comput Chem*. 2008;29(11):1859–1865. https://doi.org/10.1002/jcc.20945.
188. Jo S, Cheng X, Lee J, et al. CHARMM-GUI 10 years for biomolecular modeling and simulation. *J Comput Chem*. 2017;38(15):1114–1124. https://doi.org/10.1002/jcc.24660.
189. Brooks BR, Brooks IIICL, Mackerell Jr AD, et al. Charmm: the biomolecular simulation program. *J Comput Chem*. 2009;30(10):1545–1614. https://doi.org/10.1002/jcc.21287.

190. Abraham MJ, Murtola T, Schulz R, et al. GROMACS: high performance molecular simulations through multi-level parallelism from laptops to supercomputers. *SoftwareX*. 2015;1–2:19–25. https://doi.org/10.1016/j.softx.2015.06.001.
191. Pronk S, Páll S, Schulz R, et al. GROMACS 4.5: a high-throughput and highly parallel open source molecular simulation toolkit. *Bioinformatics*. 2013;29(7):845–854. https://doi.org/10.1093/bioinformatics/btt055.
192. Case DA, Cheatham IIITE, Darden T, et al. The Amber biomolecular simulation programs. *J Comput Chem*. 2005;26(16):1668–1688. https://doi.org/10.1002/jcc.20290.
193. Phillips JC, Braun R, Wang W, et al. Scalable molecular dynamics with NAMD. *J Comput Chem*. 2005;26(16):1781–1802. https://doi.org/10.1002/jcc.20289.
194. Jung J, Mori T, Kobayashi C, et al. Genesis: a hybrid-parallel and multi-scale molecular dynamics simulator with enhanced sampling algorithms for biomolecular and cellular simulations. *WIREs Comput Mol Sci*. 2015;5(4):310–323. https://doi.org/10.1002/wcms.1220.
195. Eastman P, Friedrichs MS, Chodera JD, et al. OpenMM 4: a reusable, extensible, hardware independent library for high performance molecular simulation. *J Chem Theory Comput*. 2013;9(1):461–469. https://doi.org/10.1021/ct300857j.
196. Arthur EJ, Brooks IIICL. Parallelization and improvements of the generalized born model with a simple switching function for modern graphics processors. *J Comput Chem*. 2016;37(10):927–939. https://doi.org/10.1002/jcc.24280.
197. Plimpton S. Fast parallel algorithms for short-range molecular dynamics. *J Comput Phys*. 1995;117(1):1–19. https://doi.org/10.1006/jcph.1995.1039.
198. Bowers KJ, Chow E, Xu H, et al. Scalable algorithms for molecular dynamics simulations on commodity clusters. In: *Proceedings of the 2006 ACM/IEEE Conference on Supercomputing*. New York, NY: ACM; 2006. https://doi.org/10.1145/1188455.1188544.
199. Lomize MA, Pogozheva ID, Joo H, Mosberg HI, Lomize AL. OPM database and PPM web server: resources for positioning of proteins in membranes. *Nucleic Acids Res*. 2011;40(D1):D370–D376. https://doi.org/10.1093/nar/gkr703.
200. Pourmousa M, Pastor RW. Molecular dynamics simulations of lipid nanodiscs. *Biochim Biophys Acta Biomembr*. 2018;1860(10):2094–2107. https://doi.org/10.1016/j.bbamem.2018.04.015.
201. Qi Y, Lee J, Klauda JB, Im W. CHARMM-GUI nanodisc builder for modeling and simulation of various nanodisc systems. *J Comput Chem*. 2019;40(7):893–899. https://doi.org/10.1002/jcc.25773.
202. Cheng X, Jo S, Lee HS, Klauda JB, Im W. CHARMM-GUI micelle builder for pure/mixed micelle and protein/micelle complex systems. *J Chem Inf Model*. 2013;53(8):2171–2180. https://doi.org/10.1021/ci4002684.
203. Rand RP, Fuller NL, Gruner SM, Parsegian VA. Membrane curvature, lipid segregation, and structural transitions for phospholipids under dual-solvent stress. *Biochemistry*. 1990;29(1):76–87. https://doi.org/10.1021/bi00453a010.
204. Chen Z, Rand RP. Comparative study of the effects of several n-alkanes on phospholipid hexagonal phases. *Biophys J*. 1998;74(2):944–952. https://doi.org/10.1016/S0006-3495(98)74017-2.
205. Sodt AJ, Pastor RW. Bending free energy from simulation: correspondence of planar and inverse hexagonal lipid phases. *Biophys J*. 2013;104(10):2202–2211. https://doi.org/10.1016/j.bpj.2013.03.048.
206. Qi Y, Cheng X, Lee J, Marrink SJ, Im W. CHARMM-GUI HMMM builder for membrane simulations with the highly mobile membrane-mimetic model. *Biophys J*. 2015;109(10):2012–2022.
207. Qi Y, Ingólfsson HI, Cheng X, Marrink SJ, Im W. CHARMM-GUI martini maker for coarse-grained simulations with the martini force field. *J Chem Theory Comput*. 2015;11(9):4486–4494. https://doi.org/10.1021/acs.jctc.5b00513.

208. Hsu PC, Bruininks BMH, Jefferies D, et al. CHARMM-GUI Martini maker for modeling and simulation of complex bacterial membranes with lipopoly-saccharides. *J Comput Chem*. 2017;38(27):2354–2363. https://doi.org/10.1002/jcc.24895.
209. de Jong DH, Singh G, Bennett WFD, et al. Improved parameters for the martini coarse-grained protein force field. *J Chem Theory Comput*. 2013;9(1):687–697. https://doi.org/10.1021/ct300646g.
210. Yesylevskyy SO, Schäfer LV, Sengupta D, Marrink SJ. Polarizable water model for the coarse-grained MARTINI force field. *PLoS Comput Biol*. 2010;6(6):1–17. https://doi.org/10.1371/journal.pcbi.1000810.
211. Arnarez C, Uusitalo JJ, Masman MF, et al. Dry martini, a coarse-grained force field for lipid membrane simulations with implicit solvent. *J Chem Theory Comput*. 2015;11(1):260–275. https://doi.org/10.1021/ct500477k.
212. Periole X, Cavalli M, Marrink SJ, Ceruso MA. Combining an elastic network with a coarse-grained molecular force field: structure, dynamics, and intermolecular recognition. *J Chem Theory Comput*. 2009;5(9):2531–2543. https://doi.org/10.1021/ct9002114.
213. Wassenaar TA, Pluhackova K, Böckmann RA, Marrink SJ, Tieleman DP. Going backward: a flexible geometric approach to reverse transformation from coarse grained to atomistic models. *J Chem Theory Comput*. 2014;10(2):676–690. https://doi.org/10.1021/ct400617g.
214. Schott-Verdugo S, Gohlke H. PACKMOL-memgen: a simple-to-use, generalized workflow for membrane-protein-lipid-bilayer system building. *J Chem Inf Model*. 2019;59(6):2522–2528. https://doi.org/10.1021/acs.jcim.9b00269.
215. Knight CJ, Hub JS. MemGen: a general web server for the setup of lipid membrane simulation systems. *Bioinformatics*. 2015;31(17):2897–2899. https://doi.org/10.1093/bioinformatics/btv292.
216. Martínez L, Andrade R, Birgin EG, Martínez JM. PACKMOL: a package for building initial configurations for molecular dynamics simulations. *J Comput Chem*. 2009;30(13):2157–2164. https://doi.org/10.1002/jcc.21224.
217. Humphrey W, Dalke A, Schulten K. VMD: visual molecular dynamics. *J Mol Graph*. 1996;14(1):33–38. https://doi.org/10.1016/0263-7855(96)00018-5.
218. Faraldo-Gómez JD, Smith GR, Sansom MS. Setting up and optimization of membrane protein simulations. *Eur Biophys J*. 2002;31(3):217–227. https://doi.org/10.1007/s00249-002-0207-5.
219. Kandt C, Ash WL, Tieleman DP. Setting up and running molecular dynamics simulations of membrane proteins. *Methods*. 2007;41(4):475–488. https://doi.org/10.1016/j.ymeth.2006.08.006.
220. Schmidt TH, Kandt C. LAMBADA and InflateGRO2: efficient membrane alignment and insertion of membrane proteins for molecular dynamics simulations. *J Chem Inf Model*. 2012;52(10):2657–2669. https://doi.org/10.1021/ci3000453.
221. Wolf MG, Hoefling M, Aponte-Santamaría C, Grubmller H, Groenhof G. g_membed: efficient insertion of a membrane protein into an equilibrated lipid bilayer with minimal perturbation. *J Comput Chem*. 2010;31(11):2169–2174. https://doi.org/10.1002/jcc.21507.
222. Javanainen M, Martinez-Seara H. Efficient preparation and analysis of membrane and membrane protein systems. *Biochim Biophys Acta Biomembr*. 2016;1858(10):2468–2482. https://doi.org/10.1016/j.bbamem.2016.02.036.
223. Ghahremanpour MM, Arab SS, Aghazadeh SB, Zhang J, van der Spoel D. MemBuilder: a web-based graphical interface to build heterogeneously mixed membrane bilayers for the GROMACS biomolecular simulation program. *Bioinformatics*. 2013;30(3):439–441. https://doi.org/10.1093/bioinformatics/btt680.

224. Wassenaar TA, Ingólfsson HI, Böckmann RA, Tieleman DP, Marrink SJ. Computational lipidomics with insane: a versatile tool for generating custom membranes for molecular simulations. *J Chem Theory Comput.* 2015;11(5):2144–2155. https://doi.org/10.1021/acs.jctc.5b00209.
225. Gutmann T, Schäfer IB, Poojari C, et al. Cryo-EM structure of the complete and ligand-saturated insulin receptor ectodomain. *J Cell Biol.* 2020;219(1). https://doi.org/10.1083/jcb.201907210.
226. Mobarak E, Håversen L, Manna M, et al. Glucosylceramide modifies the LPS-induced inflammatory response in macrophages and the orientation of the LPS/TLR4 complex in silico. *Sci Rep.* 2018;8(1):13600.
227. Kaszuba K, Grzybek M, Orłowski A, et al. N-glycosylation as determinant of epidermal growth factor receptor conformation in membranes. *Proc Natl Acad Sci.* 2015;112(14):4334–4339. https://doi.org/10.1073/pnas.1503262112.
228. Danne R, Poojari C, Martinez-Seara H, et al. doGlycans-tools for preparing carbohydrate structures for atomistic simulations of glycoproteins, glycolipids, and carbohydrate polymers for GROMACS. *J Chem Inf Model.* 2017;57(10):2401–2406. https://doi.org/10.1021/acs.jcim.7b00237.
229. Park SJ, Lee J, Qi Y, et al. CHARMM-GUI glycan modeler for modeling and simulation of carbohydrates and glycoconjugates. *Glycobiology.* 2019;29(4):320–331. https://doi.org/10.1093/glycob/cwz003.
230. Case DA, Ben-Shalom IY, Brozell SR, et al. *AMBER 2019.* San Francisco: University of California; 2019.
231. McGibbon RT, Beauchamp KA, Harrigan MP, et al. MDTraj: a modern open library for the analysis of molecular dynamics trajectories. *Biophys J.* 2015;109(8):1528–1532. https://doi.org/10.1016/j.bpj.2015.08.015.
232. Schrödinger LLC. *The PyMOL Molecular Graphics System, Version 1.8;* 2015.
233. Pettersen EF, Goddard TD, Huang CC, et al. Ucsf chimera—a visualization system for exploratory research and analysis. *J Comput Chem.* 2004;25(13):1605–1612. https://doi.org/10.1002/jcc.20084.
234. Natick M. *MATLAB version 9.1 (R2016b).* The Mathworks, Inc.; 2016.
235. Williams T. Gnuplot 4.4: an interactive plotting program; 2010. http://www.gnuplot.info/docs_4.4/gnuplot.pdf.
236. Romo TD, Grossfield A. LOOS: an extensible platform for the structural analysis of simulations. *2009 Annual International Conference of the IEEE Engineering in Medicine and Biology Society.* 2009:2332–2335. https://doi.org/10.1109/IEMBS.2009.5335065.
237. Gapsys V, de Groot BL, Rodolfo R. Computational analysis of local membrane properties. *J Comput Aided Mol Des.* 2013;27(10):845–858. https://doi.org/10.1007/s10822-013-9684-0.
238. Allen WJ, Lemkul JA, Bevan DR. GridMAT-MD: a grid-based membrane analysis tool for use with molecular dynamics. *J Comput Chem.* 2009;30(12):1952–1958. https://doi.org/10.1002/jcc.21172.
239. Carr M, MacPhee CE. Membrainy: a 'smart', unified membrane analysis tool. *Source Code Biol Med.* 2015;10(1):3.
240. Guixà-González R, Rodriguez-Espigares I, Ramírez-Anguita JM, et al. MEMBPLUGIN: studying membrane complexity in VMD. *Bioinformatics.* 2014;30(10):1478–1480. https://doi.org/10.1093/bioinformatics/btu037.
241. Buchoux S. FATSLiM: a fast and robust software to analyze MD simulations of membranes. *Bioinformatics.* 2016;33(1):133–134. https://doi.org/10.1093/bioinformatics/btw563.
242. Castillo N, Monticelli L, Barnoud J, Tieleman DP. Free energy of WALP23 dimer association in DMPC, DPPC, and DOPC bilayers. *Chem Phys Lipids.* 2013;169:95–105. https://doi.org/10.1016/j.chemphyslip.2013.02.001.

243. Michaud-Agrawal N, Denning EJ, Woolf TB, Beckstein O. MDAnalysis: a toolkit for the analysis of molecular dynamics simulations. *J Comput Chem.* 2011;32(10): 2319–2327. https://doi.org/10.1002/jcc.21787.
244. Hilger D, Masureel M, Kobilka BK. Structure and dynamics of GPCR signaling complexes. *Nat Struct Mol Biol.* 2018;25(1):4–12.
245. Sriram K, Insel PA. G protein-coupled receptors as targets for approved drugs: how many targets and how many drugs? *Mol Pharmacol.* 2018;93(4):251–258. https://doi.org/10.1124/mol.117.111062.
246. Paila YD, Chattopadhyay A. The function of G-protein coupled receptors and membrane cholesterol: specific or general interaction? *Glycoconj J.* 2008;26(6):711. https://doi.org/10.1007/s10719-008-9218-5.
247. Oates J, Watts A. Uncovering the intimate relationship between lipids, cholesterol and GPCR activation. *Curr Opin Struct Biol.* 2011;21(6):802–807. https://doi.org/10.1016/j.sbi.2011.09.007.
248. Pucadyil TJ, Chattopadhyay A. Cholesterol modulates ligand binding and G-protein coupling to serotonin1A receptors from bovine hippocampus. *Biochim Biophys Acta Biomembr.* 2004;1663(1):188–200. https://doi.org/10.1016/j.bbamem.2004.03.010.
249. Gimpl G, Fahrenholz F. Cholesterol as stabilizer of the oxytocin receptor. *Biochim Biophys Acta Biomembr.* 2002;1564(2):384–392. https://doi.org/10.1016/S0005-2736(02)00475-3.
250. Jastrzebska B, Debinski A, Filipek S, Palczewski K. Role of membrane integrity on G protein-coupled receptors: rhodopsin stability and function. *Prog Lipid Res.* 2011;50(3):267–277. https://doi.org/10.1016/j.plipres.2011.03.002.
251. Cherezov V, Rosenbaum DM, Hanson MA, et al. High-resolution crystal structure of an engineered human β2-adrenergic G protein-coupled receptor. *Science.* 2007;318(5854):1258–1265. https://doi.org/10.1126/science.1150577.
252. Michael FB. Modulation of rhodopsin function by properties of the membrane bilayer. *Chem Phys Lipids.* 1994;73(1):159–180. https://doi.org/10.1016/0009-3084(94)90180-5.
253. Manna M, Niemelä M, Tynkkynen J, et al. Mechanism of allosteric regulation of β_2-adrenergic receptor by cholesterol. *eLife.* 2016;5:e18432. https://doi.org/10.7554/eLife.18432.
254. Cang X, Yang L, Yang J, et al. Cholesterol-β1AR interaction versus cholesterol-β2AR interaction. *Proteins.* 2014;82(5):760–770. https://doi.org/10.1002/prot.24456.
255. Paila YD, Tiwari S, Sengupta D, Chattopadhyay A. Molecular modeling of the human serotonin1A receptor: role of membrane cholesterol in ligand binding of the receptor. *Mol BioSyst.* 2011;7(1):224–234. https://doi.org/10.1039/C0MB00148A.
256. Shan J, Khelashvili G, Mondal S, Mehler EL, Weinstein H. Ligand-dependent conformations and dynamics of the serotonin 5-HT2A receptor determine its activation and membrane-driven oligomerization properties. *PLoS Comput Biol.* 2012;8(4):1–15. https://doi.org/10.1371/journal.pcbi.1002473.
257. Hurst DP, Grossfield A, Lynch DL, et al. A lipid pathway for ligand binding is necessary for a cannabinoid G protein-coupled receptor. *J Biol Chem.* 2010;285(23): 17954–17964.
258. Pitman MC, Grossfield A, Suits F, Feller SE. Role of cholesterol and polyunsaturated chains in lipid-protein interactions: molecular dynamics simulation of rhodopsin in a realistic membrane environment. *J Am Chem Soc.* 2005;127(13):4576–4577. https://doi.org/10.1021/ja042715y.
259. Lee JY, Lyman E. Predictions for cholesterol interaction sites on the A_{2A} adenosine receptor. *J Am Chem Soc.* 2012;134(40):16512–16515. https://doi.org/10.1021/ja307532d.

260. Lyman E, Higgs C, Kim B, et al. A role for a specific cholesterol interaction in stabilizing the Apo configuration of the human A_{2A} adenosine receptor. *Structure*. 2009;17(12):1660–1668.
261. Marino KA, Shang Y, Filizola M. Insights into the function of opioid receptors from molecular dynamics simulations of available crystal structures. *Br J Pharmacol*. 2018;175(14):2834–2845.
262. Yuan S, Wu R, Latek D, Trzaskowski B, Filipek S. Lipid receptor S1P1 activation scheme concluded from microsecond all-atom molecular dynamics simulations. *PLoS Comput Biol*. 2013;9(10):1–8. https://doi.org/10.1371/journal.pcbi.1003261.
263. Morra G, Razavi AM, Pandey K, Weinstein H, Menon AK, Khelashvili G. Mechanisms of lipid scrambling by the G protein-coupled receptor opsin. *Structure*. 2018;26(2):356–367.
264. Montgomery D, Campbell A, Sullivan HJ, Wu C. Molecular dynamics simulation of biased agonists at the dopamine D2 receptor suggests the mechanism of receptor functional selectivity. *J Biomol Struct Dyn*. 2019;37(12):3206–3225.
265. Liu W, Chun E, Thompson AA, et al. Structural basis for allosteric regulation of GPCRs by sodium ions. *Science*. 2012;337(6091):232–236. https://doi.org/10.1126/science.1219218.
266. Guixà-González R, Albasanz JL, Rodriguez-Espigares I, et al. Membrane cholesterol access into a G-protein-coupled receptor. *Nat Commun*. 2017;8:14505.
267. Jaakola VP, Griffith MT, Hanson MA, et al. The 2.6 angstrom crystal structure of a human A2A adenosine receptor bound to an antagonist. *Science*. 2008;322(5905):1211–1217. https://doi.org/10.1126/science.1164772.
268. Ernst OP, Menon AK. Phospholipid scrambling by rhodopsin. *Photochem Photobiol Sci*. 2015;14:1922–1931. https://doi.org/10.1039/C5PP00195A.
269. Goren MA, Morizumi T, Menon I, et al. Constitutive phospholipid scramblase activity of a G protein-coupled receptor. *Nat Commun*. 2014;5:5115.
270. Menon I, Huber T, Sanyal S, et al. Opsin is a phospholipid flippase. *Curr Biol*. 2011;21(2):149–153. https://doi.org/10.1016/j.cub.2010.12.031.
271. Pomorski TG, Menon AK. Lipid somersaults: uncovering the mechanisms of protein-mediated lipid flipping. *Prog Lipid Res*. 2016;64:69–84. https://doi.org/10.1016/j.plipres.2016.08.003.
272. Bevers EM, Williamson PL. Getting to the outer leaflet: physiology of phosphatidylserine exposure at the plasma membrane. *Physiol Rev*. 2016;96(2):605–645. https://doi.org/10.1152/physrev.00020.2015.
273. Pomorski T, Menon AK. Lipid flippases and their biological functions. *Cell Mol Life Sci*. 2006;63(24):2908–2921. https://doi.org/10.1007/s00018-006-6167-7.
274. Sanyal S, Menon AK. Flipping lipids: why an' what's the reason for? *ACS Chem Biol*. 2009;4(11):895–909. https://doi.org/10.1021/cb900163d.
275. Ploier B, Menon AK. A fluorescence-based assay of phospholipid scramblase activity. *J Vis Exp*. 2016;115:1–8.
276. Alberts B, Johnson A, Lewis J, Raff M, Roberts K, Walter P. Principles of membrane transport. In: Alberts B, Johnson A, Lewis J, Raff M, Roberts K, Walter P, eds. In: *Molecular Biology of the Cell*. 4th ed. Garland Science; 2002:1585–1603.
277. Martens C, Shekhar M, Borysik AJ, et al. Direct protein-lipid interactions shape the conformational landscape of secondary transporters. *Nat Commun*. 2018;9(1):4151.
278. Quistgaard EM, Löw C, Moberg P, Trésaugues L, Nordlund P. Structural basis for substrate transport in the GLUT-homology family of monosaccharide transporters. *Nat Struct Mol Biol*. 2013;20(6):766.
279. Mirza O, Guan L, Verner G, Iwata S, Kaback HR. Structural evidence for induced fit and a mechanism for sugar/H+ symport in LacY. *EMBO J*. 2006;25(6):1177–1183. https://doi.org/10.1038/sj.emboj.7601028.

280. Khelashvili G, Stanley N, Sahai MA, et al. Spontaneous inward opening of the dopamine transporter is triggered by PIP_2-regulated dynamics of the N-Terminus. *ACS Chem Neurosci*. 2015;6(11):1825–1837. https://doi.org/10.1021/acschemneuro.5b00179.
281. Penmatsa A, Gouaux E. How LeuT shapes our understanding of the mechanisms of sodium-coupled neurotransmitter transporters. *J Physiol*. 2014;592(5):863–869. https://doi.org/10.1113/jphysiol.2013.259051.
282. Nichols CG, Lopatin AN. Inward rectifier potassium channels. *Annu Rev Physiol*. 1997;59(1):171–191. https://doi.org/10.1146/annurev.physiol.59.1.171.
283. Hibino H, Inanobe A, Furutani K, Murakami S, Findlay I, Kurachi Y. Inwardly rectifying potassium channels: their structure, function, and physiological roles. *Physiol Rev*. 2010;90(1):291–366. https://doi.org/10.1152/physrev.00021.2009.
284. Rosenhouse-Dantsker A, Noskov S, Durdagi S, Durdagi S, Logothetis DE, Levitan I. Identification of novel cholesterol-binding regions in Kir2 channels. *J Biolog Chem*. 2013;288(43):31154–31164. https://doi.org/10.1074/jbc.M113.496117.
285. Spiro RG. Glycoproteins. In: Anfinsen CB, Edsall JT, Richards FM, eds. vol. 27. *Advances in Protein Chemistry*. Academic Press; 1973:349–467. https://doi.org/10.1016/S0065-3233(08)60451-9.
286. Meyer B, Möller H. Conformation of glycopeptides and glycoproteins. In: Wittmann V, ed. *Glycopeptides and Glycoproteins: Synthesis, Structure, and Application*. Berlin, Heidelberg: Springer; 2007:187–251. https://doi.org/10.1007/128_2006_078.
287. Wormald MR, Rudd PM, Harvey DJ, Chang SC, Scragg IG, Dwek RA. Variations in oligosaccharide-protein interactions in immunoglobulin G determine the site-specific glycosylation profiles and modulate the dynamic motion of the Fc oligosaccharides. *Biochemistry*. 1997;36(6):1370–1380. https://doi.org/10.1021/bi9621472.
288. Shental-Bechor D, Levy Y. Folding of glycoproteins: toward understanding the biophysics of the glycosylation code. *Curr Opin Struct Biol*. 2009;19(5):524–533. https://doi.org/10.1016/j.sbi.2009.07.002.
289. Shental-Bechor D, Levy Y. Effect of glycosylation on protein folding: a close look at thermodynamic stabilization. *Proc Natl Acad Sci*. 2008;105(24):8256–8261. https://doi.org/10.1073/pnas.0801340105.
290. Flintegaard TV, Thygesen P, Rahbek-Nielsen H, et al. N-glycosylation increases the circulatory half-life of human growth hormone. *Endocrinology*. 2010;151(11):5326–5336. https://doi.org/10.1210/en.2010-0574.
291. Van Kooyk Y, Rabinovich GA. Protein-glycan interactions in the control of innate and adaptive immune responses. *Nat Immunol*. 2008;9(6):593.
292. Spiro RG. Protein glycosylation: nature, distribution, enzymatic formation, and disease implications of glycopeptide bonds. *Glycobiology*. 2002;12(4). https://doi.org/10.1093/glycob/12.4.43R43R–56R.
293. Shinzawa-Itoh K, Aoyama H, Muramoto K, et al. Structures and physiological roles of 13 integral lipids of bovine heart cytochrome c oxidase. *EMBO J*. 2007;26(6):1713–1725. https://doi.org/10.1038/sj.emboj.7601618.
294. Malkamäki A, Sharma V. Atomistic insights into cardiolipin binding sites of cytochrome c oxidase. *Biochim Biophys Acta*. 2019;1860(3):224–232. https://doi.org/10.1016/j.bbabio.2018.11.004.
295. Dudkina NV, Kouřil R, Peters K, Braun HP, Boekema EJ. Structure and function of mitochondrial supercomplexes. *Biochim Biophys Acta*. 2010;1797(6):664–670. https://doi.org/10.1016/j.bbabio.2009.12.013.
296. Osellame LD, Blacker TS, Duchen MR. Cellular and molecular mechanisms of mitochondrial function. *Best Pract Res Clin Endocrinol Metab*. 2012;26(6):711–723. https://doi.org/10.1016/j.beem.2012.05.003.
297. Schmidt O, Pfanner N, Meisinger C. Mitochondrial protein import: from proteomics to functional mechanisms. *Nat Rev Mol Cell Biol*. 2010;11(9):655.

298. Friedman JR, Nunnari J. Mitochondrial form and function. *Nature.* 2014;505(7483): 335–343.
299. Duchen MR. Mitochondria and calcium: from cell signalling to cell death. *J Physiol.* 2000;529(1):57–68. https://doi.org/10.1111/j.1469-7793.2000.00057.x.
300. Hill S, Remmen HV. Mitochondrial stress signaling in longevity: a new role for mitochondrial function in aging. *Redox Biol.* 2014;2:936–944. https://doi.org/10.1016/j.redox.2014.07.005.
301. Kühlbrandt W. Structure and function of mitochondrial membrane protein complexes. *BMC Biol.* 2015;13(1):89. https://doi.org/10.1186/s12915-015-0201-x.
302. Schlame M, Towbin JA, Heerdt PM, Jehle R, DiMauro S, Blanck TJJ. Deficiency of tetralinoleoyl-cardiolipin in Barth syndrome. *Ann Neurol.* 2002;51(5):634–637. https://doi.org/10.1002/ana.10176.
303. Valianpour F, Wanders RJA, Barth PG, Overmars H, van Gennip AH. Quantitative and compositional study of cardiolipin in platelets by electrospray ionization mass spectrometry: application for the identification of Barth syndrome patients. *Clin Chem.* 2002;48(9):1390–1397. http://clinchem.aaccjnls.org/content/48/9/1390.
304. Hunte C. Specific protein-lipid interactions in membrane proteins. *Biochem Soc Trans.* 2005;33(5):938–942. https://doi.org/10.1042/BST0330938.
305. Sedlák E, Robinson NC. Phospholipase A2 digestion of cardiolipin bound to bovine cytochrome c oxidase alters both activity and quaternary structure. *Biochemistry.* 1999;38(45):14966–14972. https://doi.org/10.1021/bi9914053.
306. Hardy J, Selkoe DJ. The amyloid hypothesis of Alzheimer's disease: progress and problems on the road to therapeutics. *Science.* 2002;297(5580):353–356. https://doi.org/10.1126/science.1072994.
307. Selkoe DJ. Alzheimer's disease is a synaptic failure. *Science.* 2002;298(5594):789–791. https://doi.org/10.1126/science.1074069.
308. Selkoe DJ. Alzheimer's disease: genes, proteins, and therapy. *Physiol Rev.* 2001;81(2):741–766. https://doi.org/10.1152/physrev.2001.81.2.741.
309. Yan Y, Wang C. Aβ42 is more rigid than Aβ40 at the C terminus: implications for Aβ aggregation and toxicity. *J Mol Biol.* 2006;364(5):853–862. https://doi.org/10.1016/j.jmb.2006.09.046.
310. Kayed R, Head E, Thompson JL, et al. Common structure of soluble amyloid oligomers implies common mechanism of pathogenesis. *Science.* 2003;300(5618):486–489. https://doi.org/10.1126/science.1079469.
311. Haass C, Selkoe DJ. Soluble protein oligomers in neurodegeneration: lessons from the Alzheimer's amyloid β-peptide. *Nat Rev Mol Cell Biol.* 2007;8(2):101.
312. Chimon S, Shaibat MA, Jones CR, Calero DC, Aizezi B, Ishii Y. Evidence of fibril-like β-sheet structures in a neurotoxic amyloid intermediate of Alzheimer's β-amyloid. *Nat Struct Mol Biol.* 2007;14(12):1157.
313. Straub JE, Thirumalai D. Membrane-protein interactions are key to understanding amyloid formation. *J Phys Chem Lett.* 2014;5(3):633–635. https://doi.org/10.1021/jz500054d.
314. Vestergaard M, Hamada T, Takagi M. Using model membranes for the study of amyloid beta:lipid interactions and neurotoxicity. *Biotechnol Bioeng.* 2008;99(4):753–763. https://doi.org/10.1002/bit.21731.
315. Matsuzaki K. Physicochemical interactions of amyloid β-peptide with lipid bilayers. *Biochim Biophys Acta Biomembr.* 2007;1768(8):1935–1942. https://doi.org/10.1016/j.bbamem.2007.02.009.
316. Butterfield SM, Lashuel HA. Amyloidogenic protein-membrane interactions: mechanistic insight from model systems. *Angew Chem Int Ed.* 2010;49(33):5628–5654. https://doi.org/10.1002/anie.200906670.
317. Makin OS, Serpell LC. The structure of amyloid. *Fibre Diffr Rev.* 2004;12:29–35.

318. Williams TL, Serpell LC. Membrane and surface interactions of Alzheimer's Aβ peptide–insights into the mechanism of cytotoxicity. *FEBS J.* 2011;278(20):3905–3917. https://doi.org/10.1111/j.1742-4658.2011.08228.x.
319. Poojari C, Kukol A, Strodel B. How the amyloid-β peptide and membranes affect each other: an extensive simulation study. *Biochim Biophys Acta Biomembr.* 2013;1828(2): 327–339. https://doi.org/10.1016/j.bbamem.2012.09.001.
320. Poojari C, Strodel B. Stability of transmembrane amyloid β-peptide and membrane integrity tested by molecular modeling of site-specific Aβ42 mutations. *PLoS ONE.* 2013;8(11):1–12. https://doi.org/10.1371/journal.pone.0078399.
321. Lemkul JA, Bevan DR. Lipid composition influences the release of Alzheimer's amyloid β-peptide from membranes. *Protein Sci.* 2011;20(9):1530–1545. https://doi.org/10.1002/pro.678.
322. Owen MC, Kulig W, Poojari C, Rog T, Strodel B. Physiologically-relevant levels of sphingomyelin, but not GM1, induces a β-sheet-rich structure in the amyloid-β(1-42) monomer. *Biochim Biophys Acta Biomembr.* 2018;1860(9):1709–1720. https://doi.org/10.1016/j.bbamem.2018.03.026.
323. Ntarakas N, Ermilova I, Lyubartsev AP. Effect of lipid saturation on amyloid-beta peptide partitioning and aggregation in neuronal membranes: molecular dynamics simulations. *Eur Biophys J.* 2019;48(8):813–824. https://doi.org/10.1007/s00249-019-01407-x.
324. Lu Y, Shi XF, Nguyen PH, Sterpone F, Salsbury FR, Derreumaux P. Amyloid-β(29-42) dimeric conformations in membranes rich in omega-3 and omega-6 polyunsaturated fatty acids. *J Phys Chem B.* 2019;123(12):2687–2696. https://doi.org/10.1021/acs.jpcb.9b00431.
325. Söderberg M, Edlund C, Kristensson K, Dallner G. Fatty acid composition of brain phospholipids in aging and in Alzheimer's disease. *Lipids.* 1991;26(6):421. https://doi.org/10.1007/BF02536067.
326. Wood PL. Lipidomics of Alzheimer's disease: current status. *Alzheimer's Res Ther.* 2012;4(1):5.
327. Söderberg M, Edlund C, Alafuzoff I, Kristensson K, Dallner G. Lipid composition in different regions of the brain in Alzheimer's disease/senile dementia of Alzheimer's type. *J Neurochem.* 1992;59(5):1646–1653. https://doi.org/10.1111/j.1471-4159.1992.tb10994.x.
328. Hossain S, Hashimoto M, Katakura M, Miwa K, Shimada T, Shido O. Mechanism of docosahexaenoic acid-induced inhibition of in vitro Aβ1-42 fibrillation and Aβ1-42-induced toxicity in SH-S5Y5 cells. *J Neurochem.* 2009;111(2):568–579. https://doi.org/10.1111/j.1471-4159.2009.06336.x.
329. Ionov M, Klajnert B, Gardikis K, et al. Effect of amyloid beta peptides Aβ1-28 and Aβ25-40 on model lipid membranes. *J Thermal Anal Calorim.* 2010;99(3):741–747. https://doi.org/10.1007/s10973-009-0405-9.
330. Jia Y, Qian Z, Zhang Y, Wei G. Adsorption and orientation of human islet amyloid polypeptide (hIAPP) monomer at anionic lipid bilayers: implications for membrane-mediated aggregation. *Int J Mol Sci.* 2013;14(3):6241–6258.
331. Qian Z, Jia Y, Wei G. Binding orientations and lipid interactions of human amylin at zwitterionic and anionic lipid bilayers. *J Diabetes Res.* 2016;2016:1–13.
332. Dong X, Qiao Q, Qian Z, Wei G. Recent computational studies of membrane interaction and disruption of human islet amyloid polypeptide: monomers, oligomers and protofibrils. *Biochim Biophys Acta Biomembr.* 2018;1860(9):1826–1839. https://doi.org/10.1016/j.bbamem.2018.03.006.
333. Dignon GL, Zerze GH, Mittal J. Interplay between membrane composition and structural stability of membrane-bound hIAPP. *J Phys Chem B.* 2017;121(37):8661–8668. https://doi.org/10.1021/acs.jpcb.7b05689.

334. Zhang Y, Luo Y, Deng Y, Mu Y, Wei G. Lipid interaction and membrane perturbation of human islet amyloid polypeptide monomer and dimer by molecular dynamics simulations. *PLoS ONE*. 2012;7(5):1–10. https://doi.org/10.1371/journal.pone.0038191.
335. Zhao J, Luo Y, Jang H, et al. Probing ion channel activity of human islet amyloid polypeptide (amylin). *Biochim Biophys Acta Biomembr*. 2012;1818(12):3121–3130. https://doi.org/10.1016/j.bbamem.2012.08.012.
336. Zhao J, Hu R, Sciacca MFM, et al. Non-selective ion channel activity of polymorphic human islet amyloid polypeptide (amylin) double channels. *Phys Chem Chem Phys*. 2014;16:2368–2377. https://doi.org/10.1039/C3CP53345J.
337. Quist A, Doudevski I, Lin H, et al. Amyloid ion channels: a common structural link for protein-misfolding disease. *Proc Natl Acad Sci*. 2005;102(30):10427–10432. https://doi.org/10.1073/pnas.0502066102.
338. Poojari C, Xiao D, Batista VS, Strodel B. Membrane permeation induced by aggregates of human islet amyloid polypeptides. *Biophys J*. 2013;105(10):2323–2332. https://doi.org/10.1016/j.bpj.2013.09.045.
339. Luca S, Yau WM, Leapman R, Tycko R. Peptide conformation and supramolecular organization in amylin fibrils: constraints from solid-state NMR. *Biochemistry*. 2007;46(47):13505–13522. https://doi.org/10.1021/bi701427q.
340. Bedrood S, Li Y, Isas JM, et al. Fibril structure of human islet amyloid polypeptide. *J Biol Chem*. 2012;287(8):5235–5241. https://doi.org/10.1074/jbc.M111.327817.
341. Xiao D, Fu L, Liu J, Batista VS, Yan ECY. Amphiphilic adsorption of human islet amyloid polypeptide aggregates to lipid/aqueous interfaces. *J Mol Biol*. 2012;421(4):537–547. https://doi.org/10.1016/j.jmb.2011.12.035.
342. Yasuda T, Nakata Y, Mochizuki H. α-synuclein and neuronal cell death. *Mol Neurobiol*. 2013;47(2):466–483. https://doi.org/10.1007/s12035-012-8327-0.
343. Rao JN, Dua V, Ulmer TS. Characterization of α-synuclein interactions with selected aggregation-inhibiting small molecules. *Biochemistry*. 2008;47(16):4651–4656. https://doi.org/10.1021/bi8002378.
344. Press-Sandler O, Miller Y. Molecular mechanisms of membrane-associated amyloid aggregation: computational perspective and challenges. *Biochim Biophys Acta Biomembr*. 2018;1860(9):1889–1905. https://doi.org/10.1016/j.bbamem.2018.03.014.
345. Eliezer D, Kutluay E, Bussell R, Browne G. Conformational properties of α-synuclein in its free and lipid-associated states. *J Mol Biol*. 2001;307(4):1061–1073. https://doi.org/10.1006/jmbi.2001.4538.
346. Ulmer TS, Bax A, Cole NB, Nussbaum RL. Structure and dynamics of micelle-bound human α-synuclein. *J Biol Chem*. 2005;280(10):9595–9603. https://doi.org/10.1074/jbc.M411805200.
347. Rao JN, Jao CC, Hegde BG, Langen R, Ulmer TS. A combinatorial NMR and EPR approach for evaluating the structural ensemble of partially folded proteins. *J Am Chem Soc*. 2010;132(25):8657–8668. https://doi.org/10.1021/ja100646t.
348. Li B, Ge P, Murray KA, et al. Cryo-EM of full-length α-synuclein reveals fibril polymorphs with a common structural kernel. *Nat Commun*. 2018;9(1):3609.
349. Pfefferkorn CM, Heinrich F, Sodt AJ, et al. Depth of α-synuclein in a bilayer determined by fluorescence, neutron reflectometry, and computation. *Biophys J*. 2012;102(3):613–621. https://doi.org/10.1016/j.bpj.2011.12.051.
350. Perlmutter JD, Braun AR, Sachs JN. Curvature dynamics of α-synuclein familial Parkinson disease mutants: molecular simulations of the micelle- and bilayer-bound forms. *J Biol Chem*. 2009;284(11):7177–7189. https://doi.org/10.1074/jbc.M808895200.
351. Garten M, Prévost C, Cadart C, et al. Methyl-branched lipids promote the membrane adsorption of α-synuclein by enhancing shallow lipid-packing defects. *Phys Chem Chem Phys*. 2015;17:15589–15597. https://doi.org/10.1039/C5CP00244C.

352. Viennet T, Wördehoff MM, Uluca B, et al. Structural insights from lipid-bilayer nanodiscs link α-synuclein membrane-binding modes to amyloid fibril formation. *Commun Biol.* 2018;1(1):44.
353. Bikfalvi A, Klein S, Pintucci G, Rifkin DB. Biological roles of fibroblast growth factor-2*. *Endocrine Reviews.* 1997;18(1):26–45. https://doi.org/10.1210/edrv.18.1.0292.
354. Presta M, Dell'Era P, Mitola S, Moroni E, Ronca R, Rusnati M. Fibroblast growth factor/fibroblast growth factor receptor system in angiogenesis. *Cytokine Growth Factor Rev.* 2005;16(2):159–178. https://doi.org/10.1016/j.cytogfr.2005.01.004.
355. Noh KH, Kim SH, Kim JH, et al. API5 confers tumoral immune escape through FGF2-dependent cell survival pathway. *Cancer Research.* 2014;74(13):3556–3566. https://doi.org/10.1158/0008-5472.CAN-13-3225.
356. Pardo OE, Wellbrock C, Khanzada UK, et al. FGF-2 protects small cell lung cancer cells from apoptosis through a complex involving PKCε, B-Raf and S6K2. *EMBO J.* 2006;25(13):3078–3088. https://doi.org/10.1038/sj.emboj.7601198.
357. La Venuta G, Zeitler M, Steringer JP, Müller HM, Nickel W. The startling properties of fibroblast growth factor 2: how to exit mammalian cells without a signal peptide at hand. *J Biol Chem.* 2015;290(45):27015–27020. https://doi.org/10.1074/jbc.R115.689257.
358. Nickel W. The unconventional secretory machinery of fibroblast growth factor 2. *Traffic.* 2011;12(7):799–805. https://doi.org/10.1111/j.1600-0854.2011.01187.x.
359. Nickel W, Rabouille C. Mechanisms of regulated unconventional protein secretion. *Nat Rev Mol cell Biol.* 2009;10(2):148.
360. Nickel W. Unconventional secretion: an extracellular trap for export of fibroblast growth factor 2. *J Cell Sci.* 2007;120(14):2295–2299. https://doi.org/10.1242/jcs.011080.
361. Temmerman K, Ebert AD, Müller HM, Sinning I, Tews I, Nickel W. A direct role for phosphatidylinositol-4,5-bisphosphate in unconventional secretion of fibroblast growth factor 2. *Traffic.* 2008;9(7):1204–1217. https://doi.org/10.1111/j.1600-0854.2008.00749.x.
362. Steringer JP, Bleicken S, Andreas H, et al. Phosphatidylinositol 4,5-bisphosphate (pi(4,5)p2)-dependent oligomerization of fibroblast growth factor 2 (fgf2) triggers the formation of a lipidic membrane pore implicated in unconventional secretion. *J Biol Chem.* 2012;287(33):27659–27669. https://doi.org/10.1074/jbc.M112.381939.
363. Steringer JP, Müller HM, Nickel W. Unconventional secretion of fibroblast growth factor 2-a novel type of protein translocation across membranes? *J Mol Biol.* 2015;427(6, Pt. A):1202–1210. https://doi.org/10.1016/j.jmb.2014.07.012.
364. Müller HM, Steringer JP, Wegehingel S, et al. Formation of disulfide bridges drives oligomerization, membrane pore formation, and translocation of fibroblast growth factor 2 to cell surfaces. *J Biol Chem.* 2015;290(14):8925–8937. https://doi.org/10.1074/jbc.M114.622456.
365. Steringer JP, Lange S, Čujová S, et al. Key steps in unconventional secretion of fibroblast growth factor 2 reconstituted with purified components. *eLife.* 2017;6:e28985. https://doi.org/10.7554/eLife.28985.
366. Rayne F, Debaisieux S, Bonhoure A, Beaumelle B. HIV-1 Tat is unconventionally secreted through the plasma membrane. *Cell Biol Int.* 2010;34(4):409–413. https://doi.org/10.1042/CBI20090376.
367. Debaisieux S, Rayne F, Yezid H, Beaumelle B. The ins and outs of HIV-1 Tat. *Traffic.* 2012;13(3):355–363. https://doi.org/10.1111/j.1600-0854.2011.01286.x.
368. Vitale N, Beaumelle B, Bader MF, Tryoen-Toth P. HIV-1 Tat protein perturbs diacylglycerol production at the plasma membrane of neurosecretory cells during exocytosis. *Commun Integr Biol.* 2013;6(5):e25145. https://doi.org/10.4161/cib.25145.

369. Martín-Sánchez F, Diamond C, Zeitler M, et al. Inflammasome-dependent il-1β release depends upon membrane permeabilisation. *Cell Death Differ.* 2016;23(7):1219.
370. Cho W, Stahelin RV. Membrane-protein interactions in cell signaling and membrane trafficking. *Annu Rev Biophys Biomol Struct.* 2005;34(1):119–151. https://doi.org/10.1146/annurev.biophys.33.110502.133337.
371. Di Paolo G, De Camilli P. Phosphoinositides in cell regulation and membrane dynamics. *Nature.* 2006;443(7112):651.
372. Kutateladze TG. Translation of the phosphoinositide code by PI effectors. *Nat Chem Biol.* 2010;6(7):507.
373. Stahelin RV, Scott JL, Frick CT. Cellular and molecular interactions of phosphoinositides and peripheral proteins. *Chem Phys Lipids.* 2014;182:3–18. https://doi.org/10.1016/j.chemphyslip.2014.02.002.
374. Lemmon MA. Membrane recognition by phospholipid-binding domains. *Nat Rev Mol Cell Biol.* 2008;9(2):99.
375. Morales KA, Igumenova TI. Synergistic effect of pb2+ and phosphatidylinositol 4,5-bisphosphate on c2 domain-membrane interactions. *Biochemistry.* 2012;51(16):3349–3360. https://doi.org/10.1021/bi201850h.
376. Lai CL, Landgraf KE, Voth GA, Falke JJ. Membrane docking geometry and target lipid stoichiometry of membrane-bound PKCα C2 domain: a combined molecular dynamics and experimental study. *J Mol Biol.* 2010;402(2):301–310. https://doi.org/10.1016/j.jmb.2010.07.037.
377. Ceccarelli DFJ, Blasutig IM, Goudreault M, et al. Non-canonical interaction of phosphoinositides with pleckstrin homology domains of Tiam1 and ArhGAP9. *J Biol Chem.* 2007;282(18):13864–13874. https://doi.org/10.1074/jbc.M700505200.
378. Jian X, Tang WK, Zhai P, et al. Molecular basis for cooperative binding of anionic phospholipids to the PH domain of the Arf GAP ASAP1. *Structure.* 2015;23(11):1977–1988. https://doi.org/10.1016/j.str.2015.08.008.
379. Pant S, Tajkhorshid E. Microscopic characterization of GRP1 PH domain interaction with anionic membranes. *J Comput Chem.* 2019;9999:1–11. https://doi.org/10.1002/jcc.26109.
380. Senju Y, Kalimeri M, Koskela EV, et al. Mechanistic principles underlying regulation of the actin cytoskeleton by phosphoinositides. *Proc Natl Acad Sci.* 2017;114(43):E8977–E8986. https://doi.org/10.1073/pnas.1705032114.
381. DiNitto JP, Delprato A, Lee MTG, et al. Structural basis and mechanism of autoregulation in 3-phosphoinositide-dependent Grp1 family Arf GTPase exchange factors. *Mol Cell.* 2007;28(4):569–583. https://doi.org/10.1016/j.molcel.2007.09.017.
382. Fuss B, Becker T, Zinke I, Hoch M. The cytohesin Steppke is essential for insulin signalling in Drosophila. *Nature.* 2006;444(7121):945.
383. Hafner M, Schmitz A, Grüne I, et al. Inhibition of cytohesins by SecinH3 leads to hepatic insulin resistance. *Nature.* 2006;444(7121):941.
384. Claing A, Chen W, Miller WE, et al. β-arrestin-mediated ADP-ribosylation factor 6 activation and β2-adrenergic receptor endocytosis. *J Biol Chem.* 2001;276(45):42509–42513. https://doi.org/10.1074/jbc.M108399200.
385. Clodi M, Vollenweider P, Klarlund J, et al. Effects of general receptor for phosphoinositides 1 on insulin and insulin-like growth factor I-induced cytoskeletal rearrangement, glucose transporter-4 translocation, and deoxyribonucleic acid synthesis. *Endocrinology.* 1998;139(12):4984–4990. https://doi.org/10.1210/endo.139.12.6351.
386. Frank SR, Hatfield JC, Casanova JE. Remodeling of the actin cytoskeleton is coordinately regulated by protein kinase C and the ADP-ribosylation factor nucleotide exchange factor ARNO. *Mol Biol Cell.* 1998;9(11):3133–3146. https://doi.org/10.1091/mbc.9.11.3133.

387. Blanchoin L, Boujemaa-Paterski R, Sykes C, Plastino J. Actin dynamics, architecture, and mechanics in cell motility. *Physiol Rev*. 2014;94(1):235–263.
388. Pollard TD, Cooper JA. Actin, a central player in cell shape and movement. *Science*. 2009;326(5957):1208–1212. https://doi.org/10.1126/science.1175862.
389. Saarikangas J, Zhao H, Lappalainen P. Regulation of the actin cytoskeleton-plasma membrane interplay by phosphoinositides. *Physiol Rev*. 2010;90(1):259–289. https://doi.org/10.1152/physrev.00036.2009.
390. Yin HL, Janmey PA. Phosphoinositide regulation of the actin cytoskeleton. *Annu Rev Physiol*. 2003;65(1):761–789. https://doi.org/10.1146/annurev.physiol.65.092101.142517.
391. Gorelik R, Yang C, Kameswaran V, Dominguez R, Svitkina T. Mechanisms of plasma membrane targeting of formin mDia2 through its amino terminal domains. *Mol Biol Cell*. 2011;22(2):189–201. https://doi.org/10.1091/mbc.e10-03-0256.
392. van Rheenen J, Song X, van Roosmalen W, et al. EGF-induced PIP2 hydrolysis releases and activates cofilin locally in carcinoma cells. *J Cell Biol*. 2007;179(6):1247–1259. https://doi.org/10.1083/jcb.200706206.
393. Chinthalapudi K, Rangarajan ES, Patil DN, George EM, Brown DT, Izard T. Lipid binding promotes oligomerization and focal adhesion activity of vinculin. *J Cell Biol*. 2014;207(5):643–656. https://doi.org/10.1083/jcb.201404128.
394. Rivera MG, Vasilescu D, Papayannopoulos V, Lim WA, Mayer BJ. A reciprocal interdependence between Nck and PI(4,5)P2 promotes localized N-WASp-mediated actin polymerization in living cells. *Mol Cell*. 2009;36(3):525–535. https://doi.org/10.1016/j.molcel.2009.10.025.
395. Wollman R, Meyer T. Coordinated oscillations in cortical actin and Ca2+ correlate with cycles of vesicle secretion. *Nat Cell Biol*. 2012;14(12):1261.
396. Yonezawa N, Nishida E, Iida K, Yahara I, Sakai H. Inhibition of the interactions of cofilin, destrin, and deoxyribonuclease I with actin by phosphoinositides. *J Biol Chem*. 1990;265(15):8382–8386.
397. Janmey PA, Stossel TP. Modulation of gelsolin function by phosphatidylinositol 4,5-bisphosphate. *Nature*. 1987;325(6102):362.
398. Schafer DA, Jennings PB, Cooper JA. Dynamics of capping protein and actin assembly in vitro: uncapping barbed ends by polyphosphoinositides. *J Cell Biol*. 1996;135(1):169–179. https://doi.org/10.1083/jcb.135.1.169.
399. Prehoda KE, Scott JA, Mullins RD, Lim WA. Integration of multiple signals through cooperative regulation of the N-WASP-Arp2/3 complex. *Science*. 2000;290(5492):801–806. https://doi.org/10.1126/science.290.5492.801.
400. Rohatgi R, Ho HYH, Kirschner MW. Mechanism of N-wasp activation by CDC42 and phosphatidylinositol 4,5-bisphosphate. *J Cell Biol*. 2000;150(6):1299–1310. https://doi.org/10.1083/jcb.150.6.1299.
401. Martel V, Racaud-Sultan C, Dupe S, et al. Conformation, localization, and integrin binding of talin depend on its interaction with phosphoinositides. *J Biol Chem*. 2001;276(24):21217–21227. https://doi.org/10.1074/jbc.M102373200.
402. Yamamoto M, Hilgemann DH, Feng S, et al. Phosphatidylinositol 4,5-bisphosphate induces actin stress-fiber formation and inhibits membrane ruffling in Cv1 cells. *J Cell Biol*. 2001;152(5):867–876. https://doi.org/10.1083/jcb.152.5.867.
403. Raucher D, Stauffer T, Chen W, et al. Phosphatidylinositol 4,5-bisphosphate functions as a second messenger that regulates cytoskeleton-plasma membrane adhesion. *Cell*. 2000;100(2):221–228. https://doi.org/10.1016/S0092-8674(00)81560-3.
404. Rozelle AL, Machesky LM, Yamamoto M, et al. Phosphatidylinositol 4,5-bisphosphate induces actin-based movement of raft-enriched vesicles through WASP-Arp2/3. *Curr Biol*. 2000;10(6):311–320. https://doi.org/10.1016/S0960-9822(00)00384-5.

405. Cunningham CC, Vegners R, Bucki R, et al. Cell permeant polyphosphoinositide-binding peptides that block cell motility and actin assembly. *J Biol Chem.* 2001;276(46):43390–43399. https://doi.org/10.1074/jbc.M105289200.
406. Kielian M. Mechanisms of virus membrane fusion proteins. *Annu Rev Virol.* 2014;1(1):171–189. https://doi.org/10.1146/annurev-virology-031413-085521.
407. Kim IS, Jenni S, Stanifer ML, et al. Mechanism of membrane fusion induced by vesicular stomatitis virus G protein. *Proc Natl Acad Sci.* 2017;114(1):E28–E36. https://doi.org/10.1073/pnas.1618883114.
408. Guardado-Calvo P, Atkovska K, Jeffers SA, et al. A glycerophospholipid-specific pocket in the RVFV class II fusion protein drives target membrane insertion. *Science.* 2017;358(6363):663–667. https://doi.org/10.1126/science.aal2712.
409. Mizushima N. Autophagy: process and function. *Genes Dev.* 2007;21(22):2861–2873. https://doi.org/10.1101/gad.1599207. http://genesdev.cshlp.org/content/21/22/2861.abstract.
410. Lee WS, Yoo WH, Chae HJ. ER stress and autophagy. *Curr Mol Med.* 2015;15(8):735–745.
411. Kirkin V, Lamark T, Sou YS, et al. A role for NBR1 in autophagosomal degradation of ubiquitinated substrates. *Mol Cell.* 2009;33(4):505–516. https://doi.org/10.1016/j.molcel.2009.01.020.
412. Khaminets A, Heinrich T, Mari M, et al. Regulation of endoplasmic reticulum turnover by selective autophagy. *Nature.* 2015;522:354–3581. https://doi.org/10.1038/nature14498.
413. Brady JP, Claridge JK, Smith PG, Schnell JR. A conserved amphipathic helix is required for membrane tubule formation by Yop1p. *Proc Natl Acad Sci.* 2015;112(7):E639–E648. https://doi.org/10.1073/pnas.1415882112.
414. Breeze E, Dzimitrowicz N, Kriechbaumer V, et al. A C-terminal amphipathic helix is necessary for the in vivo tubule-shaping function of a plant reticulon. *Proc Natl Acad Sci.* 2016;113(39):10902–10907. https://doi.org/10.1073/pnas.1605434113.
415. Hu J, Shibata Y, Voss C, Rist JM, Rapoport TA. Membrane proteins of the endoplasmic reticulum induce high-curvature tubules. *Science.* 2008;319(5867):1247–1250. https://doi.org/10.1126/science.1153634.
416. Voeltz GK, Prinz WA, Shibata Y, et al. A class of membrane proteins shaping the tubular endoplasmic reticulum. *Cell.* 2006;124(3):573–586. https://doi.org/10.1016/j.cell.2005.11.047.
417. Bhaskara RM, Grumati P, Garcia-Pardo J, et al. Curvature induction and membrane remodeling by FAM134B reticulon homology domain assist selective ER-phagy. *Nat Commun.* 2019;10(1):1–13.
418. Walther TC, Farese RV. Lipid droplets and cellular lipid metabolism. *Annu Rev Biochem.* 2012;81(1):687–714. https://doi.org/10.1146/annurev-biochem-061009-102430.
419. Zoni V, Campomanes P, Khaddaj R, Thiam AR, Schneiter R, Vanni S. Lipid droplet biogenesis is driven by liquid-liquid phase separation. *bioRxiv.* 2020. https://doi.org/10.1101/777466.
420. Bacle A, Gautier R, Jackson CL, Fuchs PFJ, Vanni S. Interdigitation between triglycerides and lipids modulates surface properties of lipid droplets. *Biophys J.* 2017;112(7):1417–1430. https://doi.org/10.1016/j.bpj.2017.02.032.
421. MacDermaid CM, Kashyap HK, DeVane RH, et al. Molecular dynamics simulations of cholesterol-rich membranes using a coarse-grained force field for cyclic alkanes. *J Chem Phys.* 2015;143(24):243144. https://doi.org/10.1063/1.4937153.
422. Campomanes P, Zoni V, Vanni SL. Local accumulation of diacylglycerol alters membrane properties nonlinearly due to its transbilayer activity. *Commun Chem.* 2019;2:1–8.

423. Shinoda W, DeVane R, Klein ML. Coarse-grained molecular modeling of non-ionic surfactant self-assembly. *Soft Matter*. 2008;4:2454–2462. https://doi.org/10.1039/B808701F.
424. Yan R, Qian H, Lukmantara I, et al. Human SEIPIN binds anionic phospholipids. *Dev Cell*. 2018;47(2):248–256.e4. https://doi.org/10.1016/j.devcel.2018.09.010.
425. Salo Y, Li S, Vihinen H, et al. Seipin facilitates triglyceride flow to lipid droplet and counteracts droplet ripening via endoplasmic reticulum contact. *Dev Cell*. 2019;50(4). https://doi.org/10.1016/j.devcel.2019.05.016478–493.e9.
426. Cartwright BR, Binns DD, Hilton CL, Han S, Gao Q, Goodman JM. Seipin performs dissectible functions in promoting lipid droplet biogenesis and regulating droplet morphology. *Mol Biol Cell*. 2015;26(4):726–739. https://doi.org/10.1091/mbc.E14-08-1303.
427. Nettebrock NT, Bohnert M. Born this way—biogenesis of lipid droplets from specialized ER subdomains. *Biochim Biophys Acta*. 2020;1865(1). https://doi.org/10.1016/j.bbalip.2019.04.008.
428. Dupuy AD, Engelman DM. Protein area occupancy at the center of the red blood cell membrane. *Proc Natl Acad Sci*. 2008;105(8):2848–2852. https://doi.org/10.1073/pnas.0712379105.
429. Nyholm TKM. Lipid-protein interplay and lateral organization in biomembranes. *Chem Phys Lipids*. 2015;189:48–55. https://doi.org/10.1016/j.chemphyslip.2015.05.008.
430. Contreras FX, Ernst AM, Wieland F, Brügger B. Specificity of intramembrane protein-lipid interactions. *Cold Spring Harb Perspect Biol*. 2011;3(6):1–18. https://doi.org/10.1101/cshperspect.a004705.
431. Corradi V, Mendez-Villuendas E, Ingólfsson HI, et al. Lipid-protein interactions are unique fingerprints for membrane proteins. *ACS Cent Sci*. 2018;4(6):709–717. https://doi.org/10.1021/acscentsci.8b00143.

CHAPTER EIGHT

Minimalistic coarse-grained modeling of viral capsid assembly

Szilard N. Fejer[a,b,*]
[a]Provitam Foundation, Department of Research and Development, Sfântu Gheorghe, Romania
[b]University of Pécs, Department of Chemistry, Faculty of Sciences, Pécs, Hungary
[*]Corresponding author: e-mail address: szilard.fejer@cantab.net

Contents

Abstract

The spontaneous formation of virus capsids from multiple copies of capsid proteins is a fascinating example of supramolecular self-assembling processes. Most known viruses protect their genome with icosahedral capsids, but other morphologies exist as well, including elongated, conical, tubular, head-tail structures. The mechanisms of assembly can be diverse and are still not perfectly understood. In this chapter we present theoretical models developed over the years that reproduce the basic physics of self-assembly of empty viral capsids. All these models are highly coarse-grained, as it is still not possible to access the long timescales of such processes with atomistic modeling. Very different particle-based models can result in the same overall behavior, showing that such processes are governed by the effective anisotropic interactions between protein building blocks.

1. Introduction

Viruses are highly efficient microscopic particles that hijack the machinery of infected cells to replicate, using information stored in the viral genome (RNA or DNA). Since viruses are completely inactive when not

Progress in Molecular Biology and Translational Science, Volume 170
ISSN 1877-1173
https://doi.org/10.1016/bs.pmbts.2019.12.003

inside a cell, they cannot be considered living organisms. The viral genome contains all necessary information for the virus to recreate itself using the machinery of the infected cell.

Simple viruses consist of a protective envelope (capsid) that packages the nucleic acid. As space within this envelope is limited, so is the information that can be encoded in the genome of a virus. The capsid of simple viruses consists therefore of multiple copies of a single protein, which has evolved to assemble into shells with high symmetries that protect the genome from the environment. Most known viruses have capsids with icosahedral symmetry,[1] although lower-order or asymmetrical virus capsids are known as well.[2] The size of known viruses varies from a mere 17 nm to the size of a bacterial cell (1.5 μm in length).[3,4] Larger viruses can have several different proteins in their capsids.

Caspar and Klug[5] were the first to formulate generic rules by which icosahedral virus capsids can exist. The theory of quasi-equivalence states that the same capsid proteins (CPs), due to their inherent flexibility, can assemble into various non-equivalent positions on a hexagonal lattice folded into an icosahedron. Icosahedral symmetries for virus capsids are generally described with a triangulation number, which describes the number of distinct non-equivalent environments in one triangular facet of an icosahedron. A simplified description of the triangulation number is $T = h^2 + hk + k^2$, where h and k are non-negative integers, and in order to reach a neighboring pentagonal facet in an icosahedral structure consisting of pentagons and hexagons, h steps must be taken in one direction, followed by k steps in a direction shifted by 60°. The triangulation number of a capsid can be 1, 3, 4, 7, 9, 12, 13, etc. Icosahedral capsids therefore contain 12 pentameric and $10(T-1)$ hexameric units. In recent years, with the progress in structure determination methods, more and more icosahedral viruses have been discovered that do not obey the Caspar-Klug theory, suggesting that a more general framework might exist for the description of such structures. Recently, it has been shown that by using other Archimedean lattices that contain hexagonal sublattices instead of the simplest hexagonal lattice for the Caspar-Klug construction of icosahedral architecture, various non-quasi-equivalent arrangements of CPs can be explained.[6]

In silico computer models can shed light on the processes governing virus assembly. For the simplest class of viruses, consisting of genetic material and capsid protein, the class of non-icosahedral capsids contains elongated cylindrical morphologies with quasi-spherical caps, oblate structures,[7] non-capped tubes, conical shells, or amorphous arrangements. Even more complex morphologies have evolved, such as head-tail assemblies (phages), or viruses with an external or internal lipid envelope. Explaining the spontaneous formation of such structures with coarse-grained models is a challenge, and related models will be presented in this chapter.

High-level theoretical model development would not be possible without the structure determination efforts over the last few decades (Section 2), which give us insight into the generic rules that control the final shape of capsids but does not tell us how this assembly is achieved. In fact, for some simple viruses, capsid proteins can self-assemble into empty shells without the presence of genetic material that could act as a nucleus.[8–11] Therefore, the characteristics of the final shell are somewhat encoded in the shape of the capsid proteins. The intermediates of capsid assembly are generally not well-described, and likely mechanisms are inferred through indirect observation of changes in physical characteristics of the solution. For cases in which direct observation of intermediates is possible, even a rough two-dimensional picture of an intermediate can give useful insight into the probable mechanism. For example, a possible assembly pathway of the conical HIV capsid has been shown to proceed via the curling of hexagonal sheets in a confined environment.[12] Starting from a formalism derived for aggregation of proteins into a shell-like polymer,[13] several assembly mechanisms have been identified so far, including hierarchical assembly through smaller oligomeric units (dimers, trimers, pentamers or hexamers), sequential addition of monomers/oligomers, or assembly around templates. The latter mechanism allows the efficient formation of large capsids.[14]

Universal aspects of capsid self-assembly were captured experimentally through extensive structural studies on well-behaving capsid proteins. A favorite capsid protein for experimentalists is that of the cowpea chlorotic mottle virus (CCMV),[15] which readily assembles on its own into shells with different symmetries depending on external conditions such as ionic strength, pH, or a rigid DNA template.[8,16,17] Studies have shown that the assembly of the native $T=3$ capsid needs a pentamer of dimers as a nucleus, and proceeds through addition of capsid protein dimers.[18,19] The flexibility of dimers allows the dimers to be inserted into pentagonal and hexagonal sites of the capsid. Fig. 1 shows the main building blocks relevant for CCMV assembly: the critical nucleus of pentamer of dimers, the dimer with the most stable interface,[20] and the assembled icosahedral capsid, illustrated on the structure of a point mutant protein (salt-stable CCMV,[21] PDB code 1za7).

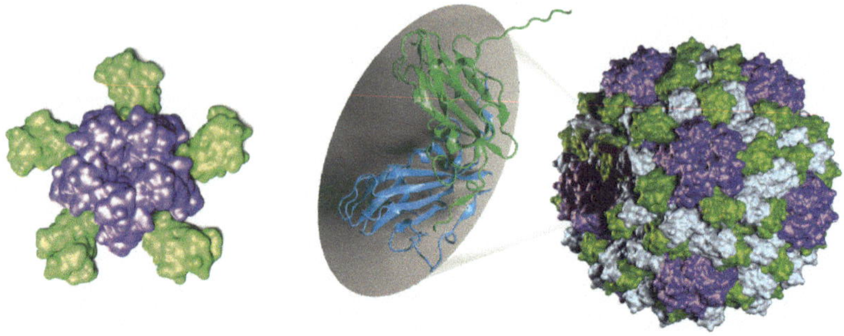

Fig. 1 Representative structures of salt-stable CCMV relevant for self-assembly: pentamer of dimers, the dimer with the most stable interface, and the assembled $T=3$ icosahedral capsid.

Because very different capsid proteins can self-assemble into capsids with similar morphologies, generic features might exist that allow such capsids to form with a high efficiency. Early attempts to explain assembly of icosahedral capsids were based on the local rule theory, which stipulated that each new protein binding to the capsid knows where to bind based on the local micro-environments caused by the different conformations already existing in the shell due to quasiequivalence.[22,23] However, to date there is no evidence that such local rules exist.

The minimal physics required for the formation of capsids can be best investigated with coarse-grained modeling. If a model has many emergent features that were not pre-programmed, the likelihood of it capturing generic aspects of the process increases.

This chapter presents an overview of minimal models that were built to explain certain universal features of self-assembling empty capsids, with emphasis on non-icosahedral morphologies. These models do not include any coarse-grained representation for the nucleic acids, although their role is important in many aspects of virus capsid stability[24,25] and assembly.[26–34]

2. Experimental structure determination methods

Cryo-electron microscopy (cryo-EM) and X-ray crystallography are the standard methods of structural determination of virus capsids. Although routinely used nowadays, cryo-EM techniques just recently reached near-atomistic resolution due to continuous development over the last 50 years. In cases where crystallization of virus capsids is possible, X-ray crystallography can provide the highest resolution images. In contrast to X-ray crystallography, cryo-EM techniques can be used to study the structure of capsids in their normal environment. Essentially, cryo-EM is an image reconstruction technique based on an iterative refinement of a structural model of a three-dimensional electron density map that can be fit to thousands of images representing two-dimensional snapshots of proteins in different orientation frozen in a matrix (a very thin layer of water) at cryogenic temperatures.

The first reconstruction of a three-dimensional structure from electron micrographs was that of the T4 phage tail, which has helical symmetry.[35] Icosahedral viruses have been extensively used in the development of cryo-EM image processing techniques, due to their high symmetry and large mass.[36] The symmetry and size of macromolecular assemblies helps the image reconstruction process, as the signal to noise ratio is proportional to the size, and symmetrical objects reduce the number of possible orientations that have to be taken into account during modeling. The development of direct electron detectors in recent years contributed significantly to the increase in signal quality.[37] With the resolution increase over the years, previously hidden asymmetric features could be resolved that were not visible before due to the overuse of symmetry averaging methods, such as the DNA packaging portal region of the herpes simplex virus capsid.[38]

Although the electron micrographs are static images, large-scale conformational changes in already formed virus capsids can be imaged. For example, the maturation of the herpes simplex virus from a spherical procapsid to the final faceted icosahedral form has been visualized using time-lapse cryo-EM.[39] Recently, partially open capsid structures with expelled pentamers due to genome release have been resolved at near-atomistic resolution as well using cryo-EM.[40] Other structures with lower-order symmetry have been resolved at high resolution recently as well, for example, that of a plant geminivirus containing two fused icosahedral capsids with overall D_5 symmetry.[2] However, the computed resolution of cryo-EM is just an approximation, and systematic modeling errors can actually improve the apparent resolution, thus overestimating the quality of the structure.[41,42] New approaches have to be employed to accurately estimate the quality of structural cryo-EM models.[43]

The most comprehensive collection of experimentally solved icosahedral virus structures can be found on the ViperDB website[44] (http://viperdb.scripps.edu), containing over 800 X-ray and cryo-EM models, corresponding to almost 600 unique structures. Although originally conceived as a database for icosahedral structures, the current database contains various non-icosahedral capsid structures as well.

Assembly dynamics can be investigated with a variety of indirect methods: time-resolved small-angle X-ray scattering,[45] light scattering,[18] mass spectrometry,[46,47] size exclusion chromatography,[48] fluorescent labeling. These methods must be complemented with theoretical models in order to understand the signals obtained from the bulk experimental system.

3. Models for assembly

Atomistic modeling of the complete capsid assembly process from basic building blocks is not yet possible, given the system size and large time scales involved (millions of atoms, seconds/minutes), therefore significant coarse-graining must be used to access such timescales. Multiscale approaches[49,50] can bridge the gap from atomistic to supramolecular length and time scales, but these are usually used to explain very specific behaviors of certain virus capsids.[51]

All coarse-grained models for capsid proteins are designed to represent the rough shape of protein monomers or oligomers and the interactions between them under experimental conditions. Given the complexity of the interactions involved, these models employ effective interaction potentials that favor aggregation of CP oligomers. To study the assembly dynamics, standard methods are adapted for the coarse-grained models: Monte-Carlo, molecular dynamics, energy landscape exploration (global optimization[52] and discrete path sampling[53]).

All models represent CPs as anisotropic particles with anisotropic interactions. The range and directionality of interactions greatly affects the aggregation behavior of CPs, and the earliest coarse-grained models could therefore reproduce only very specific capsid morphologies. Effective anisotropic interactions can also be obtained by restricting spherical particles on spherical or other curved surfaces, which gives insight into the relative stabilities of certain oligomers, but cannot describe the assembly dynamics.

The interactions between building blocks are represented with energy terms in all these models, but these are effectively free energies. It is important to note that the aggregation of virus capsid proteins is usually driven by entropy instead of strong protein-protein interactions, this is called the "weak association model" for virus capsids.[48] The free energy change of association is negative due to the entropy gain through the release of water molecules bound to the surface, which is greater than the entropy loss of the proteins due to aggregation.

There is no consensus on whether capsid assembly is reversible or irreversible under biological conditions.[54] Clearly, the virus has to be stable outside its normal environment for a very long time in order to be able to infect other host cells. Once assembled, larger capsids usually undergo maturation that ensure long-term capsid stability.[55] Maturation steps generally involve large-scale conformational rearrangements, disulfide bond formation,[56,57] and there is an interesting example of covalent bond forming between HK97 capsid proteins as well, creating a macromolecular catenane, which contributes to the increased stability of the capsid.[58,59] Phages infect the cells by attaching to the cell membrane and injecting their genetic material instead of releasing it through capsid disassembly, and have therefore evolved extremely stable protein shells.

Capsid stability is especially important in harsh environments. The thermostable phage P74-26 stands out from other related phages due to its capsid size which is able to package twice as much genetic material than the related HK97 phage, despite both having the same $T=7$ symmetry.[60] Usually, more complex capsids have larger triangulation numbers. In this case, however, the capsid protein keeps the relatively simple $T=7$ configuration by increasing the diameter of the capsid protein, relative to its homologues shown in Fig. 2. It is possible that capsid stability can be best ensured by assembly into low triangulation number capsids in the case of phages that infect bacteria living in extreme environments.[60]

Experimental evidence shows that capsid assembly kinetics is similar to crystallization, with two distinct phases: nucleation and elongation.[61] The process has been reviewed extensively by Hagan.[32] Basically, the formation of the critical nucleus is the rate determining step of the process, which is followed by the elongation phase, usually by sequential addition of small protein units (monomers, dimers, pentamers and hexamers, depending on the system). Structures smaller than the critical nucleus are predominantly unstable, while those in the elongation phase disassemble less likely.

One of the earliest models studying self-assembly of polyhedra was that of Rapaport,[62,63] obtaining self-assembly of shells with $T=1$ and 3 using triangular and trapezoidal building blocks with attractive patches, with additional specific (unphysical) assembly rules that prevented the formation of incorrectly assembled structures. Correctly formed bonds were irreversible in character. The model was subsequently refined to allow for reversible bonding between the building blocks, giving insight into the dynamics of capsid growth.[64]

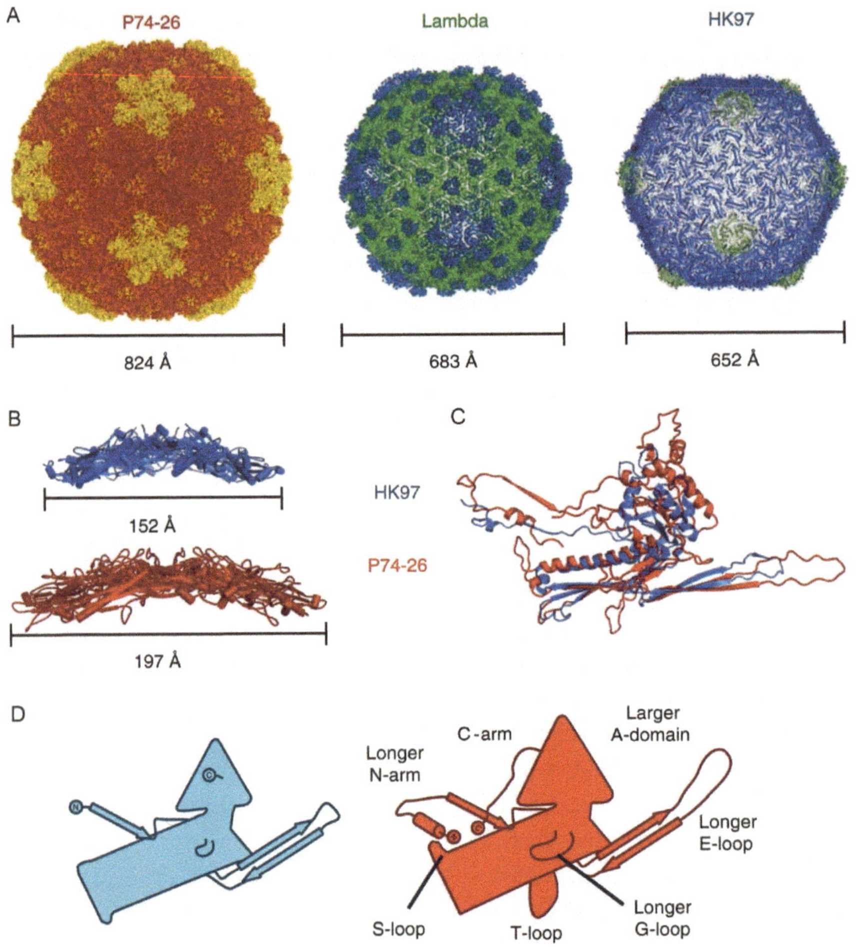

Fig. 2 (A) Comparison of capsid sizes for three $T=7$ icosahedral capsids of phages: P74-26, Lambda and HK97. (B) Comparative side view of hexons for HK97 (blue) and P74-26 (red). (C) Overlaid tertiary structure of the main capsid proteins for HK97 (blue) and P74-26 (red). (D) Schematic view of the secondary structures for the two proteins, with the features unique to P74-26. *Reprinted from Stone NP, Demo G, Agnello E, Kelch BA. Principles for enhancing virus capsid capacity and stability from a thermophilic virus capsid structure.* Nat Commun. *2019;10(1):4471. doi:10.1038/s41467-019-12341-z under the Creative Commons license. To view a copy of this license, visit http://creativecommons.org/licenses/by/4.0/.*

Hagan and Chandler[65] introduced a model based on simple pairwise interactions of spherical particles with interaction anisotropies designed to replicate protein-protein binding in $T=1$ icosahedral configurations of 60 particles. By averaging over multiple trajectories, the model could identify several mechanisms through which these capsomers can form complete shells, with reversible interactions and the model obeying detailed balance. The system, however, presented many kinetic traps that narrowed the region of parameter space for efficient forming of complete capsids.

The self-assembly kinetics of simple model capsids have been explored in detail by Nguyen et al.[66,67] by using a more physically realistic model than that of Rapaport, and discontinuous molecular dynamics (DMD). The building blocks in the earlier model[66] are triangular pyramidal or trapezoidal units, with attractive patches along the edges, making possible the forming of native and non-native contacts as well. The trapezoidal units compatible with $T=1$ and $T=3$ capsids contain more specific attractive patches, which only interact with patches of the same type in other units. The building blocks interact with simple square-well and hard-sphere potentials. DMD simulations on these systems showed that under appropriate conditions (temperature and monomer concentration), efficient self-assembly of complete icosahedral shells is possible. Nevertheless, the model yielded other important byproducts as well, including oblate, prolate capsids, partial capsids, and misaggregated "monster" particles.

Van Workum and Douglas used a model of rigidly linked multipoles to obtain symmetric self-assembling structures.[68] The model building blocks were able to form pentagonal and hexagonal subunits spontaneously, together with $T=1$ icosahedral shells, despite their simplicity (directional dipolar and isotropic Lennard-Jones interactions). Triangular building blocks with fixed dipoles on each LJ particle could spontaneously form flat aggregates with hexagonal symmetries, coexisting with pentagonal aggregates in which the triangles tilt out of plane. This remarkably simple minimalistic model therefore achieved spontaneous assembly of structures analogous to capsid protein pentamers and hexamers.

3.1 Confined models

A minimalistic model of interacting disks restricted on spherical surfaces gave great insight into the reasons for preference of certain symmetries compatible with virus capsids.[69] In many ways, this model is similar to the Thomson problem,[70] which describes the optimal arrangement of points that repel each other while being confined on a sphere, and to the Tammes problem,[71] discussing the optimal surface coverage of disks on a sphere. Additionally, an earlier model of interacting identical disks[72] could produce a phase diagram as a function of spontaneous curvature between two adjacent disks and the mole fraction of various sized capsids. The model of Zandi et al. showed that simple excluded volume interaction (for hard disks on a sphere) is not enough to ensure preferred icosahedral packing. Although the most stable configuration for $N=12$ particles is the $T=1$ capsid, for larger N of identical disks, the "magic numbers" do not always represent structures with icosahedral symmetry. Other, highly symmetrical capsid structures appeared for $N=24$ (snub cube – octahedral symmetry) and 48 (cubic symmetry), while the $N=32$ and 42 configurations were lower in symmetry (D_5 and D_{5h}, respectively). Two main factors were identified that increase the stability of icosahedral structures relative to other spherical configurations: the disks at pentameric sites have to be smaller in size, and a long-range attraction must exist between the disks.

Prolate viruses with the highest symmetry can be described as hexamers in a cylindrical arrangement, capped by two halves an icosahedron of a $T=1$ configuration containing six pentamers each, and having fivefold axial symmetry. The original Caspar-Klug formalism was extended to explain and classify such structures,[73,74] although even this classification did not include all experimentally known prolate capsids. A phenomenological model of hard spheres with pairwise attraction, confined on a prolate spheroid surface, was able to reproduce threefold and fivefold symmetries of prolate virus capsids (aberrant flock house virus, alfalfa mosaic virus mutants and bacteriophages $\Phi29$ and $T4$). Luque et al. developed a model for elongated virus capsids, which describe self-assembly of spherical particles confined on spherocylindrical surfaces.[75] By adjusting several parameters for clusters with two different morphological units (pentamers and hexamers), the most stable structures were found to be either having partial $T=1$ and $T=3$ caps with threefold or fivefold symmetries. For $T=1$ caps, the threefold symmetry was more stable, while for $T=3$ caps it was the fivefold symmetry that was the global minimum. $T=1$ caps with twofold symmetries were also found to be quite low in energy. For caps with larger triangulation number, all global minima had fivefold symmetries. If only one type of capsomer was considered, the model supported nonicosahedral caps with snub cube or cubic symmetries, additionally to the icosahedral caps corresponding to $T=1$, 3 and 7.

3.2 Interactions between capsid proteins

The range of all protein-protein interactions relevant for capsid assembly is generally classified as short,[32] although long-range electrostatic repulsive interactions might help orient the proteins[9,76] while they approach each other, or maintain the weak association between proteins that is necessary for infectivity.[77] Generally, the outer side of the capsid has a different distribution of polar residues than the inside. Considering that a typical capsid protein has a radius of gyration of about 5 nm,[78] and the Debye screening length in physiologically relevant ionic strengths is about 1 nm, we would not classify protein-protein electrostatic interactions as short ranged. In fact, the estimated surface charge density of virus capsids was found to be significantly higher than that of biomolecules.[79] High ionic strengths screen the electrostatic interactions between proteins, and, since repulsive interactions are more prevalent, capsid stability increases, as demonstrated experimentally on hepatitis B virus capsids[48] and with theoretical modeling as well.[80] The increased capsid stability may prevent disassembly in cellular environments, therefore decreasing the infectivity of the virus. Interfering with capsid stability is a potential way of action of novel antiviral drugs.[81,82]

The different components of protein-protein interaction can be grouped into two main classes: those that can be decomposed into pairwise energetic interactions (electrostatic, hydrogen bonding, van der Waals), and those that are harder to quantify due to complexity of the environment (entropic changes of the solvent during the process, depletion interaction in local environments due to crowding). Electrostatic interactions are responsible for the aggregation of various proteins,[83] but in virus capsid assembly the main driving force is thought to be the burial of hydrophobic sites. The widely accepted view is that per-subunit protein-protein association free energies are weak in the capsid.

Macromolecular crowding[84] plays a very important role in facilitating assembly of virus capsids. In cellular environments, the concentration of macromolecules is high, and, due to their size, the excluded volume is also significant, compared to the volume occupied by small molecules. This raises the effective concentration of the proteins by up to two orders of magnitude, and thus can enhance molecular recognition, depending on the shape and size of the protein. For example, HIV CA proteins can assemble into tubes composed of CA hexamers *in vitro* in high ionic strength solutions, but not under physiologically relevant ionic strengths. However, when using inert macromolecular crowding agents, the same self-assembly of tubular structures can be observed at close to physiological ionic strengths as well.[85] Currently it is possible to simulate crowding by using a rules-based discrete event stochastic simulator, coupled with regression models parameterized based on light scattering data.[86] These simulations show that the efficiency of assembly is indeed enhanced in highly crowded environments, while it directs the system into kinetic traps when assembly kinetics is not limited by nucleation. This phenomenological model also suggests that *in vitro* assembly pathways are applicable for nucleation-limited models within a broad range of macromolecular crowding.

All above interactions have to be taken into account when considering building minimalistic coarse-grained models. Using too short ranged interactions significantly complicates the energy landscape,[87,88] and makes the appearance of unphysical kinetic traps more likely. Environmental conditions must be factored into any model that does not use explicit solvent or crowding agents.

3.3 Energy landscapes of coarse-grained capsid models

Every physical or chemical process takes place on the underlying energy landscape.[89,90] A wealth of information can be extracted from these processes by exploring their energy landscape with various methods: for example the self-assembling character of the system can be estimated, kinetic traps described, rates of interconversion between different structures calculated, etc. The difficulty of the method consists of characterizing this multidimensional landscape, for which relevant regions have to be explored efficiently. For self-assembling systems, these are usually the low-energy regions of the landscape.

We characterize the energy landscapes in terms of their stationary points relevant for thermodynamics and kinetics, representing them with a network of local potential energy minima and transition states that connect these minima. Additional entropic effects due to vibration can be included by using the harmonic superposition approximation.[89] The complexity of the landscape, best described with the number of stationary points, is highly system-dependent and increases with system size. Molecular dynamics simulations can easily get stuck in certain regions of the landscape due to existing kinetic traps—these are low energy regions of the landscape separated by high barriers from other regions. Modifications of MD such as umbrella sampling or parallel tempering can overcome kinetic traps, at the expense of longer simulation times. As discrete path sampling involves finding connections between distant minima through a series of minima and transition states, it usually explores the relevant region of the energy landscape more efficiently than MD methods. The reduced representation of the landscape by disconnectivity graphs[91,92] gives visual insight into its organization, and characteristic motifs show if the system is efficiently self-assembling or not.

For simulations involving explicit solvent, MC and MD methods are generally more efficient than discrete path sampling, since the latter method needs properly converged stationary points, and gradient-based energy minimizers can reach the required convergence very slowly, if the number of degrees of freedom is large. We therefore use energy landscape methods for small to medium-sized systems, involving at most several thousand particles.

In order to be able to explore stationary points on the landscape, the potential energy function and its first derivative must be continuous. Any cutoffs routinely used in MD or MC methods have to be adapted to not have any discontinuities in energy and gradient. Clearly, this applies to any minimalistic coarse-grained potential energy functions for virus capsid building blocks. For example, it is not possible to use hard sphere potentials in conjunction with proper energy minimization.

The structures of stationary points can be well-defined only if no particle or group of particles is completely dissociated from the rest of the system. MD-based methods are therefore more appropriate to study nucleation and growth of nanostructures in bulk. It is, however, possible to include loosely bound particles as part of the cluster, if these structures are well-converged, albeit shallow, local minima on the landscape. In most cases, energy landscape exploration with discrete path sampling can take place in clusters with a fixed number of particles. The recent development of the grand and semigrand canonical basin-hopping method[93] makes it possible to investigate the thermodynamics of clusters with variable sizes as well, although the method has not been tested for virus capsid models yet.

The self-assembling behavior of simple rigid polyhedra has been studied with energy landscape methods by Wales,[90] using pentagonal pyramidal building blocks (capsomers) with Morse interaction sites and a repulsive site at the apex. He found that the global minimum of a cluster composed of 12 such building blocks is a $T=1$ structure, but the energy landscape exhibited a deep kinetic trap in which a capsomer was flipped in a closed shell. Follow-up investigation on the model with molecular dynamics confirmed the occurrence of such kinetic traps in bulk as well, and showed that for $T=3$ capsids the dynamics is very complex.[94] The addition of an extra interaction site that represents the implicit hydrophilic-hydrophobic repulsion between the inside and outside surfaces of a capsid protein[95] effectively removed such kinetic traps, improving the self-assembling character on the landscape for the studied capsid sizes ($T=1$ and 3, and a snub cube shell of 24 pentagonal pyramids) and therefore it simplified the dynamics on the landscape.

All these results showed that shape and interaction anisotropy is paramount in defining the self-assembling behavior of capsids. In order to gain further insight into the universal conditions that drive capsid assembly, we developed a model[96] that removes the spatial constraints of the previous minimalistic models, and in which the building blocks are not engineered to intrinsically form icosahedral structures either. In our model, the building blocks are rigid and consist of ellipsoids with anisotropic Lennard-Jones-like interaction between them, and two "polarized" apex sites as presented in Ref. [95], and in the inset of Fig. 3A. The potential through which the ellipsoids interact was developed by Paramonov and Yaliraki,[97] and subsequently implemented in many of our rigid-body mesoscopic models due to its versatility and efficiency which can properly describe not only the shape of particles, but their interaction anisotropies as well.[98–103]

3.3.1 "Magic number" clusters

Using basin-hopping global optimization on clusters of building blocks with the simplest parameterization we could think of, we found that the model building blocks tend to form highly symmetric shells. For the range

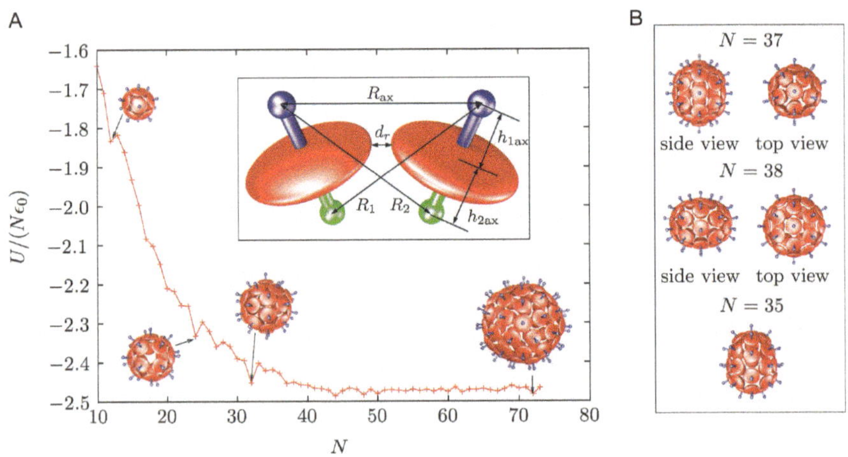

Fig. 3 Global minima for the simplest parameterization of the model: (A) Energy/particle (U/N) in units of ε_0 of the global minima as a function of cluster size. Inset: definition of interactions between two anisotropic building blocks with two repulsive sites on either side of the ellipsoid. Primary apex sites are colored in blue. (B) Conical, prolate, and oblate-shaped global minima. Secondary apex sites are only displayed in the dimer structure. *Reprinted with permission from Fejer SN, Chakrabarti D, Wales DJ. Emergent complexity from simple anisotropic building blocks: shells, tubes and spirals.* ACS Nano. *2010;4:219–228. Copyright (2010) American Chemical Society.*

$10 \leq N \leq 73$, we found seven cluster sizes which are more stable than their neighbors: 12, 24, 32, 44, 48, 72. Among these, three structures are polymorphic in character: the second-lowest minimum is very close in energy, but very different in structure for $N = 27$, 44, 48. The remaining four structures are all spherical, corresponding to capsid triangulation numbers $T = 1$, 3, 7, and a "snub cube" structure. The energy per particle diagram for the encountered global minima, as a function of cluster size, is shown in Fig. 3, together with representative structures of high symmetry. The building blocks were able to occupy both pentameric and hexameric coordination sites in the capsid, despite being of one size. The capsomers in pentavalent sites got slightly raised from the sphere defined by the centers of the hexavalent building blocks in the lowest energy structures of icosahedral symmetry. This essentially maximizes the contact of the pentavalent building blocks with their neighbors, at the expense of slightly shorter distance between primary and secondary apex sites, which in turn increases the overall repulsion. Nevertheless, this very basic parameterization did not yield a capsid size of $N = 42$ as a magic number cluster. The global minimum in this case is a D_{5h} structure, while the icosahedral $T = 4$ structure is the second lowest minimum on the landscape, very close in energy. We found that increasing the range of repulsive interactions and the different size of building blocks in pentameric and hexameric sites can increase the relative stability of the icosahedral structure compared to other low-energy spherical arrangements (Fig. 5A).[96] Other minimalistic models also had difficulties in stabilizing the $T = 4$ icosahedral structure compared to the D_{5h} structure, suggesting that some additional parameters must be necessary that allow the preference of the icosahedral structure. The model of Zandi and our model also showed that one such factor is the different size of pentamers and hexamers. However, another possibility is incorporating additional interaction terms according to the true symmetry of hexamers (C_2).[104] As hexamer complexity[105] increases, these additional terms can get stronger, therefore adding more frustration to D_5 structures, compared to their icosahedral polymorphs.

3.3.2 Non-spherical shells with polyhedral symmetries

Our model building blocks formed non-spherical shells for cluster sizes that were not compatible with spherical arrangements, whenever there was a possibility of forming shells with some symmetry. In this way, we found global minima that were oblate or prolate shells, and even shells with variable curvature. Clearly, such structures, when confined on a spherical surface, would get artificially high in energy, therefore structures that are roughly spherical are overstabilized in energy in all models where there is a spherical confinement.[69]

We also mapped the energy landscape for several cluster sizes and found that for $T=1$ and 3, the obtained disconnectivity graphs[89,91,92] were characteristic for self-assembly, having overall low downhill energy barriers from every minimum on the landscape toward the global minimum, with a perfect funnel shape. However, for $N=33$, we observed a low energy kinetic trap on the landscape, separated by a high barrier from the global minimum. The kinetic trap corresponds to a fully formed $T=3$ shell with a capsomer loosely bound to it. Inserting an extra capsomer into a stable icosahedral shell is therefore energetically not favored. We found that the insertion is in fact a single transition state rearrangement, and has an energy barrier that is about 2.5 times higher than a similar rearrangement that yields an icosahedral $T=3$ capsid. Fig. 4 shows a comparison between the energy landscapes for 32 and 33 particles, and the two single transition state pathways for incorporating a loosely bound capsomer into a shell of 31 and 32 particles, respectively.

The parameter space of our model building blocks is huge: besides the position and repulsion strength of the apex sites, we can change the size and shape of the ellipsoidal core and the attraction/repulsion strengths along its main axes. We showed in the Supplementary Information of Ref. [96] that decreasing the anisotropy of the ellipsoidal core increases the overall symmetry of the global minimum for all cluster sizes where symmetry is not already maximal (icosahedral or octahedral). Interestingly, some of the lowest energy structures found using a spherical core (essentially, a Lennard-Jones particle with two apex sites) are compatible with known non-icosahedral virus capsid morphologies: the third lowest minima for $N=22$ and 27 are both of the same symmetry as that suggested for two non-icosahedral virus capsids consisting of 120 and 150 protein subunits (12 pentamers and 10 or 15 hexamers, respectively).[106]

The kinetics of viral capsid assembly have been studied recently using a similar model, in which the preferred orientation of two spherical building blocks was adjusted with explicit angular parameters instead of repulsive apex sites.[104,107] The authors found that in order to achieve efficient assembly into shells with a given triangulation number, two factors are necessary: relatively weak interactions and preferred dimer curvatures that are smaller than that of the complete shell.

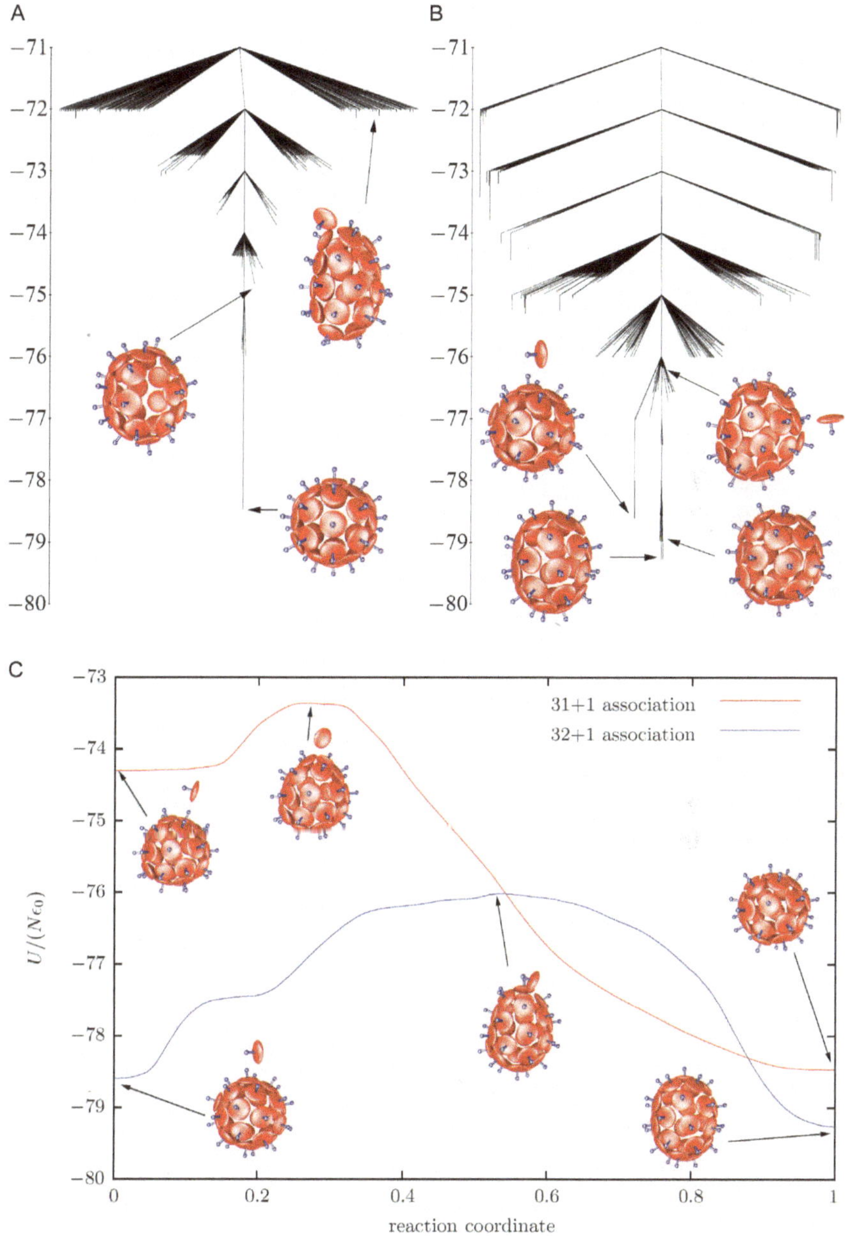

Fig. 4 Disconnectivity graphs for (A) $N=32$ and (B) $N=33$ building blocks. (C) Single transition state pathways from a capsomer loosely bound to shells containing 31 and 32 particles, forming closed shells with 32 and 33 particles, respectively. The activation energy for the capsomer insertion into an icosahedral shell (blue line) is 2.5 times higher than capsomer insertion into an incomplete shell. *Adapted with permission from Fejer SN, Chakrabarti D, Wales DJ. Emergent complexity from simple anisotropic building blocks: shells, tubes and spirals. ACS Nano. 2010;4:219–228. Copyright (2010) American Chemical Society.*

3.3.3 Open tubes

Our model highlights other universal aspects of capsid assembly as well. By increasing the side-to-side interaction strength of ellipsoids and decreasing the offset of the "polarized" apex sites, we found that higher on the landscape kinetic traps appear corresponding to open tubes of various diameters. These structures are analogous to carbon nanotubes. HIV CA and RSV capsid proteins can form *in vitro* analogous chiral tubular structures, for example.[108–110] We found open conical structures on the landscape as well, which belong to the basin of attraction of the spheroidal global minimum. The landscape is effectively split into two parts, with the structures that are open tubes being well-separated from the rest. Achiral and chiral tubular structures of different diameters are illustrated in Fig. 5C for a cluster of 60 particles.

As tubes with various diameter and chiral pitch exist on the landscape, their stability is also different. We identified a simple mechanism through which more strained tubes relax into less strained structures. These are cooperative single transition state rearrangements, with successive diamond-square-diamond type motions along the length of the tube. The identified "sliding" mechanism can act as a simple way through which misassembled tubes with large strain energy can relax into less strained tubes that can act as nuclei for tubular assembly, without the need of dissociation and reassembly from scratch.

It is useful to note in our capsid model there is an inherent curvature between two building blocks, this being the preferred orientation in a dimer configuration, irrespectively if the building block is considered to be a pentamer or a hexamer. Many oligomers of capsid proteins have this innate curvature, which helps stabilize certain curved configurations. For example, high-level coarse-grained simulations have shown such curvature to spontaneously arise in the HIV CA hexamer surrounded by six CA trimers,[111] while this structure acts as a nucleus for assembly. Also, a recent experimental example shows that even a point mutation can affect the preferred angle of capsid protein dimers: the mutation enhanced hydrogen bonding between two regions, which in turn slightly lowered the angle between dimers, thus increasing the curvature, and the mutant protein self-assembled into a $T=1$ structure instead of the $T=3$ capsid of the wild-type protein.[112]

Although the hexamers in the model of van Workum and Douglas[68] were not observed to spontaneously aggregate into tubular structures, these assemblies could readily be obtained using small tubular seeds in the simulation box.

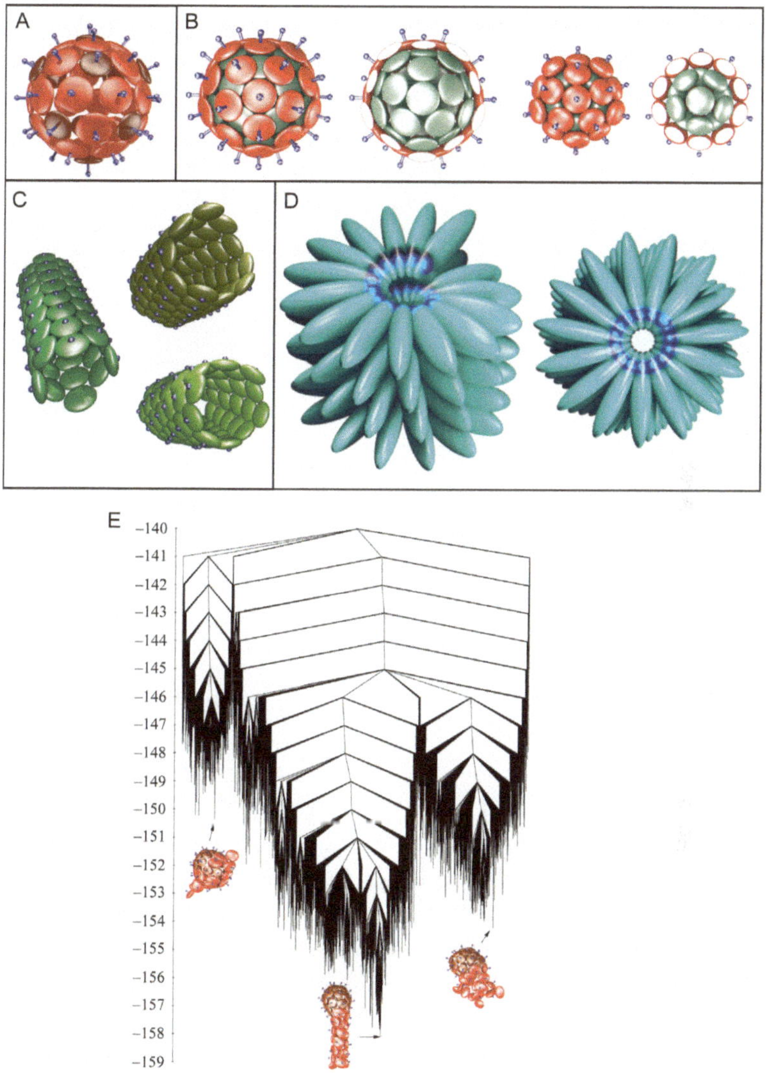

Fig. 5 (A) $T=4$ global minimum for a binary system with 12 smaller and 30 original capsomers. (B) Double-shell icosahedral structures ($T=4$, 3 and $T=3$, 1). (C) Examples of chiral tubular kinetic traps on the landscape. (D) Low-energy spiral structure similar to the TMV coat protein. (E) Disconnectivity graph for a binary system featuring low-energy head-tail assemblies. *Panels (B–D): Adapted from Fejer SN, Chakrabarti D, Wales DJ. Emergent complexity from simple anisotropic building blocks: shells, tubes and spirals.* ACS Nano. *2010;4:219–228. Copyright (2010) American Chemical Society.*

3.3.4 Scaffolding

Many large capsids can be assembled only with the help of scaffolding proteins or a spherical core which helps select capsid size. This templated assembly can be modeled with an appropriate number of building blocks of two different types. In our approach, these building blocks had the same ellipsoidal core and apex sites, but the apex sites between the two types did not interact. In this way, both building blocks preferred attaching to building blocks of the other type, due to less pairwise repulsion. Global optimization showed that in a binary system with 42 type A particles and 32 type B particles, the most stable structure is that of a double shell, with the type B particles forming an inner $T=3$ icosahedral shell, and the 42 type A particles being in a $T=4$ configuration, with the two shells aligned perfectly along the fivefold symmetry axes of the icosahedron. Other double-shelled structures are also possible, with the same overall icosahedral symmetry. The greater size mismatch between a $T=1$ and $T=3$ shell can be compensated by using thicker uniaxial ellipsoids than those for the basic parameterization. Fig. 5B shows the two double-shelled global minima. Interestingly, such structures have been shown to exist *in vitro* for RSV CA proteins.[110] Functional double-shelled viruses exist as well, e.g., in the *Reoviriade* and *Cystoviriade* families, in which the number of protein units is not compatible with fully formed icosahedral shells.[113–116] These observed "symmetry anomalies" do not affect the overall symmetry of the double shells: in all cases, the fivefold axes of the icosahedron coincide for the outer and inner shell. The anomaly is due to missing proteins from certain positions in the capsid, which was explained in terms of compatibility of protein units with the local symmetry in the shell: asymmetric proteins cannot occupy positions with non-trivial symmetries.[117]

It can be expected that the above shown scaffolding is mechanistically different from that around an already preformed template or a shell with nonspecific attraction toward capsid proteins. For this latter case, a minimalistic model has been developed in the group of Zandi,[14] showing with continuum elasticity theory that a weak nonspecific attraction toward a spherical template is enough under certain conditions to cause disclinations in the hexagonal sheet assembling around it, *in the exact places of what become the vertices of an icosahedron* once assembly is complete.

3.3.5 Helical capsids

The helical symmetry of virus capsids is not as common as icosahedral symmetry, but nevertheless, it is a common motif in plant viruses, filamentous phages, and even more complex virus structures. About half of plant virus species are helical.[118] As in case of icosahedral virus capsids, very different capsid protein sequences can give structures with the same overall symmetry, suggesting that the overall protein shape is important in this case as well.

Helical capsids are morphologically different from the chiral tubes discussed previously. In order to classify something as helical, a one-dimensional string of proteins has to be assigned unambiguously that wraps around an axis. For this to occur, two different length scales are necessary, these control the radius of the capsid and the helical pitch. These length scales can be easily tuned in a coarse-grained model by changing the preferred interaction strengths and the preferred curvature between building blocks. Our ellipsoidal capsid model[96] presented two ways that give rise to helically self-assembling morphologies: (1) changing the shape of the oblate ellipsoids so that they do not retain a rotational symmetry along the axis defined by the apex sites, and (2) with an alternative model with a triaxial ellipsoidal repulsive core and a patchy attractive interaction along the longest axis. In both cases, we found competing low-energy structures which could be best described as spirals (Fig. 5D) and stacked rings. Since the building blocks were achiral, left- and right-handed helices have the same energy in this model. Stronger interactions along the direction of the strand favor the emergence of helical order if the energy loss due to the pitch is smaller than the cost of accommodating a ring with a larger than ideal curvature. The selection of rings or spirals as global minima depends on the cluster size as well. For example, for the parameterization considered in Ref. [96], the global minimum for a 27-particle cluster is a spiral, while that for a 32-particle cluster is two stacked rings of 16 particles each.

Changing solvent conditions affects protein-protein interaction strengths, and this can lead to preference of stacked rings over spirals in the most studied helical virus, tobacco mosaic virus (TMV).[119,120] It can be expected that for strong side-by-side contacts, sequential addition of proteins should lead to spiral assemblies instead of stacked rings after a small spiral is formed as a critical nucleus. In TMV, this critical nucleus is facilitated by the binding of RNA between the rings in the "disk," which is thought to consist of two stacked rings. The interaction with RNA changes the two rings into a proto-helix that facilitates elongation in two different ways along the two ends of RNA: by sequential addition of disks toward the 5′-end, and addition of smaller protein aggregates toward the 3′-end.[121] However, some experimental evidence points to the possibility that the "disk" is in fact a small helix of a bit more than two turns.[122]

3.3.6 Head-tail assemblies

As of now, our understanding about the evolution of viruses is far from complete. Since no virus fossils exist, only genetic analysis can be used to get very limited insight into their evolution. The appearance of complex capsid structures is very intriguing. Viruses with a distinct head and tail region usually infect bacteria. The assembly pathways of such bacteriophages are well-described through experimental studies,[123] but their assembly from coarse-grained building blocks has not been replicated yet, due to the complexity of the system. Nevertheless, an intriguing result of the ellipsoidal capsid model was that of obtaining low-energy structures with a distinct "head" and "tail" region for binary systems in which the two types of building blocks preferred to form spherical shells and helical tubular structures. We found that the spontaneous emergence of the tubular structure attached to a spheroidal "head" as low-energy structures occurred without any fine tuning of interaction parameters, therefore it is likely a generic process that can spontaneously happen under the right conditions. We speculate that this is a possible route for the emergence of the very first head-tail phages during evolution, which likely occurred through horizontal gene transfer between two viruses with different coat morphology that infected the same host cell, and the resulting head-tail capsid could accommodate the proper combination of their genes. To the best of our knowledge, this minimalistic capsid model[96] is the first that provides computational support to the well-established theory of modular evolution of bacteriophages.[124] The triple helical tail itself has interesting geometrical characteristics that can be best described as a simple superhelix, since the building blocks of each strand are twisted around the imaginary axis of the strand, which is itself wrapped around the helical axis of the tail. The two helical properties are in phase. The disconnectivity graph of this system (Fig. 5E) shows a complex underlying energy landscape, in which high energy regions where the two building blocks mix are well-separated from the regions supporting a distinct head and an amorphous and well-ordered tail, respectively.

3.3.7 Hierarchical self-assembly of addressable capsids

We have used the self-assembling virus capsid model to investigate possible theoretical routes for efficient hierarchical self-organization of addressable particles.[125] Reducing the magnitude of non-nearest neighbor attraction between capsomers with respect to an icosahedral $T=1$ reference structure significantly decreased the complexity and enhanced the self-assembling character of the landscape, driving a cluster of 12 addressable particles toward the icosahedral global minimum in which every particle occupies its predefined (addressed) place in the shell. We found that for larger clusters, individual, correctly addressed $T=1$ shells form, which interact with much weaker forces with each other. These loosely bound clusters of shells then behave analogously to Lennard-Jones particles, forming aggregates with high symmetry: 156 particles form 13 icosahedral shells organized into an icosahedron themselves. Clearly, the proposed interactions are not possible among virus capsid proteins, but we can envisage DNA-coated nanoparticles that could support such analogous interactions.

4. Conclusions and outlook

All minimalistic models highlight the fact that capsid self-assembly is primarily controlled by building block shape and interaction anisotropies, which can only be roughly approximated with coarse-graining. The quality of these approximations is still not ideal, but the good correspondence between various particle-based coarse-grained methods shows that the specific interactions are not important for assembly, if the overall effective interactions are captured by the model.

Considerable challenges still remain in the field of computer modeling for a complete picture of mechanisms relevant for capsid assembly and maturation. Particularly tricky is including the effects of the environment (solvent, macromolecular crowding) in an effective potential between minimalistic building blocks. Although several different models are now generic enough to support self-assembly of icosahedral capsids from "pentameric" and "hexameric" building blocks, the spontaneous formation of capsids with larger triangulation numbers (4, 7, 9, 12, 13, etc.) is still not possible from *generic* coarse-grained model building blocks designed to represent protein monomers. Patchy particle models can be a promising route toward such models, although current approaches support only preprogrammed triangulation numbers. It is not trivial to encode preference for different hexameric environments necessary for larger triangulation numbers with different hexamer complexities.[105] As our coarse-grained models based on ellipsoids could produce virtually all morphologies relevant in virus capsid assembly, we propose that inclusion of such ellipsoidal shapes could be useful for representing monomers as well that can be stable in multiple environments. Models based on monomers would be indispensable for proper explanation of capsid maturation mechanisms. We are yet to see a buckling transition pathway for a complete virus capsid in computer simulations, in which both the initial and final structures can form spontaneously using coarse-graining of the same level.

Acknowledgments

This work was supported by a grant of the Romanian Ministry of Research and Innovation, CNCS-UEFISCDI, project number PN-III-P1-1.1-TE-2016-1279, within PNCDI III.

References

1. Johnson JE, Speir JA. Quasi-equivalent viruses: a paradigm for protein assemblies. 11 Edited by T. Richmond, *J Mol Biol.* 1997;269(5):665–675. https://doi.org/10.1006/jmbi.1997.1068.
2. Hesketh EL, Saunders K, Fisher C, et al. The 3.3 Å structure of a plant geminivirus using cryo-EM. *Nat Commun.* 2018;9(1):2369. https://doi.org/10.1038/s41467-018-04793-6.
3. Todd D, Niagro FD, Ritchie BW, et al. Comparison of three animal viruses with circular single-stranded DNA genomes. *Arch Virol.* 1991;117(1):129–135. https://doi.org/10.1007/BF01310498.
4. Legendre M, Bartoli J, Shmakova L, et al. Thirty-thousand-year-old distant relative of giant icosahedral DNA viruses with a pandoravirus morphology. *Proc Natl Acad Sci USA.* 2014;111(11):4274. https://doi.org/10.1073/pnas.1320670111.
5. Caspar DL, Klug A. Physical principles in the construction of regular viruses. *Cold Spring Harb Symp Quant Biol.* 1962;27:1–24.
6. Twarock R, Luque A. Structural puzzles in virology solved with an overarching icosahedral design principle. *Nat Commun.* 2019;10(1):4414. https://doi.org/10.1038/s41467-019-12367-3.
7. Luque A, Reguera D. The structure of elongated viral capsids. *Biophys J.* 2010; 98(12):2993–3003. https://doi.org/10.1016/j.bpj.2010.02.051.
8. Fox JM, Wang G, Speir JA, et al. Comparison of the native CCMV virion with in vitro assembled CCMV virions by cryoelectron microscopy and image reconstruction. *Virology.* 1998;244(1):212–218. https://doi.org/10.1006/viro.1998.9107.
9. Garcea R, Salunke D, Caspar D. Site-directed mutation affecting polyomavirus capsid self-assembly in vitro. *Nature.* 1987;329(6134):86–87. https://doi.org/10.1038/329086a0.
10. Salunke DM, Caspar DL, Garcea RL. Polymorphism in the assembly of polyomavirus capsid protein VP1. *Biophys J.* 1989;56(5):887–900.
11. Flint SJ. *Principles of Virology: Molecular Biology, Pathogenesis, and Control.* Washington, DC: ASM Press; 2000.
12. Yu Z, Dobro MJ, Woodward CL, et al. Unclosed HIV-1 capsids suggest a curled sheet model of assembly. *J Mol Biol.* 2013;425(1):112–123. https://doi.org/10.1016/j.jmb.2012.10.006.
13. Zlotnick A. To build a virus capsid: an equilibrium model of the self assembly of polyhedral protein complexes. *J Mol Biol.* 1994;241(1):59–67. https://doi.org/10.1006/jmbi.1994.1473.
14. Li S, Roy P, Travesset A, Zandi R. Why large icosahedral viruses need scaffolding proteins. *Proc Natl Acad Sci USA.* 2018;115(43):10971. https://doi.org/10.1073/pnas.1807706115.
15. Bancroft JB, Hiebert E, Rees MW, Markham R. Properties of cowpea chlorotic mottle virus, its protein and nucleic acid. *Virology.* 1968;34(2):224–239. https://doi.org/10.1016/0042-6822(68)90232-8.

16. Bancroft JB, Hills GJ, Markham R. A study of the self-assembly process in a small spherical virus formation of organized structures from protein subunits in vitro. *Virology*. 1967;31(2):354–379. https://doi.org/10.1016/0042-6822(67)90180-8.
17. Burns K, Mukherjee S, Keef T, Johnson JM, Zlotnick A. Altering the energy landscape of virus self-assembly to generate kinetically trapped nanoparticles. *Biomacromolecules*. 2010;11(2):439–442. https://doi.org/10.1021/bm901160b.
18. Zlotnick A, Aldrich R, Johnson JM, Ceres P, Young MJ. Mechanism of capsid assembly for an icosahedral plant virus. *Virology*. 2000;277(2):450–456.
19. Tang J, Johnson JM, Dryden KA, Young MJ, Zlotnick A, Johnson JE. The role of subunit hinges and molecular "switches" in the control of viral capsid polymorphism. *J Struct Biol*. 2006;154(1):59–67.
20. Antal Z, Szoverfi J, Fejer SN. Predicting the initial steps of salt-stable cowpea chlorotic mottle virus capsid assembly with atomistic force fields. *J Chem Inf Model*. 2017;57(4):910–917. https://doi.org/10.1021/acs.jcim.7b00078.
21. Bancroft JB, Rees MW, Johnson MW, Dawson JRO. A salt-stable mutant of cowpea chlorotic mottle virus. *J Gen Virol*. 1973;21(3):507–513.
22. Berger B, Shor PW, Tucker-Kellogg L, King J. Local rule-based theory of virus shell assembly. *Proc Natl Acad Sci USA*. 1994;91(16):7732–7736. https://doi.org/10.1073/pnas.91.16.7732.
23. Schwartz R, Shor PW, Prevelige Jr PE, Berger B. Local rules simulation of the kinetics of virus capsid self-assembly. *Biophys J*. 1998;75(6):2626–2636. https://doi.org/10.1016/S0006-3495(98)77708-2.
24. Arkhipov A, Freddolino PL, Schulten K. Stability and dynamics of virus capsids described by coarse-grained modeling. *Structure*. 2006;14(12):1767–1777. https://doi.org/10.1016/j.str.2006.10.003.
25. Roos WH, Bruinsma R, Wuite GJL. Physical virology. *Nat Phys*. 2010;6(10):733–743. https://doi.org/10.1038/nphys1797.
26. Choi YG, Rao ALN. Packaging of tobacco mosaic virus subgenomic RNAs by brome mosaic virus coat protein exhibits RNA controlled polymorphism. *Virology*. 2000;275(2):249–257. https://doi.org/10.1006/viro.2000.0532.
27. Elrad OM, Hagan MF. Encapsulation of a polymer by an icosahedral virus. *Phys Biol*. 2010;7(4):045003. https://doi.org/10.1088/1478-3975/7/4/045003.
28. ElSawy KM, Caves LSD, Twarock R. The impact of viral RNA on the association rates of capsid protein assembly: bacteriophage Ms2 as a case study. *J Mol Biol*. 2010;400:935.
29. Morton VL, Dykeman EC, Stonehouse NJ, Ashcroft AE, Twarock R, Stockley PG. The impact of viral RNA on assembly pathway selection. *J Mol Biol*. 2010;401:298.
30. Cadena-Nava RD, Comas-Garcia M, Garmann RF, Rao ALN, Knobler CM, Gelbart WM. Self-assembly of viral capsid protein and RNA molecules of different sizes: requirement for a specific high protein/RNA mass ratio. *J Virol*. 2012; 86(6):3318. https://doi.org/10.1128/JVI.06566-11.
31. Perlmutter JD, Qiao C, Hagan MF. Viral genome structures are optimal for capsid assembly. *Elife*. 2013;2:e00632. https://doi.org/10.7554/eLife.00632.
32. Hagan MF. Modeling viral capsid assembly. *Adv Chem Phys*. 2014;155:1–68. https://doi.org/10.1002/9781118755815.ch01.
33. Li S, Erdemci-Tandogan G, Wagner J, van der Schoot P, Zandi R. Impact of a non-uniform charge distribution on virus assembly. *Phys Rev E*. 2017;96(2):022401. https://doi.org/10.1103/PhysRevE.96.022401.
34. van Galen M, Higler R, Sprakel J. Allosteric pathway selection in templated assembly. *Sci Adv*. 2019;5(10):eaaw3353. https://doi.org/10.1126/sciadv.aaw3353.
35. de Rosier DJ, Klug A. Reconstruction of three dimensional structures from electron micrographs. *Nature*. 1968;217(5124):130–134. https://doi.org/10.1038/217130a0.

36. Jiang W, Tang L. Atomic cryo-EM structures of viruses. *Curr Opin Struct Biol.* 2017;46:122–129. https://doi.org/10.1016/j.sbi.2017.07.002.
37. Li X, Mooney P, Zheng S, et al. Electron counting and beam-induced motion correction enable near-atomic-resolution single-particle cryo-EM. *Nat Methods.* 2013;10(6):584–590. https://doi.org/10.1038/nmeth.2472.
38. Goetschius DJ, Parrish CR, Hafenstein S. Asymmetry in icosahedral viruses. *Curr Opin Virol.* 2019;36:67–73. https://doi.org/10.1016/j.coviro.2019.05.006.
39. Heymann JB, Cheng N, Newcomb WW, Trus BL, Brown JC, Steven AC. Dynamics of herpes simplex virus capsid maturation visualized by time-lapse cryo-electron microscopy. *Nat Struct Mol Biol.* 2003;10(5):334–341. https://doi.org/10.1038/nsb922.
40. Buchta D, Füzik T, Hrebík D, et al. Enterovirus particles expel capsid pentamers to enable genome release. *Nat Commun.* 2019;10(1):1138. https://doi.org/10.1038/s41467-019-09132-x.
41. Subramaniam S, Earl LA, Falconieri V, Milne JL, Egelman EH. Resolution advances in cryo-EM enable application to drug discovery. *Curr Opin Struct Biol.* 2016;41:194–202. https://doi.org/10.1016/j.sbi.2016.07.009.
42. Baker M. Cryo-electron microscopy shapes up. *Nature.* 2018;561:565–567. https://doi.org/10.1038/d41586-018-06791-6.
43. Herzik Jr MA, Fraser JS, Lander GC. A multi-model approach to assessing local and global cryo-EM map quality. *Structure.* 2019;27(2):344–358.e3. https://doi.org/10.1016/j.str.2018.10.003.
44. Ho PT, Montiel-Garcia DJ, Wong JJ, et al. VIPERdb: a tool for virus research. *Annu Rev Virol.* 2018;5(1):477–488. https://doi.org/10.1146/annurev-virology-092917-043405.
45. Chevreuil M, Law-Hine D, Chen J, et al. Nonequilibrium self-assembly dynamics of icosahedral viral capsids packaging genome or polyelectrolyte. *Nat Commun.* 2018;9(1):3071. https://doi.org/10.1038/s41467-018-05426-8.
46. Uetrecht C, Barbu IM, Shoemaker GK, van Duijn E, Heck AJR. Interrogating viral capsid assembly with ion mobility–mass spectrometry. *Nat Chem.* 2010;3:126.
47. Snijder J, Rose RJ, Veesler D, Johnson JE, Heck AJR. Studying 18 MDa virus assemblies with native mass spectrometry. *Angew Chem Int Ed.* 2013;52(14):4020–4023. https://doi.org/10.1002/anie.201210197.
48. Ceres P, Zlotnick A. Weak protein-protein interactions are sufficient to drive assembly of hepatitis B virus capsids. *Biochemistry.* 2002;41(39):11525–11531.
49. Ayton GS, Noid WG, Voth GA. Multiscale modeling of biomolecular systems: in serial and in parallel. *Curr Opin Struct Biol.* 2007;17:192.
50. May ER, Arora K, Mannige RV, Nguyen HD, Brooks CL. Multiscale modeling of virus structure, assembly, and dynamics. In: Dokholyan NV, ed. *Computational Modeling of Biological Systems: From Molecules to Pathways.* Boston, MA: Springer US; 2012:167–189. https://doi.org/10.1007/978-1-4614-2146-7_7.
51. Wang B, Zhang J, Wu Y. A multiscale model for the self-assembly of coat proteins in bacteriophage MS2. *J Chem Inf Model.* 2019;59(9):3899–3909. https://doi.org/10.1021/acs.jcim.9b00514.
52. Wales DJ, Scheraga HA. Global optimization of clusters, crystals and biomolecules. *Science.* 1999;285:1368–1372.
53. Wales DJ. Discrete path sampling. *Mol Phys.* 2002;100:3285–3306.
54. Zlotnick A. Distinguishing reversible from irreversible virus capsid assembly. *J Mol Biol.* 2007;366(1):14–18. https://doi.org/10.1016/j.jmb.2006.11.034.
55. Steven AC, Heymann JB, Cheng N, Trus BL, Conway JF. Virus maturation: dynamics and mechanism of a stabilizing structural transition that leads to infectivity. *Curr Opin Struct Biol.* 2005;15(2):227–236. https://doi.org/10.1016/j.sbi.2005.03.008.

56. Buck CB, Thompson CD, Pang Y-YS, Lowy DR, Schiller JT. Maturation of papillomavirus capsids. *J Virol.* 2005;79(5):2839–2846. https://doi.org/10.1128/JVI.79.5.2839-2846.2005.
57. Mukherjee S, Thorsteinsson MV, Johnston LB, DePhillips PA, Zlotnick A. A quantitative description of in vitro assembly of human papillomavirus 16 virus-like particles. *J Mol Biol.* 2008;381(1):229–237. https://doi.org/10.1016/j.jmb.2008.05.079.
58. Duda RL. Protein chainmail: catenated protein in viral capsids. *Cell.* 1998;94(1):55–60. https://doi.org/10.1016/S0092-8674(00)81221-0.
59. Wikoff WR, Liljas L, Duda RL, Tsuruta H, Hendrix RW, Johnson JE. Topologically linked protein rings in the bacteriophage HK97 capsid. *Science.* 2000;289(5487):2129. https://doi.org/10.1126/science.289.5487.2129.
60. Stone NP, Demo G, Agnello E, Kelch BA. Principles for enhancing virus capsid capacity and stability from a thermophilic virus capsid structure. *Nat Commun.* 2019;10(1). 4471https://doi.org/10.1038/s41467-019-12341-z.
61. Prevelige Jr PE, Thomas D, King J. Nucleation and growth phases in the polymerization of coat and scaffolding subunits into icosahedral procapsid shells. *Biophys J.* 1993;64(3):824–835.
62. Rapaport DC, Johnson JE, Skolnick J. Supramolecular self-assembly: molecular dynamics modeling of polyhedral shell formation. *Comput Phys Commun.* 1999; 121–122:231–235. https://doi.org/10.1016/S0010-4655(99)00319-7.
63. Rapaport DC. Self-assembly of polyhedral shells: a molecular dynamics study. *Phys Rev E.* 2004;70(5) 051905.
64. Rapaport DC. Role of reversibility in viral capsid growth: a paradigm for self-assembly. *Phys Rev Lett.* 2008;101(18): 186101. https://doi.org/10.1103/PhysRevLett.101.186101.
65. Hagan MF, Chandler D. Dynamic pathways for viral capsid assembly. *Biophys J.* 2006;91(1):42–54.
66. Nguyen HD, Reddy VS, Brooks III CL. Deciphering the kinetic mechanism of spontaneous self-assembly of icosahedral capsids. *Nano Lett.* 2007;7(2):338–344.
67. Nguyen HD, Reddy VS, Brooks III CL. Invariant polymorphism in virus capsid assembly. *J Am Chem Soc.* 2009;131(7):2606–2614.
68. Van Workum K, Douglas JF. Symmetry, equivalence, and molecular self-assembly. *Phys Rev E.* 2006;73(3):1–17.
69. Zandi R, Reguera D, Bruinsma RF, Gelbart WM, Rudnick J. Origin of icosahedral symmetry in viruses. *Proc Natl Acad Sci USA.* 2004;101(44):15556. https://doi.org/10.1073/pnas.0405844101.
70. Thomson JJ. On the structure of the atom: an investigation of the stability and periods of oscillation of a number of corpuscles arranged at equal intervals around the circumference of a circle; with applications of the results to the theory of atomic structure. *Philos Mag.* 1904;7(39):237.
71. Tammes PML. On the origin of number and arrangement of the places of exit on the surface of pollen-grains. *Recl Trav Bot Néerl.* 1930;27(1):1–84.
72. Bruinsma RF, Gelbart WM, Reguera D, Rudnick J, Zandi R. Viral self-assembly as a thermodynamic process. *Phys Rev Lett.* 2003;90(24) 248101.
73. Moody MF. The shape of the T-even bacteriophage head. *Virology.* 1965; 26(4):567–576. https://doi.org/10.1016/0042-6822(65)90319-3.
74. Moody MF. Geometry of phage head construction. *J Mol Biol.* 1999;293(2):401–433. https://doi.org/10.1006/jmbi.1999.3011.
75. Luque A, Zandi R, Reguera D. Optimal architectures of elongated viruses. *Proc Natl Acad Sci USA.* 2010;107(12):5323. https://doi.org/10.1073/pnas.0915122107.

76. Karlin S, Brendel V. Charge configurations in viral proteins. *Proc Natl Acad Sci USA*. 1988;85(24):9396–9400. https://doi.org/10.1073/pnas.85.24.9396.
77. del Álamo M, Mateu MG. Electrostatic repulsion, compensatory mutations, and long-range non-additive effects at the dimerization interface of the HIV capsid protein. *J Mol Biol*. 2005;345(4):893–906. https://doi.org/10.1016/j.jmb.2004.10.086.
78. Lošdorfer Božič A, Šiber A, Podgornik R. Statistical analysis of sizes and shapes of virus capsids and their resulting elastic properties. *J Biol Phys*. 2013;39(2):215–228. https://doi.org/10.1007/s10867-013-9302-3.
79. Lošdorfer Božič A, Siber A, Podgornik R. How simple can a model of an empty viral capsid be? Charge distributions in viral capsids. *J Biol Phys*. 2012;38(4):657–671. https://doi.org/10.1007/s10867-012-9278-4.
80. Siber A, Podgornik R. Role of electrostatic interactions in the assembly of empty spherical viral capsids. *Phys Rev E*. 2007;76 061906.
81. Bourne CR, Finn MG, Zlotnick A. Global structural changes in hepatitis B virus capsids induced by the assembly effector HAP1. *J Virol*. 2006;80(22):11055–11061. https://doi.org/10.1128/JVI.00933-06.
82. Zlotnick A, Lee A, Bourne CR, Johnson JM, Domanico PL, Stray SJ. In vitro screening for molecules that affect virus capsid assembly (and other protein association reactions). *Nat Protoc*. 2007;2(3):490–498.
83. Zhou H-X, Pang X. Electrostatic interactions in protein structure, folding, binding, and condensation. *Chem Rev*. 2018;118(4):1691–1741. https://doi.org/10.1021/acs.chemrev.7b00305.
84. Minton AP. The effect of volume occupancy upon the thermodynamic activity of proteins: some biochemical consequences. *Mol Cell Biochem*. 1983;55(2):119–140. https://doi.org/10.1007/BF00673707.
85. del Alamo M, Rivas G, Mateu MG. Effect of macromolecular crowding agents on human immunodeficiency virus type 1 capsid protein assembly in vitro. *J Virol*. 2005;79(22):14271–14281. https://doi.org/10.1128/JVI.79.22.14271-14281.2005.
86. Smith GR, Xie L, Lee B, Schwartz R. Applying molecular crowding models to simulations of virus capsid assembly in vitro. *Biophys J*. 2014;106(1):310–320. https://doi.org/10.1016/j.bpj.2013.11.022.
87. Doye JPK, Wales DJ, Berry RS. The effect of the range of the potential on the structures of clusters. *J Chem Phys*. 1995;103(10):4234–4249. https://doi.org/10.1063/1.470729.
88. Doye JPK, Wales DJ. The effect of the range of the potential on the structure and stability of simple liquids: from clusters to bulk, from sodium to C60. *J Phys B Atomic Mol Phys*. 1996;29(21):4859–4894. https://doi.org/10.1088/0953-4075/29/21/002.
89. Wales DJ. *Energy Landscapes: Applications to Clusters, Biomolecules and Glasses*. Cambridge University Press; 2003.
90. Wales DJ. The energy landscape as a unifying theme in molecular science. *Phil Trans R Soc A*. 2005;363:357–377.
91. Becker OM, Karplus M. The topology of multidimensional potential energy surfaces: theory and application to peptide structure and kinetics. *J Chem Phys*. 1997;106(4):1495–1517.
92. Wales DJ, Miller MA, Walsh TR. Archetypal energy landscapes. *Nature*. 1998;394:758.
93. Calvo F, Schebarchov D, Wales DJ. Grand and semigrand canonical basin-hopping. *J Chem Theory Comput*. 2016;12(2):902–909. https://doi.org/10.1021/acs.jctc.5b00962.
94. Johnston IG, Louis AA, Doye JPK. Modelling the self-assembly of virus capsids. *J Phys Condens Matter*. 2010;22(10):104101. https://doi.org/10.1088/0953-8984/22/10/104101.

95. Fejer SN, James TR, Hernández-Rojas J, Wales DJ. Energy landscapes for shells assembled from pentagonal and hexagonal pyramids. *Phys Chem Chem Phys*. 2009;11: 2098–2104.
96. Fejer SN, Chakrabarti D, Wales DJ. Emergent complexity from simple anisotropic building blocks: shells, tubes and spirals. *ACS Nano*. 2010;4:219–228.
97. Paramonov L, Yaliraki SN. The directional contact distance of two ellipsoids: coarse-grained potentials for anisotropic interactions. *J Chem Phys*. 2005;123(19) 194111.
98. Fejer SN, Wales DJ. Helix self-assembly from anisotropic molecules. *Phys Rev Lett*. 2007;99(8) 086106.
99. Fejer SN, Chakrabarti D, Wales DJ. Self-assembly of anisotropic particles. *Soft Matter*. 2011;7:3553–3564.
100. Olesen SW, Fejer SN, Chakrabarti D, Wales DJ. A left-handed building block self-assembles into right- and left-handed helices. *RSC Adv*. 2011;3:3553–3564.
101. Forman CJ, Fejer SN, Chakrabarti D, Barker PD, Wales DJ. Local frustration determines molecular and macroscopic helix structures. *J Phys Chem B*. 2013;117(26): 7918–7928. https://doi.org/10.1021/jp4040503.
102. Fejer SN, Chakrabarti D, Kusumaatmaja H, Wales DJ. Design principles for Bernal spirals and helices with tunable pitch. *Nanoscale*. 2014;6:9448–9456.
103. Fejer SN, Wales DJ. Design of a Kagome lattice from soft anisotropic particles. *Soft Matter*. 2015;11(33):6663–6668. https://doi.org/10.1039/C5SM01191D.
104. Llorente JMG, Hernández-Rojas J, Bretón J. A minimal representation of the self-assembly of virus capsids. *Soft Matter*. 2014;10(20):3560–3569. https://doi.org/10.1039/C4SM00087K.
105. Mannige RV, Brooks III CL. Periodic table of virus capsids: implications for natural selection and design. *PLoS One*. 2010;5(3):e9423. https://doi.org/10.1371/journal.pone.0009423.
106. Cusack S, Oostergetel GT, Krijgsman PCJ, Mellema JE. Structure of the top a-t component of alfalfa mosaic-virus—a non-icosahedral virion. *J Mol Biol*. 1983; 171(2):139–155.
107. Reguera D, Hernández-Rojas J, Llorente JMG. Kinetics of empty viral capsid assembly in a minimal model. *Soft Matter*. 2019;15(36):7166–7172. https://doi.org/10.1039/C9SM01593K.
108. Li S, Hill CP, Sundquist WI, Finch JT. Image reconstructions of helical assemblies of the HIV-1 CA protein. *Nature*. 2000;407(6802):409–413.
109. Kingston RL, Fitzon-Ostendorp T, Eisenmesser EZ, et al. Structure and self-association of the Rous sarcoma virus capsid protein. *Structure*. 2000;8(6):617–628.
110. Cardone G, Purdy JG, Cheng N, Craven RC, Steven AC. Visualization of a missing link in retrovirus capsid assembly. *Nature*. 2009;457(7230):694–698.
111. Grime JMA, Dama JF, Ganser-Pornillos BK, et al. Coarse-grained simulation reveals key features of HIV-1 capsid self-assembly. *Nat Commun*. 2016;7(1):11568. https://doi.org/10.1038/ncomms11568.
112. Asensio MA, Morella NM, Jakobson CM, et al. A selection for assembly reveals that a single amino acid mutant of the bacteriophage MS2 coat protein forms a smaller virus-like particle. *Nano Lett*. 2016;16(9):5944–5950. https://doi.org/10.1021/acs.nanolett.6b02948.
113. Grimes JM, Burroughs JN, Gouet P, et al. The atomic structure of the bluetongue virus core. *Nature*. 1998;395(6701):470–478. https://doi.org/10.1038/26694.
114. Rossmann MG, Tao Y. Courageous science: structural studies of bluetongue virus core. *Structure*. 1999;7(3):R43–R46. https://doi.org/10.1016/S0969-2126(99)80031-8.
115. Nakagawa A, Miyazaki N, Taka J, et al. The atomic structure of rice dwarf virus reveals the self-assembly mechanism of component proteins. *Structure*. 2003;11(10): 1227–1238. https://doi.org/10.1016/j.str.2003.08.012.

116. Iwasaki K, Miyazaki N, Hammar L, et al. Pleomorphic configuration of the trimeric capsid proteins of rice dwarf virus that allows formation of both the outer capsid and tubular crystals. *J Mol Biol*. 2008;383(1):252–265.
117. Rochal SB, Konevtsova OV, Myasnikova AE, Lorman VL. Hidden symmetry of small spherical viruses and organization principles in "anomalous" and double-shelled capsid nanoassemblies. *Nanoscale*. 2016;8(38):16976–16988. https://doi.org/10.1039/C6NR04930C.
118. Stubbs G, Kendall A, Viruses H. In: Rossmann MG, Rao VB, eds. *Viral Molecular Machines*. Boston, MA: Springer US; 2012:631–658. https://doi.org/10.1007/978-1-4614-0980-9_28.
119. Durham ACH, Finch JT, Klug A. States of aggregation of tobacco mosaic virus protein. *Nat New Biol*. 1971;229(2):37.
120. Klug A. The tobacco mosaic virus particle: structure and assembly. *Philos Trans R Soc B*. 1999;354(1383):531–535.
121. Butler PJG. Self-assembly of tobacco mosaic virus: the role of an intermediate aggregate in generating both specificity and speed. *Philos Trans R Soc B*. 1999;354:537.
122. Correia JJ, Shire S, Yphantis DA, Schuster TM. Sedimentation equilibrium measurements of the intermediate-size tobacco mosaic virus protein polymers. *Biochemistry*. 1985;24(13):3292–3297. https://doi.org/10.1021/bi00334a033.
123. Aksyuk AA, Rossmann MG. Bacteriophage assembly. *Viruses*. 2011;3(3):172–203. https://doi.org/10.3390/v3030172.
124. Botstein D. A theory of modular evolution for bacteriophages. *Ann N Y Acad Sci*. 1980;354:484–491.
125. Fejer SN, Mantell RG, Wales DJ. Designing hierarchical molecular complexity: icosahedra of addressable icosahedra. *Mol Phys*. 2018;116(21 – 22):2954–2964. https://doi.org/10.1080/00268976.2018.1439190.

CHAPTER NINE

Aggregation of disease-related peptides

Phuong H. Nguyen[a,b], Fabio Sterpone[a,b], Philippe Derreumaux[c,d,*]

[a]CNRS, Université de Paris, UPR 9080, Laboratoire de Biochimie Théorique, Paris, France
[b]Institut de Biologie Physico-Chimique-Fondation Edmond de Rothschild, PSL Research University, Paris, France
[c]Laboratory of Theoretical Chemistry, Ton Duc Thang University, Ho Chi Minh City, Vietnam
[d]Faculty of Pharmacy, Ton Duc Thang University, Ho Chi Minh City, Vietnam
*Corresponding author: e-mail address: philippe.derreumaux@tdtu.edu.vn

Contents

Abstract

Protein misfolding and aggregation of amyloid proteins is the fundamental cause of more than 20 diseases. Molecular mechanisms of the self-assembly and the formation of the toxic aggregates are still elusive. Computer simulations have been intensively used to study the aggregation of amyloid peptides of various amino acid lengths related to neurodegenerative diseases. We review atomistic and coarse-grained simulations of short amyloid peptides aimed at determining their transient oligomeric structures and the early and late aggregation steps.

Progress in Molecular Biology and Translational Science, Volume 170
ISSN 1877-1173
https://doi.org/10.1016/bs.pmbts.2019.12.002

1. Introduction

Numerous diseases affecting either the central nervous system or a variety of peripheral tissues result from the self-assembly of amyloid proteins. Disorders range from Alzheimer's disease (AD, implication of Aβ proteins of 39–43 amino acids and tau protein of 441 amino acids), Parkinson (α-synuclein protein of 140 amino acids) to amyotrophic lateral sclerosis or ALS (superoxide dismutase, SOD of 154 amino acids) and type II diabetes (IAPP or islet amyloid polypeptide of 37 amino acids). All these proteins differ in amino acid sequence and length, yet they all form amyloid fibrils with cross-β structure.[1] This propensity to self-assembly into amyloid fibrils under given conditions is also observed for heptapeptides (e.g., Aβ16–22),[2] tetrapeptides (e.g., KFFE),[3] dipeptides[4] and even by single amino acids.[5]

Experimentally, aggregation kinetics of amyloid proteins display a sigmoidal curve with a lag phase, during which monomers self-assembly into oligomers and undergo structural rearrangements, until the growth phase where fibril elongation and primary/secondary nucleation occur, followed by the saturation phase where the system is in equilibrium between fibrils and a small concentration of monomers. Note that our understanding of amyloid aggregation kinetics goes much beyond classical nucleation theory where primary nucleation event is sufficient to fit the experimental curves.[1]

Structural determination of all species along amyloid fibril formation pathways is challenging by experimental means because of their transient character despite the use of a large variety of biophysical techniques.[6] Complexity comes from the metastable character of each species (a very large number of conformations for each aggregate), but also from the sensitivity of the kinetics to experimental conditions (solution pH, temperature, salt concentration, agitation, ions) and external conditions (presence of membrane, crowding, etc.).[7]

To add further complexity, amyloid aggregation kinetics is also modulated by the exact composition of the lipid bilayers. For example, although Aβ1–40, Aβ1–42 and tau oligomers are key players in AD,[8] dietary PUFA (polyunsaturated fatty acids) supplementation change molecular phospholipids, and there is evidence that increased intake of omega-3 PUFA slows the progression of AD, while omega-6 PUFA is linked to higher risk of AD.[9] Another example of the modulation of the aggregation upon membrane presence can be taken from experiments on α-synuclein. On the one hand,

DPLS lipid bilayers significantly augment its aggregation rate, while DOPE lipid bilayers have no impact on its aggregation rate.[10]

Simulations on amyloids at different time and length scales can complement experiment, but require accurate potential energy models ranging from atomistic in explicit aqueous solution/lipid bilayer, coarse-grained with implicit solvent/membrane to mesoscopic or super-mesoscopic representations.[11–14] Molecular dynamics (MD), replica exchange molecular dynamics (REMD) simulations and other sampling methods are often used to generate the conformational ensemble of intrinsically disordered proteins (IDPs). These methods are described in Section 2 and we will see that their results vary with the protein force field used. Section 3 focuses on the application of computer simulations to a better understanding of small oligomers of short amyloid peptides. While an implicit solvent representation reduces the number of degrees of freedom, hydrodynamics, which deals with the motion of fluids and the forces acting on solid bodies immersed in fluids and in motion relative to them, plays a significant role in the early aggregation steps of amyloids using simplified models (Section 4). Section 5 reports on the simulations aimed at understanding the primary nucleation and the surface-catalyzed secondary nucleation. The final section reports on recent advances in the determination of oligomer structures of Aβ40/42 peptides in aqueous solution and lipid bilayers based on computer simulations. Our review will be mainly centered on Aβ peptides. For computer simulations of tau, hIAAP and synuclein, see Ref. [14].

2. Computer simulation models for amyloid protein aggregation

Numerous conformational sampling methods are used to study amyloid proteins. We will not describe them in detail but rather summarize their main features. Atomistic molecular dynamics (MD) simulations in explicit environment offer the most detailed dynamic and energetic pictures of protein folding and aggregation. The longest simulation on the fastest computer (Anton) does not exceed, however, 1 ms in explicit solvent, sufficient for sampling the monomeric state of amyloid proteins, but far too short for exploring the aggregation during the experimental lag phase.[15] To overcome the multiple minima problem and converge faster to equilibrium properties, replica exchange MD (REMD) simulations, which run multiple MD simulations at different temperatures in parallel and exchange consecutive replicas according to the Metropolis criterion, are commonly used as their implementations are easy. Interestingly, it is possible to determine peptide dimerization-dissociation rates from REMD simulations.[16] Recent modeling methods based on Markovian descriptions of conformational states have also led to frameworks allowing accurate description of dynamics. In conjunction with REMD simulations, these frameworks allow systematic and accurate determination of transition probabilities between the corresponding states, in the case of Markov state models, and of transition rates, in the case of the corresponding coarse master equations.[17]

Among other sampling techniques, Monte Carlo, replica exchange MC (REMC), replica exchange with solute tempering, simulated tempering, discrete or discontinuous MD, and metadynamics are also employed.

On-lattice (where the position of the particles are limited to nodes) and off-lattice Monte Carlo simulations of models having a small number of internal degrees of freedom have been used to investigate the nucleation mechanisms.[18–20] On-lattice REMC simulations were also used to determine the free energy surfaces of various oligomers of the Aβ16–22 and Aβ35–42 peptides[21,22] and the nucleus size of Aβ42.[23]

A drawback of REMD is its sensitivity to the overlap between the replicas total energies, and to the system size, the energy overlap scaling with the number of particles as $N^{-1/2}$. Optimal exchange rate between replicas is thus limited by the number of solvent molecules. Because protein-protein and protein-solvent interactions are dominant in protein recognition,[24] it is interesting to focus on these interactions and simulation schemes based on potential energy rescaling and Hamiltonian replica exchange exploit this feature. Replica exchange simulation with solute tempering (REST) belongs to this class of schemes.[25–27] Smith et al. performed REST and standard REMD simulations to study the Aβ25–35 peptide in water and a system consisting of two Aβ10–40 peptides binding to a DMPC lipid bilayer.[27] Their study led to three conclusions. First, the number of replicas in REST simulations can be reduced four to five times. Second, REST produces much fewer conformational states than REMD, but the number of unique states is very similar between the REST and REMD simulations. Third comparing REST and REMD equilibrium conformational ensembles, REST reproduces REMD data extremely well for the system of two Aβ peptides bound to the DMPC lipid bilayer, and REST demonstrates much better convergence for the Aβ1–40 system bound to the lipid bilayer rather than for the small unstructured Aβ25–35 peptide in water.[27]

In simulated tempering, temperature is a dynamical variable with discrete values T_n. Simulated tempering scheme requires the determination of a priori unknown weight parameters to ensure a random walk in temperature space, namely, the Helmholtz free energies at T_n. We have developed a simulated tempering scheme with on-the-fly weight determination. The weights are self-updated via a trapezoid rule during the simulation, eliminating therefore the need for trial simulations.[28] The advantage of this simulated tempering method over REMD was illustrated on the equilibrium states of Aβ16–22 trimer using the OPEP CG force field and the folding of the 20-residue Trp-cage using atomistic force field starting from randomly chosen states.[29] Of interest is that it is possible to include different atomistic force fields into a single simulated tempering run.[30]

Unlike classical MD simulations, which integrate the equation of motion with a time step of 1–2 ns, DMD (discontinuous MD) is event-driven and keeps track of particle positions and velocities only at collision times. As a result, DMD is several orders of magnitude faster than the traditional continuous MD if coarse-graining is used. The main disadvantage of DMD is that pairs of particles must interact by means of spherically symmetric potentials that consist of square wells. The DMD scheme has been coupled to all-atom model with implicit solvent,[31] and various CG models such as PRIME20.[32]

Finally in metadynamics, enhanced conformational sampling is realized by an external history-dependent bias potential affecting few selected degrees of freedom, referred to as collective variables. The bias is adaptively constructed as a sum of Gaussians deposited along the system trajectory to prevent the system from revisiting previous conformational regions. Although metadynamics requires a limited yet effective set of collective variables, this is not a trivial task even for the monomeric state of long amyloid peptides,[33] and of course for amyloid fibril formation.[34]

There is a constant effort in improving atomistic force fields in explicit water. Many versions of CHARMM and AMBER with modified water force fields have been designed over the years to reproduce experimental data on folded proteins (early versions) and both folded and disordered proteins (CHARMM36m-TIP3P modified,[35] and AMBERsb99-disp,[15] where disp is a variant of TIP4P water model). Other current force fields include OPLS-AA and GROMOS force fields.[36] We summarize some recent results on short amyloid peptides and the Aβ40 and Aβ42 peptides.

Following a previous study on Aβ16–22 dimers and trimers using three force fields (AMBER99, GROMOS96 and OPLS-AA/TIP3P),[37] Nguyen et al. studied the effects of 17 widely used atomistic molecular force fields on the structures and kinetics of the Aβ16–22 dimer in aqueous solution, for a total MD time of 0.34 ms.[38] While the AMBER94, AMBER99 and

AMBER12SB force fields fail to predict β-sheet, the AMBER96, GROMOS45a3, GROMOS53a5, GROMOS53a6, GROMOS43a1, GROMOS43a2, and GROMOS54a7 force fields form β-sheets very rapidly. In contrast, the AMBER99-ILDN, AMBER14SB, CHARMM22* and CHARMM36m force fields are the best candidates for studying amyloid peptide assembly, as they provide good balances in terms of structures, thermodynamics and kinetics (assembly and dissociation).

In the same line, Strodel et al. investigated the effect of five different force fields (GROMOS54a7, OPLS-AA, CHARMM22*, AMBER99SB*-ILDN and AMBER03WS) on the hexamer formation of the wild-type Aβ16–22 peptide and two mutants: Aβ16–22 (with F19V and F20V) which does not form fibrils, and Aβ16–22 (F19L) which forms fibrils faster than the wild-type.[39] Note that the AMBER03WS force field reduces the over-stabilization of protein-protein interactions generated by older force fields.[36] They found that the structural properties are independent of the force field, but oligomer formation kinetics depends strongly on the force field. It is very important to note that the kinetics generated by the various force fields differ more from each other than the kinetics between amyloid and non-amyloid peptides simulated with a single force field.[39]

Siwy et al. performed comparative analysis of the REMD conformational ensembles of Aβ10–40 monomer produced by five force fields combining CHARMM36, CHARMM22*, CHARMM22/cmap, and OPLS-AA and two water models (standard and modified TIP3P).[40] As expected, all force fields predict that Aβ10–40 remains unfolded dominated by turn and random coil structure. The secondary structures generated by CHARMM36 and CHARMM22/cmap simulations are qualitatively similar, but their tertiary interactions show little consistency. Among all force fields, CHARMM22* differs the most from CHARMM36, and based on the analysis of 3JHNHα-coupling and RDC constants, the CHARMM36 force field with standard TIP3P model produces the most accurate representation of Aβ10–40 conformational ensemble.[40]

The Aβ42 peptide has also been subject of particular attention over the years.[41–43] Strodel et al. used REMD simulations on the monomeric form with the five OPLS, AMBER99SB, AMBER99SB*ILDN, AMBER99SBILDN-NMR and CHARMM22* force fields and compared the equilibrium ensembles to nuclear magnetic resonance (NMR) experimental data. They found that all force fields except AMBER99SBILDN-NMR reproduce local NMR observables, and CHARMM22* is slightly better than the other force fields.[42]

Nguyen et al. generated the Aβ42 dimer ensembles obtained by four force fields (OPLS-AA, CHARMM22*, AMBER99sb-ildn and AMBERsb14 with TIP3P model) coupled to REMD simulations.[44] The simulation results were compared to experimental circular dichroism (CD) and ion-mobility mass-spectrometry data. On the basis of 144 μs, the four force fields lead to random coil ensembles with cross-collision sections, hydrodynamics properties, and small-angle X-ray scattering profiles independent of the force field. Overall, there are significant differences in secondary, tertiary and quaternary conformations among the four force fields, for example, the intra-molecular β-hairpin content spanning residues 17–21 and 30–36 varying between 1.5% and 13%.

Finally, the stability of tetrameric Aβ40 and Aβ42 β-barrel structures was studied considering four atomistic force field-water model combinations: Amber99SB-ILDN/TIP3P, OPLS/TIP3P, CHARMM36m/TIP3P-modified, and Amber99sb-disp. For each force field, Nguyen et al. found that Aβ42 samples barrel structures in aqueous solution, while this is less the case for Aβ40. It is to be noted that the authors used the two latest force fields that were developed for IDPs (CHARMM36m and AMBER99SB-disp) and spent a lot of computing time to converge the sampling by using REMD.[45]

There is also a constant effort in designing coarse-grained models with implicit solvent for amyloid aggregation. Using a two-state model for each peptide and Langevin dynamics, Caflisch et al. found that the nucleus size varies between 4 and 35 depending on the energy difference between the amyloid-competent and amyloid-protected minima. This simple model is able to generate fibril topologies with twist and multifilament composition as observed experimentally.[46]

Shea and co-workers designed a three-bead model (two for the backbone, and one for the side-chain) that reproduces the polar, hydrophobic and charged nature of each residue. Using a simple energy function that controls the β-strand propensity of each peptide and Langevin dynamics simulations,[47] they found that, when the β-strand propensity is not too high, the oligomeric states consist of disordered aggregates, multiple β-sheet layers with different orientations including the cross-β structure, and β-barrels consistent other coarse grained simulations.[48–50]

Frenkel and co-workers describe the monomers as single spherocylinders able to coexist in two states: a soluble state and an amyloid-like state. Using this model and Monte Carlo simulations, they estimated a critical nucleus size of four peptides for Aβ42. They also established the concentration, temperature and interpeptide interactions conditions upon which aggregation follows the one-step nucleation (where two peptides in the amyloid-competent state binds) and the two-step nucleation (where the amorphous aggregates reorganize into β-rich nuclei).[19,20]

The four-bead "Urbanc" and PRIME20 models consist of three spheres for the backbone (N, Ca, and CO) and one bead for the side chain. Using DMD, Urbanc et al. reproduced differences in oligomer size distributions between the Aβ40 and Aβ42 peptides consistent with experimental data.[51] Using PRIME20/DMD, Hall reproduced the propensity of a series of hexapeptides known to form amyloid and non-amyloid fibrils in vitro and studied the mixing of Aβ40 and Aβ16–22 peptides.[32]

Wolynes et al. also designed the AWSEM (associative memory, water mediated, structure and energy model) three-bead model with Ca, Cb and O and applied it to the aggregation of Aβ40 and Aβ42 peptides.[52,53] Scheraga et al. used their CG united-residue UNRES force field with MD to study the aggregation of 5 Aβ1–28 peptides,[54] and of the repeat domain of tau in the presence of Aβ40 fibrils.[55] Derreumaux and co-workers designed the OPEP (optimized potential for efficient peptide structure prediction) model, where each amino acid is represented by six beads: one bead for each side-chain and the atomic resolution is used for the backbone (N, H, Ca, C and O).[56–58] This model has been used for on-lattice and off-lattice simulations of many short-amyloid peptides, and the Aβ40 and Aβ42 peptide.[59,60] Note that it is beyond the scope of this review to describe the potential energy functions used by all models.

3. Structures of small aggregates

Among all amyloid sequences, the Aβ16–22 peptide has been the most studied. Due to its simplicity, its amyloid fibril at neutral pH consists of antiparallel β-strands and parallel β-sheets,[2] and the amino-acids 17–21 constitute the main driving force for the aggregation of the full-length Aβ peptides.[8] Based on MD and REMD simulations with all force fields except AMBER99sb-disp, it is clear that the probability of Aβ16–22 fibril formation following the one-step nucleation is very small.

Using the modified AMBER94 force field and REMD, Nussinov et al. found six configurations including shifted parallel strand and parallel loop, antiparallel strand, parallel strand, shifted antiparallel strand, cross, and tight cross/lock.[61] Employing the OPEP coarse-grained force field, in-register and out-of-register parallel and antiparallel strands were generated along with amorphous states.[62] Nguyen et al. demonstrated with 0.34 ms MD that the probability of the amyloid-competent state is on the order of at most 10% using the most recent atomistic force fields. Of interest is that, in principle there are nine out-register antiparallel β-sheets in addition to the native in-register β-sheet, but only five are populated at the dimer level.[38] This feature is important as experiments and simulations have shown that large aggregates consist of out-of-register β-strands and transition to the fibril state involves the reptation of the β-strands at high concentration and detachment/binding of the peptides at low concentration.[63,64] A second important aspect from the Aβ16–22 dimer simulations is that the parallel β-sheet conformation is a state of low probability. This feature explains why the E22Q mutation can induce antiparallel to parallel β-sheet transition of Aβ16–22 fibrils as observed by solid-state NMR and all-atom simulation studies.[65]

Simulations of Aβ16–22 trimer reveal that the main configurations found at the dimer level remain and we have a low population of mixed antiparallel—parallel β-sheets.[63] In terms of dynamics, we can infer a very likely scenario from other peptides known to form fibrils with antiparallel β-strands. By studying the aggregation of the KFFE peptide up to 4-mers by REMD and Markov models, Sengupta et al. found that disordered and β-sheet oligomers do not interconvert, leading to separate pathways for their formations.[66]

De Groot et al. used unbiased explicit solvent MD simulations to investigate the structural and dynamical features of aggregates formed by Aβ16–22 at atomistic resolution for 2 μs. Simulations of 12 peptides were performed at 2 and 20 mM peptide concentration.[67] To characterize the structures, they computed collision cross sections of the individual aggregates sampled from the MD trajectories and found heterogeneous ensemble of aggregates lacking a defined quaternary structure, fully consistent with the same simulations of other amyloid peptides ranging from 6- to 12-mers. In particular, they found a very large variety of ordered intermolecular *β*-structure motifs including distinct fibril-like configurations and *β*-barrel oligomers of various sizes already prevalent in smaller aggregates. This conformational ensemble is fully consistent with all-atom Aβ16–22 DMD simulations of 6–12 peptides where *β*-barrel oligomers of various sizes were found with a probability varying between 2% and 4%,[31] CG Langevin dynamics simulation results of Shea et al.,[47] and a multitude of all-atom and CG simulations of other sequences.[26,68,69]

The second peptide we want to discuss is the Aβ37–42 peptide because, in contrast to many short-amyloid peptides, its fibril structure displays parallel β-strands, as formed by the full length Aβ40/42 peptides. Nguyen et al. performed a REMD simulation of 16 all-atom Aβ37–42 peptides in explicit water starting from dispersed orientations.[70] The total simulation time of 23 μs allows them to obtain the conformational distribution of oligomers and β-sheet sizes at 300 K and a peptide concentration of 12 mM. Self-assembly is described by the condensation-polymerization conversion from micelle to high β-sheet structures. When equilibrium is reached, the oligomer distribution consists of large aggregates (70%) and free monomers (30%). The population of 4–5 fully parallel β-strands consistent with the native fibril is low, the population of 4–5 fully antiparallel β-strands is non-negligible, and the system has a large population of mixed parallel/antiparallel β-strands in equilibrium with amorphous oligomers. This distribution is in agreement with all-atom metadynamics results of Laio et al.[34] We now report on Aβ16–22 simulations of unprecedented sizes that further support the second-step nucleation.

4. Exploring the early aggregates of amyloid peptides at quasi-atomic level with hydrodynamics

Coarse-grained simplified molecular models with implicit solvent have been extensively used to explore the aggregation process of amyloid systems and to inspect the impact of the peptide β-propensity and amino acid sequence, peptide-peptide interactions, concentration, temperature,[46,47,52,53,71,72] and crowding.[73] However, these simulations based on simplified models neglect the effect of solvent mediated interactions on the kinetic behavior of the system and the diffusion limited aggregation process. It is therefore important to include these hydrodynamic interactions (HI). This can be done using specialized simulation techniques, e.g., Brownian dynamics with HI, the stochastic rotational method, and the lattice Boltzmann molecular dynamics (LBMD).

Recently we implemented an effective LBMD approach to study biomolecular systems based on the OPEP CG model. This technique, which includes naturally hydrodynamics in implicit solvent coarse-grained molecular dynamics of biomolecules, has been used to explore the mobility of proteins in a crowded solution, the unfolding and dynamics of proteins and lipid systems under shear flow, and the multi-scale dynamics of amyloid peptides aggregation.[22,74–76] Before describing the results of interest, we pause to describe the methodology, see Fig. 1A.

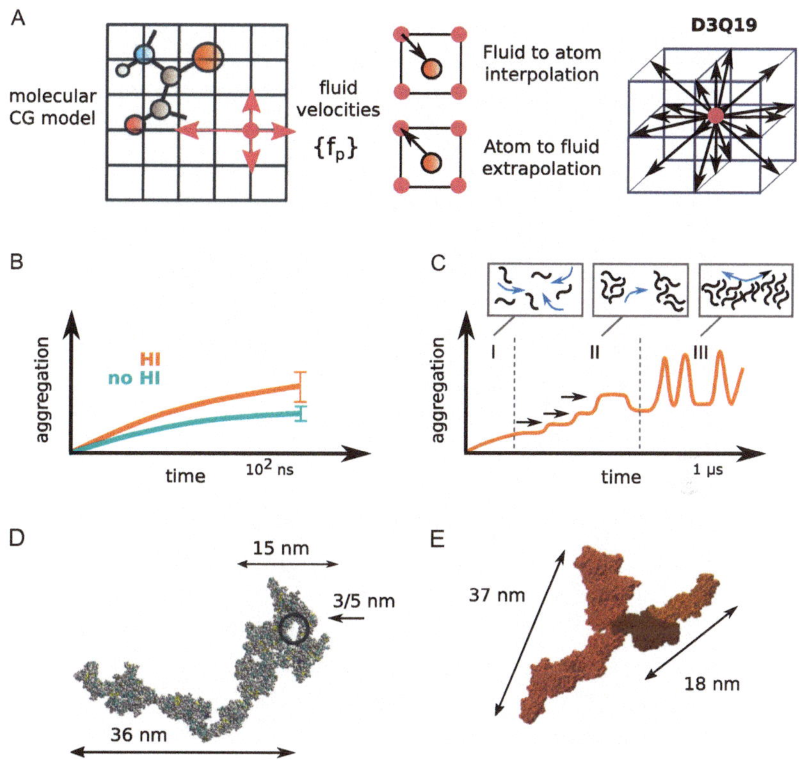

Fig. 1 LBMD study of amyloid aggregation. (Panel A): Scheme describing the LBMD technique with particles embedded in a solvent. The set of velocities of the fluid (fp) is represented by displacement in the lattice space (pink arrows). The drag force acting on the particle and its feedback on the evolution of the fluid velocity field require an interpolation/extrapolation communication strategy. In lattice Boltzmann, the typical 3D set of displacements is represented by 18 directions and the zero displacement (D3Q19). (Panel B): simulations of 100 Aβ16–22 peptides. The curves represents the growth of the largest cluster during the simulation time including or not HI. The vertical bars schematically indicate the fluctuations of the cluster size. (Panel C): simulations of 1000 Aβ16–22 peptides. The growth of the largest cluster manifests three regimes: individual peptide inclusion, cluster fusion, and fragmentation. (Panel D): amyloid amorphous protofibril sampled in LBMD, capped by an annular pore. (Panel E): Aggregation of a large amorphous protofibril can occur via lateral branching.

In LBMD, the particle-based dynamics is coupled to a kinetic representation of the solvent, simulated via the Lattice Boltzmann (LB) technique.[77] The coupling between the particles and the solvent has the form of a Stokes-like drag force acting on a per-particle basis:

$$\vec{F}_i^D = -\gamma\left(\vec{v}_i - \vec{u}_i\right) \tag{1}$$

where the first term is the velocity of *i*-th particle, the second term is the fluid velocity smeared over a finite extension of the *i*-th particle, and γ is the friction, an adjustable parameter in the methodology taken to be equal for all particles of a given molecule. The drag force adds up to the usual conservative forces derived from the Hamiltonian of the system and to a random white noise that represents thermal fluctuations. The reader can find more technical details in Refs. [74, 78].

In our simulations, the LB evolution was solved by using the Bhatnagar-Gross-Krook (BGK) collisional operator, with a lattice grid spacing that was varied depending on the size of the system and the wished resolution (1–5 Å), and the bulk water kinematic viscosity.[74,78] When using the flexible OPEP force field,[79] the molecular and LB dynamics were evolved synchronously using a time step of 1.5 fs, a value defined by the properties of the molecular mechanics force field. When focusing on the mesoscale behavior of aggregate growth, an elastic model was used in tandem with a time step of 10 fs. The friction γ was optimized in order to reproduce the diffusion of an isolated peptide in dilute condition.

We first considered a system composed of 100 Aβ16–22 peptides placed in a box corresponding to a concentration of about 60 mM.[78] The first issue we addressed was to estimate the characteristic time of complete aggregation with HI compared to a Langevin simulation. It turns out, as already anticipated by Skolnick and co-workers in the study of lipid assembly,[80] that the presence of HI speeds up the aggregation process. In the LBMD simulation, after 100 ns we observe the formation of a large aggregate containing about 80% of the peptides. At the same time scale and in the absence of HI, the aggregation proceeds more slowly, and only 50% of peptides form the largest cluster, see Fig. 1B. The result can be understood considering that by including HI the diffusivity of peptide aggregates scales more favorably with the aggregate size than in their absence. Also, when focusing on the fluctuations of the size of the formed aggregates, it results that the presence of HI increases the fast exchange of peptides.

The aggregation process of Aβ16–22 was further investigated by considering a system of much larger size, 1000 peptides placed in a box of L = 300 Å, corresponding to a concentration of 60 mM.[81] Using atomistic and explicit solvent, this system would count for 2.7 millions of particles, becoming prohibitive for classical MD simulations.

The aggregation process was followed at the microsecond time scale, and for the first time three different regimes were clearly individuated. At the short timescale, $<10^2$ ns, the growth of peptides aggregates proceeds continuously in time, with progressive absorption of single molecules. When in the solution the presence of finite size oligomers (10–20 peptides in each) becomes relevant, the aggregation proceeds via the fusion of these entities, and the growth mechanism is discontinuous manifesting sudden size jumps (see the arrows in Fig. 1C). Several fusion mechanisms were individuated, from head-tail elongation to branching, involving coherently from two up to four aggregates. Interestingly, the time evolution of the aggregate size distribution shows the emergence of a shoulder around the pentamer and decamer states, a signature that can be related to the presence of a critical nucleus size. Finally, the system reaches a steady regime for time >350 ns, with the formation of a large amorphous protofibril and three minor ones, no residual monomers are present in the system at this timescale. The progress of the amorphous protofibril elongation is limited by the diffusion of these very large (massive) clusters. To be noted that very large size fluctuations of the largest amorphous protofibril are visible and are caused by fragmentation events.

Having individuated these relevant timescales it is worth to stress that the simulated system is at a very high concentration. If one considers a lower concentration regime, e.g., reproducing the upper limit of experiments, μM, the early phase of the aggregation would be on the microsecond timescale and would approach the second timescale at in vivo concentrations (nM).

Looking at the early phase of aggregation, following the formation of the small oligomers, the peptides acquire in part β-sheet organization, going up to 15% of the whole system. The further conversion is then rate-limited by the size of the formed aggregates that constraints the monomer conformational sampling. The possibility to explore the growth of very large amorphous protofibril at the quasi-atomistic resolution allowed to individuate critical structural features like the formation of a large asymmetric pore (dimensions of 3–5 nm) at the edge of the structure, see Fig. 1D. It is interesting that amyloid plaques with annular pores have been detected in the brain of Alzheimer's disease patients by antibodies and electron microscopy experiments.[82]

Evolution of the amorphous protofibril elongation was further explored by deploying an ad hoc multi-scale strategy. The four final clusters individuated in the LBMD simulation based on the fully flexible OPEP force field were converted into an elastic-network and placed in much larger simulation box (60 nm) to avoid finite-size artifact. The elastic network allowed the use of a larger integration time-step for the particle dynamics, and scaling arguments were applied to ensure a kinetic coherence when changing fluid and particle resolutions. With this tool in hands, it was possible to follow the hydrodynamic sustained formation of a larger branched amorphous protofibril, and the critical role of the amorphous protofibril surface acting as an entropically favorable seed for the formation of branched structures via aggregate fusion, see Fig. 1E.

5. Primary and secondary nucleation from simulations

The foundation of the nucleation for amyloid fibrils dates from the polymerization studies of actin by Oosawa[83] and deoxyhemoglobin by Weaton.[84] They theoretically explained the nucleation and subsequent polymerization processes by employing the homogeneous nucleation theory developed for vapor condensation. Their theories have served as a background for the development of kinetic and thermodynamic analyses for amyloid nucleation.[85]

Basically, in the first step, the fluctuations in the system induce spontaneous association of monomers, leading to the formation of small clusters of monomers. Due to small sizes, these clusters are thermodynamically unstable with high positive interfacial free energy. With increasing size, the intra-cluster interactions become more significant, and the clusters become more stable, the bulk free energy is negatively decreased, resulting in the decrease in the total free energy. A smallest cluster, whose total free energy is highest along the nucleation pathway, is usually defined as a critical nucleus. The probability of forming a larger cluster started from a nucleus is higher than the probability of dissociation of the nucleus. The initial formation of nuclei in a system where there are still no formed aggregates present is called primary nucleation. The primary nucleation may occur in homogeneous solution, called homogeneous nucleation, or at a surface, called heterogeneous nucleation. A surface can be the wall of the vessel, an air-water interface or the wall of a lipid vesicle etc. The secondary nucleation is induced by the surface of existing aggregates, referred to as surface-catalyzed nucleation.[86]

From the experimental side, the chemical kinetic and thermodynamic analyses play an important role in revealing the mechanism of nucleation of protein aggregation.[87] These analyses were applied to the aggregation of IAPP, alpha-synuclein and, notably the Aβ peptides with wild-type sequences and many variants.[7,88–90] These studies revealed the role of mutations and truncations at the N-terminus on the concentration-dependent time course of Aβ fibril formation and lag phase from a solution of monomers.

From the theoretical side, master equations have been used to interpret experimental data. Basically, rate equations with increasing complexity are fitted to the experimental sigmoidal curve. It has been shown that the best fit can only be obtained if all processes, including primary nucleation, fibril growth and secondary nucleation are taken into account. The use of only the first two processes can fail to reproduce the experimental curve.[87] The master equations do not provide, however, information on the nucleus. As a remedy, simulations have been carried out, employing various protein models so as to provide insights into the structure and size of the primary nucleus.[6,91]

At the coarse-grained level, Li et al. developed an on-lattice protein model, where each residue is entered on a site of a cubic lattice.[23,92] MC simulations of multiple Aβ42 chains, each chain consisting of eight residues, show that the nucleus size N* is 11 chains. Irback et al. developed another lattice model, where a residue is represented by an unit vector, located on a site of a cubic lattice, and interacts with other peptides by hydrogen bonding and hydrophobic forces. The simulation of 256 peptides shows that fibril formation occurs with a sigmoidal kinetics shape, and the width of the aggregate prior to nucleation is 3.5.[18]

Abeln and coworkers developed a more sophisticated lattice model by considering the formation of hydrogen bonds, the directions of side-chains and pairwise interaction energies between the 20 amino acids. REMC simulations captured the amyloid formation with $N^*=10$ for a designed 7-residue TFTFTFT peptide at low concentration.[93] We have developed further this model by incorporating the OPEP force field and notably the four-body hydrogen bonding interaction so as to explore the critical nucleus size of fibril formation for real amyloid peptides. REMC simulations of oligomers formed by 10–20 chains of Aβ16–22 or Aβ37–42 peptides allowed us to characterize the structure, thermodynamics and nucleus sizes.[21,22] We show that Aβ16–22 decamer forms a stable fibril with two anti-parallel five-stranded-sheets, consistent with the NMR fibril structure, indicating that the nucleus size of Aβ16–22 is about 10 chains. For a system of 20 Aβ37–42 peptides, Fig. 2 shows the population of one, two and three layers of β-sheets composed of n-strands. Overall, the system is mainly disordered with a low population of two and three layers of long n-strand β-sheets. It is to be noted that the free energy surface changes for 30 Aβ37–42 peptides with the presence of well-defined native fibrils (with parallel β-strands) indicating that $20 < N^* < 30$.[22]

In addition to the Caflish's and Frenkel's models described in Section 2, Linse et al. developed an algorithm for MC simulation of amyloid formation

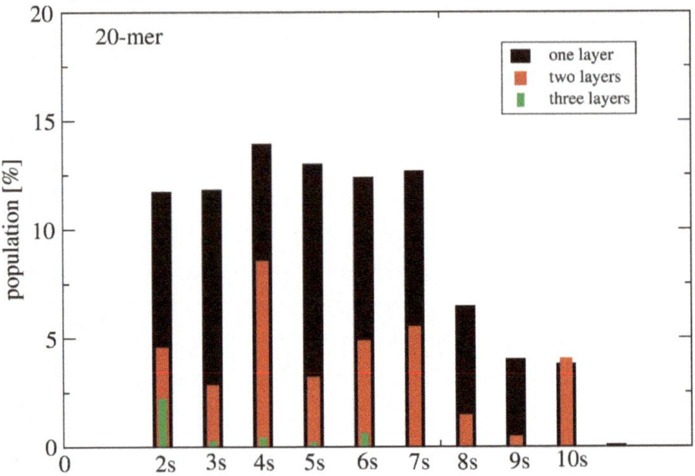

Fig. 2 The population of one (black), two (red) and three (green) β-sheet layers as a function of the number of β-strands. Shown is the REMC result of 20-mer Aβ37–42 below the melting temperature of the aggregates.

from flexible peptides. The algorithm is based on kinetic discrimination among fibrillar and non-fibrillar contacts. The simulations of up to 500 hexapeptides show that the fibrillation follows a sigmoidal curve, but there is no requirement for any complicated structural pathway to explain the lag phase and sigmoidal shape. This shape arises from cooperative among multiple interactions within each pair of peptides in a fibril.[94] Another kinetic model developed by Hsieh et al. also shows a two-step nucleation for Aβ16–22.[95] Using PRIME20 model with DMD method, Hall et al. showed that at low temperatures, the aggregation of 48 Aβ16–22 chains follows a primary nucleation mechanism with $N^* = 8$.[96]

Finally, Laio et al. investigated the all-atom aggregation of 18 Val8 and 18 Aβ35–40 peptides employing the bias-exchange metadynamics method in explicit solvent with the Amber99 force field. The simulations show that the aggregation follows a primary nucleated conformational conversion mechanism and the nucleus size is on the order of 14. The most populated aggregate is amorphous. The free energy landscape projected on appropriate variables reveals that the maximum free energy is characterized by a transition from mixed parallel/antiparallel to parallel β-strand orientations and only when a sufficient number of parallel sheets are formed that the free energy decreases to a free energy minimum with fully parallel β-sheets.[34,97]

To study the secondary nucleation mechanisms, large system sizes are necessary. However, due to the limitation of current computer power, most atomistic simulations start from preformed fibrils and free monomers. Strodel et al. carried out 1 μs all-atom simulations of six Aβ42 monomers placed near the surface of a preformed fibril fragment composed of 20 Aβ17–42 peptides. During the simulations, the monomers slide on the fibril surface, leading to oligomer formation. The results show that the hydrophobic surface at the C-terminal of the fibril causes the unfolding of the monomers which catalyzes the formation of β-sheet rich oligomers on the fibril surface.[98] This suggests one possible molecular basis of the secondary nucleation pathway.

Zacharias et al. carried out atomistic simulations using the CHARMM27 force field to study the association of monomer, dimer and larger oligomers to the surface of a Aβ9–40 fibril model. The results show that the association during the secondary nucleation follows a dock-lock mechanism consisting of a fast initial docking phase and a slow structural phase, and the hydrophobic effect is a major driving force for the association. Also, the surface induced association process is faster than the elongation at the fibril ends, indicating that secondary nucleation process can become the dominant process of amyloid formation in agreement with experiment.[99]

Very interestingly, Wilson and colleagues studied the co-aggregation of Aβ40 and Aβ16–22 to obtain molecular insights into the surface-catalyzed secondary nucleation of Aβ40 by Aβ16–22.[32] Employing DMD with the CG PRIME20 force field, simulations were performed on different systems: six Aβ40 monomers, six Aβ16–22 monomers and six Aβ40 monomers, and six Aβ40 peptides in the presence of preformed two, three and four β-sheet Aβ16–22 protofilaments. The simulations show that while all primary/secondary nucleation and growth processes occur at the same time, secondary nucleation is dominant in Aβ40 fibril formation kinetics during co-assembly with Aβ16–22. Interestingly, although the fibril surface of Aβ16–22 catalyzes the assembly of Aβ40, the assembly mechanism resembles that of Aβ40 alone in solution.[32]

6. Recent advances in structures of Aβ40/42 oligomers from simulations

In this last section, we would like to review some recent and significant contributions in the field of Aβ40/42 peptide simulations. We already discussed the simulation results of Aβ42 dimer in aqueous solution (see Section 2).

In the case of Aβ40 dimer in aqueous solution, Nguyen et al. performed atomistic REMD simulations on the wild-type (WT) sequence, the A2V/A2V mutant and the mixed WT/A2V mutant.[100–102] Experimentally, the A2V mutation protects from Alzheimer's disease in its heterozygous form and leads to early AD in its homozygous form. Biophysical experimental showed that the aggregation rate follows the order A2V > WT > A2V-WT. It is found by REMD that the secondary structure content (18% β-strand and 10% α-helix) is invariant, but the patterns of intra-molecular and intermolecular conformations drastically change upon single and double A2V mutation. In all sequences, the antiparallel and perpendicular peptide orientations are preferred over the parallel organization. However, upon single A2V mutation, the intermolecular potential energies are reduced, and the population of intra-molecular three-stranded β-sheets and of multiple β-hairpins spanning the residues 17–21 and 30–36 is increased, providing an explanation for its slower aggregation kinetics. Similar computational results were obtained for the Aβ40 WT/A2T dimer and the Aβ42 A2T/A2T dimer, A2T mutation known to reduce Aβ aggregation kinetics and the risk of AD.[102,103] Of particular interest is the application of a Markov state model to the transition pathways and associated kinetics resulting from a hybrid-resolution model and adaptive sampling technique that explores over 2.7 ms of simulations of formation of Aβ40 dimers. Indeed β-hairpin-structures are formed by direct binding of soluble Aβ in β-hairpin conformations, whereas formation of parallel β-structures resembling the U-fibril state results from Aβ monomers in arbitrary conformations and occurs 100-fold more slowly.[104]

For larger aggregates, two computational studies in aqueous solution are worth discussing. First, using the predictive coarse-grained protein AWSEM force field, Wolynes et al. calculated and compared the free energy landscapes and relative stabilities of Aβ42 and Aβ40 in their monomeric up to octameric forms. At the same concentration, the aggregation free energy profile of Aβ42 is more downhill and Aβ42 has a computed solubility 10 times smaller than that of Aβ40. While, at a concentration of 40 μM, there is a clear free energy barrier between the pre-fibril tetramer form and the fibril pentamer in the Aβ40 aggregation landscape, this is not the case for Aβ42. It is to be noted that the fibril state depicts the U-shape. Using oligomerization maps that capture the paths of conversion between similar states of oligomers, several key differences are revealed. Moreover, the two C-terminal residues stabilize the oligomeric structures of Aβ42, and facilitate the conversion from pre-fibril trimers to fibril tetramers.[52,53]

Second, atomistic MD simulations of 20 Aβ disordered chains in implicit solvent followed by transition networks analysis showed that pathways of Aβ40 and Aβ42 aggregation depend on oligomer shape, with compact and extended configurations and different solvent-exposures of hydrophobic residues.[105]

Numerous experiments indicate that toxicity could involve pore-forming Aβ oligomers in membranes, but aggregate heterogeneity has prevented experimental high-resolution structure determination. Nussinov and coworkers have modeled annular pores of 1.7–2.5 nm inner diameters consistent with atomic force microscopy experiments.[106] Their channels were built from the assembly of the U-shape fibril conformation of Aβ peptide. Recent biophysical experiments in membrane-mimicking environments also suggest Aβ42 pores with different oligomer sizes, and notably tetramers[107] and hexamers.[108] Using atomistic REMD simulations, low-resolution data obtained in lipid bilayers, and other theoretical factors, Derreumaux and coworkers designed 3D structures of Aβ40 and Aβ42 β-barrels in a bilayer mimicking a neuronal membrane.[109] The tetrameric model with two distinct β-hairpin motifs, eight antiparallel β-strands and an inner pore diameter of 0.7 nm is highly stable in all-atom MD and REMD simulations for Aβ42 and is much less likely for Aβ40.[109] Overall, this result sheds light on the amyloid pore hypothesis and explains the higher toxicity of Aβ42. The authors also reported novel findings as the stability of this Aβ40/42 β-barrel in aqueous solution has not been tested before at the atomistic level, and is very different from previous computational studies.[45] We cannot ignore, however, that pores may be formed of alpha-helical topologies.[110,111]

7. Conclusions

Understanding how amyloid aggregates actually become toxic is truly a real challenge in developing a treatment for neurodegenerative diseases, as only monomers are nontoxic. Thus far, all molecules (antibodies and drugs) targeting amyloid-β oligomers have failed to pass clinical trials. Many reasons have been put forward to explain this repetitive failure.[112,113]

Atomistic and coarse-grained simulations with increased computer efficiency, improved force field accuracy,[15,114] coupling to machine learning[115] and integration of the main cellular partners[116–118] are likely to lead to a better molecular understanding of all events involved in amyloid fibril formation and structural characterization of the Aβ and tau proteins aggregates in

Alzheimer's disease. These molecular insights coupled to experiments such as microbubbles in combination with focused ultrasound to deliver drugs through blood-brain barrier[119] may help design more efficient inhibitors of AD toxicity.

Conflict of interest

The authors declare no competing financial interest.

Acknowledgments

We acknowledge support by the "Initiative d'Excellence" program from the French State (Grant "DYNAMO," ANR-11-LABX-0011-01, and "CACSICE," ANR-11-EQPX-0008). PhD thanks Université de Paris, CNRS and PSL.

References

1. Chiti F, Dobson CM. Protein misfolding, amyloid formation, and human disease: a summary of progress over the last decade. *Annu Rev Biochem*. 2017;86:27–68.
2. Petkova AT, Buntkowsky G, Dyda F, et al. Solid state NMR reveals a pH-dependent antiparallel β-sheet registry in fibrils formed by a β-amyloid peptide. *J Mol Biol*. 2004;335:247–260.
3. Melquiond A, Boucher G, Mousseau N, et al. Following the aggregation of amyloid-forming peptides by computer simulations. *J Chem Phys*. 2005;122:174904.
4. Gazit E. Reductionist approach in peptide-based nanotechnology. *Annu Rev Biochem*. 2018;87:533–553.
5. Gour N, Kanth PC, Koshti B, et al. Amyloid-like structures formed by single amino acid self-assemblies of cysteine and methionine. *ACS Chem Neurosci*. 2019;10:1230–1239.
6. Nasica-Labouze J, Nguyen PH, Sterpone F, et al. Amyloid beta protein and Alzheimer's disease: when computer simulations complement experimental studies. *Chem Rev*. 2015;115:3518–3563.
7. Meisl G, Yang X, Frohm B, et al. Quantitative analysis of intrinsic and extrinsic factors in the aggregation mechanism of Alzheimer-associated Aβ-peptide. *Sci Rep*. 2016;6:18728.
8. Selkoe DG, Hardy J. The amyloid hypothesis of Alzheimer's disease at 25 years. *EMBO Mol Med*. 2016;8:595–608.
9. Lu Y, Shi XF, Nguyen PH, et al. Amyloid-β(29-42) dimeric conformations in membranes rich in omega-3 and omega-6 polyunsaturated fatty acids. *J Phys Chem B*. 2019;123:2687–2696.
10. Galvagnion C, Brown JW, Ouberai MM, et al. Chemical properties of lipids strongly affect the kinetics of the membrane-induced aggregation of α-synuclein. *Proc Natl Acad Sci USA*. 2016;113:7065–7070.
11. Nguyen PH, Tarus B, Derreumaux P. Familial Alzheimer A2V mutation reduces the intrinsic disorder and completely changes the free energy landscape of the Aβ1-28 monomer. *J Phys Chem B*. 2014;118:501–510.
12. Levine ZA, Shea JE. Simulations of disordered proteins and systems with conformational heterogeneity. *Curr Opin Struct Biol*. 2017;43:95–103.
13. Zhang M, Ren B, Chen H, et al. Molecular simulations of amyloid structures toxicity and inhibition. *Isr J Chem*. 2017;57:586–601.

14. Ilie IM, Caflisch A. Simulation studies of amyloidogenic polypeptides and their aggregates. *Chem Rev*. 2019;119:6956–6993. https://doi.org/10.1021/acs.chemrev.8b00731.
15. Robustelli P, Piana S, Shaw DE. Developing a molecular dynamics force field for both folded and disordered protein states. *Proc Natl Acad Sci USA*. 2018;115:E4758–E4766.
16. Leahy CT, Kells A, Hummer G, et al. Peptide dimerization-dissociation rates from replica exchange molecular dynamics. *J Chem Phys*. 2017;147:152725.
17. Narayan B, Herbert C, Yuan Y, et al. Conformational analysis of replica exchange MD: temperature-dependent Markov networks for FF amyloid peptides. *J Chem Phys*. 2018;149:072323.
18. Irback A, Jonsson S, Linnemann N, et al. Aggregate geometry in amyloid fibril nucleation. *Phys Rev Lett*. 2013;110:058101.
19. Šarić A, Michaels TCT, Zaccone A, et al. Kinetics of spontaneous filament nucleation via oligomers: insights from theory and simulation. *J Chem Phys*. 2016;145:211926.
20. Bieler NS, Knowles TP, Frenkel D, et al. Connecting macroscopic observables and microscopic assembly events in amyloid formation using coarse grained simulations. *PLoS Comput Biol*. 2012;8:e1002692.
21. Tran TT, Nguyen PH, Derreumaux P. Lattice model for amyloid peptides: OPEP force field parametrization and applications to the nucleus size of Alzheimer's peptides. *J Chem Phys*. 2016;144:205103.
22. Sterpone F, Doutreligne S, Tran TT, et al. Multi-scale simulations of biological systems using the OPEP coarse-grained model. *Biochem Biophys Res Commun*. 2018;498: 296–304.
23. Li MS, Klimov DK, Straub JE, et al. Probing the mechanisms of fibril formation using lattice models. *J Chem Phys*. 2008;129:175101.
24. Tuffery P, Derreumaux P. Flexibility and binding affinity in protein-ligand, protein-protein and multi-component protein interactions: limitations of current computational approaches. *J R Soc Interface*. 2012;9:20–33.
25. Stirnemann G, Sterpone F. Recovering protein thermal stability using all-atom Hamiltonian replica-exchange simulations in explicit solvent. *J Chem Theory Comput*. 2015;11:5573–5577.
26. Nasica-Labouze J, Meli M, Derreumaux P, et al. A multiscale approach to characterize the early aggregation steps of the amyloid-forming peptide GNNQQNY from the yeast prion sup-35. *PLoS Comput Biol*. 2011;7(5):e1002051.
27. Smith AK, Lockhart C, Klimov DK. Does replica exchange with solute tempering efficiently sample Aβ peptide conformational ensembles? *J Chem Theory Comput*. 2016;12: 5201–5214.
28. Nguyen PH, Okamoto Y, Derreumaux P. Communication: simulated tempering with fast on-the-fly weight determination. *J Chem Phys*. 2013;138:061102.
29. Zhang T, Nguyen PH, Nasica-Labouze J, et al. Folding atomistic proteins in explicit solvent using simulated tempering. *J Phys Chem B*. 2015;119:6941–6951.
30. Hoang Viet M, Derreumaux P, Nguyen PH. Communication: multiple atomistic force fields in a single enhanced sampling simulation. *J Chem Phys*. 2015;143:021101.
31. Ge X, Sun Y, Ding F. Structures and dynamics of β-barrel oligomer intermediates of amyloid-β16-22 aggregation. *Biochim Biophys Acta Biomembr*. 2018;1860:1687–1697.
32. Bunce SJ, Wang Y, Stewart KL, et al. Molecular insights into the surface-catalyzed secondary nucleation of amyloid-β_{40} (Aβ_{40}) by the peptide fragment A$\beta_{16\text{-}22}$. *Sci Adv*. 2019;5(6):eaav8216.
33. Granata D, Baftizadeh F, Habchi J, et al. The inverted free energy landscape of an intrinsically disordered peptide by simulations and experiments. *Sci Rep*. 2015;5:15449.
34. Baftizadeh F, Biarnes X, Pietrucci F, et al. Multidimensional view of amyloid fibril nucleation in atomistic detail. *J Am Chem Soc*. 2012;134:3886–3894.

35. Huang J, Rauscher S, Nawrocki G, et al. CHARMM36m: an improved force field for folded and intrinsically disordered proteins. *Nat Methods*. 2017;14:71–73.
36. Best RB. Atomistic force fields for proteins. *Methods Mol Biol*. 2019;2022:3–19.
37. Nguyen PH, Li MS, Derreumaux P. Effects of all-atom force fields on amyloid oligomerization: replica exchange molecular dynamics simulations of the Aβ(16-22) dimer and trimer. *Phys Chem Chem Phys*. 2011;13:9778–9988.
38. Man VH, He X, Derreumaux P, et al. Effects of all-atom molecular mechanics force fields on amyloid peptide assembly: the case of $A\beta_{16-22}$ dimer. *J Chem Theory Comput*. 2019;15:1440–1452.
39. Carballo-Pacheco M, Ismail AE, Strodel B. On the applicability of force fields to study the aggregation of amyloidogenic peptides using molecular dynamics simulations. *J Chem Theory Comput*. 2018;14:6063–6075.
40. Siwy CM, Lockhart C, Klimov DK. Is the conformational ensemble of Alzheimer's Aβ10-40 peptide force field dependent? *PLoS Comput Biol*. 2017;13:e1005314.
41. Rosenman DJ, Wang C, García AE. Characterization of Aβ monomers through the convergence of ensemble properties among simulations with multiple force fields. *J Phys Chem B*. 2016;120:259–277.
42. Carballo-Pacheco M, Strodel B. Comparison of force fields for Alzheimer's Aβ42: a case study for intrinsically disordered proteins. *Protein Sci*. 2017;26:174–185.
43. Weber OC, Uversky VN. How accurate are your simulations? Effects of confined aqueous volume and AMBER FF99SB and CHARMM22/CMAP force field parameters on structural ensembles of intrinsically disordered proteins: Amyloid-β_{42} in water. *Intrinsically Disord Proteins*. 2017;5:e1377813.
44. Man VH, Nguyen PH, Derreumaux P. High-resolution structures of the amyloid-β 1-42 dimers from the comparison of four atomistic force fields. *J Phys Chem B*. 2017;121:5977–5987.
45. Nguyen PH, Campanera JM, Ngo ST, et al. Tetrameric Aβ40 and Aβ42 β-barrel structures by extensive atomistic simulations. II. In aqueous solution. *J Phys Chem B*. 2019;123:6750–6756.
46. Pellarin R, Caflisch A. Interpreting the aggregation kinetics of amyloid peptides. *J Mol Biol*. 2006;360:882–892.
47. Bellesia G, Shea J-E. Effect of β-sheet propensity on peptide aggregation. *J Chem Phys*. 2009;130:145103.
48. Melquiond A, Mousseau N, Derreumaux P. Structures of soluble amyloid oligomers from computer simulations. *Proteins*. 2006;65:180–191.
49. Song W, Wei G, Mousseau N, Derreumaux P. Self-assembly of the β2-microglobulin NHVTLSQ peptide using a coarse-grained protein model reveals a beta-barrel species. *J Phys Chem B*. 2008;112:4410–4418.
50. Lu Y, Derreumaux P, Guo Z, et al. Thermodynamics and dynamics of amyloid peptide oligomerization are sequence dependent. *Proteins*. 2009;75:954–963.
51. Urbanc B, Betnel M, Cruz L, et al. Elucidation of amyloid β-protein oligomerization mechanisms: discrete molecular dynamics study. *J Am Chem Soc*. 2010;132: 4266–4280.
52. Weihua Zheng MC, Tsai MY, Wolynes PG. Exploring the aggregation free energy landscape of the amyloid-β protein (1–40). *Proc Natl Acad Sci USA*. 2016;113: 11835–11840.
53. Zheng W, Tsai M-Y, Wolynes PG. Comparing the aggregation free energy landscapes of amyloid β(1-42) and amyloid β(1-40). *J Am Chem Soc*. 2017;139:16666–16676.
54. Rojas AV, Liwo A, Scheraga HA. A study of the α-helical intermediate preceding the aggregation of the amino-terminal fragment of the β amyloid peptide (Aβ(1-28)). *J Phys Chem B*. 2011;115:12978–12983.

55. Rojas AV, Maisuradze GG, Scheraga HA. Dependence of the formation of tau and Aβ peptide mixed aggregates on the secondary structure of the N-terminal region of Aβ. *J Phys Chem B*. 2018;122:7049–7056.
56. Maupetit J, Tuffery P, Derreumaux P. A coarse-grained protein force field for folding and structure prediction. *Proteins*. 2007;69:394–408.
57. Derreumaux P. Generating ensemble averages for small proteins from extended conformations by Monte Carlo simulations. *Phys Rev Lett*. 2001;85:206–209.
58. Wei GH, Derreumaux P, Normand M. Sampling the complex energy landscape of a simple β-hairpin. *J Chem Phys*. 2003;119:6403–6406.
59. Sterpone F, Melchionna S, Tuffery P, et al. The OPEP protein model: from single molecules, amyloid formation, crowding and hydrodynamics to DNA/RNA Systems. *Chem Soc Rev*. 2014;43:4871–4893.
60. Côté S, Laghaei R, Derreumaux P, et al. Distinct dimerization for various alloforms of the amyloid-β protein: Aβ(1-40), Aβ(1-42), and Aβ(1-40)(D23N). *J Phys Chem B*. 2012;116:4043–4055.
61. Gnanakaran S, Nussinov R, García AE. Atomic-level description of amyloid beta-dimer formation. *J Am Chem Soc*. 2006;128:2158–2159.
62. Santini S, Wei G, Mousseau N, Derreumaux P. Pathway complexity of Alzheimer's Aβ16-22 peptide assembly. *Structure*. 2004;12:1245–1255.
63. Santini S, Mousseau N, Derreumaux P. In silico assembly of Alzheimer's Aβ16-22 peptide into β-sheets. *J Am Chem Soc*. 2004;126:11509–11516.
64. Petty SA, Decatur SM. Experimental evidence for the reorganization of β-strands within aggregates of the Aβ (16-22) peptide. *Proc Natl Acad Sci USA*. 2005;102:14272–14277.
65. Li X, Lei J, Qi R, et al. Mechanistic insight into E22Q-mutation-induced antiparallel-to-parallel β-sheet transition of $A\beta_{16-22}$ fibrils: an all-atom simulation study. *Phys Chem Chem Phys*. 2019;21:15686–15694.
66. Sengupta U, Carballo-Pacheco M, Strodel B. Automated Markov state models for molecular dynamics simulations of aggregation and self-assembly. *J Chem Phys*. 2019;150:115101.
67. Matthes D, Gapsys V, Brennecke JT, de Groot BL. An atomistic view of amyloidogenic self-assembly: structure and dynamics of heterogeneous conformational states in the pre-nucleation phase. *Sci Rep*. 2016;6:33156.
68. Sun Y, Ge X, Xing Y, Wang B, Ding F. β-barrel oligomers as common intermediates of peptides self-assembling into cross-β aggregates. *Sci Rep*. 2018;8:10353.
69. De Simone A, Derreumaux P. Low molecular weight oligomers of amyloid peptides display β-barrel conformations: a replica exchange molecular dynamics study in explicit solvent. *J Chem Phys*. 2010;132:165103.
70. Nguyen PH, Derreumaux P. Conformational ensemble and polymorphism of the all-atom Alzheimer's Aβ(37–42) amyloid peptide oligomers. *J Phys Chem B*. 2013;117:5831–5840.
71. Barz B, Olubiyi O, Strodel B. Early amyloid β-protein aggregation precedes conformational change. *Chem Commun*. 2014;50:5373–5375.
72. Saric A, Chebaro YC, Knowles TPJ, et al. Crucial role of non-specific interactions in amyloid nucleation. *Proc Natl Acad Sci USA*. 2014;111:17869–17874.
73. Latshaw DC, Cheon M, Hall CK. Effects of macromolecular crowding on amyloid β(16–22) aggregation using coarse-grained simulations. *J Phys Chem B*. 2014;118:13513–13526.
74. Sterpone F, Derreumaux P, Melchionna S. Protein simulations in fluids: coupling the OPEP coarse-grained force field with hydrodynamics. *J Chem Theory Comput*. 2015;11:1843–1853.

75. Languin-Cattoën O, Melchionna S, Derreumaux P, et al. Three weaknesses for three perturbations: comparing protein unfolding under shear, force, and thermal stresses. *J Phys Chem B.* 2018;122:11922–11930.
76. Brandner AF, Timr S, Melchionna S, et al. Modelling lipid systems in fluid with Lattice Boltzmann Molecular Dynamics simulations and hydrodynamics. *Sci Rep.* 2019;9: 16450.
77. Succi S. *The Lattice Boltzmann Equation for Fluid Dynamics and Beyond.* Clarendon Press; 2001.
78. Chiricotto M, Melchionna S, Derreumaux P, et al. Hydrodynamic effects on β-amyloid(16–22) peptide aggregation. *J Chem Phys.* 2016;145:035102.
79. Chebaro Y, Pasquali S, Derreumaux P. The coarse-grained OPEP force field for non-amyloid and amyloid proteins. *J Phys Chem B.* 2012;116:8741–8752.
80. Ando T, Skolnick J. On the importance of hydrodynamic interactions in lipid membrane formation. *Biophys J.* 2013;104:96–105.
81. Chiricotto M, Melchionna S, Derreumaux P, et al. Multiscale aggregation of the amyloid Aβ16–22 peptide: from disordered coagulation and lateral branching to *Amorphous prefibrils*. *J Phys Chem Lett.* 2019;10:1594–1599.
82. Lasagna-Reeves CA, Glabe CG, Kayed R. Amyloid-β annular protofibrils evade fibrillar fate in Alzheimer disease brain. *J Biol Chem.* 2011;286:22122–22130.
83. Oosawa F, Kasai M. A theory of linear and helical aggregations of macromolecules. *J Mol Biol.* 1962;4:10–21.
84. Hofricher J, Ross PD, Eaton WA. Kinetics and mechanisms of deoxyhemoglobin S gelation: a new approach to understanding sickle cell disease. *Proc Natl Acad Sci USA.* 1974;71:4864–4868.
85. Ferrone F. Analysis of protein aggregation kinetics. *Methods Enzymol.* 1999; 309:256–274.
86. Tornquist M, Michaels TCT, Sanagavarapu K, et al. Secondary nucleation in amyloid formation. *Chem Commun.* 2018;54:8667–8684.
87. Linse S. Mechanism of amyloid protein aggregation and the role of inhibitors. *Pure Appl Chem.* 2019;91:211–229.
88. Meisl G, Yang X, Hellstrand E, et al. Differences in nucleation behavior underlie the contrasting aggregation kinetics of the Aβ40 and Aβ42 peptides. *Proc Natl Acad Sci USA.* 2014;111:9384–9389.
89. Dammers C, Schwarten M, Buell AK, et al. Pyroglutamate-modified Aβ(3-42) affects aggregation kinetics of Aβ(1-42) by accelerating primary and secondary pathways. *Chem Sci.* 2017;8:4996.
90. Yang X, Meisl G, Frohm B, et al. On the role of sidechain size and charge in the aggregation of Aβ42 with familial mutations. *Proc Natl Acad Sci USA.* 2018;115: E5849–E5858.
91. Nguyen PH, Derreumaux P. Understanding amyloid fibril nucleation and Aβ oligomer/drug interactions from computer simulations. *Acc Chem Res.* 2014;47: 603–611.
92. Co NT, Li MS. New method for determining size of critical nucleus of fibril formation of polypeptide chains. *J Chem Phys.* 2012;137:095101.
93. Abeln S, Vendruscolo M, Dobson C, Frenkel D. A simple lattice model that captures protein folding aggregation and amyloid formation. *PLoS One.* 2014;9:e85185.
94. Linse B, Linse S. Monte Carlo simulations of protein amyloid formation reveal origin of sigmoidal aggregation kinetics. *Mol Biosyst.* 2011;7:2296–2303.
95. Hsieh MC, Lynn DG, Grover MA. Kinetic model for two-step nucleation of peptide assembly. *J Phys Chem B.* 2017;121:7401–7411.
96. Cheon M, Chang I, Hall CK. Spontaneous formation of twisted $A\beta_{16-22}$ fibrils in large-scale molecular dynamics simulations. *Biophys J.* 2011;101:2493–2501.

97. Baftizadeh F, Pietrucci F, Biarnes X, Laio A. Nucleation process of a fibril precursor in the C-terminal segment of amyloid-β. *Phys Rev Lett.* 2013;110:168103.
98. Barz B, Strodel B. Understanding amyloid-β oligomerization at the molecular level: the role of the fibril surface. *Chem A Eur J.* 2016;22:8768–8772.
99. Schwierz N, Frost CV, Geissler PL, et al. From Aβ filament to fibril: molecular mechanism of surface activated secondary nucleation from all-atom MD simulations. *J Phys Chem B.* 2017;121:671–682.
100. Tarus B, Tran TT, Nasica-Labouze J, et al. Structures of the Alzheimer's wild-type Aβ1-40 dimer from atomistic simulations. *J Phys Chem B.* 2015;119: 10478–10487.
101. Nguyen PH, Sterpone F, Campanera JM, et al. Impact of the A2V mutation on the heterozygous and homozygous Aβ1-40 dimer structures from atomistic simulations. *ACS Chem Neurosci.* 2016;7:823–832.
102. Nguyen PH, Sterpone F, Pouplana R, et al. Dimerization mechanism of Alzheimer $A\beta_{40}$ peptides: the high content of intrapeptide-stabilized conformations in A2V and A2T heterozygous dimers retards amyloid fibril formation. *J Phys Chem B.* 2016;120:12111–12126.
103. Das P, Chacko AR, Belfort G. Alzheimer's protective cross-interaction between wild-type and A2T variants alters $A\beta_{42}$ dimer structure. *ACS Chem Neurosci.* 2017;8:606–618.
104. Cao Y, Jiang X, Han W. Self-assembly pathways of β-sheet-rich amyloid-β(1-40) dimers: Markov state model analysis on millisecond hybrid-resolution simulations. *J Chem Theory Comput.* 2017;13:5731–5744.
105. Barz B, Liao Q, Strodel B. Pathways of amyloid-β aggregation depend on oligomer shape. *J Am Chem Soc.* 2018;140:319–327.
106. Jang H, Arce FT, Ramachandran S, et al. Disordered amyloidogenic peptides may insert into the membrane and assemble into common cyclic structural motifs. *Chem Soc Rev.* 2014;43:6750–6764.
107. Serra-Batiste M, Ninot-Pedrosa M, Bayoumi M, et al. Aβ42 assembles into specific β-barrel pore-forming oligomers in membrane-mimicking environments. *Proc Natl Acad Sci USA.* 2016;113:10866–10871.
108. Österlund N, Moons R, Ilag LL, et al. Native ion mobility-mass spectrometry reveals the formation of β-Barrel shaped amyloid-β hexamers in a membrane-mimicking environment. *J Am Chem Soc.* 2019;141:10440–10450.
109. Nguyen PH, Campanera JM, Ngo ST, et al. Tetrameric Aβ40 and Aβ42 β-barrel structures by extensive atomistic simulations. I. In a bilayer mimicking a neuronal membrane. *J Phys Chem B.* 2019;123:3643–3648.
110. Diaz JC, Linnehan J, Pollard H, et al. Histidines 13 and 14 in the Aβ sequence are targets for inhibition of Alzheimer's disease Aβ ion channel and cytotoxicity. *Biol Res.* 2006;39:447–460.
111. Ngo ST, Derreumaux P, Vu VV. Probable transmembrane amyloid α-helix bundles capable of conducting Ca^{2+} ions. *J Phys Chem B.* 2019;123:2645–2653.
112. Doig AJ, Derreumaux P. Inhibition of protein aggregation and amyloid formation by small molecules. *Curr Opin Struct Biol.* 2015;30:50–56.
113. Doig AJ, Del Castillo-Frias MP, Berthoumieu O, et al. Why is research on amyloid-β failing to give new drugs for Alzheimer's disease? *ACS Chem Neurosci.* 2017;8: 1435–1437.
114. Sterpone F, Nguyen PH, Kalimeri M, et al. Importance of the ion-pair interactions in the OPEP coarse-grained force field: parametrization and validation. *J Chem Theory Comput.* 2013;9:4574–4584.
115. Noe F, Olsson S, Kohler J, et al. Boltzmann generators: sampling equilibrium states of many-body systems with deep learning. *Science.* 2019;365:eaaw1147.

116. Menon S, Sengupta N. Influence of crowding and surfaces on protein amyloidogenesis: a thermo-kinetic perspective. *Biochim Biophys Acta Proteins Proteomics*. 2019;1867: 941–953.
117. Owen MC, Gnutt D, Gao M, et al. Effects of in vivo conditions on amyloid aggregation. *Chem Soc Rev*. 2019;48:3946–3996.
118. Chiricotto M, Sterpone F, Derreumaux P, et al. Multiscale simulation of molecular processes in cellular environments. *Philos Transact A Math Phys Eng Sci*. 2016; 374(2080):2080–2092.
119. Man VH, Truong PM, Li MS, et al. Molecular mechanism of the cell membrane pore formation induced by bubble stable cavitation. *J Phys Chem B*. 2019;123:71–78.

CHAPTER TEN

Computational studies of protein aggregation mediated by amyloid: Fibril elongation and secondary nucleation

Yang Cao[†], Xuan Tang[†], Miao Yuan[†], Wei Han[*,††]
State Key Laboratory of Chemical Oncogenomics, School of Chemical Biology and Biotechnology, Peking University Shenzhen Graduate School, Shenzhen, China
[*]Corresponding author: e-mail address: hanw@pkusz.edu.cn

Contents

Abstract

Amyloid formation by proteins and peptides is the hallmark of many diseases. Growing evidence suggests that oligomeric species arising during aggregation are toxic, but the molecular mechanism of aggregation and oligomer generation remains unclear. Recent discoveries that amyloid fibrils can convert soluble proteins into oligomeric nuclei to facilitate aggregation highlight the role played by fibrils in protein aggregation. We review here computational studies conducted to elucidate the molecular mechanism of two fibril-dependent processes during protein aggregation, namely, secondary nucleation and fibril elongation. Secondary nucleation occurs on the lateral surface of a fibril to generate nuclei while fibril elongation, through addition of proteins to the ends of fibrils increases the lateral surface of the fibril. We summarize the molecular

[†] These authors contributed equally.
[††] Senior author

Progress in Molecular Biology and Translational Science, Volume 170
ISSN 1877-1173
https://doi.org/10.1016/bs.pmbts.2019.12.008

insights into each process unraveled by computational methods at levels ranging from coarse-grained to atomic and discuss the connection between these insights and experimental observations. The computational challenges faced by these studies and their solutions are also discussed. Finally, we propose possible computational studies that could shed light on the mechanistic aspects of secondary nucleation and fibril elongation that have been unaddressed.

1. Introduction

Growing evidence suggests that many diseases are related to amyloid aggregation by certain proteins. These diseases include major neural degenerative diseases (Alzheimer's disease, Parkinson's disease), non-neuropathic systemic diseases (AA Amyloidosis and Senile Systematic Amyloidosis) and non-neuropathic localized diseases (Type 2 Diabetes and Arterial Amyloidosis).[1–3] Based on much genetic and mutagenetic evidence,[4–6] it has been hypothesized that the end products and intermediates may play critical roles in causing cytotoxicity and thus act as key molecular players in the pathology of diseases.[4,7,8] Although great efforts have been invested in modulating protein aggregation as means of therapeutic management of the diseases, success in clinical trials with this approach has not been achieved.[9–11] One of the main reasons is the lack of understanding of the molecular nature of toxic species and how they are generated during the aggregation process.

After years of biophysical and biochemical studies,[8,12,13] it is now known that protein aggregation is a complex process involving a myriad of transient intermediates (Fig. 1A).[16–22] It begins with soluble peptides that are either natively folded or intrinsically disordered, proceeds through a series of oligomeric intermediates varying in size with largely unknown structures, and eventually gives rise to fibrillar aggregates in which the peptides adopting special "cross-β" conformations are aligned periodically along the fibril axis. The key to understanding the mechanism of protein aggregation is to elucidate the molecular details of elementary processes involving interconversion between different intermediates, and the detailed role that these intermediates play.[13,23,24]

A large body of experimental literature has pinpointed several elementary processes essential for the progression of the protein aggregation.[14,19,22,25–27] At the very beginning of the aggregation process, proteins in solution self-assemble into oligomeric nuclei that are capable of growing into fibrillar

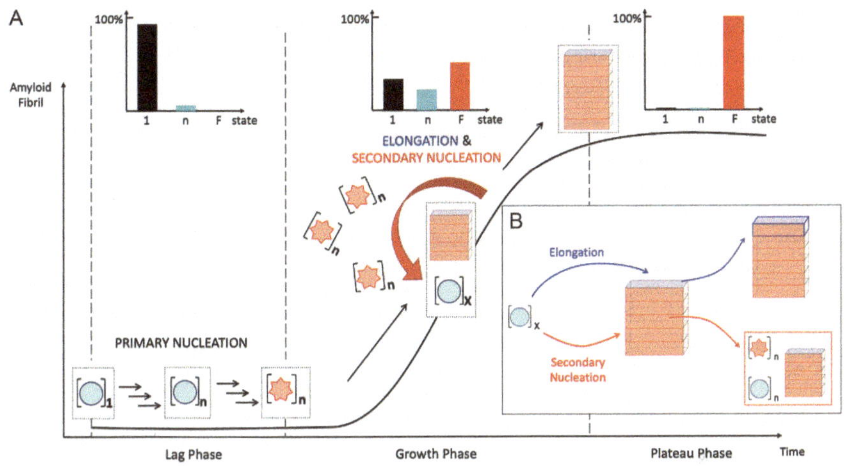

Fig. 1 Schematic view of amyloid aggregation process. (A) Macroscopic kinetic curve of amyloid aggregation and microscopic elementary processes. The vertical axis is the signal in proportion to quantity of amyloid fibrils. The horizontal axis is time evolution and can be divided into three phases: lag phase in which the amyloid signal rarely changes at the beginning, growth phase where the amyloid curve elevates sharply, and plateau phase where the amyloid signal reaches its maximum and remains steady. Three microscopic elementary processes are denoted in capitals, namely, primary nucleation process, elongation process and secondary nucleation process. The red curved arrow denotes the positive feedback of amyloid fibril due to its secondary catalytic effects. (B) Schemes of the elongation process (pathway connected by blue arrows) and the secondary nucleation process (pathway connected by red arrows). Shown in this figure are the different forms of amyloidogenic proteins, including monomers (either disordered or folded, represented as cyan cycles subscripted with 1), m-sized oligomer intermediates (either partially folded or misfolded, represented as cyan cycles subscripted with m), oligomeric nuclei (red asteroids subscripted with n), and amyloid fibrils (assemblies of cubic building block with red surface and blue tip). The cyan circle subscripted with x in panel (B) denotes reactants in elongation and secondary nucleation, where it has been proposed to be a monomer ($x=1$) in both cases[14,15] while we can't exclude the possibility that the reactant is an oligomer neither.[15] Shown as bar-charts above each phase are schematic views of percentage distribution of amyloid states, where horizontal axis indicates different amyloid states, namely, monomer state (black bar and denoted by 1), oligomer state (cyan bar and denoted by n) and fibril state (red bar and denoted by F).

aggregates. The fibrillar aggregates, once formed, can also use their lateral surface to convert the soluble proteins into nuclei, and, moreover, can do so at a much faster pace than is enjoyed by nucleation in solution (Fig. 1B). These two processes, termed primary and secondary nucleation, represent the major elementary processes that take place in the initial lag phase of the protein aggregation commonly seen experimentally (Fig. 1A). While primary nucleation is critical for triggering the entire cascade of the protein aggregation reactions, secondary nucleation is thought to be responsible for the generation of the majority of oligomeric nuclei.[25,27] A third elementary process is the elongation of the existing fibrillar aggregates by addition of soluble proteins onto the growing ends of these aggregates (Fig. 1B).[14,28] When significant amounts of nuclei have accumulated, this process occurs much more rapidly than the nucleation processes, and accounts for the fast growth phase of the protein aggregation that is observed experimentally.

There is mounting evidence showing that the oligomers arising transiently from the protein aggregation are probably the actual molecular species which are toxic to cells and animal models.[8,18,21,23,29–31] Experimental and theoretical studies have suggested that all the three aforementioned elementary processes play important but distinct roles in generation of the toxic oligomers.[13,32,33] The oligomer level could be altered differently when each of these elementary processes is modulated by other molecules such as small compounds and large protein molecules.[34–38] Collectively, these findings highlight the need to characterize the molecular details of these elementary processes.

Experimental characterization of the elementary processes of protein aggregation remains challenging. Current characterization techniques cannot provide information with sufficient spatial and temporal details. Complementary to the experimental methods, computer simulations studying protein aggregation have been useful, and have begun to unravel the critical details inaccessible to experimental efforts. In the past decades, extensive computational studies have been conducted to understand the mechanism of the primary nucleation process. There have been excellent reviews that cover this topic,[39,40] including a part of Chapter 11 of this series by Nguyen et al. Of particular interest to the present review is the mechanism of secondary nucleation and fibril elongation. Remarkably, both processes share the same catalyst, namely, fibrils, involving the transformation of the same substrate, namely, soluble proteins/peptides but their outcomes are markedly different. We will thus focus here on the recent computational advances in deciphering the molecular basis underlying secondary nucleation and elongation processes, paying particular attention to the roles played by fibrils.

2. Computational insights into fibril elongation

2.1 "Fast-deposition" versus "lock-and-dock" mechanisms

Fibrillar aggregates, once formed *via* the nucleation process, can elongate rapidly and continuously by addition of proteins or peptides to the growing ends of the aggregates. In most of the cases, an apparent first-order kinetics of fibril elongation was experimentally observed,[25,41–43] indicating that it is most likely that monomers rather than oligomers are added to the fibril ends. Some of aggregation-prone proteins or peptides such as Aβ, hIAPP or α-synuclein are disordered in their soluble forms while others like TTR and prions have native conformations which differ considerably from those present in the fibrils. Generally, this elongation process must involve (1) the association of soluble proteins with the fibril ends and (2) the structural transition of these proteins, and is reminiscent of the folding and binding process associated with protein-protein interactions.[44,45] A key mechanistic aspect of this type of process is the timing of the folding event and of the binding event.[44] The soluble monomers may undergo a conformational transition before or after their attachment to the ends of the existing fibrils. Two possible mechanisms have been proposed to describe the elongation process (Fig. 2): a unimolecular "fast-deposition" mechanism and a two-step "dock-and-lock" mechanism.[28,46] For the "dock-and-lock" mechanism, there must be a separation of the timescales between the binding of proteins

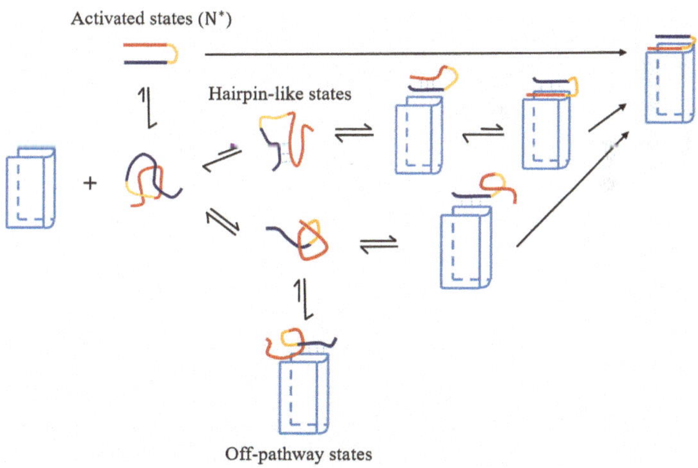

Fig. 2 The possible pathways of fibril elongation. In the "fast-deposition" pathway, the monomer in activated conformations (N*) attaches to the fibril tips quickly to extend the fibrils. In the "dock-and-lock" pathway, the monomer first quickly attaches to the fibril end. The docked monomer is likely to assume heterogeneous structures, some of which, such as hairpin-like states, have been thought to be important. Subsequently, the bound monomer undergoes structural rearrangement, during which native contacts between the monomer and the fibril end are formed at the cost of disruption of intramolecular and misaligned interactions. Some docked states, such as antiparallel states shown in this figure, may be off the pathway, retarding the process of fibril elongation and causing the "stop" phase seen in experiments.

and the structural rearrangement of bound proteins, and the latter process should be significantly slower.

In the "fast-deposition" mechanism, the soluble peptides assume various conformations (N) but must reorganize into a set of activated conformations (N*) (Fig. 2) that are prone to aggregation. These activated peptides can rapidly deposit onto the fibril ends elongating the fibrils. The "dock-and-lock" mechanism, on the other hand, suggests that regardless of the conformations that they take, the soluble peptides can always bind to the fibril end, giving rise to a covered fibril end structurally distinct from the fibril. In a second step, structural reorganization of either the attached monomer or the entire covered end takes place, completing the elongation process.

In an early theoretical work, Massi and Straub developed an analytical model of the kinetics of fibril elongation by both mechanisms.[46] The relationship between elementary processes of the fibril elongation and overall elongation kinetics could be interrogated with this model. The main conclusion of this study was that the overall elongation kinetics could be influenced by different elementary processes of different elongation mechanisms. The elongation rate of the fast-deposition mechanism may largely be affected by how fast the soluble peptide can diffuse to and form contacts with the fibril ends and by the possibility that the peptide happens to be in an activated conformation. If on the other hand, the elongation follows the "dock-and-lock" mechanism, the elongation rate is probably affected by the height of free energy barriers of reorganization of bound peptides or covered ends, as well as the binding equilibrium between the soluble peptides and the fibril ends. These analytical models were used to fit experimental kinetic data of aggregation of Aβ and its mutants, supporting the "dock-and-lock" mechanism.

While analytical models are useful for determining possible elementary processes during protein aggregation and associated rate constants, simulations have been proved valuable for interrogating the molecular basis of these processes. The simulations of fibril elongation have, however, been limited by the need to follow the slow elongation processes that take place on timescales between milliseconds and seconds.[47,48] Coarse-grained models at various levels of details have been developed and employed to overcome this computational challenge.[40] The minimalist models include the cuboid model developed by Muthukumar et al.,[49] the lattice model by Frenkel et al.,[50] the lattice model by Thirumalai and coworkers,[51] and several off-lattice models such as the 10-site models by Caflisch et al.,[52,53] the low-resolution model by Shea and coworkers,[54] the tube model by

Auer et al.,[55] the OPEP model by Derreumaux et al.[56,57] and the PRIME model by Chang et al.[58] These models were designed to be simple enough for efficient simulations of protein aggregation but still capable of capturing the basic physics of protein aggregation for investigation of the generic mechanism of the aggregation processes. The application of the minimalist models in studying the aggregation mechanism has been well reviewed by Wu and Shea,[59] and we discuss here only the insights into fibril elongation derived from these studies.

Both the "fast-deposition" pathway and the "dock-and-lock" pathway suggested by Straub were observed in the off-lattice simulations of fibril elongation by Auer, Caflisch and Shea.[52,54,55,60] In particular, the simulations of Caflisch and coworkers showed that when the aggregation-prone conformations of monomers become more populated, the elongation rate of fibril will also increase.[52,60] These studies suggest that the increased population of aggregation-prone conformations plays an important role not only in the "fast-deposition" scheme, but also in the "dock-and-lock" mechanism. Moreover, using a Markov chain formalism to analyze elongation kinetics, the authors estimated the partition coefficient between the two types of elongation pathways as a function of the probability of occurrence of the aggregation-prone conformations.[60] They found that the probability of the elongation by the "fast-deposition" mechanism indeed increases with improved aggregation propensity of monomers, but the "dock-and-lock" mechanism remained always as the dominant elongation pathway.

Other evidence in support of the "dock-and-lock" mechanism arose from the analysis of elongation timescales based on simulations at higher resolutions. For instance, Head-Gordon et al. developed an α-carbon-only model with anisotropic hydrogen-bonding (HB).[61] With this model, these authors studied the elongation of fibrils by Aβ40.[62,63] They observed a timescale gap ($\tau_{\text{lock}}/\tau_{\text{dock}}$) of ~4–20 between the initial binding of Aβ and subsequent structural rearrangement, an indication of the "dock-and-lock" mechanism. Especially, even in the best scenario, that incoming monomers tended toward the fibril-like conformations (N*) to favor the "fast-deposition" mechanism, the lock phase still lasts four times longer than the dock phase. Han and Schulten developed a hybrid-resolution model called PACE for protein folding.[64] This model retains most of atomic details of proteins but simplifies the representation of the solvent to accelerate simulations. The elongation of Aβ17–42 fibrils was simulated with this model.[65] The analysis of the simulation results with the transition path theory suggested that $\tau_{\text{lock}}/\tau_{\text{dock}}$ is about 10^3–10^4, consistent with the experimental finding of Maggio et al. that the timescale of the lock step is at least two orders of magnitude longer than that of the dock step.[26]

Apart from CG models, all-atom models with explicit solvent representation are arguably the most accurate for modeling aggregation processes. Due to the significant computational cost, early studies with atomistic simulations were limited to the fibril elongation of short peptide segments such as Aβ16–22, Aβ15–28, Aβ35–40 and Aβ37–42,[66–69] as well as segment GNNQQNY taken from Sup35.[68] In these studies, the β-strand content of the incoming monomer as a function of time [β(t)] was monitored during the addition of the monomer to a fibril end. The quantity β(t) was found to change significantly only after the monomer docked onto the fibril. The timescale of the locking process was around 10 times longer than that of the binding of the monomer. Recent works employed advanced techniques for the enhanced sampling of rare events and the effective extraction of kinetic information from simulations to permit the investigation of fibril elongation of full-length proteins/peptides with atomistic simulations.[70,71] In particular, Zacharias et al. employed an umbrella sampling technique to construct the free energy profile of the association of Aβ with the fibril end, and then derived the elongation kinetics from this profile.[70] Their study revealed a timescale gap ($\tau_{\text{lock}}/\tau_{\text{dock}}$) of $>$100-fold, which is in excellent agreement with both experiments and CG simulations.[26,65]

Collectively, the "dock-and-lock" mechanism is a consensus elongation mechanism observed in a myriad of simulations conducted with various levels of the details. This mechanism seems robust regardless of amino acid sequence and conformational preference of aggregating proteins/peptides. Hence, we will review here the recent computational insights into this elongation mechanism.

2.2 Insights into the "dock-and-lock" mechanism of fibril elongation

According to Straub et al., the kinetics of the "dock-and-lock" elongation process are affected by both the equilibrium between docking and undocking of the monomer and the subsequent post-absorption rearrangement of the monomer. Thus, the thermodynamic characters of these elementary processes and their molecular basis are crucial information for understanding the elongation mechanism. There have been numerous experimental studies conducted to examine the enthalpy and entropy components of the elongation kinetics.[14,72–74] Although the measured data only pertain to the overall elongation kinetics, they revealed that the elongation barriers of various types of aggregating proteins/peptides had a large unfavorable enthalpic contribution and a favorable entropic contribution.[14,72,75] Several possible explanations have been postulated to understand these results, including the release of bound water and/or a large unfolding of the monomer needed for the activation which breaks favorable interactions, increasing its flexibility.[14,72] In the following paragraphs, we will summarize the computational discoveries of the dock-and-lock phases of fibril elongation that assisted an understanding of the molecular basis of activation barriers of fibril elongation.

2.2.1 Initial docking of peptides driven by water release

The absorption equilibrium of soluble proteins/peptides, often unstructured onto fibril tips is determined by the free energy difference between unbound and bound monomers. As fibril ends exhibit a large number of unpaired HB donors and acceptors as well as exposed hydrophobic, hydrophilic and charged amino acid sidechains, various types of non-bonded interactions could arise. A key question is what are the main forces that could overcome unfavorable loss of translational and conformational entropy of incoming monomers and so drive this absorption process. The free energy change of the absorption contributes also to the overall activation barrier, and it is thus important to address this question. By analogy with protein folding, one may expect that the bound monomers would gain enthalpy through interactions with the fibril end to compensate the loss of entropy. Surprisingly, the experiments showed that the equilibrium constant of the absorption increases with temperature, suggesting that the absorption process is entropy-driven.[48]

There have been numerous computational studies aimed at explaining this mechanistic puzzle. Through atomistic simulations of Aβ10–35 and analysis of its hydration shell, Thirumalai et al. proposed that water could play a key role.[76,77] By assuming a U-shaped fibril-like conformation as the activated conformation (N*) for fibril formation, they found that only a small fraction of the conformational ensemble of Aβ10–35 samples the activated conformation. As in the activated conformation, D23 and K28 need be in close contact to form a salt bridge interaction while they are bound with multiple water molecules in non-activated conformations. Thus, the formation of fibrils involves expulsion of these hydration water molecules, leading to a favorable entropy change. Following this reasoning, the authors also examined through simulations, the hydration of a Sup35 peptide (GNNQQNY) during its addition to fibril ends.[68] They found that the release of hydration water concurred with the docking of the peptide onto the fibril end and the subsequent lock phase.

Recently, Su and Dias performed a systematic computational analysis of contribution of various types of interactions to the overall docking energetics.[78] In this study, the free energy change of the binding of two, three, and four Ala10 in β-strand conformations was calculated through the umbrella sampling method. In all cases, the binding process is spontaneous. However, the energy decomposition analysis revealed that during the peptide binding, the energy gain from peptide/peptide and water/water hydrogen bonds (HB) approximately cancels out the loss caused by the rupture of peptide/water HBs. The overall Lennard-Jones energy contributions, on the other hand, are favorable for peptide aggregation. The authors pointed out that although the HBs formed between peptides do not drive the binding process, they are essential to compensate unfavorable dehydration of peptides, which is approximately ~9.4 kJ/mol/residue. A similar analysis has also been conducted by Zacharias and coworkers for the fibril elongation by Aβ9–40.[70] They found that the docking of this peptide to the fibril end caused only slight change in total number of HBs in the system as well as in the enthalpy ($\Delta H \approx 0$). Instead, they observed a large gain of entropy resulting from the release of ~60 hydration water molecules during the docking process.

Still, the entropy loss of peptide conformations needs to be taken into account to evaluate the overall change of entropy in the dock phase. Although estimation of this quantity remains challenging, the extensive sampling of Aβ bound on the fibril end revealed that the bound peptides adopt a large number of distinct conformations and interact with the fibril end in rather heterogeneous manners.[65,79–81] This is also consistent with the analysis of kinetic data of Aβ fibril elongation.[48] These findings imply that the entropy loss caused by the docking of peptides on the fibril end could be much smaller than what would be normally expected for protein folding. Therefore, the entropy gained from water release is likely to exceed the loss of conformational entropy, resulting in a net gain of entropy in the dock phase. Collectively, the majority of simulation results obtained so far suggest that the initial docking of proteins/peptides on the fibril ends is driven mainly by the release of hydration water molecules.

2.2.2 Structural rearrangement in the locking step

The lock phase involves the conformational transition of bound monomers (and covered fibril ends) from the docked conformational states to the locked conformational states. Numerous computational studies have been conducted to address several questions regarding the mechanism of this locking transition. These studies have been mainly focused on the evolution of conformation of docked monomers and their interactions with the fibril ends during the locking transition.

The conformational complexity associated with the locking transition grows exponentially with sequence length, and many early simulation studies employed short peptides as model systems. These peptides include Aβ16–22,[66,82] Sup35 segment GNNQQNY,[83] Aβ37–42[68] and Aβ35–40,[69] most of which are also the model system for experimental study of fibril formation (Fig. 3A). Due to the short length of the peptides studied, it was feasible to observe the docking/locking transition through unbiased simulations with atomistic models. In these simulations, a peptide was often observed to encounter an existing fibril end at random, often forming incorrect contacts with the fibril end that are different from parallel in-register contacts seen in canonical fibril structures. These contacts can be rectified by the dissociation of the peptide into solution. The lock stage starts only when at least one of the fibrillar contact between the peptide and the fibril end is formed correctly. At this point, the peptide gradually forms more fibrillar HB interactions with the fibril end. Although the models are simple, the results are significant. They suggest that due to the stochastic nature of the docking process, the incoming peptide can dock onto the fibril end in heterogeneous poses, only a select few of which, however, are productive for subsequent locking.

Bolhuis and coworkers performed atomistic simulations of the fibril elongation by ^{11}LVEALYL^{17} from the insulin B-chain.[87] They employed transition path sampling (TPS) techniques[88] to construct the ensemble of docking/locking transition paths, which allowed for evaluation of kinetic mechanism of the elongation process. They found that when the peptide formed extensive antiparallel HB contacts or parallel but out-of-register contacts with the fibril end, it will not be able to proceed to undergo the locking transition. Instead, the peptide needs to await slow dissociation. Thus, these docked states appear to be kinetic traps off the locking transition pathway. It has been observed experimentally that during the fibril elongation there are intermittent periods during which the elongation is halted.[48,89–93] The halt of the elongation is likely caused by the docked monomer trapped in the off-pathway states that disable the fibril end from further growth.

Compared with the short peptide systems, real aggregation-prone peptides are much longer (≥40 amino acids). Not only do these peptides have

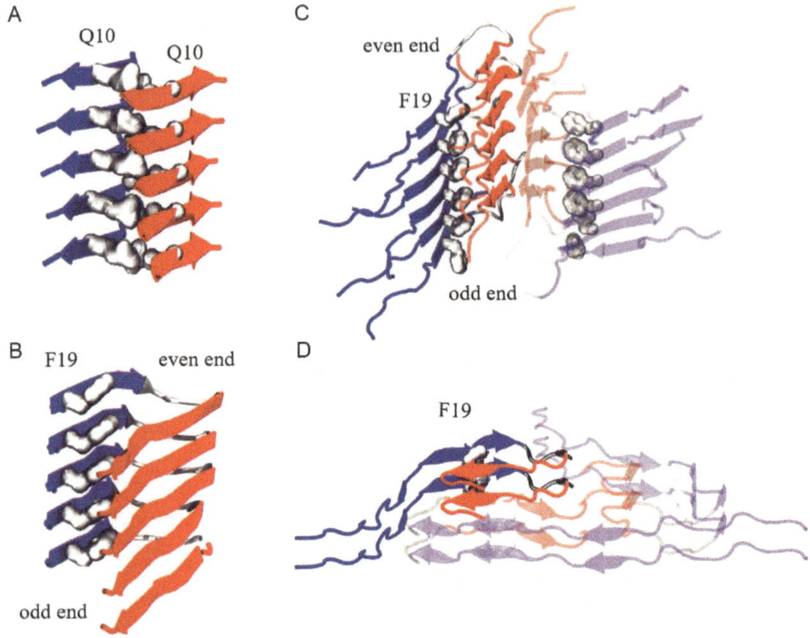

Fig. 3 Fibril structural models employed in the computational studies of fibril elongation discussed in this review. (A) Architecture of the ^{7}GNNQQNY^{13} peptide from Sup35 (PDB ID: 1YJP), which forms parallel, in-register stacks between two β-sheets by a network of hydrogen bonds.[83] (B) Structure of Aβ17–42 fibrils (PDB ID: 2BEG) comprising a strand-turn-strand motif.[84] The regions of 17–26, 27–30 and 31–42 are shown in blue, white and red, respectively. The even fibril tip, F19 in N-terminal was exposed to solvent. (C) Structure model of Aβ9–40 fibril with twofold symmetry along the fibril axis (PDB ID: 2LMN),[85] which has opposite staggering direction to the structure of Aβ17–42 fibrils. The regions of 9–24, 25–29 and 30–40 are shown in blue, white and red, respectively. (D) The threefold symmetric model of Aβ1–40 fibril (PDB ID: 2M4J).[86] The regions of 1–24, 25–29 and 30–40 are shown in blue, white and red, respectively. Except for the charged residues D23 and K28 which form a salt bridge in the loop region, the residues located in the regions other than the N-terminus are mostly neutral and hydrophobic.

much larger conformational space than the short peptides, but they may also be more easily trapped into deeper free energy minima during the docking/locking transition as more misaligned interactions could arise. Therefore, the sampling of such systems presents a huge computational challenge, which, as discussed in the following paragraph, can be handled by the application of various advanced simulation techniques.

A further complexity is that the longer peptides do not form a simple parallel in-register β-sheet in which the entire peptide chain assumes a straight β-strand structure. Instead, depending on the amino acid sequences, the long peptide has intermittent short β-strand regions, forming multiple parallel in-register β-sheets that can further assemble into unique architectures (Fig. 3B–D).[86,94,95] For example, an NMR model (PDB ID: 2LMN) of Aβ40 fibrils (containing the Aβ9–40 segment) shows that the fibril is comprised of an N-terminal β-sheet and a C-terminal β-sheet (Fig. 3B).[85] The peptides in the fibrils have a U-shaped topology, bringing the two sheets into an anti-parallel packing. Moreover, there is a perpendicular "stagger" between the two sheets. As a result, the C-terminal edge of Aβ40 is exposed on one end of the Aβ40 fibril (called an odd end by Lührs et al.[84]) while on the other end (the even end), the N-terminal edge is exposed. Similarly, the structural difference of two fibril ends is also observed in a NMR model (PDB ID: 2BEG) of Aβ42 fibril (containing Aβ17–42 segment) while the direction of staggering is opposite to that of the Aβ40 fibril (Fig. 3C).[84] It has been suggested that the structural difference between two fibril ends may account for the asymmetric growth of Aβ fibrils[62,84] although the corresponding molecular basis remains elusive.

The computational studies have been primarily focused on the fibril elongation of Aβ due to its moderate length and the wealth of experimental data available to date. These studies will be reviewed in this chapter. Progress has also been made recently in computational study of fibril elongation by much longer peptides such as α-synulein (~140 amino acid residues).[96,97] These works have been covered in an excellent review by Ilie and Caflisch.[40]

Scheraga and coworkers studied the elongation of Aβ9–40 fibrils.[98] They examined the equilibrium of binding of a free monomer to a fibrillar seed with a united-residue CG force field called UNRES.[99,100] The replica exchange molecular dynamics (REMD) technique was employed to keep the system from kinetic traps to boost the sampling.[101] The monomer was found to bind to both fibril ends and there was a ~1% chance of finding the monomer locked in the fibril state. Although the REMD simulations perturb dynamics and cannot be used to derive kinetic information, the authors tracked the sequence of events of the locking transitions observed in their simulation trajectories. This revealed a two-step locking mechanism. In the first step, the monomer docked onto either N- or C-terminal edge of the fibril. Non-native HBs formed between the monomer and the fibril end in this step needed to be removed by dissociation, similar to what was found in the previous studies of the short peptide systems. Once native HBs were formed, the transition proceeded in the second step, gradually spreading native HB contacts in the rest of the monomer.

Gurry and Stultz employed an atomistic model with implicit solvent to characterize the elongation with atomic details.[102] They constructed the free energy surface (FES) of the elongation through extensive simulations of unbinding of a monomer coupled with umbrella sampling.[103] The binding of a monomer to the odd ends of both Aβ17–42 and Aβ9–40 fibrils were investigated. The transition paths were determined as the minimum free energy pathways connecting the unbound and locked states. In these pathways, the β1-strand (18–26) of the monomer first needs to form native contacts with the N-terminal edge of the fibril end and then the β2-strand (31–42) of the monomer interacts with β1-strand, forming a β-hairpin as an obligatory structural intermediate of the bound monomer. Finally, the β2-strand of the monomer forms native contacts with the C-terminal edge at the cost of the intramolecular HBs. This locking mechanism appeared to be generic for the elongation of both Aβ40 and Aβ42 fibrils, and the observation of a β-hairpin intermediate was consistent with experimental data.[104–106] Intriguingly, a similar β-hairpin intermediate has also been observed in the atomistic simulations of both monomers and oligomers.[107,108] The stabilization of the loop of the β-hairpin by introducing a lactam bond between D23 and K28 greatly accelerated the fibril growth.[106] On the other hand, the sequestration of this β-hairpin conformation was found to slow down the aggregation.[104]

To systematically evaluate the elongation pathways of Aβ17–42 fibrils, Han and Schulten combined umbrella sampling techniques and REMD simulations, the former method used to probe the monomer binding and the latter method to explore extensively the conformational fluctuation of the incoming monomer coupled with its binding.[65] Based on over 1 ms simulations with the hybrid-resolution model PACE,[64,109] the pathway ensemble and associated kinetics were derived by kinetic network analysis[110] using the transition path theory.[111,112] The simulations showed that consistent with other studies, the monomer bound on both fibril ends can adopt a wide spectrum of binding poses involving parallel but out-of-register, anti-parallel and native HB interactions with the fibril ends.[79–81] However, the productive transition pathways always involve initial native interactions of the monomer either with the N-terminal edge of the fibril end or occasionally with the C-terminal edge. In particular, the high hydrophobicity of ^{17}LVFF^{20} in β1 renders this region much more likely to form native contacts than could be expected from a random encounter between the monomer and the fibril end. The hydrophilic loop was the last to form the native contacts. The β-hairpin of the bound monomer, similar to that found by Gurry and Stultz, was observed on the productive pathways for both fibril ends with a >50% chance, indicating that this hairpin is an important but not required intermediate for the locking transition. The authors also showed that the elongation occurred ~40 times faster at the even end than at the odd end. The activation enthalpy of the elongation at the even end was calculated to be ~22 kcal/mol, close to the experimental values (~23 and ~15.8 kcal/mol).[14,72] This indicated that favorable intra-/intermolecular interactions are disrupted in the locking transition states. The transition states for the elongation at the odd end are less stable by ~12 kcal/mol. The authors attributed the faster elongation rate at the even end to the exposed N-terminal edge. The hydrophobic residues such as F19 contained in this edge were likely to provide transient stabilization to the bound monomer which needs to reorganize its structure. These data were consistent with the asymmetric growth of Aβ fibrils observed experimentally[43,92] although the odd end of Aβ17–42 was proposed by Lührs et al. to be the faster growing end based on the presence of a hydrophobic cleft at this odd end.[84]

A different approach by Schwierz et al. of modeling fibril elongation[70] is first to use umbrella sampling to calculate both profiles of free energy and diffusion constants for the detachment of a monomer from a fibril end and then study pathways and kinetics based on these profiles with one-dimensional Fokker-Planck equation.[113,114] They investigated the elongation of Aβ9–40 fibrils. In agreement with the results of Han and Schulten,[65] they found that various non-native contacts existed in the docked states and the incoming monomer had to fold from a partially disordered structure in solution into a β-hairpin structure in the bound states. The difference is that during the locking transitions the native contacts were formed at the cost of non-native contacts. The need to break these non-native contacts is a major factor retarding the locking process. In addition, they showed that the fibril grew about twofold faster at the even end (originally named the odd end by Schwierz et al.) than at the odd end, consistent with the results of Han and Schulten.[65]

As the simulation studies described above employed umbrella sampling or REMD to enhance sampling, positional restraints were often needed to maintain the structural integrity of the fibril parts during the simulations. In addition, only the core fragments visible in experimental structural models were studied while the remaining parts that appeared disordered were ignored. In the recent work of Bacci et al., the fibril elongation of Aβ42 fibril (built from the Aβ17–42 fibril) was studied through atomistic simulations free of such restraints.[71] These simulations were carried out with an adaptive sampling algorithm called progress index-guided sampling (PIGS) that permits efficient sampling of transition paths from the locked state.[115] The path ensembles, obtained with the Markov State Model,[111] revealed that the mechanism of the locking transition of Aβ42 at the odd fibril end bore several similarities to those proposed previously, such as the structural disorder of the docked states,[65,70] the prominent role of ^{17}LVFF^{20} in initial native contacts[65] and the transfer from intra- to intermolecular contacts in the locking pathways.[70] However, the β-hairpin was not found in the locking pathways. Other new insights were also revealed. Apart from the incoming peptide, the penultimate peptide was also found to participate in non-native contacts, raising the possibility of structural rearrangement of the fibril end as a part of the locking mechanism. The flexible N-terminal part of Aβ42 can shield the fibril end in some of the docked states and potentially block the locking transition, highlighting the importance of considering this part in the study of fibril elongation.

The free energy landscape of the locking transition of Aβ17–42 was recently explored by Roder and Wales by implicit solvent simulations.[116] A large number of docked minima and transition states between these minima were systematically sampled by discrete path sampling.[117] Again, no restraints were applied in their simulations. The transition landscape was obtained through the disconnectivity graph analysis.[118] The docked states with non-native contacts can lie in the funnels separated from the native funnel by very high energy barriers. In the native funnel, the locking transition occurred by a mechanism similar to those reported previously, except that the β-hairpin was not observed. In other funnels, the docked states disfavored the locking transition due to the misaligned tight contacts with the fibril end or the large entropy of the hydrophilic loop. Based on their observation of the greater flexibility at the even fibril end, the authors hypothesized that the elongation at this end would suffer a greater loss of entropy and, thus was slower than the elongation at the opposite end. This conclusion agreed with that of Lührs et al.[84]

Head-Gordon and coworkers investigated the addition of Aβ40 to a fibril template with a pathological morphology (PDB ID: 2M4J, see also Fig. 3D).[62,63,86] Using a united-residue model,[61] they performed very long (~65 ms) unbiased, restraint-free simulations. The monomer was observed to deposit onto both ends of the fibril at a similar rate, which failed to support the asymmetric growth. By perturbing the conformational distribution of the incoming monomers, the authors examined the effect of conformational preference of the incoming monomer on the elongation kinetics. They found that the monomer in a U-shaped conformation was locked onto the fibril end twice as fast as in other conformations. This result supported the notion that certain conformational states of the incoming monomer could be key intermediates for the locking transition.

In summary, the computational studies suggest a consensus mechanism of the elongation of Aβ fibrils (Fig. 2), and (1) the docked states are heterogeneous and involved in misaligned contacts with the fibril ends, giving rise to a rugged free energy landscape; (2) the locking transition is initiated by the formation of native contacts at one of β-strands of Aβ and proceeds with more native contacts spreading to the other β-strand; (3) formation of native contacts relies on the rupture of intramolecular and misaligned intermolecular contacts, which slows the transition. However, there is still discrepancy among these studies regarding more specific mechanistic questions as to whether the fibril growth is asymmetric, which of the fibril ends grows faster, and whether a particular conformation such as the β-hairpin of the incoming monomer is a key intermediate in the elongation process.

This discrepancy may result from several factors. First, the use of different force fields and sampling protocols in these studies may cause the difference in calculation of the energetics of intra- and intermolecular interactions. This energetic difference would in turn impact the calculated rates of the growth at both ends. In addition, Aβ modeled with different force fields showed vastly different tendencies to adopt the β-hairpin conformation,[119,120] which may also affect the results of the elongation simulations. Second, there are multiple resolved structural models that differ in the count of filaments and the molecular packing within the filaments.[84–86,94,95] The structural features of the fibril ends vary between the different models. Of note, for most of the simulations, the fibril parts studied were single filaments extracted from the experimental fibril models and, thus can only be regarded as a model system.

Finally, as demonstrated in recent works,[71,116] whether or not to consider the flexibility of fibril ends can make a difference in the simulation results. For example, the simulations of Roder and Wales[116] showed that due to conformational flexibility, addition of a monomer in the β-hairpin conformation to a tetrameric stack of Aβ17–42 appeared to be prohibited, but would otherwise be observed in the simulations with rigidified fibrils.[65,70,102] Although this highlights the importance of including a flexible model of fibrils in the simulations of fibril elongation, caution is still needed in doing so. The works mentioned above used a tetrameric stack as fibril models which may not be sufficient to maintain the structural stability of fibrils.[63] A model with more peptide chains may give rise to more stable fibril ends, as shown by Okumura and Itoh.[121] Moreover, the instability of fibril ends may be overestimated by current non-polarizable force fields. In their simulations of a fibril model composed of seven hIAPP20–29s,[122] Zhang et al. showed that the fibril structures were well maintained in the simulations using polarizable force fields[123] but became highly twisted or had their edges ripped off if modeled with non-polarizable force fields. This study and another by Lemkul and coworkers[124] showed that the electrostatic polarization between different peptide chains plays a critical role in the stability of the fibrils. Therefore, a fibril template containing enough peptide chains modeled with polarizable force field would be considered in further computational study of the elongation mechanism.

3. Computational insights into the mechanism of secondary nucleation

3.1 Experimental background

Secondary nucleation was originally referred to as an auto-catalytic crystallization process where production of nascent crystal nuclei is efficiently catalyzed by the surfaces of parental crystals.[125,126] In 1985, Ferrone et al. showed that aggregation kinetics of sick-cell hemoglobin was consistent with an auto-catalysis of oligomeric nuclei on the surfaces of existing hemoglobin amyloid fibrils.[127,128] In recent years, several independent studies on other amyloidogenic proteins confirmed the conclusion of Ferrone et al., indicating that secondary nucleation at the fibril surface is common in protein aggregation.[25,27,129] In several diseases, oligomers are identified as potential pathogenic factors,[21,130–132] and secondary nucleation catalyzed by existing fibrillation aggregates is responsible for rapid generation of these oligomers.[133]

In general, it is believed that the secondary nucleation starts with the adsorption of amyloidogenic peptides on the fibril surface (Fig. 4), followed by a series of structural conversions into nuclei.[33] Diffusion and coalescence of the adsorbates may also be needed for the conversion process to take place. Lastly, the resulting nuclei need to detach from the surface. Recently, Linse and coworkers conducted extensive experiments to investigate kinetics of Aβ aggregation and its dependence upon temperature.[134] They found that the activation barrier of the fibril-dependent secondary nucleation is mainly composed of the free energy changes of absorption of proteins and subsequent structural reorganization. Although a similar composition of the

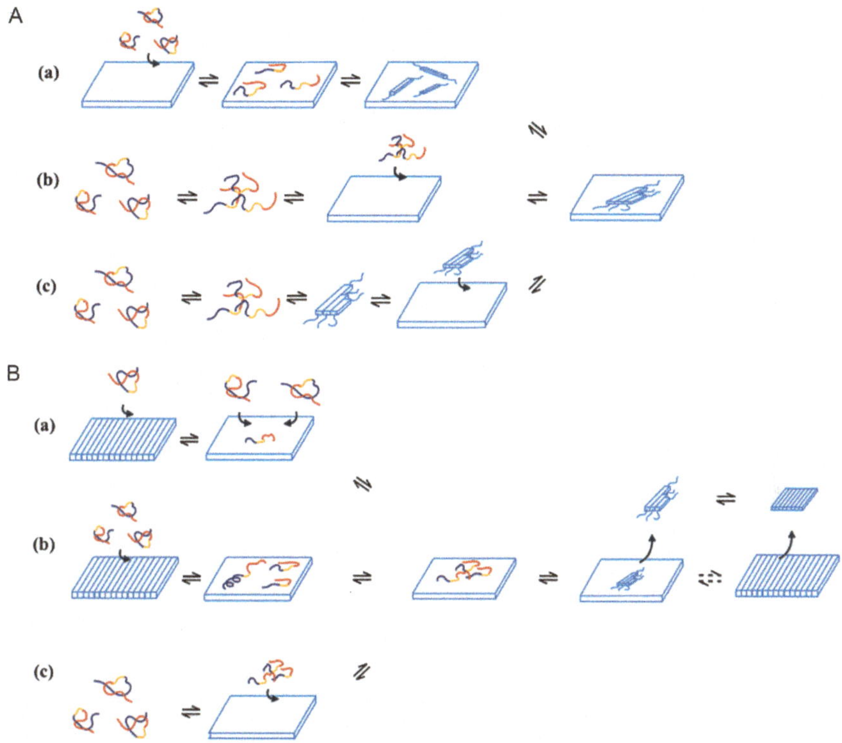

Fig. 4 Possible pathways of nucleation processes on a surface learnt from computational studies. (A) and fibril-dependent secondary nucleation processes (B). Shown in (A) are the three types of general pathways of nucleation induced by a surface. (a) Individual peptides are adsorbed on the surface and are transformed into β-strands and then diffuse and assemble into nuclei, reminiscent of the canonical nucleation-polymerization mechanism. (b) Oligomers are first formed in solution, then adsorbed on the surface, and finally converted into nuclei by the surface. (c) Oligomers are formed and converted into ordered β-sheet structures in solution, and then these β-sheet structures bind to the surface. (B) Possible pathways of secondary nucleation catalyzed by a fibril surface. (a) Individual peptides bind to the peptides that are already attached on the fibril surface to form bound oligomers, which subsequently rearrange their structures to form intermediate oligomers that are presumably easier to nucleate. The oligomers will be released to solution and then evolve into nuclei. (b) Individual peptides are first adsorbed on the fibril surface and then diffuse on the surface and coalesce into disordered monomers. A method with which these oligomers could nucleate has not been observed computationally. (c) It is also possible that oligomers are directly formed in solution and then attach on the fibril surface. Again, the mechanism of subsequent nucleation of the bound oligomers remains unclear.

activation barrier was also found for the fibril elongation, the thermodynamic signature of the secondary nucleation is different. The absorption of the peptides on the fibril surface causes a large enthalpy gain and a large entropy loss, which are opposite to what was observed for the docking step of the fibril elongation. The overall barrier of the secondary nucleation has a moderate favorable enthalpic contribution and an unfavorable entropic contribution. These results imply that the molecular basis of the secondary nucleation is unique. Moreover, this study illustrated also the similarity between the secondary nucleation and classical heterogeneous catalysis. For such a process to be efficient, there has to be a facile release of nuclei so that the catalytic surface can be recycled.[135]

Clearly, the experimental investigations have shed light on important microscopic features of fibril-dependent secondary nucleation, but they also raise questions as to molecular details of absorption of proteins, nucleated conversion on fibril surface and detachment of nuclei, as well as how the synergy of these processes results in an accelerated nucleation. In contrast to the fibril elongation, the discovery of the fibril-dependent secondary nucleation is relatively new, but there have been a number of computational studies of the molecular basis of the process. More generally, the secondary nucleation bears many similarities to protein aggregation on a surface, which is already well known and has been studied by numerous simulations to gain insights into the general mechanism of protein aggregation on surfaces. These general insights are also helpful for understanding the fibril-dependent nucleation. Consequently, in the following paragraphs, we will first cover the studies of surface-mediated aggregation and then discuss the insights into the fibril-dependent secondary nucleation that have been derived from recent computational and experimentation studies. Due to the space limitations, we try to exercise restraint on the results for the general insights into surface-mediated aggregation without covering too much details for particular systems, especially for the membrane-mediated aggregation which by itself, deserves a comprehensive review and has been covered recently by Derreumaux et al.[39]

3.2 Insights into surface-induced nucleation of peptides

Surface-mediated aggregation is an important pathway of protein aggregation.[136–140] It can take place at protein concentrations much lower than are normally required for homogeneous aggregation in solution. The lag phase of aggregation is usually greatly reduced,[136,141] and the surface-mediated aggregation could be peculiarly relevant to *in vivo* conditions where the concentration of amyloidogenic peptides is extremely low.[142] Various types of surfaces have been reported to promote aggregation, including those of nanoparticles (NP),[136,138] liquid-air interfaces[143] and membranes.[144] It has been proposed that the acceleration of the aggregation could result from high local concentration of peptides because of the absorption and preferred orientations and conformations of the absorbed peptides ready for nucleation,[145,146] but the molecular basis of the aggregation acceleration remains largely unclear.

3.2.1 Protein/peptide-surface interactions

Numerous experimental studies have suggested that the chemical nature of both peptides/proteins and surfaces play a critical role in modulating the protein aggregation. As an example, it has been shown that a type of hydrophobic copolymeric NPs can promote the nucleation of β-sheet-rich β_2-microglobin but inhibit the nucleation of Aβ and hIAPP.[136,137,141] The lag phase of Aβ aggregation can, however, be shortened by another type of hydrophobic nanoparticles.[147] Moreover, other surface types, such as inorganic TiO_2 NPs and hydrophobic, hydrophilic and charged self-assembled mono-layer (SAM) have all been shown to promote the nucleation of Aβ.[148,149]

Atomistic simulations have been used to probe the peptide-surface interactions that drive the absorption process as this type of simulation includes full atomic details of both peptides and surfaces and an understanding of the distinct nucleation behaviors observed for diverse peptides and surfaces. For instance, Zheng and coworkers simulated the binding between several types of SAMs and pentameric Aβ42 in the fibrillar conformation.[149] On the hydrophobic SAM, the bound Aβ42 pentamer was oriented to form tight contacts between its C-terminal β-sheet and the surface whereas no particular orientation was preferred on the hydrophilic or charged SAMs. Although hydrophobic interactions were essential for all the Aβ-SAM systems studies, hydrophobic and electrostatic interactions were found to contribute differently to different systems. Zheng et al. investigated later the initial absorption of fibrillar dimer to hexamer on the SAMs.[150] Upon binding, the dimer quickly lost its fibrillar structures but the fibrillar structures of the larger oligomers remained stable, indicating that the trimer could be the smallest seed for aggregation on the SAMs.

Inorganic NPs represent another type of surfaces that can promote or inhibit peptide aggregation. Through atomic simulations, Corni and coworkers investigated the interactions between Aβ40/42 and the surface of Au NPs.[151] The T-REMD and H-REMD methods were used to enhance sampling of peptide conformations in solution and on surfaces, respectively. Their results showed a considerable structural change of both peptides upon the absorption on the surface. A large fraction of intramolecular contacts formed in solution was replaced with contacts between the peptides and the surface. The aromatic and HB interactions between the peptides and the surface allowed segments 17–21 and 30–40/42 to adopt stand-alone β-strand structures. There is in fact a higher chance to find fibrillar conformations of the peptides on the surface than in solution, and this could explain why the Au NPs can facilitate the Aβ aggregation. On the other hand, Mani and coworkers showed that positively charged Au NPs had no effect on Aβ aggregation while negatively charged NPs inhibited the process.[152] To provide a molecular explanation, the authors conducted the simulations of binding of Aβ40 to these two types of Au NPs. It turned out that only the negatively charged Au NPs can form salt bridge interactions with K16 and K28 of Aβ, sequestering ^{16}KLVFF^{20} from a β-strand structure which otherwise forms the core of fibrils.

Membranes formed by lipid molecules are ubiquitous in biological systems and provide surfaces for aggregation of proteins and peptides.[144] As such, the interactions between peptides and lipid membranes have also been extensively studied by molecular simulations. Davis and Berkowitz performed atomic simulations to examine the absorption of Aβ42 on the membrane surface of zwitterionic lipid (DPPC) and an anionic lipid (DOPS). They found that Aβ is bound tightly to the DPPC membrane, partly embedding itself into the membrane to enable hydrophobic interactions with the tail region of the membrane. On the other hand, Aβ formed less tight electrostatic contacts with the DOPS membrane and remained more exposed. The authors hypothesized that the peptide bound on the DOPS membrane would interact more easily with other peptides than on the DPPC membrane, which is consistent with the greater ability of the DOPS membrane to accelerate the Aβ aggregation.[153] Subsequently, these authors and others showed that the absorbed Aβ40/42 tended to adopt coiled structures due to strong peptide-lipid interactions.[154,155] However, a helix-turn-helix conformation of the absorbed Aβ40/42 was also proposed by Thirumalai and coworkers based on their REMD simulations with an implicit membrane model.[156] In addition, Zheng and coworkers showed through atomic simulations that hIAPP prefers helical structures on the surfaces of POPC and POPC/POPE to allow its basic sidechains to interact favorably with phosphate groups of the lipid molecules.[157]

3.2.2 Insights from simulations of nucleation processes on surfaces

Although the simulations of absorption of proteins and peptides gave first glimpses of the mechanism of the surface-mediated protein aggregation, it is necessary to model the assembly process on the surfaces to obtain a complete mechanistic picture.

As nucleation of peptides like Aβ occurs on a long timescale (>ms) and could involve complex conformational rearrangements, atomic simulations of this process is still challenging. Despite this difficulty, atomic simulations have been used to probe very early stage of the peptide aggregation on surfaces. In their studies of trimer formations of Aβ42 on the DPPC membrane, Mu and coworkers started simulations with three peptide chains in a NMR helical conformation that were placed on the membrane surface and separated by a short distance.[158] After 1 μs simulation, the helix in the C- and N-terminal parts of the peptides was unfolded and an intermolecular β-sheet was found in these parts. In another simulation study, three Aβ42 were added sequentially into the simulation, letting each one diffuse freely and bind to the membrane surface.[159] A trimer structure similar to that defined by Mu et al. was observed.

Alternatively, short peptides have also been used as simple model systems with which to simulate events beyond initial coalescence. Lyubchenko and coworkers studied the dimerization of Aβ14–23 in the presence of mica and DLPE membrane surfaces, both of which had promoted the aggregation of the peptide in experiments.[160] A similar dimerization path was observed in both cases: one peptide was first absorbed rapidly on the surface, followed by a transition from a short helix into a bend structure; the other peptide was later recruited onto the surface and interacted with the first peptide to form a dimer. To understand why citrate-covered Au NPs can promote aggregation of Sup35, Risselada et al. conducted simulations of oligomerization of a ∼10 segment of Sup35 (GNNQQNY) in the presence of a citrate-covered Au surface.[161] They found that the peptides were initially absorbed on the surface through electrostatic interactions between the citrate groups and the N-terminus of the peptides. As a result, the bound peptides were aligned parallel to each other and subsequently gave rise to parallel β-sheet structures. This result supports the notion that the surface can reduce the entropy cost of nucleation by properly orientating the absorbed peptides.

The work by Pomes and coworkers is particularly interesting.[162] They simulated the aggregation of $(GA)_4$ and $(GV)_4$ on the water-octane interface, a model system simple enough for them to observe the assembly process of tens of peptides through atomic simulations over hundreds of nanoseconds. Two nucleation pathways were identified (Fig. 4A): (1) at low concentrations, the aggregation on the surface followed a mechanism reminiscent of the canonical nucleation polymerization,[163] proceeding through initial absorption of individual peptides and their structural transition into β-strands, followed by diffusion and association of the absorbed peptides; (2) at high concentrations, the disordered oligomers formed in solution were instead absorbed first on the interface and then converted into nuclei by the surface.

CG models have been employed to explore the mechanism of surface-mediated protein aggregation on much longer timescales. Auer and coworkers employed a tube model to simulate the aggregation of peptides in the presence of a spherical NP.[164] Through ms-long simulations, they found that the aggregation was catalyzed by the NP *via* a condensation-ordering mechanism, similar to the second mechanism proposed by Pomes et al.[162] In the CG study by Shea and coworkers,[165] a united-residue model[166] was employed to investigate peptide aggregation on a solid-like surface composed of attractive particles arranged in a square lattice. Both (I) the surface-mediated nucleation-polymerization and (II) the condensation-ordering pathways were observed. In addition, the authors reported a third pathway (III) in which an ordered β-sheet arose in solution and bound to the surface (Fig. 4A). Moreover, the model employed in this study had two key parameters that could be adjusted to probe how the strength of peptide-surface interactions and conformational stiffness of peptides affected aggregation. The authors found that the peptides with a stiffer backbone preferred pathway (III) over (II), and where there was a stronger interaction between the peptides and the surface, pathway (I) was preferred over (II) and (III). This was consistent with the experimental observation that the nature of surface affected aggregation.[139,140,167] Using the same model, Shea and coworkers investigated the peptide aggregation on the CG lipid bilayer.[168] The observed assembly pathways were similar to those of aggregation on a solid surface. However, the thermal fluctuation of the membrane led to the formation of β-barrel intermediates which were absent on the solid surface. Also, the lack of structural anisotropy on the membrane rendered the aggregation much less cooperative and slower than on the solid surface.

The surface effect on aggregation kinetics of peptides was also probed in the CG studies of Vácha et al. and Ding et al.[169] Vácha et al. modeled amyloidogenic peptides as two-state patchy spherocylinders,[170] one representing coil state and the other representing aggregation-prone β-state. The fluctuation between these two states was sampled with a dynamic Monte Carlo method.[171] With this model, they showed that the aggregation rate depended on the strength of attraction between surface and the peptides. The nucleation and elongation were accelerated on a strongly attractive surface but, surprisingly, a weakly-attractive surface could retard aggregation. The authors explained that when the peptides had a high propensity for aggregation, the weakly attractive surface did not allow intermolecular aggregation but instead slowed down bulk-phase aggregation by depletion of the peptides in solution. This computational rationalization was further validated by these authors with the measured aggregation kinetics of Aβ/α-synuclein-surface systems. On the other hand, the peptide aggregation could also be retarded when the attraction between the peptides and surface was too strong, as demonstrated by the CG study of Ding et al.[172] Using an intermediate-level CG model, they showed that the surface with too strong an attraction could decrease stability of the fibrils and, more importantly, reduce the lateral diffusion of the peptides. This finding was in line with the experimental observation that the sufficient translational mobility of peptides on surface is a key requirement for efficient surface-mediated aggregation.[173–175]

The simulation studies so far have depicted a complex mechanistic picture of peptide nucleation on surfaces in which the nucleation pathways and kinetics are determined by various factors such as intrinsic aggregation propensity of peptides, peptide-surface interactions, orientational and conformational preferences of adsorbed peptides and their mobility. These factors can be sensitive to the chemical nature of both surfaces and peptides. Such mechanistic complexity is probably inherited by secondary nucleation on the surface of fibrils.

3.3 Computational insights into fibril-dependent secondary nucleation

3.3.1 *Insights from CG simulations of secondary nucleation*

Despite the aforementioned similarities, there are some basic differences between nucleation of peptides/protein on surfaces and fibril-dependent secondary nucleation. First, nuclei and fibrils that provide surfaces are both made of the same type of molecules. Once nuclei are formed on the fibril surface, they need to be released into solution to grow into new fibrils, providing significantly more catalytic surfaces. Thus, the detachment of nuclei from the fibril is a key step of secondary nucleation that must be considered in mechanistic studies. In addition, with the cross-β arrangement of constituent peptides, the surface fibrils display unique structural features (Fig. 3): (a) in general, a fibril has a prism shape with multiple lateral faces, each of which extends in the direction of the fibril axis but is narrow (several nm) in width; (b) there is roughly a 5 Å periodicity of the fibril surface in the direction of the fibril axis due to the repeated β-sheet packing of peptide chains in the fibril; (c) the parallel, in-register stack of peptides leads to the presence on the surface of columns of side chains from the same amino acid types, aligned in the direction of the fibril axis. It is crucial to understand how these structural features of fibril surface could dictate the nucleation behavior of peptides.

To probe the mechanism of secondary nucleation in general, Anwar and coworkers conducted simulations with solute and solvent molecules represented by a single LJ particle model.[176,177] Their simulations started with a crystal slab of the solute in a solution of the same solute at different degrees of supersaturation. The secondary nucleation was not observed either at low supersaturation or at high supersaturation. Only at an intermediate level of supersaturation could the nucleation occur on the slab surface *via* the association of spherical amorphous clusters of solute with the slab and subsequent structural change of the clusters into crystallites upon contacts with the slab, a mechanism reminiscent of the condensation ordering. Moreover, these newly formed crystallites could also catalyze the nucleation of other spherical clusters in the solution. Intriguingly, the contact area between the slab surface and the newly-formed crystallites and between the seed and child crystallites was small, implying that these nuclei, once formed, were bound weakly and could be readily released by shear force in agitated fluid, consistent with earlier studies.[178] These computational findings were also validated by the authors through the AFM studies of seeded crystallization of the drug bicalutamide.[177]

Recently, Frenkel and coworkers developed a minimal model to study the fibril-dependent secondary nucleation.[179] Similar to the work of Vácha et al.,[169] they employed the patchy spherocylinders to represent peptides. What was new in their model was the introduction of an intermediate state with an energy level between those of the coil and β-states, representing a conformation of the bound peptides. In its intermediate state, the peptide has a self-interaction stronger than its interaction with the fibril to ensure that oligomers can be detached from the surface, as was observed experimentally. The authors found that the soluble peptides could be attached to the surface of a pre-existing fibril where they could grow into a cluster. When the cluster reached a critical size, it turned into an oligomer with the intermediate states and then detached into solution, eventually forming a nucleus in the β-state (Fig. 4B). There exists a narrow range of interaction strength between the peptides and the fibrils for the secondary nucleation to take place. When the peptide-surface interaction was too weak, the primary nucleation events were dominant; when it was too strong, it was more difficult for the bound peptides to transform into the intermediate state because they needed to form a critical cluster with a bigger size, which was, however, limited by the geometry of the fibril surface. As a result, the fibril surface was saturated with bound peptides in the coil state. The results presented in this study explained why the kinetics of secondary nucleation was sensitive to mutations and external conditions such as pH and salt concentrations that may alter peptide-surface interactions.[25,179–181]

It is clear that the CG simulation studies have been capable of shedding light on the fundamental mechanistic picture enabling us to understand various observations about the fibril-dependent secondary nucleation. Nevertheless, it is still necessary to characterize in greater detail, this process which relies on simulations with higher resolutions.

3.3.2 Peptide-fibril interactions characterized by simulations at high resolution

Given the crucial role of interactions between amyloidogenic peptides and fibril surfaces that is played in the secondary nucleation, there have been atomic simulation studies aimed at understanding the molecular basis of peptide-fibril interactions. The main mechanistic questions of interest in these studies usually included the fibril-binding region of the peptide, the peptide-binding site on the fibril, the key interaction types that drive the peptide-fibril binding and the conformational preference of the bound peptide. Just as in the studies of the fibril elongation, the computational simulations of the secondary nucleation need to deal with the complex conformational space of the amyloidogenic peptides both in solution and on the surface. Moreover, compared to the fibril ends, the lateral surface of a fibril has a much larger area that could harbor the potential substrate-binding sites but little is known about the location and the nature of these sites. Hence, simulations of peptide binding to the fibril surface represent an even greater sampling challenge.

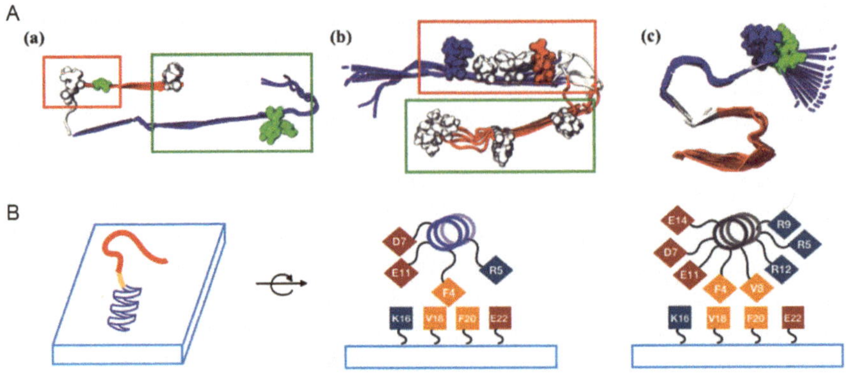

Fig. 5 (A) Various peptide-binding sites of Aβ fibril predicted by computational studies using different fibril models. (a) Two predicted peptide-binding sites using Aβ17–42 fibrils (PDB ID: 2BEG). Shown in the red box is the A30-L34 region that, as suggested by Barz and Strodel, recognized Aβ42[182] while the green box denotes Y10 and Y41 which were found to produce tau fragments by Qi et al. who used an extended Aβ42 fibril model based on the Aβ17–42 fibril.[183] (b) Two predicted binding sites on the Aβ9–40 fibril surface (PDB ID: 2LMN). The red box denotes K16, V18, F20 and E22 of the N-terminal fibril that was found to be the Aβ40-binding site in the work of Han et al.[184] The green box denotes I31, M35 and V40 of the C-terminal fibril that were shown to interact with Aβ9–40 monomer, dimer, trimer and tetramer in fibril-like conformations in the work of Schwierz et al.[185] (c) The Aβ42-binding sites on Aβ11–42 fibril surface (PDB ID: 2MXU) predicted by Bellaiche and Best.[186] The key binding residues on the surface including Q15 and K16 are shown in green and blue ellipsoids, respectively. (B) Schematic illustration of the binding poses of the N-terminal region of Aβ40 and the designed inhibitor on the Aβ9–40 fibril surface according to the work of Han et al.[184,187] The N-terminal region of Aβ40 is shown as a blue spring. The other parts of Aβ40 are shown as pink and red threads. The inhibitor is shown as a gray spring.

So far, the interactions between Aβ and its fibrils have been most extensively studied with simulations (Fig. 5A). One of the earliest studies is the work of Barz and Strodel,[182] who conducted multiple microsecond atomic simulations with an implicit solvent representation to investigate how free Aβ42 monomers are adsorbed onto an Aβ17–42 fibril segment (PDB ID: 2BEG). They found that the Aβ peptides could either assemble into oligomers (i.e., dimers to tetramers) first and attached to the fibril surface or individually bind to the surface first and then coalesce with other adsorbed peptides to give rise to bound oligomers (Fig. 4B). These observed pathways were consistent with the previous CG studies of the early stage of peptide aggregation on fibril or other types of surfaces (Fig. 4A).[165,179] Both the monomers and the oligomers were found to bind predominantly to the C-terminal part of the fibril, especially on the surface of segments 30–34.

Binding of the oligomers to the loop segment of the fibril was also observed. The Aβ peptides used mainly their C-terminal regions to form hydrophobic contacts with the fibril surface. Although the limitation on simulation length did not allow observation of the actual conformational rearrangement required for nucleation, the monomer was observed to tend to be stretched in order to form more extensive interactions with the fibril surface. In addition, the bound monomer, when assuming a compact shape, was observed to slide on the fibril surface. Its diffusion constant was close to that of free diffusion in bulk phase, suggesting that the bound monomer can be mobile on the surface, which was thought to be a crucial condition for efficient surface-induced nucleation.[173–175]

To gain insight into the thermodynamic signature of lateral association between Aβ and its fibril, Schwierz et al. carried out umbrella sampling simulations with atomic force field to evaluate the free energy change of detachment of Aβ monomer to tetramer from the Aβ9–40 fibril (Fig. 5A).[185] These simulations started with the monomer and oligomers that adopted the U-shaped fibrillar shape and were attached to the C-terminal part of the fibril as in the Aβ9–40 fibril model containing two filaments (PDB: 2LMN). The simulations showed that the U-shaped fibrillar conformation of the monomer was not stable and adopted collapsed-coils in solution and stretched structures on the fibril surface, consistent with the observation of Barz and Strodel.[182] The fibrillar conformation of the dimer, trimer and tetramer were stable both on the fibril surface and in solution. Free energy calculations showed that the dimer has the weakest affinity for the fibril surface, suggesting that the dimer could be readily removed from the fibril surface. The affinities of the trimer and the tetramer, on the other hand, were too strong and their detachment was thought to be extremely unlikely. The decomposition of the binding free energy further suggested that the binding of Aβ and the C-terminal surface of the fibril was driven by the gain of entropy from the release of hydration water molecules, which was, however, opposite to the experimental conclusion based on kinetic analysis that the adsorption of Aβ42 on its own fibril surface was driven by a large gain of enthalpy.[134]

To improve the sampling efficiency, Han et al. employed the hybrid-resolution model PACE to conduct hundreds of microseconds of the binding of Aβ40 to the Aβ9–40 fibril.[184] With the temperature-based REMD technique, they examined the conformational spaces of the bound and unbound peptide and the interactions between the peptide and different parts of the fibril surface. The simulations revealed primarily two Aβ-binding sites, one located in region ^{30}AIIGL^{34} that had also been found by Braz and Strodel,[182] and the other located in region ^{16}KLVFFAE^{22}. The Aβ peptide was five times more likely to bind to region ^{16}KLVFFAE^{22} than to the other region. The standard binding free energy was calculated to be ~33 kJ/mol, which predicted a half saturation concentration of 2–5 μM, consistent with the corresponding experimental value (~6 μM).[188] The decomposition of the free energy showed that the binding enthalpy (ΔH) and entropy (TΔS) were approximately −65 and ~33 kJ/mol, respectively, which also agreed well with the recent experimental data (−67 and 40 kJ/mol, respectively) suggesting that the Aβ-fibril binding was enthalpy driven.[134] The major binding site consisted of four columns of side chains of K16, V18, F20 and E22. When bound with this site, the Aβ was also observed to be stretched although strikingly, its N-terminal region (3–14) adopted a helical structure. The bound peptide formed both electrostatic interactions with the K16 and E22 columns and hydrophobic interactions with the V18 and F20 columns. This result highlighted the role of electrostatic interactions in stabilizing the binding between Aβ and the fibril.

The further support to the importance of electrostatic interactions for the Aβ-fibril binding comes later from the work of Bellaiche and Best who investigated the binding of Aβ42 to its fibril (Fig. 5A).[186] In particular, they employed a recently derived NMR Aβ42 fibril model[189] that differs from that of Aβ17–42 used by Barz and Strodel.[182] With the Hamiltonian-REMD method,[190] they explored the binding of Aβ42 on different sites of this fibril. The Aβ42 was found to predominantly attach to the N-terminal part of the fibril, which is different from the Aβ binding on the C-terminal part of the Aβ17–42 fibril observed by Barz and Strodel, suggesting that the Aβ-fibril interactions may be sensitive to fibril polymorphism. The Aβ-fibril interactions were formed mainly between the side chain columns of Q15 and K16 of the fibril and E22 and D23 of the peptide, indicating that the electrostatic interactions played a major role. Indeed, the free energy calculation further showed that the binding enthalpy and entropy were approximately −90 and ~70 kJ/mol, respectively, suggesting again that the Aβ42-fibril binding was not driven by the entropy effect. These results agree qualitatively with the work of Han and coworkers.[184] Recent kinetic studies of Aβ aggregation suggested that the disease-related mutations that alter the charge states of residues 22 and 23 could significantly accelerate the secondary nucleation process.[181] The results from this work as well as that of Han and coworkers may provide an intriguing explanation: if the desorption of Aβ, as shown in previous studies at high salt concentrations or low pH,[191] is rate limiting, the weakened electrostatic interactions between Aβ and the fibril may lead to a faster turn-over, thereby facilitating the secondary process.[179]

Amyloid cross-seeding, a phenomenon in which the amyloid of one protein nucleates the aggregation of another, has important implications in the pathology of multiple diseases.[192] The Aβ fibril is known to be capable of seeding the nucleation of tau proteins, inducing tau pathology.[193] To understand the molecular basis of this cross-seeding effect, Qi et al. conducted REMD simulations to investigate the binding between the Aβ1–42 fibril (constructed based on the Aβ17–42 fibril) and a 120–160 amino acid fragment of tau corresponding to the core of tau amyloid.[183] Unlike the Aβ-fibril binding motif shown earlier, this tau fragment was found to simultaneously bind to both the N- and C-terminal parts of the fibrils. In particular, the specific interactions of the peptide around the columns of Y10 and I41 contributed greatly to the overall stability of the binding. The peptide, once bound, was also observed to unfold and adopt stretched conformations, exposing its two key regions (^{275}VQIINK^{280} and ^{306}VQIVYK^{311}) that were crucial for the nucleation of tau. In addition, the bound peptide displayed a much higher content of helix than β-sheet, which corroborated the α-helix formation of the bound Aβ that was observed by Han and coworkers.[184] The α-helix has been suggested to be an important intermediate of protein nucleation in solution and on surfaces.[144,194] The authors hypothesized that the exposure of the nucleating region and the presence of transient helical structures would lead to a faster nucleation.

More recently, the cross-seeding of Aβ40 by the fibrils formed by its fragment, Aβ16–22, was reported in a combined experimental and computational study by Bunce et al.[195] To gain molecular insights into the cross-seeding mechanism, the authors simulated the binding of Aβ40 to a Aβ16–22 fibril. An intermediate-resolution model, namely, the PRIME force field,[196] was employed together with discrete molecular dynamics technique to accelerate simulations, extending simulations on timescale of around 100 μs. According to their simulations, soluble Aβ40 oligomers that have largely random coils quickly associated with the fibril surface and the C-terminus of Aβ40 exhibited a greater tendency to contact directly with the fibril surface. The bound oligomers were gradually converted into structures rich in anti-parallel β-sheet. This "bind and reorganization" mechanism is largely consistent with what was reported by Barz and Strodel,[182] and Schwierz et al.[185] Nonetheless, the extended simulations allowed the authors of this work to observe events that took place on longer timescales.

The β-sheet oligomers on the fibril surface could further recruit monomers from solution and convert them into β-sheet structures, leading to a further increase in β-sheet contents of Aβ40.

Given the importance of peptide-fibril interactions for the secondary nucleation process, they have been advocated as intervening in these interactions to control the generation of toxic oligomers.[33,181] To this end, an effective approach is to use certain key structural motifs of bound peptides found at the binding interface as templates with which to design inhibitors.[197] The discovery of such motifs, in theory, could be facilitated with computational studies such as those reviewed here. Although in most of these studies the bound peptides were usually stretched and lacking in specific structural features, the α-helix emerges as an intriguing recognition motif as demonstrated by the studies of Aβ/tau-fibril interactions by Qi et al. and Han et al.[183,184] The importance of the α-helical motif is also supported by a recent experimental and computational study by Baram et al.[198] who showed that the insulin, a natural aggregation inhibitor of amylin, used a helical segment (^{10}HLVEALYLV^{19}) of its B chain to recognize the surface of amylin fibrils.

As a part of the work by Han and coworkers,[184] the ability of the N-terminal region (^{3}EFRHDSGYEVHH^{14}) of Aβ to recognize the fibril was further investigated (Fig. 5B). The analysis revealed that this region, when adopting an α-helix, arranged its acidic side chains (D7 and E11) on one side of the helix to interact with the K16 column of the fibril, and its basic side chain (R5) on the opposite side to interact with the E22 column. The hydrophobic side chain F4 was placed in between the acidic and the basic sides, forming contacts with the V18 and F20 columns. Through a series of *in silico* mutations in this helical region and the binding simulations with atomic force fields, it was demonstrated that a short peptide (Ac-AFRADVRAERAE-NMe) could recognize the fibril surface in the same way as the N-terminal region of Aβ40 with a greatly enhanced affinity (Fig. 5B). Subsequently, Jiang et al. employed the proposed peptide to design inhibitors of the nucleation of Aβ40.[187] They postulated that if an α-helix was critical, the restriction of the peptide in this conformation by means of chemical modification would reduce the entropy cost of the folding of the α-helix and thus promote its binding to the fibril surface. In accordance with their expectations, once a helix-promoting N-capping group was introduced, the peptide displayed improved helicity and affinity for the Aβ40 fibril. Furthermore, this peptide was shown to be able to specifically retard the secondary nucleation of Aβ40, reduce its oligomer level and to mitigate cytotoxicity caused by its aggregation.

In summary, the computational studies so far have characterized the adsorption of peptides on fibril surface in great detail. Remarkably, even in the case of peptides binding to the Aβ fibrils, there is a large variation in bound structures of the peptides, the location of peptide-binding sites, the key peptide-fibril interactions and the thermodynamic signature of the binding for different Aβ isomers and fibril morphologies. The differences in the initial peptide-fibril interactions might further extend to the later steps of the secondary nucleation, resulting in distinct nucleation behaviors. This is in line with the general notion that the surface-induced aggregation is sensitive to the chemical nature of surfaces and amyloidogenic peptides. This sensitivity implies that the peptide-fibril interactions might be specific to both peptide sequences and fibril structures, the latter of which not only depend on peptide sequences but also on external conditions. Notably, recent kinetic experiments demonstrated that despite the sequence similarity between Aβ40 and Aβ42, Aβ42, which contained only two additional amino acids at its C-terminus, exhibited much faster kinetics of secondary nucleation. Moreover, neither Aβ42 fibrils were able to catalyze the nucleation of Aβ40, nor Aβ40 fibrils were able to catalyze the nucleation of Aβ42.[199] Given that Aβ42 is much more prone to aggregate and more toxic than Aβ40, there is a great need to understand the molecular basis of difference in secondary nucleation behaviors between the two isomers. Computational studies to this end are still lacking. Comparative computational studies of Aβ40/42-fibril interactions may provide clues to solve this mechanistic puzzle.

4. Summary and outlook

Secondary nucleation and fibril elongation are two essential processes of protein/peptide aggregation, the former process, with fibril surface as catalyst, producing efficiently aggregation nuclei and the latter process responsible for growth of catalytic surfaces. Given the importance of these two processes, it is important to understand their molecular basis. With recent development in the experimental techniques of characterization and the kinetic theory of protein aggregation, the thermodynamic and kinetic signatures of these processes have been individually characterized, but there is still a gap that needs to be filled in with atomic details. The computational studies presented herein have provided valuable atomic insights in this regard.

For fibril elongation, a consensus mechanism supported by the extensive computational studies is the "dock-and-lock" mechanism in which soluble monomers, driven mainly by an entropy effect, are rapidly adsorbed on the fibril ends, followed by a slow structural rearrangement of the bound monomers, which is hindered by the need to break interactions within the monomers or misaligned monomer-fibril interactions. As for the fibril-dependent secondary nucleation that has been discovered more recently, its mechanism is less well investigated computationally. The simulation studies so far have unraveled the key factors that could be essential for an efficient secondary nucleation, including a proper interaction strength between the peptides and the fibril surface, the preference of the bound peptides for amyloidogenic conformations, the unhindered movement of the peptides on the fibril surface and their facile detachment for a fast turn-over. All these factors are sensitive to peptide sequences and fibril structures. Many of these mechanistic features are consistent with the more general mechanism of surface-induced nucleation which has been extensively studied.

Despite the advances in understanding the basic mechanism of secondary nucleation and fibril elongation, they raise also more mechanistic questions that are worthy of future investigation. In the majority of computational studies reported so far, a single filament from the early models of fibrils was often chosen as a model system with which to investigate fibril-dependent processes. However, a growing number of amyloid structures resolved recently have illustrated much more diverse and complex architectures of amyloids than previously expected. This amyloid polymorphism has been regarded as a fundamental structural feature of amyloid fibrils that is of both biological and pathological significance. It is thus suggested that various recently-resolved fibril models should be included to obtain more realistic insights into the processes. Also, more attention may need to be paid in future to large proteins like tau and α-synuclein. With few exceptions, the studies that have been conducted focused on short amyloidogenic peptides such as Aβ and Amylin whose conformations can be sampled more easily. Nonetheless, the study of the larger proteins can not only provide insights into the aggregation mechanisms of other disease-causing proteins but also into the generic mechanism of protein aggregation. Besides, the computational studies of secondary nucleation have only revealed the atomic details of adsorption of monomers or ordered oligomers that take place very early during the secondary nucleation process. The full atomic details of the entire process are yet to be determined, and are indispensable for understanding secondary nucleation. Finally, protein aggregation takes place on mesoscopic scales in crowded environment. In such systems, hydrodynamic interactions[200] and crowing effect[201,202] play essential roles in determining kinetics. Thus, taking into account these factors in modeling of fibril elongation and secondary nucleation would furnish more realistic insights into these processes.

The ultimate goal of computational studies is more than merely providing a molecular explanation of experimental observations regarding aggregation mechanism. More importantly, the key mechanistic findings revealed from these studies should be able to aid experimental researchers in developing new means of modulating protein aggregation. It is our hope that this review would inspire more future computational works toward this goal.

Acknowledgments

We thank the National Science Foundation of China (21977011 and 21673013) and the Shenzhen STIC (KQTD2015032709315529 and JCYJ20170818085409785) for financial support.

References

1. Chiti F, Dobson CM. Protein misfolding, functional amyloid, and human disease. *Annu Rev Biochem*. 2006;75(1):333–366.
2. Chiti F, Dobson CM. Protein misfolding, amyloid formation, and human disease: a summary of progress over the last decade. *Annu Rev Biochem*. 2017;86(1):27–68.
3. Selkoe DJ. Folding proteins in fatal ways. *Nature*. 2003;426(6968):900–904.
4. Westermark GT, Fändrich M, Westermark P. AA amyloidosis: pathogenesis and targeted therapy. *Annu Rev Pathol*. 2015;10(1):321–344.
5. Selkoe DJ, Hardy J. The amyloid hypothesis of Alzheimer's disease at 25 years. *EMBO Mol Med*. 2016;8(6):595–608.
6. Mukherjee A, Morales-Scheihing D, Butler PC, Soto C. Type 2 diabetes as a protein misfolding disease. *Trends Mol Med*. 2015;21(7):439–449.
7. Eisenberg D, Jucker M. The amyloid state of proteins in human diseases. *Cell*. 2012;148(6):1188–1203.
8. Hamley IW. The amyloid B peptide: a chemist's perspective. role in Alzheimer's and fibrillization. *Chem Rev*. 2012;112(10):5147–5192.
9. Bachmann MF, Jennings GT, Vogel M. A vaccine against Alzheimer's disease: anything left but faith? *Expert Opin Biol Ther*. 2019;19(1):73–78.
10. Mehta D, Jackson R, Paul G, Shi J, Sabbagh M. Why do trials for Alzheimer's disease drugs keep failing? A discontinued drug perspective for 2010-2015. *Expert Opin Investig Drugs*. 2017;26(6):735–739.
11. Doig AJ, del Castillo-Frias MP, Berthoumieu O, et al. Why is research on Aβ failing to give new drugs for Alzheimer's disease? *ACS Chem Nerosci*. 2017;8(7):1435–1437.
12. Hayden EY, Teplow DB. Amyloid β-protein oligomers and Alzheimer's disease. *Alzheimers Res Ther*. 2013;5(6):60.
13. Michaels TCT, Šarić A, Habchi J, et al. Chemical kinetics for bridging molecular mechanisms and macroscopic measurements of amyloid fibril formation. *Annu Rev Phys Chem*. 2018;69(1):273–298.

14. Kusumoto Y, Lomakin A, Teplow DB, Benedek GB. Temperature dependence of amyloid β-protein fibrillization. *Proc Natl Acad Sci U S A*. 1998;95(21):12277–12282.
15. Linse S. Monomer-dependent secondary nucleation in amyloid formation. *Biophys Rev*. 2017;9(4):329–338.
16. Kjaergaard M, Dear AJ, Kundel F, et al. Oligomer diversity during the aggregation of the repeat region of Tau. *ACS Chem Nerosci*. 2018;9(12):3060–3071.
17. Sang JC, Lee J-E, Dear AJ, et al. Direct observation of prion protein oligomer formation reveals an aggregation mechanism with multiple conformationally distinct species. *Chem Sci*. 2019;10(17):4588–4597.
18. Fändrich M. Oligomeric intermediates in amyloid formation: structure determination and mechanisms of toxicity. *J Mol Biol*. 2012;421(4):427–440.
19. Bernstein SL, Dupuis NF, Lazo ND, et al. Aβ protein oligomerization and the importance of tetramers and dodecamers in the aetiology of Alzheimer's disease. *Nat Chem*. 2009;1:326.
20. Benilova I, Karran E, De Strooper B. The toxic Aβ oligomer and Alzheimer's disease: an emperor in need of clothes. *Nat Neurosci*. 2012;15(3):349–357.
21. Haass C, Selkoe DJ. Soluble protein oligomers in neurodegeneration: lessons from the Alzheimer's amyloid β-peptide. *Nat Rev Mol Cell Biol*. 2007;8(2):101–112.
22. Ono K, Condron MM, Teplow DB. Structure–neurotoxicity relationships of amyloid β-protein oligomers. *Proc Natl Acad Sci U S A*. 2009;106(35):14745–14750.
23. Cline E, Assunção Bicca M, Viola K, Klein W. The Aβ oligomer hypothesis: beginning of the third decade. *J Alzheimers Dis*. 2018;64:s567–s610.
24. Aguzzi A, O'Connor T. Protein aggregation diseases: pathogenicity and therapeutic perspectives. *Nat Rev Drug Discov*. 2010;9(3):237–248.
25. Buell AK, Galvagnion C, Gaspar R, et al. Solution conditions determine the relative importance of nucleation and growth processes in α-synuclein aggregation. *Proc Natl Acad Sci U S A*. 2014;111(21):7671–7676.
26. Esler WP, Stimson ER, Jennings JM, et al. Alzheimer's disease amyloid propagation by a template-dependent dock-lock mechanism. *Biochemistry*. 2000;39(21):6288–6295.
27. Cohen SI, Linse S, Luheshi LM, et al. Proliferation of amyloid-β42 aggregates occurs through a secondary nucleation mechanism. *Proc Natl Acad Sci U S A*. 2013; 110(24):9758–9763.
28. Straub JE, Thirumalai D. Toward a molecular theory of early and late events in monomer to amyloid fibril formation. *Annu Rev Phys Chem*. 2011;62:437–463.
29. Teplow DB. On the subject of rigor in the study of amyloid β-protein assembly. *Alzheimers Res Ther*. 2013;5(4):39.
30. Kalia LV, Kalia SK, McLean PJ, Lozano AM, Lang AE. α-Synuclein oligomers and clinical implications for Parkinson disease. *Ann Neurol*. 2013;73(2):155–169.
31. Haataja L, Gurlo T, Huang CJ, Butler PC. Islet amyloid in type 2 diabetes, and the toxic oligomer hypothesis. *Endocr Rev*. 2008;29(3):303–316.
32. Michaels TCT, Lazell HW, Arosio P, Knowles TPJ. Dynamics of protein aggregation and oligomer formation governed by secondary nucleation. *J Chem Phys*. 2015;143(5):054901.
33. Tornquist M, Michaels TCT, Sanagavarapu K, et al. Secondary nucleation in amyloid formation. *Chem Commun*. 2018;54(63):8667–8684.
34. Cohen SIA, Arosio P, Presto J, et al. A molecular chaperone breaks the catalytic cycle that generates toxic Aβ oligomers. *Nat Struct Mol Biol*. 2015;22:207.
35. Joshi P, Chia S, Habchi J, Knowles TPJ, Dobson CM, Vendruscolo M. A fragment-based method of creating small-molecule libraries to target the aggregation of intrinsically disordered proteins. *ACS Comb Sci*. 2016;18(3):144–153.
36. Habchi J, Chia S, Limbocker R, et al. Systematic development of small molecules to inhibit specific microscopic steps of Aβ42 aggregation in Alzheimer's disease. *Proc Natl Acad Sci U S A*. 2017;114(2):E200–E208.

37. Aprile F, Sormanni P, Perni M, et al. Selective targeting of primary and secondary nucleation pathways in Aβ42 aggregation using a rational antibody scanning method. *Sci Adv*. 2017;3:e1700488.
38. Habchi J, Arosio P, Perni M, et al. An anticancer drug suppresses the primary nucleation reaction that initiates the production of the toxic Aβ42 aggregates linked with Alzheimer's disease. *Sci Adv*. 2016;2(2):e1501244.
39. Nasica-Labouze J, Nguyen PH, Sterpone F, et al. Amyloid β protein and Alzheimer's disease: when computer simulations complement experimental studies. *Chem Rev*. 2015;115(9):3518–3563.
40. Ilie IM, Caflisch A. Simulation studies of Amyloidogenic polypeptides and their aggregates. *Chem Rev*. 2019;119(12):6956–6993.
41. Naiki H, Hashimoto N, Suzuki S, Kimura H, Nakakuki K, Gejyo F. Establishment of a kinetic model of dialysis-related amyloid fibril extension in vitro. *Amyloid*. 1997;4(4):223–232.
42. Esler WP, Stimson ER, Ghilardi JR, et al. In vitro growth of Alzheimer's disease β-amyloid plaques displays first-order kinetics. *Biochemistry*. 1996;35(3):749–757.
43. Ban T, Hoshino M, Takahashi S, et al. Direct observation of Aβ amyloid fibril growth and inhibition. *J Mol Biol*. 2004;344(3):757–767.
44. Dyson HJ, Wright PE. Coupling of folding and binding for unstructured proteins. *Curr Opin Struct Biol*. 2002;12(1):54–60.
45. Sugase K, Dyson HJ, Wright PE. Mechanism of coupled folding and binding of an intrinsically disordered protein. *Nature*. 2007;447(7147):1021–1025.
46. Massi F, Straub J. Energy landscape theory for Alzheimer's amyloid β-peptide fibril elongation. *Proteins*. 2001;42:217–229.
47. Ban T, Yamaguchi K, Goto Y. Direct observation of amyloid fibril growth, propagation, and adaptation. *Acc Chem Res*. 2006;39(9):663–670.
48. Qiang W, Kelley K, Tycko R. Polymorph-specific kinetics and thermodynamics of β-amyloid fibril growth. *J Am Chem Soc*. 2013;135(18):6860–6871.
49. Zhang J, Muthukumar M. Simulations of nucleation and elongation of amyloid fibrils. *J Chem Phys*. 2009;130(3):035102.
50. Abeln S, Vendruscolo M, Dobson CM, Frenkel D. A simple lattice model that captures protein folding, aggregation and amyloid formation. *PLoS One*. 2014;9(1):e85185.
51. Li MS, Klimov DK, Straub JE, Thirumalai D. Probing the mechanisms of fibril formation using lattice models. *J Chem Phys*. 2008;129(17):175101.
52. Pellarin R, Caflisch A. Interpreting the aggregation kinetics of amyloid peptides. *J Mol Biol*. 2006;360(4):882–892.
53. Pellarin R, Schuetz P, Guarnera E, Caflisch A. Amyloid fibril polymorphism is under kinetic control. *J Am Chem Soc*. 2010;132:14960–14970.
54. Bellesia G, Shea JE. Effect of β-sheet propensity on peptide aggregation. *J Chem Phys*. 2009;130(14):145103.
55. Auer S, Meersman F, Dobson CM, Vendruscolo M. A generic mechanism of emergence of amyloid protofilaments from disordered oligomeric aggregates. *PLoS Comput Biol*. 2008;4(11):e1000222.
56. Sterpone F, Doutreligne S, Tran TT, et al. Multi-scale simulations of biological systems using the OPEP coarse-grained model. *Biochem Biophys Res Commun*. 2018;498(2):296–304.
57. Sterpone F, Melchionna S, Tuffery P, et al. The OPEP protein model: from single molecules, amyloid formation, crowding and hydrodynamics to DNA/RNA systems. *Chem Soc Rev*. 2014;43(13):4871–4893.
58. Cheon M, Hall CK, Chang I. Structural conversion of Aβ17-42 peptides from disordered oligomers to U-shape protofilaments via multiple kinetic pathways. *PLoS Comput Biol*. 2015;11(5):e1004258.

59. Wu C, Shea JE. Coarse-grained models for protein aggregation. *Curr Opin Struct Biol.* 2011;21(2):209–220.
60. Pellarin R, Guarnera E, Caflisch A. Pathways and intermediates of amyloid fibril formation. *J Mol Biol.* 2007;374(4):917–924.
61. Brown S, Fawzi NJ, Head-Gordon T. Coarse-grained sequences for protein folding and design. *Proc Natl Acad Sci U S A.* 2003;100(19):10712–10717.
62. Fawzi NL, Okabe Y, Yap EH, Head-Gordon T. Determining the critical nucleus and mechanism of fibril elongation of the Alzheimer's Aβ(1-40) peptide. *J Mol Biol.* 2007;365(2):535–550.
63. Sasmal S, Schwierz N, Head-Gordon T. Mechanism of nucleation and growth of Aβ40 fibrils from all-atom and coarse-grained simulations. *J Phys Chem B* 2016;120(47): 12088–12097.
64. Han W, Schulten K. Further optimization of a hybrid united-atom and coarse-grained force field for folding simulations: Improved backbone hydration and interactions between charged side chains. *J Chem Theory Comput.* 2012;8(11):4413–4424.
65. Han W, Schulten K. Fibril elongation by Aβ(17-42): kinetic network analysis of hybrid-resolution molecular dynamics simulations. *J Am Chem Soc.* 2014;136(35): 12450–12460.
66. Nguyen PH, Li MS, Stock G, Straub JE, Thirumalai D. Monomer adds to preformed structured oligomers of Aβ-peptides by a two-stage dock–lock mechanism. *Proc Natl Acad Sci U S A.* 2007;104(1):111–116.
67. Baumketner A, Krone MG, Shea J-E. Role of the familial Dutch mutation E22Q in the folding and aggregation of the 15-28 fragment of the Alzheimer amyloid-β protein. *Proc Natl Acad Sci U S A.* 2008;105(16):6027–6032.
68. Reddy G, Straub JE, Thirumalai D. Dynamics of locking of peptides onto growing amyloid fibrils. *Proc Natl Acad Sci U S A.* 2009;106(29):11948–11953.
69. O'Brien EP, Okamoto Y, Straub JE, Brooks BR, Thirumalai D. Thermodynamic perspective on the dock-lock growth mechanism of amyloid fibrils. *J Phys Chem B* 2009;113(43):14421–14430.
70. Schwierz N, Frost CV, Geissler PL, Zacharias M. Dynamics of seeded Aβ40-fibril growth from atomistic molecular dynamics simulations: kinetic trapping and reduced water mobility in the locking step. *J Am Chem Soc.* 2016;138(2):527–539.
71. Bacci M, Vymetal J, Mihajlovic M, Caflisch A, Vitalis A. Amyloid β fibril elongation by monomers involves disorder at the tip. *J Chem Theory Comput.* 2017;13(10):5117–5130.
72. Buell AK, Dhulesia A, White DA, Knowles TP, Dobson CM, Welland ME. Detailed analysis of the energy barriers for amyloid fibril growth. *Angew Chem Int Ed Engl.* 2012;51(21):5247–5251.
73. Carrotta R, Manno M, Bulone D, Martorana V, San Biagio PL. Protofibril formation of amyloid β-protein at low pH via a non-cooperative elongation mechanism. *J Biol Chem.* 2005;280(34):30001–30008.
74. Skerget K, Vilfan A, Pompe-Novak M, et al. The mechanism of amyloid-fibril formation by stefin B: temperature and protein concentration dependence of the rates. *Proteins.* 2009;74(2):425–436.
75. Knowles TPJ, Shu W, Devlin GL, et al. Kinetics and thermodynamics of amyloid formation from direct measurements of fluctuations in fibril mass. *Proc Natl Acad Sci U S A.* 2007;104(24):10016–10021.
76. Tarus B, Straub JE, Thirumalai D. Dynamics of Asp23 – Lys28 salt-bridge formation in Aβ10-35 monomers. *J Am Chem Soc.* 2006;128(50):16159–16168.
77. Thirumalai D, Reddy G, Straub JE. Role of water in protein aggregation and amyloid polymorphism. *Acc Chem Res.* 2012;45(1):83–92.
78. Su Z, Dias CL. Driving β-strands into fibrils. *J Phys Chem B.* 2014;118(37): 10830–10836.

79. Takeda T, Klimov DK. Probing energetics of Aβ fibril elongation by molecular dynamics simulations. *Biophys J.* 2009;96(11):4428–4437.
80. Rojas A, Maisuradze N, Kachlishvili K, Scheraga HA, Maisuradze GG. Elucidating important sites and the mechanism for amyloid fibril formation by coarse-grained molecular dynamics. *ACS Chem Nerosci.* 2017;8(1):201–209.
81. Han M, Hansmann UH. Replica exchange molecular dynamics of the thermodynamics of fibril growth of Alzheimer's Aβ42 peptide. *J Chem Phys.* 2011;135(6):065101.
82. Hwang W, Zhang S, Kamm RD, Karplus M. Kinetic control of dimer structure formation in amyloid fibrillogenesis. *Proc Natl Acad Sci U S A.* 2004;101(35): 12916–12921.
83. Nelson R, Sawaya MR, Balbirnie M, et al. Structure of the cross-β spine of amyloid-like fibrils. *Nature.* 2005;435(7043):773–778.
84. Lührs T, Ritter C, Adrian M, et al. 3D structure of Alzheimer's amyloid-β(1-42) fibrils. *Proc Natl Acad Sci U S A.* 2005;102(48):17342–17347.
85. Petkova AT, Ishii Y, Balbach JJ, et al. A structural model for Alzheimer's β-amyloid fibrils based on experimental constraints from solid state NMR. *Proc Natl Acad Sci U S A.* 2002;99(26):16742–16747.
86. Lu JX, Qiang W, Yau WM, Schwieters CD, Meredith SC, Tycko R. Molecular structure of β-amyloid fibrils in Alzheimer's disease brain tissue. *Cell.* 2013;154(6): 1257–1268.
87. Schor M, Vreede J, Bolhuis PG. Elucidating the locking mechanism of peptides onto growing amyloid fibrils through transition path sampling. *Biophys J.* 2012;103(6): 1296–1304.
88. Dellago C, Bolhuis PG, Geissler PL. Transition path sampling. *Adv Chem Phys.* 2002; 123:1–78.
89. Hoyer W, Cherny D, Subramaniam V, Jovin TM. Rapid self-assembly of α-synuclein observed by in situ atomic force microscopy. *J Mol Biol.* 2004;340(1):127–139.
90. Wordehoff MM, Bannach O, Shaykhalishahi H, et al. Single fibril growth kinetics of α-synuclein. *J Mol Biol.* 2015;427:1428–1435.
91. Ferkinghoff-Borg J, Fonslet J, Andersen CB, et al. Stop-and-go kinetics in amyloid fibrillation. *Phys Rev E.* 2010;82(1):010901.
92. Kellermayer MSZ, Karsai A, Benke M, Soós K, Penke B. Stepwise dynamics of epitaxially growing single amyloid fibrils. *Proc Natl Acad Sci U S A.* 2008;105(1): 141–144.
93. Watanabe-Nakayama T, Ono K, Itami M, Takahashi R, Teplow DB, Yamada M. High-speed atomic force microscopy reveals structural dynamics of amyloid β1-42 aggregates. *Proc Natl Acad Sci U S A.* 2016;113(21):5835–5840.
94. Colvin MT, Silvers R, Ni QZ, et al. Atomic resolution structure of monomorphic Aβ42 amyloid fibrils. *J Am Chem Soc.* 2016;138(30):9663–9674.
95. Fitzpatrick AWP, Falcon B, He S, et al. Cryo-EM structures of tau filaments from Alzheimer's disease. *Nature.* 2017;547(7662):185–190.
96. Ilie IM, den Otter WK, Briels WJ. The attachment of α-synuclein to a fiber: a coarse-grain approach. *J Chem Phys.* 2017;146(11):115102.
97. Ilie IM, Nayar D, den Otter WK, van der Vegt NFA, Briels WJ. Intrinsic conformational preferences and interactions in α-synuclein fibrils: insights from molecular dynamics simulations. *J Chem Theory Comput.* 2018;14(6):3298–3310.
98. Rojas A, Liwo A, Browne D, Scheraga HA. Mechanism of fiber assembly: treatment of Aβ peptide aggregation with a coarse-grained united-residue force field. *J Mol Biol.* 2010;404(3):537–552.
99. Liwo A, Czaplewski C, Pillardy J, Scheraga HA. Cumulant-based expressions for the multibody terms for the correlation between local and electrostatic interactions in the united-residue force field. *J Chem Phys.* 2001;115(5):2323–2347.

100. Rojas AV, Liwo A, Scheraga HA. Molecular dynamics with the United-residue force field: ab initio folding simulations of multichain proteins. *J Phys Chem B.* 2007;111(1):293–309.
101. Sugita Y, Okamoto Y. Replica-exchange molecular dynamics method for protein folding. *Chem Phys Lett.* 1999;314(1):141–151.
102. Gurry T, Stultz CM. Mechanism of amyloid-β fibril elongation. *Biochemistry.* 2014;53(44):6981–6991.
103. Kumar S, Rosenberg JM, Bouzida D, Swendsen RH, Kollman PA. THE weighted histogram analysis method for free-energy calculations on biomolecules. I. The method. *J Chem Theory Comput.* 1992;13(8):1011–1021.
104. Hoyer W, Grönwall C, Jonsson A, Ståhl S, Härd T. Stabilization of a β-hairpin in monomeric Alzheimer's Aβ peptide inhibits amyloid formation. *Proc Natl Acad Sci U S A.* 2008;105(13):5099–5104.
105. Sandberg A, Luheshi LM, Sollvander S, et al. Stabilization of neurotoxic Alzheimer amyloid-β oligomers by protein engineering. *Proc Natl Acad Sci U S A.* 2010;107(35): 15595–15600.
106. Sciarretta KL, Gordon DJ, Petkova AT, Tycko R, Meredith SC. Aβ40-lactam(D23/K28) models a conformation highly favorable for nucleation of amyloid. *Biochemistry.* 2005;44(16):6003–6014.
107. Cao Y, Jiang X, Han W. Self-assembly pathways of β-sheet-rich Aβ(1–40) dimers: markov state model analysis on millisecond hybrid-resolution simulations. *J Chem Theory Comput.* 2017;13(11):5731–5744.
108. Man VH, Nguyen PH, Derreumaux P. High-resolution structures of the Aβ 1–42 dimers from the comparison of four atomistic force fields. *J Phys Chem B.* 2017;121(24):5977–5987.
109. Han W, Schulten K. Characterization of folding mechanisms of Trp-cage and WW-domain by network analysis of simulations with a hybrid-resolution model. *J Phys Chem B.* 2013;117(42):13367–13377.
110. Zheng W, Gallicchio E, Deng N, Andrec M, Levy RM. Kinetic network study of the diversity and temperature dependence of Trp-Cage folding pathways: combining transition path theory with stochastic simulations. *J Phys Chem B.* 2011;115(6):1512–1523.
111. Berezhkovskii A, Hummer G, Szabo A. Reactive flux and folding pathways in network models of coarse-grained protein dynamics. *J Chem Phys.* 2009;130(20):205102.
112. Noé F, Schütte C, Vanden-Eijnden E, Reich L, Weikl TR. Constructing the equilibrium ensemble of folding pathways from short off-equilibrium simulations. *Proc Natl Acad Sci U S A.* 2009;106(45):19011–19016.
113. Zhu F, Hummer G. Theory and simulation of ion conduction in the pentameric GLIC channel. *J Chem Theory Comput.* 2012;8(10):3759–3768.
114. Hummer G. Position-dependent diffusion coefficients and free energies from Bayesian analysis of equilibrium and replica molecular dynamics simulations. *New J Phys.* 2005;7:34.
115. Bacci M, Vitalis A, Caflisch A. A molecular simulation protocol to avoid sampling redundancy and discover new states. *Biochim Biophys Acta.* 2015;1850(5):889–902.
116. Roder K, Wales DJ. Energy landscapes for the aggregation of Aβ17-42. *J Am Chem Soc.* 2018;140(11):4018–4027.
117. Wales D. Discrete path sampling. *Mol Phys.* 2002;100:3285–3305.
118. Krivov SV, Karplus M. Free energy disconnectivity graphs: application to peptide models. *J Chem Phys.* 2002;117(23):10894–10903.
119. Rosenman DJ, Wang C, Garcia AE. Characterization of Aβ monomers through the convergence of ensemble properties among simulations with multiple force fields. *J Phys Chem B.* 2016;120(2):259–277.

120. Tran L, Ha-Duong T. Exploring the Alzheimer amyloid-β peptide conformational ensemble: a review of molecular dynamics approaches. *Peptides*. 2015;69:86–91.
121. Okumura H, Itoh SG. Structural and fluctuational difference between two ends of Aβ amyloid fibril: MD simulations predict only one end has open conformations. *Sci Rep*. 2016;6:38422.
122. Li Y, Ji C, Xu W, Zhang JZ. Dynamical stability and assembly cooperativity of β-sheet amyloid oligomers—effect of polarization. *J Phys Chem B*. 2012;116(45):13368–13373.
123. Ji C, Mei Y, Zhang JZ. Developing polarized protein-specific charges for protein dynamics: MD free energy calculation of pKa shifts for Asp26/Asp20 in thioredoxin. *Biophys J*. 2008;95(3):1080–1088.
124. Davidson DS, Brown AM, Lemkul JA. Insights into stabilizing forces in amyloid fibrils of differing sizes from polarizable molecular dynamics simulations. *J Mol Biol*. 2018;430(20):3819–3834.
125. Larson MA. Secondary nucleation: an analysis. *Chem Eng Commun*. 1981;12(1–3): 161–169.
126. Davey RJ. Crystallization: how come you look so good? *Nature*. 2004;428:374–375.
127. Ferrone FA, Hofrichter J, Eaton WA. Kinetics of sickle hemoglobin polymerization. I. Studies using temperature-jump and laser photolysis techniques. *J Mol Biol*. 1985;183(4):591–610.
128. Ferrone FA, Hofrichter J, Sunshine HR, Eaton WA. Kinetic studies on photolysis-induced gelation of sickle cell hemoglobin suggest a new mechanism. *Biophys J*. 1980;32:361–380.
129. Ruschak AM, Miranker AD. Fiber-dependent amyloid formation as catalysis of an existing reaction pathway. *Proc Natl Acad Sci U S A*. 2007;104(30):12341–12346.
130. Kayed R, Head E, Thompson JL, et al. Common structure of soluble amyloid oligomers implies common mechanism of pathogenesis. *Science*. 2003;300:486–489.
131. Bucciantini M, Giannoni E, Chiti F, et al. Inherent toxicity of aggregates implies a common mechanism for protein misfolding diseases. *Nature*. 2002;416:507–511.
132. Walsh DM, Klyubin I, Fadeeva JV, et al. Naturally secreted oligomers of amyloid β protein potently inhibit hippocampal long-term potentiation in vivo. *Nature*. 2002;416:535–539.
133. Arosio P, Knowles TP, Linse S. On the lag phase in amyloid fibril formation. *Phys Chem Chem Phys*. 2015;17(12):7606–7618.
134. Cohen SIA, Cukalevski R, Michaels TCT, et al. Distinct thermodynamic signatures of oligomer generation in the aggregation of the amyloid-β peptide. *Nat Chem*. 2018;10(5):523–531.
135. Roduner E. Understanding catalysis. *Chem Soc Rev*. 2014;43(24):8226–8239.
136. Linse S, Cabaleiro-Lago C, Xue WF, et al. Nucleation of protein fibrillation by nanoparticles. *Proc Natl Acad Sci U S A*. 2007;104(21):8691–8696.
137. Cabaleiro-Lago C, Quinlan-Pluck F, Lynch I, et al. Inhibition of amyloid β protein fibrillation by polymeric nanoparticles. *J Am Chem Soc*. 2008;130:15437–15443.
138. Pronchik J, He X, Giurleo JT, Talaga DS. In vitro formation of amyloid from α-synuclein is dominated by reactions at hydrophobic interfaces. *J Am Chem Soc*. 2010;132:9797–9803.
139. Keller A, Fritzsche M, Yu Y-P, et al. Influence of hydrophobicity on the surface-catalyzed assembly of the islet amyloid polypeptide. *ACS Nano*. 2011;5:2770–2778.
140. Kowalewski T, Holtzman DM. In situ atomic force microscopy study of Alzheimer's β-amyloid peptide on different substrates: new insights into mechanism of β-sheet formation. *Proc Natl Acad Sci U S A*. 1999;96:3688–3693.
141. Cabaleiro-Lago C, Lynch I, Dawson KA, Linse S. Inhibition of IAPP and IAPP(20-29) fibrillation by polymeric nanoparticles. *Langmuir*. 2010;26(5):3453–3461.

142. Seubert P, Vigo-Pelfrey C, Esch F, et al. Isolation and quantification of soluble Alzheimer's β-peptide from biological fluids. *Nature*. 1992;359:325–327.
143. Campioni S, Carret G, Jordens S, Nicoud L, Mezzenga R, Riek R. The presence of an air-water interface affects formation and elongation of α-Synuclein fibrils. *J Am Chem Soc*. 2014;136(7):2866–2875.
144. Butterfield SM, Lashuel HA. Amyloidogenic protein-membrane interactions: mechanistic insight from model systems. *Angew Chem Int Ed Engl*. 2010;49(33):5628–5654.
145. Lin Y-C, Petersson EJ, Fakhraai Z. Surface effects mediate self-assembly of amyloid-β peptides. *ACS Nano*. 2014;8:10178–10186.
146. Gao G, Zhang M, Lu P, Guo G, Wang D, Sun T. Chirality-assisted ring-like aggregation of Aβ(1-40) at liquid-solid interfaces: a stereoselective two-step assembly process. *Angew Chem Int Ed Engl*. 2015;54(7):2245–2250.
147. Cabaleiro-Lago C, Quinlan-Pluck F, Lynch I, Dawson KA, Linse S. Dual effect of amino modified polystyrene nanoparticles on amyloid β protein fibrillation. *ACS Chem Nerosci*. 2010;1(4):279–287.
148. Wu W-H, Sun X, Yu Y-P, et al. TiO2 nanoparticles promote β-amyloid fibrillation in vitro. *Biochem Biophys Res Commun*. 2008;373(2):315–318.
149. Wang Q, Shah N, Zhao J, et al. Structural, morphological, and kinetic studies of β-amyloid peptide aggregation on self-assembled monolayers. *Phys Chem Chem Phys*. 2011;13(33):15200–15210.
150. Zhao J, Wang Q, Liang G, Zheng J. Molecular dynamics simulations of low-ordered Alzheimer β-amyloid oligomers from dimer to hexamer on self-assembled monolayers. *Langmuir*. 2011;27(24):14876–14887.
151. Bellucci L, Bussi G, Di Felice R, Corni S. Fibrillation-prone conformations of the amyloid-β-42 peptide at the gold/water interface. *Nanoscale*. 2017;9(6):2279–2290.
152. Sudhakar S, Kalipillai P, Santhosh PB, Mani E. Role of surface charge of inhibitors on amyloid β fibrillation. *J Phys Chem C*. 2017;121(11):6339–6348.
153. Lindberg DJ, Wesen E, Bjorkeroth J, Rocha S, Esbjorner EK. Lipid membranes catalyse the fibril formation of the amyloid-β (1-42) peptide through lipid-fibril interactions that reinforce secondary pathways. *Biochim Biophys Acta Biomembr*. 2017;1859(10): 1921–1929.
154. Davis CH, Berkowitz ML. Structure of the Aβ (1-42) monomer absorbed to model phospholipid bilayers: a molecular dynamics study. *J Phys Chem B*. 2009;113: 14480–14486.
155. Lemkul JA, Bevan DR. Perturbation of membranes by the amyloid β-peptide—a molecular dynamics study. *FEBS J*. 2009;276(11):3060–3075.
156. Miyashita N, Straub JE, Thirumalai D. Structures of β-amyloid peptide 1-40, 1-42, and 1-55-the 672-726 fragment of APP-in a membrane environment with implications for interactions with γ-secretase. *J Am Chem Soc*. 2009;131:17843–17852.
157. Zhang M, Ren B, Liu Y, et al. Membrane interactions of hIAPP monomer and oligomer with lipid membranes by molecular dynamics simulations. *ACS Chem Nerosci*. 2017;8(8):1789–1800.
158. Zhao LN, Chiu SW, Benoit J, Chew LY, Mu Y. Amyloid β peptides aggregation in a mixed membrane bilayer: a molecular dynamics study. *J Phys Chem B*. 2011; 115(42):12247–12256.
159. Hoshino T, Mahmood MI, Mori K, Matsuzaki K. Binding and aggregation mechanism of amyloid β-peptides onto the GM1 ganglioside-containing lipid membrane. *J Phys Chem B*. 2013;117(27):8085–8094.
160. Banerjee S, Hashemi M, Lv Z, Maity S, Rochet J-C, Lyubchenko YL. A novel pathway for amyloids self-assembly in aggregates at nanomolar concentration mediated by the interaction with surfaces. *Sci Rep*. 2017;7(1):45592–45602.

161. Gladytz A, Abel B, Risselada HJ. Gold-induced fibril growth: the mechanism of surface-facilitated amyloid aggregation. *Angew Chem Int Ed Engl.* 2016;55(37): 11242–11246.
162. Nikolic A, Baud S, Rauscher S, Pomes R. Molecular mechanism of β-sheet self-organization at water-hydrophobic interfaces. *Proteins.* 2011;79(1):1–22.
163. Serio TR, Cashikar AG, Kowal AS, et al. Nucleated conformational conversion and the replication of conformational information by a prion determinant. *Science.* 2000; 289(5483):1317–1321.
164. Auer S, Trovato A, Vendruscolo M. A condensation-ordering mechanism in nanoparticle-catalyzed peptide aggregation. *PLoS Comput Biol.* 2009;5(8):e1000458.
165. Morriss-Andrews A, Shea JE. Kinetic pathways to peptide aggregation on surfaces: the effects of β-sheet propensity and surface attraction. *J Chem Phys.* 2012;136(6):065103.
166. Liwo A, Khalili M, Scheraga HA. Ab initio simulations of protein-folding pathways by molecular dynamics with the united-residue model of polypeptide chains. *Proc Natl Acad Sci U S A.* 2005;102(7):2362–2367.
167. Ha C, Park CB. Ex situ atomic force microscopy analysis of β-amyloid self-assembly and deposition on a synthetic template. *Langmuir.* 2006;22:6977–6985.
168. Morriss-Andrews A, Brown FLH, Shea J-E. A coarse-grained model for peptide aggregation on a membrane surface. *J Phys Chem B.* 2014;118(28):8420–8432.
169. Vácha R, Linse S, Lund M. Surface effects on aggregation kinetics of amyloidogenic peptides. *J Am Chem Soc.* 2014;136(33):11776–11782.
170. Vácha R, Frenkel D. Relation between molecular shape and the morphology of self-assembling aggregates: a simulation study. *Biophys J.* 2011;101(6):1432–1439.
171. Shakhnovich E, Farztdinov G, Gutin AM, Karplus M. Protein folding bottlenecks: a lattice Monte Carlo simulation. *Phys Rev Lett.* 1991;67(12):1665–1668.
172. Radic S, Davis TP, Ke PC, Ding F. Contrasting effects of nanoparticle-protein attraction on amyloid aggregation. *RSC Adv.* 2015;5(127):105498.
173. Shen L, Adachi T, Vanden Bout D, Zhu XY. A mobile precursor determines Aβ peptide fibril formation at interfaces. *J Am Chem Soc.* 2012;134(34):14172–14178.
174. Shezad K, Zhang K, Hussain M, et al. Surface roughness modulates diffusion and fibrillation of Aβ peptide. *Langmuir.* 2016;32(32):8238–8244.
175. Lin YC, Li C, Fakhraai Z. Kinetics of surface-mediated fibrillization of amyloid-β (12-28) peptides. *Langmuir.* 2018;34(15):4665–4672.
176. Anwar J, Boateng PK, Tamaki R, Odedra S. Mode of action and design rules for additives that modulate crystal nucleation. *Angew Chem Int Ed Engl.* 2009;48(9):1596–1600.
177. Anwar J, Khan S, Lindfors L. Secondary crystal nucleation: nuclei breeding factory uncovered. *Angew Chem Int Ed Engl.* 2015;54(49):14681–14684.
178. Powers HEC. Sucrose crystals. *Nature.* 1960;188:289–291.
179. Saric A, Buell AK, Meisl G, et al. Physical determinants of the self-replication of protein fibrils. *Nat Phys.* 2016;12(9):874–880.
180. Meisl G, Yang X, Frohm B, Knowles TP, Linse S. Quantitative analysis of intrinsic and extrinsic factors in the aggregation mechanism of Alzheimer-associated Aβ-peptide. *Sci Rep.* 2016;6:18728–18739.
181. Yang X, Meisl G, Frohm B, Thulin E, Knowles TPJ, Linse S. On the role of sidechain size and charge in the aggregation of Aβ42 with familial mutations. *Proc Natl Acad Sci U S A.* 2018;115(26):E5849–E5858.
182. Barz B, Strodel B. Understanding amyloid-β oligomerization at the molecular level: the role of the fibril surface. *Chem A Eur J.* 2016;22(26):8768–8772.
183. Qi R, Luo Y, Wei G, Nussinov R, Ma B. Aβ "stretching-and-packing" cross-seeding mechanism can trigger tau protein aggregation. *J Phys Chem Lett.* 2015;6(16): 3276–3282.

184. Jiang X, Cao Y, Han W. In silico study of recognition between Aβ40 and Aβ40 fibril surfaces: an N-terminal helical recognition motif and its implications for inhibitor design. *ACS Chem Nerosci*. 2018;9(5):935–944.
185. Schwierz N, Frost CV, Geissler PL, Zacharias M. From Aβ filament to fibril: molecular mechanism of surface-activated secondary nucleation from all-atom MD simulations. *J Phys Chem B*. 2017;121(4):671–682.
186. Bellaiche MMJ, Best RB. Molecular determinants of Aβ42 adsorption to amyloid fibril surfaces. *J Phys Chem Lett*. 2018;9(22):6437–6443.
187. Jiang Y, Jiang X, Shi X, et al. α-Helical motif as inhibitors of toxic Aβ oligomer generation via highly specific recognition of amyloid surface. *iScience*. 2019;17: 87–100.
188. Meisl G, Yang X, Hellstrand E, et al. Differences in nucleation behavior underlie the contrasting aggregation kinetics of the Aβ40 and Aβ42 peptides. *Proc Natl Acad Sci U S A*. 2014;111(26):9384–9389.
189. Xiao Y, Ma B, McElheny D, et al. Aβ(1–42) fibril structure illuminates self-recognition and replication of amyloid in Alzheimer's disease. *Nat Struct Mol Biol*. 2015;22(6): 499–505.
190. Oleinikovas V, Saladino G, Cossins BP, Gervasio FL. Understanding cryptic pocket formation in protein targets by enhanced sampling simulations. *J Am Chem Soc*. 2016;138(43):14257–14263.
191. Meisl G, Yang X, Dobson CM, Linse S, Knowles TPJ. Modulation of electrostatic interactions to reveal a reaction network unifying the aggregation behaviour of the Aβ42 peptide and its variants. *Chem Sci*. 2017;8(6):4352–4362.
192. Morales R, Moreno-Gonzalez I, Soto C. Cross-seeding of misfolded proteins: implications for etiology and pathogenesis of protein misfolding diseases. *PLoS Pathog*. 2013;9(9):e1003537.
193. Gotz J, Chen F, van Dorpe J, Nitsch RM. Formation of neurofibrillary tangles in P301L tau transgenic mice induced by Aβ42 fibrils. *Science*. 2001;293:1491–1495.
194. Misra P, Kodali R, Chemuru S, Kar K, Wetzel R. Rapid α-oligomer formation mediated by the Aβ C terminus initiates an amyloid assembly pathway. *Nat Commun*. 2016;7:12419.
195. Bunce S, Wang Y, Stewart K, et al. Molecular insights into the surface-catalyzed secondary nucleation of amyloid-β40 (Aβ40) by the peptide fragment Aβ16-22. *Sci Adv*. 2019;5:eaav8216.
196. Cheon M, Chang I, Hall CK. Influence of temperature on formation of perfect tau fragment fibrils using PRIME20/DMD simulations. *Protein Sci*. 2012;21(10): 1514–1527.
197. Azzarito V, Long K, Murphy NS, Wilson AJ. Inhibition of α-helix-mediated protein-protein interactions using designed molecules. *Nat Chem*. 2013;5(3):161–173.
198. Baram M, Gilead S, Gazit E, Miller Y. Mechanistic perspective and functional activity of insulin in amylin aggregation. *Chem Sci*. 2018;9(18):4244–4252.
199. Cukalevski R, Yang X, Meisl G, et al. The Aβ40 and Aβ42 peptides self-assemble into separate homomolecular fibrils in binary mixtures but cross-react during primary nucleation. *Chem Sci*. 2015;6(7):4215–4233.
200. Chiricotto M, Melchionna S, Derreumaux P, Sterpone F. Multiscale aggregation of the amyloid Aβ16–22 peptide: from disordered coagulation and lateral branching to amorphous prefibrils. *J Phys Chem Lett*. 2019;10(7):1594–1599.
201. Latshaw David C, Hall CK. Effects of hydrophobic macromolecular crowders on amyloid β (16–22) aggregation. *Biophys J*. 2015;109(1):124–134.
202. Owen MC, Gnutt D, Gao M, et al. Effects of in vivo conditions on amyloid aggregation. *Chem Soc Rev*. 2019;48(14):3946–3996.

CHAPTER ELEVEN

Aggregation and coacervation with Monte Carlo simulations

Sandipan Mohanty*
Institute for Advanced Simulation, Jülich Supercomputing Centre, Forschungszentrum Jülich, Jülich, Germany
*Corresponding author: e-mail address: s.mohanty@fz-juelich.de

Contents

Abstract

Protein aggregation into oligomeric species has been linked to a number of neurodegenerative diseases. The ability to assemble into ordered fibril-like forms under certain conditions is now regarded as a very common property of polypeptide chains. A different form of assembly is known for some proteins in which protein molecules sequester into liquid-like droplets. Understanding the biophysical mechanisms behind these phenomena is of great relevance not only for deeper insights into the associated disorders but also for any protein-based therapeutics. Here we review a few examples of the use of Markov chain Monte Carlo simulations in the study of these two forms of protein assembly.

1. Introduction

Protein aggregation is the self-assembly of proteins into supra-molecular structures ranging from dimers to highly structured amyloid fibrils. While functional forms of fibrils are known,[1,2] much of the research on formation and characterization of amyloid fibrils stems from the fact that biophysical research over many decades has linked the process of aggregation with a variety of diseases such as Alzheimer's, Parkinson's, and Huntington's diseases,

Progress in Molecular Biology and Translational Science, Volume 170
ISSN 1877-1173
https://doi.org/10.1016/bs.pmbts.2019.12.005

type II diabetes and spongiform encephalopathies.[3–6] Detailed causal chains connecting molecular self-assembly events to disease pathology remain obscure. But it is believed that a clearer atomic picture will be a very useful aid in understanding and hopefully treating these diseases. Another related reason is that protein aggregation constitutes a common and recalcitrant obstacle in the highly target specific protein-based therapeutics.[7]

The ability to form aggregates is not just a special property of rare or disease related proteins, but rather a more general property of polypeptide chains.[8,9] Different biophysical strategies have evolved in cells to oppose protein aggregation.[10] An understanding of the biophysical mechanisms driving protein misfolding and aggregation is therefore expected to inform therapeutic strategies.

Proteins involved in aggregation related diseases such as α-synuclein are often intrinsically disordered proteins (IDP).[11,12] IDPs do not have a compact native fold, but rather populate a wide range of structures at physiologically relevant conditions. Some IDPs are also known to undergo a process called liquid–liquid phase separation (LLPS) or coacervation, in which concentrated liquid droplets of proteins develop inside the solution.[13–15] These droplets exhibit a continuous, fluid like, internal reorganization of the member molecules instead of a more ordered, solid-like, cross-β arrangement typical of fibrils.

It has been suggested that LLPS can facilitate fibril formation for some proteins[16] or present an alternative to aggregation. Functional roles for the liquid droplets have also been proposed where they serve as membrane-less organelles.[17] The phenomenon of LLPS has been studied using different theoretical and computational techniques, such as random phase approximation,[18] field theoretical simulations[19] and lattice Monte Carlo simulations.[20]

Computer simulations have long been used to provide a conceptual framework to understand biophysical phenomena. Because of the high degree of complexity arising from the large number of interacting components, they constitute an essential class of tools in interpreting experimental studies, connecting them to fundamental physics, organizing our knowledge and asking new questions based on an ever improving picture. Simulation methods used for biomolecular systems are very diverse. Most computational studies of biomolecular systems use the molecular dynamics (MD) method. For a recent review of atomistic MD simulations used to study protein aggregation, see Ref. 21. For an overview of field theoretical simulations for LLPS, see Ref. 19. A completely different computational approach is the Monte Carlo method. The Markov Chain Monte Carlo (MC) method was developed in the 1940s in the Los Alamos National Laboratory[22] in connection with nuclear weapons research. Because of the generality of the technique and its scaling properties for high-dimensional problems, the MC method has found increasing use since its inception in diverse fields of inquiry ranging from fundamental high energy physics to finance and risk evaluation. Although in biomolecular simulations the use of MC is less common than MD, for some problems they offer an alternative to MD at a significantly lower computational cost. In this chapter, we will summarize a few applications of MC simulations to explore self-assembly of proteins.

2. Markov chain Monte Carlo simulations

Thermodynamic Monte Carlo (MC) simulations sample random structures from an ensemble which is either directly a Boltzmann ensemble, or has a rigorous exact mapping to such an ensemble. From such a random sample, statistical averages can be estimated. The class of MC algorithms most commonly used for molecular simulations is called a Markov Chain Monte Carlo, and in the following, we will refer to this type of MC simulations as simply MC. In this method, the simulated system performs a random walk in the conformation space. At every step, we propose a random change to the conformation, and accept or reject the change according to an acceptance probability. This simple recipe of random moves with accept/reject decisions is repeated billions of times and sample points for statistical analysis are collected at regular intervals from the resulting Markov Chain. The principle of detailed balance is a convenient way to ensure that the sample points reach a stationary distribution with the correct relative statistical weights. If $P(A)$ and $P(B)$ are the equilibrium probabilities of states A and B, and p_{AB} is the probability to propose a move from A to B (and likewise for p_{BA}), detailed balance can be expressed as the requirement $P(A)p_{AB} = P(B)p_{BA}$. Conformation updates satisfying detailed balance do not cause a net flow of probabilities. For proper sampling, it is also important that the entire conformation space remains connected under the conformation updates.

The form of the acceptance probability depends on the type of ensemble we intend to generate. If $P(A)$ represents the weight of a state A in the target ensemble, the acceptance probability for a conformation change to a state B

is $p = \min\left(1, \frac{P(B)}{P(A)}\right)$. In the simplest case of a canonical ensemble, the microscopic weight of a state A is $P(A) \propto \exp(-\beta E_A)$, where E_A is the energy of state A, $\beta = \frac{1}{k_B T}$ is the inverse temperature, and k_B is the Boltzmann's constant. Each state with an energy E will have a weight of $\exp(-\beta E)$, and the total probability of all states with energy E is proportional to $g(E)\exp(-\beta E)$, where $g(E)$ is called the density of states, which counts the number of states at energy E. For a simple Metropolis MC simulation sampling the canonical ensemble, the acceptance probability takes the form $p \propto \min(1, \exp(-\beta \Delta E))$, ΔE being the change in energy due to the conformation change. Long Markov chains with this acceptance probability will sample each state of energy E with weight $\exp(-\beta E)$, so that with $g(E)$ states at energy E, the sampled probability for energy E will be an approximation of $g(E)\exp(-\beta E)$. In general, $g(E)$ is an unknown function, and the histogram of energy from a long Markov chain is one way to estimate it. This is one of the important goals of thermodynamic MC simulations, as a good estimate of $g(E)$ along with correlations of various state properties with energy can be used to calculate thermodynamic averages.

The canonical MC recipe described above can be improved in many ways for faster convergence of statistical averages. The $\exp(-\beta \Delta E)$ factor in the acceptance probability suppresses positive energy changes, which impedes the escape from local energy minima. Since the global energy minimum of a system is usually unknown at the start, it is important that the simulations are able to escape local energy minima, and search for the low lying states in the entire energy landscape. Sophisticated techniques have been developed to help the MC simulations navigate rough energy landscapes and converge faster. These include simulated tempering,[23] parallel tempering,[24,25] multicanonical method,[26] and Wang–Landau simulations.[27,28]

2.1 Small update MC simulations

Unlike in MD simulations, in MC, we do not integrate Newton's equations of motion. There is no restriction on how big or small a single step along the Markov Chain can be, and this property can be exploited for very fast importance sampling from a high-dimensional conformation space. In many studies, the random updates are restricted to make small overall changes to the system so that after an MC step, the system retains much of its structure from before the step, with small modifications. A Markov Chain generated in this way vaguely mimics a trajectory, i.e., continuously changing coordinates as a function of time, as seen for instance, in MD simulations. The MC simulations, however, do not contain a real "time" variable, and can not reveal information about physical time scales of processes. In our own research, we have found that small update MC simulations sampling the canonical ensemble yield pseudo-trajectories with easily rationalizable qualitative behavior. This is not entirely unexpected, as, despite not following a path set by Newton's equations of motion, the Markov Chain stochastically follows paths defined by the energy landscape. Let us compare how coordinates are updated in a small update MC simulation and an MD simulation using the same force field. The conformation updates in the MC leading in the opposite direction relative to the acceleration in the corresponding MD simulation (i.e., updates along the positive gradient of the potential energy) are unfavorable due to the $\min(1, \exp(-\beta\Delta E))$ acceptance probability. MC updates leading roughly in the direction of the acceleration on the other hand, are accepted. The MC simulation treats all possible changes which reduce the potential energy equivalently, whereas the MD will pick one of them definitively. But the barriers in the energy landscape which slow down a particular transition in an MD will have a similar effect on a small update MC, because of the exponential suppression of the positive ΔE in the acceptance probability. Larger changes, unlike larger time steps in an MD do not result in larger numeric errors in an MC simulation, but suppress their superficial resemblance to MD trajectories. Therefore, with the understanding that no time scales can be read out from the simulations, small update MC simulations have often been used to study kinetic processes because of their much smaller computational cost compared to a equivalent MD simulations.[29–36]

2.2 Folding and aggregation using all-atom MC simulations

In this section, we will review a few examples of the use of the MC method to study peptide aggregation. Although the methods discussed above are very general, currently we focus on studies conducted using a specific implicit solvent physics based model. In this model, we explicitly represent all atoms in the protein molecules, including all hydrogen atoms, as interacting entities. We ignore fluctuations in lengths of covalent bonds as well as the bond angles between converging bonds, so that all sample states considered preserve the values for these geometrical parameters. This leaves only torsional rotation around bonds inside a single molecule and global translations and rotations as degrees of freedom. The conformation update in every step of the MC simulation is chosen randomly from a small set of move types.

This set includes rigid body translations and rotations of one or more molecules, random changes to a single randomly chosen backbone or side chain torsion angle, or a concerted rotation of up to 8 backbone angles affecting a local deformation of a single protein chain.[37] The random single angle updates as well as the rigid body translations and rotations can create very large displacements of large numbers of atoms. If we worked with explicitly represented solvent molecules, most of these updates would result in steric clashes and our approach will lose much of its ability for fast exploration. We therefore developed an implicit solvent force field to model protein interactions.

Our interaction potential consists of four terms: $E = E_{exv} + E_{loc} + E_{hb} + E_{sc}$. Excluded volume effects are represented by E_{exv}. E_{loc} represents local electrostatic effects along the protein backbone. E_{hb} represents backbone and side chain hydrogen bonds. E_{sc} encapsulates side chain charge–charge interactions as well as the hydrophobic effect. Detailed mathematical forms of each of these terms can be found elsewhere.[38] Here we only note that in this reductive physical approach, the interaction parameters are tied to atoms or groups of atoms, and are regarded as universal. For instance, every phenylalanine residue has exactly the same interaction capacity with respect to each of the terms in our force field, irrespective of its position in a protein sequence or structure. This also means that in simulations with many molecules, intramolecular and intermolecular interactions necessarily have the same functional form and parameters. The form and parameters of this force field have been developed through thermodynamic folding simulations of small polypeptide chains[38,39] from randomly initialized chains. The approximations of this force field developed through studies of small peptides of sequence length 10–30 residues have occasionally proven to be sufficient to describe reversible folding transitions of somewhat larger sequences, such as Top7-CFr, several three-helix bundle proteins of about 70 residues, and even a mixed 92 residue α/β protein Top7[38,40,41] with very slow folding times of about 1 second in experiments (see Fig. 1).

The ability of simulations with this model to find the native folded states of some proteins starting from completely randomized initial configurations, without biasing the simulations with any prior information on the folded states, is a very nontrivial attribute. Calculated temperature dependence of secondary structure propensities and native populations often agree well with suitable experimental results. An implementation of the protein model, force field and MC simulation tools is available as an open source software package called ProFASi.[42]

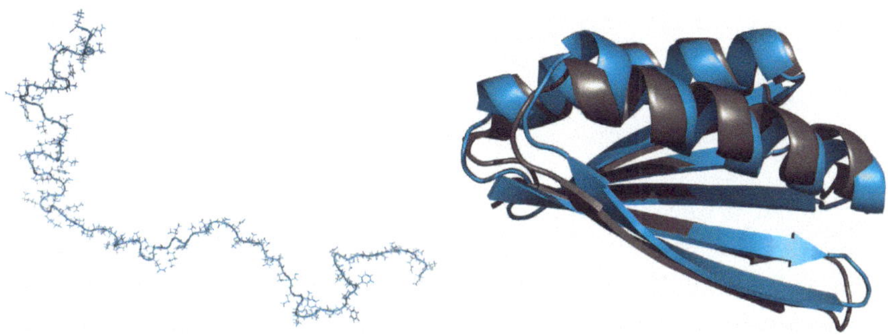

Fig. 1 The protein model highlighted in this chapter in the context of atomistic MC simulations is capable of folding many small proteins starting from completely random conformations. Here we show two snapshots from a simulation of a 92 residue α/β protein Top7 for which the native state contains two α-helices and a 5 stranded β-sheet. *Left*: an example structure near the beginning of the simulation. *Right*: a snapshot from the simulation (*blue*) representing the free energy minimum at 273 K aligned with the PDB structure 1QYS (*gray*).

We have used this all-atom protein model to explore protein aggregation and related phenomena using MC simulations. In 2004, after developing an early version of the model, we applied it to study the oligomerization of $A\beta_{16-22}$ peptides (sequence: KLVFFAE).[43] From that proof-of-concept study with only up to 6 chains of $A\beta_{16-22}$ peptides, we learned that systems of many protein chains can show interesting physical behavior in MC simulations. The MC simulations sampled multiple transitions between isolated monomers and several different kinds of oligomers. Monomer simulations showed that the isolated $A\beta_{16-22}$ behaves as a random coil. In multichain simulations, the statistical secondary structure propensities were consistent with single chain simulations at high temperatures, but with decreasing temperatures the β-sheet content smoothly increased, exhibiting a sigmoid temperature dependence. Analysis of the structures obtained at the low temperatures showed that they were predominantly oligomeric, and the β-sheet secondary structure was exclusively from the oligomers present in the structures.

In Ref. 30, we compared two truncated segments $A\beta_{16-22}$ and $A\beta_{25-35}$ (sequence: GSNKGAIIGLM) of the Alzheimer's $A\beta$ peptide with small update MC simulations with ProFASi at different concentrations and temperatures. It turned out that the two peptides show different aggregation behavior. The $A\beta_{16-22}$ system showed a rapid hydrophobic collapse to large clusters of unstructured aggregates with ample chain mobility. These structures then gradually transitioned into β-sheet rich oligomers due to the formation of interchain hydrogen bonds, which curiously exposed some hydrophobic groups. In contrast, the $A\beta_{25-35}$ showed little tendency to collapse into disordered aggregates, but rather formed β-sheet hydrogen bonds early in the process.

In Ref. 32, we examined aggregation behavior of a small segment AcPHF6 (Ac-VQIVYK-NH2) of the tau protein, also connected with the Alzheimer's disease. We analyzed the structure and organization of the oligomers forming in the simulations. For instance, we kept track of the relative orientation of the neighboring chains in a β-sheet and the interchain hydrogen bonds. We found that a variety of very small oligomers emerged quickly with no preference for orientation or registry between neighboring chains. These small oligomers formed and dissolved rapidly resulting in an equilibrium with the solution. This stochastic process sometimes results in oligomers of a critical size, which do not dissolve as easily. Of these relatively stable oligomers, we found that only a subset continued to grow into long β-sheets and two sheet oligomers (Fig. 2). The subset of the minimal stable oligomers which managed to grow in our simulations were those with the greatest internal order, i.e., those with the highest percentage of parallel in-register strand organization. Even the largest simulations in Ref. 32 had only 36 peptides in them, which makes multilayered structures less probable. The simulations did, however, result in many double layered aggregates exhibiting a steric zipper interface (see Fig. 2) between the two sheets with the V1, I3, and Y5 side chains packed in the sheet–sheet

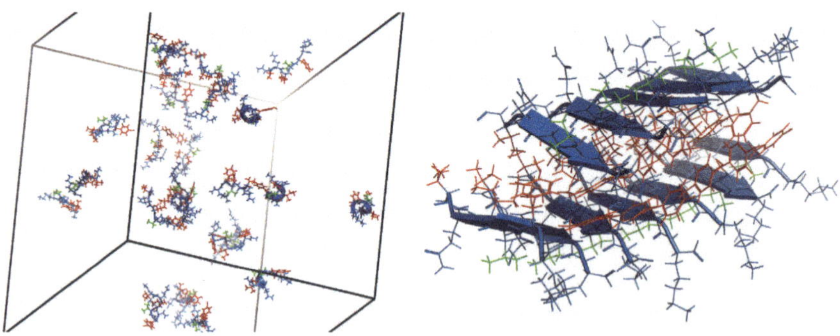

Fig. 2 Example snapshots from MC simulations of 24 chains of Ac-PHF6-NH2 at 308 Kelvin in a 95 Å periodic box. Residues V^1, I^3, and Y^5, which appear in the dry-steric zipper interface in two layer aggregates have been marked red, while V^4, another hydrophobic residue which does not get buried in the dry interface is marked *green*. *Left*: a snapshot from near the beginning of the simulation. *Right*: closer view of a large oligomer toward the end of the simulation.

interface. Such a dry-steric zipper interface has been proposed for a variety of small fibril forming peptides including PHF6 using X-ray diffraction.[44]

The above mentioned simulations are examples of small update MC simulations where the Markov chains resemble kinetic trajectories. In Ref. 31, Irbäck and Mitternacht again explored oligomerization of $A\beta_{16-22}$ using equilibrium simulations. For this purpose they used our all-atom model and its implementation in ProFASi, but did not restrict the runs to small updates. They also tested experimental modifications to the force field which contributed to the next iteration of the default interaction calculations in ProFASi. These simulations provided a different view of the oligomerization of $A\beta_{16-22}$. The temperature dependence of properties such as the size of the largest cluster in the system and the secondary structure content were calculated with excellent statistical errors. They showed that with decreasing temperature, the size of the largest cluster starts to increase a little before the β-sheet secondary structure does. This is in perfect agreement with the more kinetic simulations[30] which showed that the disordered "molten" oligomers formed before the ordered β-sheets in MC time. In addition, they found that the simulations resulted in the spontaneous formation of tight, extraordinarily long-lived β-barrels.

Aggregating systems are often characterized by strongly bimodal energy distributions. Sparse conformations without aggregates have energies corresponding to multiple free monomers. Once an aggregating nucleus forms, there is a rapid transition to a low energy oligomer rich state. For such systems, the canonical probability of intermediate states connecting the compact low energy states with the sparse higher energy states is very small. Therefore, MC simulations based on the canonical ensemble only rarely visit the intermediate states, and because of this, they may struggle to cross back and forth between the low and high energy basins in free energy. Methods such as multicanonical or Wang–Landau simulations can alleviate the difficulty of crossing the free energy barrier. In Ref. 45, we applied a slightly modified form of the Wang–Landau method to a system of 8 chains of a 7 residue fragment (sequence : GIIFNEQ) of the Cu/Zn superoxide dismutase 1 protein (SOD1). This system exhibited an aggregated and an unaggregated phase, which coexisted at the mid-point temperature with a free energy barrier of height 2.7 k_BT. The intermediate states for this system had low canonical probabilities at the mid-point temperature (approximately $\frac{1}{15}$ of the probability of the aggregated and unaggregated states). Analogous to the multicanonical method, our simulations sampled a noncanonical ensemble which amplifies these intermediate states facilitating repeated sampling of the high and low energy basins in free energy. This is achieved while maintaining a well defined mapping to the canonical ensemble so that the correct thermodynamic averages can be calculated during post-processing. Using direct comparison, we showed that results from our modified Wang–Landau method agree with those from canonical simulations, but have smaller statistical errors per consumed computing resources.

2.3 Exploring fibril formation with lattice models

While atomic resolution simulations, like those discussed above, have been used for studies of small oligomeric systems, it is easier to obtain general insights on much larger scale processes by using reduced representations of the molecules. Lattice models represent one popular class of reduced representation. Each amino acid is represented by one or a few beads, and each bead exclusively occupies one location in a grid of possible positions. A peptide chain in such a model is a sequence of connected beads. In Ref. 46, Li et al. developed a model in which an amino acid is represented as a hydrophobic, polar, or charged bead, with electrostatic and hydrophobic interactions with other beads. Their MC moves consisted of global translations and rotations of a randomly chosen peptide, and local moves such as tail rotations and crank shaft rotations. Since the set of states on a spatial grid is countable, it is possible to perform exact enumeration of all possible conformations for small system sizes. Despite the simplicity of this model, it was able to capture several interesting characteristics of peptide systems. Li et al. found a nondegenerate ground state clearly separated from degenerate higher energy conformations for a chosen peptide sequence. In simulations with multiple chains, they observed self-assembly into anti-parallel β-sheet-like structures, with the strands arranged perpendicular to the fibril axis. In Ref. 47, this model was further used to study the effect of macro molecular crowders on peptide aggregation. Crowders were modeled as impenetrable cubical entities spanning several grid positions in each dimension. The crowders limited access to a part of the spatial grid, but did not otherwise interact with the peptide chains. Using MC simulations with this model, they were able to reconcile experimental observations regarding both increase[48] and decrease[49] of fibril formation rate in the presence of cosolutes. For large crowders, the model showed a decrease in fibril formation time with increasing crowder concentrations, while for small crowders, it showed the opposite effect. This result illustrates how MC simulations with a good physical model can sometimes help clarifying very nontrivial experimental observations.

In Ref. 50, Irbäck et al. explored the thermodynamics of fibril formation using a stick model for small peptides in a lattice approximation. The peptides were modeled as unit length sticks, which were allowed to occupy any available positions on a three-dimensional cubic lattice. The hydrogen bonding direction was assumed to be perpendicular to the chain (stick) direction, and the hydrophobic interactions were assumed to be in the direction perpendicular to both of these. This is inspired by the typical geometry of a cross-β structure of a fibril, with hydrogen bonds running perpendicular to the chain axis and hydrophobic interactions bind multiple sheets together. By using a reduced model and strategies to update large clusters together, they were able to study systems consisting of up to 131,072 peptides. They found that the simulated system shows sigmoidal kinetics characteristic of fibril formation in experiments.[51] The fibril-like structures found in these simulations had an average length of around 210 and comprised 7 layers on average. Similar to the above mentioned all-atom simulations of AcPHF6 peptide, they observed an initial waiting phase where small aggregates form and dissolve until by random chance a stable growth capable aggregate forms. After that there is rapid growth in the mass of the aggregate until the depletion of available monomers slows the kinetics down. In the stick model simulation, the distinguishing characteristics of the critical aggregate was its width in terms of the number of β-sheet layers in the aggregate. Most growth capable aggregates had at least four layers.

A different type of model was developed by Abeln et al.[52] to study protein folding and aggregation on three-dimensional cubic lattices. Instead of grouping amino acids into broad categories such as hydrophobic and polar, they used 20 different kinds of beads to represent the 20 different naturally occurring amino acids. The state of each amino acid in this model consists of a position at one of the lattice points, a secondary structure marker (strand or coil) and a side chain direction marker (which could point to one of the neighboring sites not occupied by the backbone). The potential energy function was written to mimic the interactions of real proteins, including terms to account for hydrogen bonds, interactions with the solvent and steric terms for the side chains. The idea was to infuse a lattice model with enough details to capture the most essential physical aspects of protein behavior. Despite these increased details, this is still a lattice model with orders of magnitude lower computational complexity than a typical atomic resolution model. Abeln et al. designed a protein sequence for a target 3D structure and simulated the designed sequence with their model using MC simulations. The simulations showed a sharp folding transition as a function of temperature, with a rapid increase in the number of native contacts. Control simulations performed for a random sequence with a similar amino acid content did not show any similar transition to a hydrogen bond rich compact state. They also performed simulations of multichain systems starting with no contact among different chains. Spontaneous formation of small oligomers as well as fibril-like structures was observed. The observed intermolecular contacts in these aggregates are consistent with the cross-β geometry of fibrils. The thermal stability of the multilayered fibril-like structures was found to be maximum for sequences with the ability to pack hydrophobic groups between β-sheets.

2.4 MC simulations of liquid–liquid phase separation

MC simulations on cubic lattices have also been used to study liquid–liquid phase separation in intrinsically disordered proteins. In Ref. 20, Das et al. modeled protein chains as self-avoiding walks on a cubic lattice. Each monomer along the sequence exclusively occupies a lattice site and is connected to its sequence neighbors which must occupy a nearest neighbor lattice site. The monomers in each chain interact with other monomers in any chain with electrostatic interactions. Using MC simulations with 300 chains of 50 monomers, several box sizes and temperatures, they compared the temperature and concentration dependence of polymer density for different charge patterns along the sequence. Comparing sequences with zero net charge but different distribution of charge along the sequence, they found that sequences with greater clustering of like charges showed a greater tendency to phase separate.

3. Conclusion

In this short review we have visited a few examples of the use of MC simulations for the study of self-assembly of proteins. Aggregation of proteins into amyloid fibrils spans many orders of magnitude in length and time scales, and models with different levels of detail have been used to create complementary insights. The models span a range of resolutions from highly detailed all-atom descriptions of the protein chains to stick models representing entire protein chains as unit sticks on a lattice. We have used atomically detailed models for systems consisting of a few chains to few hundred short chains, and such simulations can capture essential sequence specific behavior, such as different aggregation mechanisms between two fragments taken from the Aβ peptide. They also support the idea that the tendency to form fibril-like structures is an inherent property of polypeptide chains, because interaction models developed to describe protein folding lead to spontaneous aggregation for certain sequences when many chains are present. Despite the significant speed benefits of the MC procedure, however, such models have a very high computational cost when applied to the formation of amyloid fibrils. General principles governing processes at larger length scales have been more successfully explored with simpler models, at the cost of finer sequence level details. Bead chain models on a lattice with different levels of details have been shown to capture the essential features of the folding and aggregation processes. In studies of liquid–liquid phase separation, lattice models with only electrostatic interactions have revealed that particular charge distributions are more prone to phase separate than others. Similar studies with HP models, treating each amino acid bead as either hydrophobic (H) or polar (P), are currently underway. Even simpler models with peptides as unit length sticks have been used in simulations of over 10^5 chains to gain insights about the kinetics of fibril formation. The complete picture of protein assembly into liquid-like droplets or ordered solid-like fibrils requires atomic resolution insights as well as physics at a coarser resolution. The MC technique has proven useful at each of these resolutions and will continue to play a role, alongside other theoretical, computational, and experimental methods in the immediate future.

References

1. Chiti F, Dobson CM. Protein misfolding, functional amyloid, and human disease. *Ann Rev Biochem*. 2006;75(1):333–366. https://doi.org/10.1146/annurev.biochem.75.101304.123901.
2. Greenwald J, Riek R. Biology of amyloid: structure, function, and regulation. *Structure*. 2010;18(10):1244–1260. https://doi.org/10.1016/j.str.2010.08.009.
3. Selkoe DJ. Folding proteins in fatal ways. *Nature*. 2003;426(6968):900–904. https://doi.org/10.1038/nature02264.
4. Roberson ED, Mucke L. 100 years and counting: prospects for defeating Alzheimer's disease. *Science*. 2006;314(5800):781–784. https://doi.org/10.1126/science.1132813.
5. Lansbury PT, Lashuel HA. A century-old debate on protein aggregation and neurodegeneration enters the clinic. *Nature*. 2006;443(7113):774–779. https://doi.org/10.1038/nature05290.
6. Invernizzi G, Papaleo E, Sabate R, et al. Protein aggregation: mechanisms and functional consequences. *Int J Biochem Cell Biol*. 2012;44(9):1541–1554. https://doi.org/10.1016/j.biocel.2012.05.023.
7. Zambrano R, Jamroz M, Szczasiuk A, et al. AGGRESCAN3D (A3D): server for prediction of aggregation properties of protein structures. *Nucleic Acids Res*. 2015;43(W1):W306–W313. https://doi.org/10.1093/nar/gkv359.
8. Dobson CM. Protein folding and misfolding. *Nature*. 2003;426(6968):884–890. https://doi.org/10.1038/nature02261.

9. Knowles TPJ, Vendruscolo M, Dobson CM. The amyloid state and its association with protein misfolding diseases. *Nat Rev Mol Cell Biol*. 2014;15(6):384–396. https://doi.org/10.1038/nrm3810.
10. Monsellier E, Chiti F. Prevention of amyloid-like aggregation as a driving force of protein evolution. *EMBO Rep*. 2007;8(8):737–742. https://doi.org/10.1038/sj.embor.7401034.
11. Uversky VN. What does it mean to be natively unfolded? *Eur J Biochem*. 2002;269(1):2–12. https://doi.org/10.1046/j.0014-2956.2001.02649.x.
12. Uversky VN. Natively unfolded proteins: a point where biology waits for physics. *Protein Sci*. 2002;11(4):739–756. https://doi.org/10.1110/ps.4210102.
13. Brangwynne CP, Eckmann CR, Courson DS, et al. Germline P granules are liquid droplets that localize by controlled dissolution/condensation. *Science*. 2009;324(5935):1729–1732. https://doi.org/10.1126/science.1172046.
14. Uversky VN, Kuznetsova IM, Turoverov KK, et al. Intrinsically disordered proteins as crucial constituents of cellular aqueous two phase systems and coacervates. *FEBS Lett*. 2015;589(1):15–22. https://doi.org/10.1016/j.febslet.2014.11.028.
15. Molliex A, Temirov J, Lee J, et al. Phase separation by low complexity domains promotes stress granule assembly and drives pathological fibrillization. *Cell*. 2015;163(1):123–133. https://doi.org/10.1016/j.cell.2015.09.015.
16. Wegmann S, Eftekharzadeh B, Tepper K, et al. Tau protein liquid-liquid phase separation can initiate tau aggregation. *EMBO J*. 2018;37(7):e98049. https://doi.org/10.15252/embj.201798049.
17. Nott TJ, Petsalaki E, Farber P, et al. Phase transition of a disordered nuage protein generates environmentally responsive membraneless organelles. *Mol Cell*. 2015;57(5):936–947. https://doi.org/10.1016/j.molcel.2015.01.013.
18. González-Mozuelos P, de la Cruz MO. Random phase approximation for complex charged systems: application to copolyelectrolytes (polyampholytes). *J Chem Phys*. 1994;100(1):507–517. https://doi.org/10.1063/1.466965.
19. McCarty J, Delaney KT, Danielsen SPO, et al. Complete phase diagram for liquid-liquid phase separation of intrinsically disordered proteins. *J Phys Chem Lett*. 2019;10(8):1644–1652. https://doi.org/10.1021/acs.jpclett.9b00099.
20. Das S, Eisen A, Lin YH, et al. A lattice model of charge-pattern-dependent polyampholyte phase separation. *J Phys Chem B*. 2018;122(21):5418–5431. https://doi.org/10.1021/acs.jpcb.7b11723.
21. Carballo-Pacheco M, Strodel B. Advances in the simulation of protein aggregation at the atomistic scale. *J Phys Chem B*. 2016;120(12):2991–2999. https://doi.org/10.1021/acs.jpcb.6b00059.
22. Metropolis N, Ulam S. The Monte Carlo method. *J Am Stat Assoc*. 1949;44(247):335–341. https://doi.org/10.1080/01621459.1949.10483310.
23. Marinari E, Parisi G. Simulated tempering: a new Monte Carlo scheme. *EPL (Europhysics Letters)*. 1992;19(6):451. https://doi.org/10.1209/0295-5075/19/6/002.
24. Swendsen RH, Wang JS. Replica Monte Carlo simulation of spin-glasses. *Phys Rev Lett*. 1986;57(21):2607–2609. https://doi.org/10.1103/PhysRevLett.57.2607.
25. Hukushima K, Nemoto K. Exchange Monte Carlo method and application to spin glass simulations. *J Phys Soc Jpn*. 1996;65(6):1604–1608. https://doi.org/10.1143/JPSJ.65.1604.
26. Berg BA, Neuhaus T. Multicanonical algorithms for first order phase transitions. *Phys Lett B*. 1991;267(2):249–253. https://doi.org/10.1016/0370-2693(91)91256-U.
27. Wang F, Landau DP. Efficient, multiple-range random walk algorithm to calculate the density of states. *Phys Rev Lett*. 2001;86(10):2050–2053. https://doi.org/10.1103/PhysRevLett.86.2050.

28. Wang F, Landau DP. Determining the density of states for classical statistical models: a random walk algorithm to produce a flat histogram. *Phys Rev E*. 2001;64(5): 056101. https://doi.org/10.1103/PhysRevE.64.056101.
29. Shimada J, Kussell EL, Shakhnovich EI. The folding thermodynamics and kinetics of crambin using an all-atom Monte Carlo simulation. *J Mol Biol*. 2001;308(1):79–95. https://doi.org/10.1006/jmbi.2001.4586.
30. Cheon M, Chang I, Mohanty S, et al. Structural reorganisation and potential toxicity of oligomeric species formed during the assembly of amyloid fibrils. *PLoS Comput Biol*. 2007;3(9):e173. https://doi.org/10.1371/journal.pcbi.0030173.
31. Irbäck A, Mitternacht S. Spontaneous β-barrel formation: an all-atom Monte Carlo study of Aβ16–22 oligomerization. *Proteins*. 2008;71(1):207–214. https://doi.org/10.1002/prot.21682.
32. Li DW, Mohanty S, Irbäck A, et al. Formation and growth of oligomers: a Monte Carlo study of an amyloid tau fragment. *PLoS Comput Biol*. 2008;4(12):e1000238. https://doi.org/10.1371/journal.pcbi.1000238.
33. Irback A, Mitternacht S, Mohanty S. Dissecting the mechanical unfolding of ubiquitin. *Proc Natl Acad Sci*. 2005;102(38):13427–13432.
34. Mitternacht S, Luccioli S, Torcini A, et al. Changing the mechanical unfolding pathway of FnIII10 by tuning the pulling strength. *Biophys J*. 2009;96(2):429–441. https://doi.org/10.1016/j.bpj.2008.09.043.
35. Jónsson SÆ, Mitternacht S, Irbäck A. Mechanical resistance in unstructured proteins. *Biophys J*. 2013;104(12):2725–2732. https://doi.org/10.1016/j.bpj.2013.05.003.
36. Bille A, Jensen KS, Mohanty S, et al. Stability and local unfolding of SOD1 in the presence of protein crowders. *J Phys Chem B*. 2019;123(9):1920–1930. https://doi.org/10.1021/acs.jpcb.8b10774.
37. Favrin G, Irbäck A, Sjunnesson F. Monte Carlo update for chain molecules: biased Gaussian steps in torsional space. *J Chem Phys*. 2001;114(18):8154–8158. https://doi.org/10.1063/1.1364637.
38. Irback A, Mitternacht S, Mohanty S. An effective all-atom potential for proteins. *PMC Biophys*. 2009;2(1):2. https://doi.org/10.1186/1757-5036-2-2.
39. Irbäck A, Mohanty S. Folding thermodynamics of peptides. *Biophys J*. 2005;88(3): 1560–1569. https://doi.org/10.1529/biophysj.104.050427.
40. Mohanty S, Meinke JH, Zimmermann O, et al. Simulation of Top7-CFr: a transient helix extension guides folding. *Proc Natl Acad Sci*. 2008;105:8004–8007. https://doi.org/10.1073/pnas.0708411105.
41. Mohanty S, Meinke JH, Zimmermann O. Folding of Top7 in unbiased all-atom Monte Carlo simulations. *Proteins*. 2013;81(8):1446–1456. https://doi.org/10.1002/prot.24295.
42. Irback A, Mohanty S. PROFASI: a Monte Carlo simulation package for protein folding and aggregation. *J Comput Chem*. 2006;27:1548–1555.
43. Favrin G, Irback A, Mohanty S. Oligomerization of amyloid Aβ16–22 peptides using hydrogen bonds and hydrophobicity forces. *Biophys J*. 2004;87(6):3657–3664.
44. Sawaya MR, Sambashivan S, Nelson R, et al. Atomic structures of amyloid cross-beta spines reveal varied steric zippers. *Nature*. 2007;447(7143):453–457. https://doi.org/10.1038/nature05695.
45. Jónsson SÆ, Mohanty S, Irbäck A. Accelerating atomic-level protein simulations by flat-histogram techniques. *J Chem Phys*. 2011;135(12):125102. https://doi.org/10.1063/1.3643328.
46. Li MS, Klimov DK, Straub JE, et al. Probing the mechanisms of fibril formation using lattice models. *J Chem Phys*. 2008;129(17):175101. https://doi.org/10.1063/1.2989981.

47. Co NT, Hu CK, Li MS. Dual effect of crowders on fibrillation kinetics of polypeptide chains revealed by lattice models. *J Chem Phys*. 2013;138(18):185101. https://doi.org/10.1063/1.4804299.
48. White DA, Buell AK, Knowles TPJ, et al. Protein aggregation in crowded environments. *J Am Chem Soc*. 2010;132(14):5170–5175. https://doi.org/10.1021/ja909997e.
49. Cabaleiro-Lago C, Quinlan-Pluck F, Lynch I, et al. Inhibition of amyloid β protein fibrillation by polymeric nanoparticles. *J Am Chem Soc*. 2008;130(46):15437–15443. https://doi.org/10.1021/ja8041806.
50. Irbäck A, Jónsson SÆ, Linnemann N, et al. Aggregate geometry in amyloid fibril nucleation. *Phys Rev Lett*. 2013;110(5):058101. https://doi.org/10.1103/PhysRevLett.110.058101.
51. Hellstrand E, Boland B, Walsh DM, et al. Amyloid β-protein aggregation produces highly reproducible kinetic data and occurs by a two-phase process. *ACS Chem Neurosci*. 2010;1(1):13–18. https://doi.org/10.1021/cn900015v.
52. Abeln S, Vendruscolo M, Dobson CM, et al. A simple lattice model that captures protein folding, aggregation and amyloid formation. *PLoS ONE*. 2014;9(1):e85185. https://doi.org/10.1371/journal.pone.0085185.

Index

Note: Page numbers followed by "*f*" indicate figures and "*t*" indicate tables.

B

F

G

N

O

P

W

X

Z

Kolibri 360 GmbH
Ferdinand-Jühlke-Straße 7
99095 Erfurt, Deutschland
produktsicherheit@kolibri360.de